Proceeding of the VI International Ship Design & Naval Engineering Congress (CIDIN) and XXVI Pan-American Congress of Naval Engineering, Maritime Transportation and Port Engineering (COPINAVAL)

Vice Admiral Jorge Enrique Carreño Moreno ·
Adan Vega Saenz · Luis Carral Couce ·
Jymmy Saravia Arenas
Editors

Proceeding of the VI International Ship Design & Naval Engineering Congress (CIDIN) and XXVI Pan-American Congress of Naval Engineering, Maritime Transportation and Port Engineering (COPINAVAL)

Springer

Editors
Vice Admiral Jorge Enrique Carreño Moreno
Instituto Panamericano de Ingeniería Naval
Bogotá D.C., Colombia

Luis Carral Couce
Escola Politécnica Superior
University of A Coruña
Ferrol, La Coruña, Spain

Adan Vega Saenz
Technological University of Panama
Panama City, Panama

Jymmy Saravia Arenas
COTECMAR
Cartagena, Colombia

ISBN 978-3-030-35962-1 ISBN 978-3-030-35963-8 (eBook)
https://doi.org/10.1007/978-3-030-35963-8

This Springer imprint is published by the registered company Springer Nature Switzerland AG
The registered company address is: Gewerbestrasse 11, 6330 Cham, Switzerland

Foreword

The Pan-American Institute of Naval Engineering (IPIN) is a nonprofit institution intended to promote progress in naval architecture and marine engineering and water and water transportation in the Americas. IPIN holds a Pan-American Congress in one of the member countries every two years. The Congress allow the rejoining of members to discuss specific problems, to exchange valuable information and experiences and to live in a very intense way, the Pan-American confraternization.

The present edition issues some of the main papers that were discussed in XXVI COPINAVAL, held in Cartagena de Indias, Republic of Colombia, last March 2019.

The papers presented at this IPIN Congress include topics such as: ship deck design; building and exploitation technologies of most types of vessels; different port enjoyment for cargoes load and unload, like the equipment and technologies used in shipbuilding and naval industry now-a-day. In last years, the antipollution studies, technologies and new development equipment had been very interesting themes for the international community, and the IPIN members had not ignored this very important problem of the humanity.

Finally, I promise you will read a very interesting technical article and Why not? I hope to see you in next IPIN XXVII Congress in Havana in 2021.

José González Cobas
President of IPIN Américas

Foreword

One of the purposes of the Pan American Institute of Naval Engineering, established in its Statutes, is to promote internationally and especially in the Americas, the Progress of engineering and naval techniques, navigation, shipbuilding and port activities in all related sectors, including parallel and subsidiary activities. From all of the above, the COPINAVAL congresses contribute in a extraordinary way and since 1966 have become the meeting forum, of the creation of bonds of sympathy, friendship and affection, between professionals of these aspects of work and between countries. The gold brooch of the XXVI COPINAVAL held in Cartagena de Indias is reflected in this publication which has the best technical level, due to a periodic character, celebrating every two years in a different country and publishing by a publisher of recognized prestige.

José Ángel Fraguela-Formoso
University of A Coruña

Preface

In this opportunity, the *International Ship Design and Naval Engineering Congress—CIDIN*—joined with the experience of *Pan-American Congress of Naval Engineering, Maritime Transportation and Port Engineering—COPINAVAL*, to promote the development of the naval, maritime and river industry, within the framework of an academic setting, where industry, universities, companies and other institutions shared knowledge and experiences on progress of scientific and technological development in the sector, while creating cooperative networks.

Different activities were developed in this academic space: 60 scientific papers, 9 technical papers, 5 keynote lectures and 3 large forums that contributed to the construction of knowledge about the impact of the military industry on economic improvement, strategies for the development of the blue economy and the future of the Pan-American shipyard industry. Different academic lectures were highlighted on issues like: vibration analysis, risk, modularity in warships, autonomous ships, propulsion systems, hydrodynamics of ships, maritime logistics, multimodal transport and good practices and trends in shipbuilding.

Among the main conclusions of the event, we can detach the invitation to the researchers to think about sustainability, work on the preservation of seas and water bodies, focus on social and economic development, and reflect that the future of Pan-American shipyard industry is in alliances and technological cooperation for the development of transnational projects.

I take this opportunity to thank the Colombian National Navy, the members, presidents and general secretaries of the different chapters of the Pan-American Naval Engineering Institute—IPIN, for promoting this academic scenario around the naval and maritime industry.

Rear Admiral Oscar Darío Tascón Muñoz
President of COTECMAR

Acknowledgments

The publication of these congress proceedings shows the success of the union of the VI CIDIN and XXVI COPINAVAL congresses held in Cartagena de Indias,-Colombia, from March 13 to 15, 2019. We thank all the members of the Pan-American Naval Engineering Institute (IPIN) and members of the Corporation of Science and Technology for the Development of the Naval, Maritime and Fluvial Industry, COTECMAR, whose effort, motivation and commitment made this great event possible, allowing the generation and transfer of knowledge in the areas of design, naval, maritime and port engineering. The contributions of this congress are aimed at generating an impact on the development of the Pan-American naval industry, its companies and institutions and the consolidation of an economic growth platform for our countries.

In the same way, we especially thank the authors of each chapter, who attended our call for the conformation of the academic agenda, the young people who participated in the student competition, the exhibitors and in general the audience, whose interest and participation enriched the purposes of the congress.

Finally, I would like to highlight the excellent work of the organizers, which resulted in the satisfaction of the attendees and the generation of fruitful knowledge exchange spaces that strengthen our industry.

Vice Admiral Jorge Enrique Carreño Moreno
President of IPIN Colombia

Contents

Argentine Shipbuilding Industry 100 Years (1937–2036): Industry 1
Raúl Rafael Podetti and Raúl Eugenio Podetti

Argentine Shipbuilding Industry 100 Years (1937–2036): Policies 11
Raúl Rafael Podetti and Raúl Eugenio Podetti

Argentine Shipbuilding Industry 100 Years (1937–2036): Future 21
Raúl Rafael Podetti and Raúl Eugenio Podetti

Future-Proofing Navies for Uncertainty 30
Geoffrey Till

Why Should Fishing Research Vessels (FRV) Be Certified? Their Technological Role in the Current Challenge to Reduce the Underwater Signature of Commercial and Navy Ships 43
Publio Beltrán Palomo

Risk Based Design with Goal Based Standards and Safety Level Approach. An Example Based on a Spanish Fishing Vessel Fleet 61
M. J. Núñez-Sánchez and Luis Pérez-Rojas

eMaster: Ship's Longitudinal Resistance 79
J. Freiria and E. Philippi

Analysis of Bow Shapes Proposed on a High-Speed Displacement Hull 89
Franklin J. Domínguez Ruiz, Lino Andrés García Jaramillo, and Patrick Roger Townsend Valencia

Harmonization in the Multimodal Transport of Chemical Substances and Mixtures 104
José Ángel Fraguela-Formoso, Asunción López-Arranz, Almudena Filgueira-Vizoso, and José de Troya-Calatayud

The Potential for the Growth of Maritime Transport in Brazil: Focus on Cabotage/Short Sea Shipping 113
Delmo Alves De Moura and Rui Carlos Botter

Modeling of the Learning Curve Corresponding to the Transit Through the New Expanded Panama Canal 124
Luis Manuel Carral Couce, Javier Tarrío-Saavedra, Aladar Rodríguez-Díaz, and Salvador Naya

Hydrodynamic Analysis of Inland Self-propelled Vessel for Cargo Transportation in the Magdalena River 134
Oscar D. Acosta Lopera, Kazuo Nishimoto, and Jairo H. Cabrera Tovar

Essential Aspects in the Supervision of the Ship Construction Projects 153
Pedro P. Martínez Villa and Romilio G. Carralero Garcés

Noise and Vibration Control Program for Warship: The New Spanish Frigate F110 163
J. M. Riola, J. C. Díaz-Cuadra, and Publio Beltrán

A Study on the Effect on Fatigue Damage Due to Different Shipping Routes 175
Luis C. De Gracia, Hitoi Tamaru, Naoki Osawa, Adan Vega, and Hector O. Ruiz

Offshore Patrol Vessel (OPV) Interceptors Evaluation Using Computational Fluid Dynamics (CFD) 185
E. Marcelo Cali Y. and Marcos A. Salas

Evolution and Present of Modularity in Warships 201
Luis Andrés Piñeros Bello and Camilo Ernesto Segovia Forero

Evaluation Methodology for the Selection of the Combined Propulsion System for an Offshore Patrol Vessel (OPV93C) 211
Eduardo Insignares, Bharat Verma, and David Fuentes

A Practical Approach for Decision-Making on Preliminary Naval Ship Cost Estimating Using Multiple Cost Estimation Methods 223
Kevin E. Patrón, Luis D. Leal, and Omar D. Vasquez

The Viability of Implementing a Container Terminal at Port of Açu – A Preliminary Analysis 233
Luiz Felipe Bretas Rozo and Luiz Felipe Assis

Analysis of the Performance of FLNG Vessels According to Basic Design Parameters 246
Daniel P. Vieira, Kam Yu Kang, and Claudio M. P. Sampaio

Optimization of the Emissions Profile of a Marine Propulsion System Using a Shaft Generator with a MMPT Based Control System and the Use of EEDI and EEOI 258
Joel R. Pérez Osses and Carlos A. Reusser Franck

Knowledge Management Applied to Productive Process 275
Pedro Gómez and Marlis Angulo

Comparison of Serrated Helical Strakes in Suppressing the Vortex-Induced Vibrations of a Circular Cylinder 287
Gustavo R. S. Assi and Tommaso Crespi

A Practical Analysis of the Influence of Tool Positioning During High-Frequency Mechanical Impact Treatment on a Welded Joint 298
Hector O. Ruiz, Naoki Osawa, Adan Vega, and Luis C. De Gracia

Fiberglass Stiffened Panels: Estimation of Natural Vibration Frequency 308
Franklin J. Domínguez Ruiz and Luis Manuel Carral Couce

Synthesis of a Modified Polyurethane Paint with Better Properties of Corrosion Resistance Applicable to Vessels and Naval Artifacts 321
Gabriel Peluffo Martínez, Karen Patricia Domínguez Martínez, and Eduardo Espinosa-Fuentes

Methodology for the Early-Stage Design of Electrical Power Systems in a Future Generation of Colombian Warships 334
Gabriela Salas-Berrocal and Richard Bucknall

Evaluation of the Damage in Composite Materials Modified with Viscoelastic Layers for the Hull of Boats Subjected to Slamming Impacts 347
Patrick Roger Townsend Valencia, Juan Carlos Suárez Bermejo, Paz Pinilla Cea, and Estela Sanz Horcajo

Numerical Model to Analyze an Artificial Reef 357
Maria Isabel Lamas Galdo, Patrick Roger Townsend Valencia, Maria Jesus Rodriguez Guerreiro, Laura Castro Santos, and Luis Manuel Carral Couce

Application of a General Pattern of Colonization of a "Verde-PROARR"-Artificial Reef in the Ría de Ares-Betanzos (Spain) 365
Maria Jesus Rodriguez Guerreiro, Jose Carlos Alvarez-Feal, José Ángel Fraguela-Formoso, Laura Castro Santos, and Luis Manuel Carral Couce

Virtual Laboratory with Algodoo for the Determination of the Metacentric Height of a Rectangular Vessel 373
Roberto A. Del Carpio Minaya and Yalmar Ponce Atencio

Optimization of the Life Cycle in the Warships: Maintenance Plan and Monitoring for Costs Reduction 391
Antonio Fernández Jove, Antonio Mackinlay, and José María Riola

Environmental Impact of Ship Emissions Based on AIS Big Data for the Port of Rio de Janeiro 402
Maricruz A. Fun Sang Cepeda, Gabriel P. Monteiro, João V. M. de Oliveira Moita, Luiz Felipe Assis, and Jean-David Caprace

Proactive Maintenance in the Azimuth Marine Propulsion: SCHOTTEL Condition Monitoring Solutions 413
Edgard Mulford, Julio Carrasquilla, and Gonzalo Moreno

Essential Factors for the Implementation of Sustainable Port Logistics Operations 424
Delmo Alves De Moura and Rui Carlos Botter

Implementation of Hydrodynamic Protuberances in the Design of Stabilizing Blades in a Human-Powered Submarine 434
Bryant E. Castañeda, Jorge Romero, Mariana Silva-Ortega, Aldo Barradas, and Mariano A. Hernández

Author Index 445

About the Editors

Vice Admiral Jorge Enrique Carreño Moreno is Naval Engineer from the Naval Cadet School "Almirante Padilla." He holds a master's degree in defense and national security, as well as a doctorate in naval architecture from the Polytechnic University of Madrid. His research interests focus on technological development and innovation in the naval industry, ship dynamics, shallow water maneuverability and experimental hydrodynamics.

He has worked in the shipbuilding industry since 1998, participated in the creation of the Corporation of Science and Technology for the Development of the Naval, Maritime and Fluvial Industry, COTECMAR, in 2000, leading the technology and design management for more than 10 years. He served as President of this corporation from December 27, 2013, until February 2018.

Dr. Adan Vega Saenz is Mechanical Engineer with a master's degree in materials science and another one in plant engineering, all from the Technological University of Panama. He holds a Ph.D. in marine engineering. He is Author of more than fifty scientific articles and a textbook. His professional experience includes, among others, being a university Professor for fifteen years at the Technological University of Panama and Visiting Professor at the Technological University of Bolivar and Uninorte, both in Colombia. He is Consultant of several national and international companies.

Dr. Luis Carral Couce holds a Ph.D. in naval architecture and marine engineering from the Universidad Politécnica de Madrid (Technical University of Madrid) and a master's degree in maritime law and port administration from the University of La Coruna, Spain. He previously worked at the former Spanish Navy shipyard of Bazan (now Navantia) prior to joining the faculty as a professor at the EPS-Ferrol, which is part of the University of La Coruna, Spain, where he continues to teach and conduct research. As a faculty specialist in maritime transport and ship auxiliary systems, he is part of the university's Multidisciplinary Research Group (GEM). He has developed several postgraduate degree programs at the university, including design, production and inspection of sailing and recreational watercraft and maritime logistics and transport. In addition, he has been Chairman of the Spanish Association of Naval Architects in the region of Galicia for several years.

Dr. Jymmy Saravia Arenas is a third year PhD student of the program of Technology and Innovation Management, MSc in Innovation Management from the Universidad Tecnológica de Bolívar (Colombia) and Systems and Technology Management Specialist. Since 2009, he has performed the roles of chief of Technology and Knowledge Management and chief of the Innovation Management Department of the Science and Technology Corporation for the Development of Naval, Maritime and Riverine Industries (COTECMAR). With more of 13 years of professional experience, he has worked leading and executing research, development and innovation projects in different sectors. He has as main knowledge areas technology surveillance, technology transfer, intellectual property, formulation and Research, Development and Innovation (R+D+i) project management, spin-off company creation, information system management and audit of management systems.

Argentine Shipbuilding Industry 100 Years (1937–2036)

Industry

Raúl Rafael Podetti and Raúl Eugenio Podetti(✉)

Asociación Argentina de Ingeniería Naval – AAIN, Tucumán 1438, piso 7 (704) - CABA, C1050AAD Buenos Aires, Argentina
podettiraul@gmail.com

Abstract. Authors Raúl Podetti Sr. (87) and Jr. (60) add up to a century devoted to the shipbuilding industry, in which they have pursued their career as naval architects and marine engineers. Here they contribute their view as active players in the sector, having managed private and state-owned shipyards, as well as professional and industrial business entities.

This paper is about shipbuilding policies in Argentina. It is divided in three chapters: Industry, Policies and Future.

The first one, Industry, describes its evolution and the main players providing a thorough qualitative and quantitative analysis of the sector's performance.

Keywords: Shipbuilding · Marine industrial policy in Argentina · Shipyards

1 Introduction

Several are the reasons that led to the inclusion of the concept of "one century" in this paper's title: 100 years have recently passed since the famous publication by Cap. Storni, on "Argentine Maritime Interests". In 1917 started the first shipbuilding union organization. One hundred years is the period analyzed in this paper. Starting in 1937 – going back eighty years in history to the beginning of modern shipbuilding industry with the construction of a series military vessels by the local industry. Reaching up to 2036 – projecting twenty years into the future in order to achieve a consolidation of the nation's industrial activity at levels of employment and exports similar to the maximums reached in the past.

One hundred years is additionally the time added up by the authors in their professional maritime dedication to this engrossing activity.

This industry is very broad and in order to analyze it, it can be divided along at least three dimensions. According to the type of service, into Construction or Repairs; according to the type of vessels, into Heavy or Light; and according to position within the industrial chain, into Shipyards or Parts/Services Suppliers to the shipbuilding process. The focus of this paper is on the Heavy Construction industry of ships, from the standpoint of the Shipyards.

V. A. J. E. Carreño Moreno et al. (Eds.): CIDIN 2019/COPINAVAL 2019, pp. 1–10, 2020.
https://doi.org/10.1007/978-3-030-35963-8_1

2 Industry

The first shipyards in the region, gradually established along the basin of the River Plate, as well as inland, to make use of the rivers to sail their products down to Buenos Aires, and in the far southern region, to repair those ships that were wrecked while trying to make their way through the extreem South, around to the Pacific Ocean [1].

3 Industry Evolution

The last eighty years (1937–2016) of Argentine Shipbuilding Industry history fall into four stages. They are clearly seen when graphing (at Contract Dates) the values (MMusd - Millions of dollars) for newbuildings in argentine yards during the last 80 years (Fig. 1).

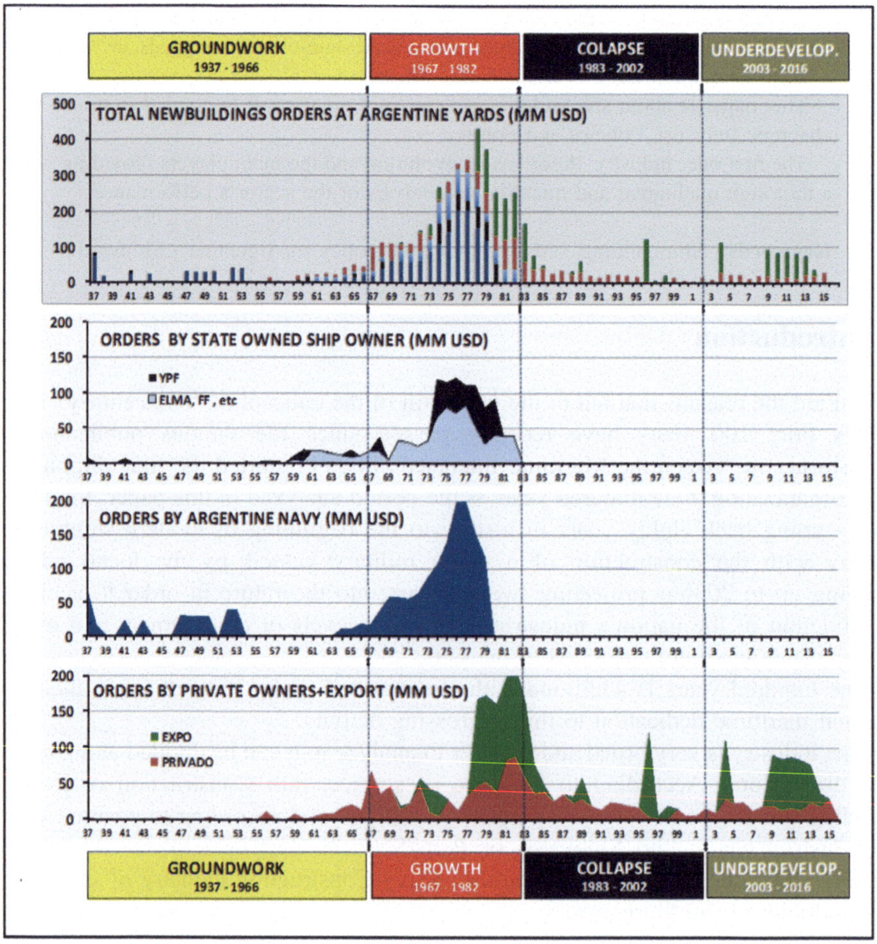

Fig. 1. Argentine newbuilding values at contratct date

First graph shows Total Newbuilding orders and in the others are shown the components of Carrier State (ELMA, EFFEA, YPF), Navy, and private and foreign shipowners. This helps to understand the classification into historical stages and the impact of the demise of the "Shipowning State", which altogether amounted to 45% of the total industrial value. The maximum commissioned rate was reached in the late 1970s, totaling over 400 MM US dollars annually.

The **Groundwork** phase starts with the first significant military shipbuilding project, in 1937, commissioned by the Navy from national private and public shipyards. On the basis of the Navy's attitude, momentum gathered, and infrastructure and institutions were created, laying the foundations for future development.

Only in 1966 did the then prevailing development minded intentions strengthen and result in suitable governmental policies for the times, which enabled a steady progress towards the second, **Growth** stage of the activity. Following a worldwide trend, promotion and protection schemes resulted in a great demand for ships from governmental and private shipowners, supported by ample public financing, similar to that in other countries, and with subsidies to match those received by foreign shipyards in their attempt to enter the Argentine market. It was a time of great development, which ended in the early 1980s.

Due to external reasons, internal crises and changes in policies, the government stopped buying ships in the country and suspended shipbuilding financing. Thus started the stage of the **Collapse** of this industry, during which the largest private shipyards died out. What's worse, as of the 1990s, a most dubious policy promoted the tax-free import of second-hand ships, which became the rule – and the most lethal sectorial policy – in the early years of the new century.

This course means that even to this date, the sector remains deeply sunk in undeserved **Underdevelopment** – the name of the fourth phase – from which this paper attempts to provide escape routes.

4 Value and Employment

In a distinctive contribution, the authors have worked on statistical shipbuilding archaeology, to identify the building data on thirteen hundred ships throughout the country since 1937 – excluding smaller (L < 20 m) and pleasure units. The work resulted in the industry's first ship production database in Argentina [2].

Using technical parameters, assessment was made of the Value (in US dollars of today) and the Direct (Adjusted Standard) Construction Man Hours applicable to each unit, which in turn made it possible to classify information under two criteria (Value and Employment) as applied to shipyards and vessels so greatly different from one another.

From the analysis of the constructed value by shipyard type, the authors found that the state-owned Río Santiago shipyard (ARS, in Spanish) delivered 40% of the total value, although in the last decade it contributed only 15%.

The large private shipyards delivered 24% of the total during their short lifespan and middle-sized shipyads have grown the most in relative contribution to the total value percentage, from 36% of the total to 85% this last decade (Fig. 2).

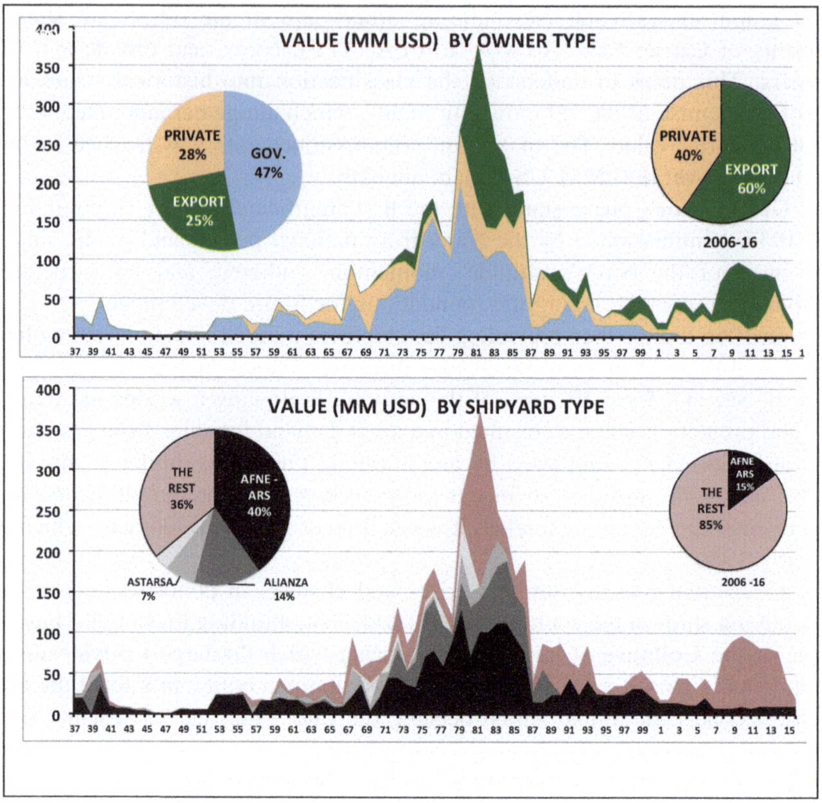

Fig. 2. Value of Argentine newbuildings

When combining this value contribution to the generation of employment, it is concluded that, at present, the shipbuilding industry increasingly depends on middle-sized private yards.

As regards the employment criterion, each ship was assigned the standard direct production hours adjusted by type of shipyard and construction period. This made it possible to reach an estimate of the necessary amount of productive staff needed for ships actually built at a given time, which is here defined as "Direct Newbuilding Employment".

"Total Employment in Shipyards" was reached by combining sources [3] (National Economic census, Shipyards publications, etc.) and also includes production workers in excess of actual building demand (for instance, idle or working in shiprepair), and indirect personnel. Direct construction employment necessary to produce what was really built reached a peak of 9,000 workers in 1979 (Fig. 3).

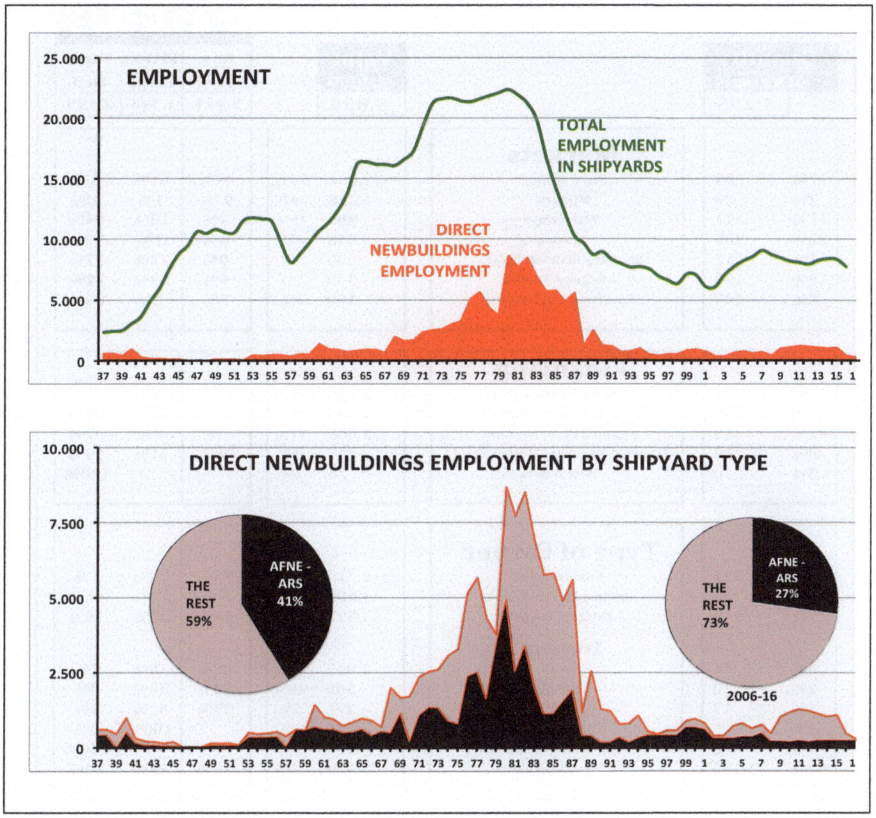

Fig. 3. Employment in Argentine shipyards

ARS provided 41% of the total productive workforce in the whole period, but only 27% in the last decade.

Total Employment in shipyards is much higher than Direct Construction Employment, for repair work has always been very sizable and requires large amounts of labor.

The ratio of Direct Jobs/MMusd produced is of certain use to compare productivity as it shows the amount of people "required" to produce a certain value of ship. This ratio is 23,2 for the whole period and 14,5 for the last decade.

It follows a statistical summary of Industry performance from 1937 to 2016 (Fig. 4).

Quantity of Vessels (%)	Quantity of Vessels 1.295		Value MMUSD 5.855	Value (%)	Gov. yard 2.351	Private Yards Large 1.395	Private Yards Med. 2.109
		Markets					
7%	95	Sea transport	2.096	36%	44%	51%	5%
2%	29	Military	1.388	24%	97%	1%	1%
47%	611	Waterways	895	15%	4%	13%	84%
22%	286	Fishing	518	9%	0%	1%	99%
1%	12	Special (Offshore, Polar)	500	9%	0%	22%	77%
12%	155	Dredges / Towing	308	5%	9%	19%	71%
8%	107	Tourism / Passengers	150	3%	7%	7%	86%
		Financing					
17%	221	State Shipowner	2.718	46%	75%	20%	5%
27%	349	External Funds/ Export	1.537	26%	14%	27%	59%
42%	545	National Prive Funding	1.008	17%	6%	7%	87%
8%	110	National Merchant Marine Fund	517	9%	9%	71%	19%
5%	70	State Banks	75	1%			100%
		Type of Owner					
17%	221	Government	2.718	46%	75%	20%	5%
56%	725	National Private Owner	1.600	27%	7%	28%	66%
27%	349	External Owner	1.537	26%	14%	27%	59%
		Owners					
4%	46	NAVY	1.545	26%	88%	10%	1%
2%	30	ELMA	586	10%	70%	30%	0%
1%	19	YPF	375	6%	59%	31%	9%
1%	8	POLSKA ZM	249	4%		100%	0%
0%	2	READING & BATES (USA)	180	3%			100%
92%	1.190	Other 24 owners	2.920	50%	12%	24%	64%
		Type of Shipyard					
9%	114	State Owned (5)	2.351	40%			
11%	138	Large Private (3)	1.395	24%			
81%	1.043	Medium Private (54)	2.109	36%			
		Shipyards					
6%	73	AFNE- ARS	2.296	39%			
3%	45	ALIANZA	820	14%			
5%	68	CORRIENTES	465	8%			
3%	40	ASTARSA	390	7%			
18%	238	PUNTA ALVEAR	375	6%			
9%	115	SANYM	250	4%			
4%	53	PRINCIPE	185	3%			
51%	663	Other 55 shipyards	1.074	18%			

Fig. 4. Argentine shipbuilding statistics (1937–2016)

5 Argentine Shipyards

Most of the shipyards in Argentina are privately owned. Up to the 80's there were three large private yards (Astarsa, Principe and Alianza) and nowadays there are 16 private medium size yards.

The state owned yards are Astillero Rio Santiago (ARS), Storni and the shiprepairers Tandanor and ANPB (Base de Puerto Belgrano).

5.1 Astillero Rio Santiago-ARS

The Río Santiago shipyard was founded in, by president Peron as part of the *State Shipyards and Naval Factories* (AFNE, in Spanish).

Its concept, design and management was undertaken by the Navy as part of the leadership exercised by the naval force regarding the shipbuilding industry in foundational and development times.

ARS was the industrial expression of a time marked by the Soviet model of overstaffed mega-factories, with maximum vertical integration (from rivets to large marine engines), which also fulfilled a strategic role in training naval architects, marine engineers, technicians, craftmen and workers who would continue the industry over the years.

Originally, ARS played a military role, soon diversifying its focus in order to meet the needs of the state owned merchant marine fleet.

It built 73 significant vessels, standouts among them being the training tall ship ARA *Libertad*, virtually all the nationally built navy fleet, YPF's largest oil tankers, and merchant ships for ELMA. In the last years, construction was of bulk carriers and oil tankers, exclusively for export.

Traditionally, ARS was a major employer in the shipbuilding industry, representing 25% of total historical employment in shipyards, with a maximum record of eight thousand employees in 1970 and a minimum of eleven hundred in 1992. Remarkably, over the last decade it represented 47% of total shipyards employment, with some 3400 agents, despite its production being much reduced – providing only 15% of the period's total industrial value.

Fifty-nine percent of the shipbuilding value was made up by military vessels, followed by overseas merchant vessels for the state-owned ELMA and YPF.

Considering current shipbuilding operations, its exports amount to 9% value of its overall historical production, having also worked for private shipowners (4%).

From 1976 to 1983, it added value exceeding 100 MM dollars annually, understanding that in this analysis, value is split over the term of construction, as is employment.

The relationship between Total Employment and Value produced is a ratio (the lower, the better) that shows a shipyard's productive features. It is worth noting that from the mid-1950s through the early 1970s, for ARS, this index increased to high levels – 230 jobs per million dollars of value – on account of its high vertical integration, the training of apprentices, and the shipyard facilities construction by its own personnel. Then the ratio was reduced to a normal value under 50. But as of the "provincialization" process, in 1993, this index increased markedly, reaching the exorbitant level of 300 jobs per million dollars of value, thus pointing to a productivity ten times worse than the rest of the state-owned shipyards of the region. This resulted from the unfettered increase in personnel not related to production (the effect of the shipyard's politicization) and the reduction in real productive activity. This situation is deemed to cause a loss of 170 MM dollars annually, a situation that has lasted for over two decades, independently of the kind of government in charge of the shipyard (Fig. 5).

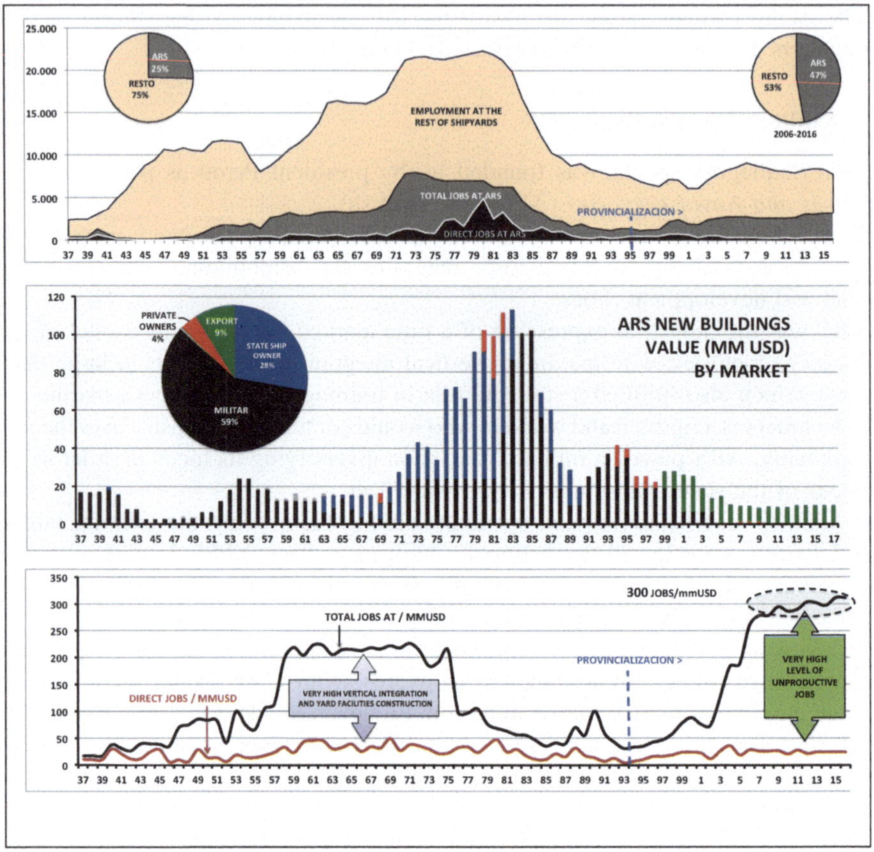

Fig. 5. Employment and value at Astillero Rio Santiago

5.2 Cinar: Tandanor + Storni

In 1972, many of the state-owned workshops and shiprepair docks were merged into one entity, named Tandanor. Six years later, it started work to install the largest ship lift system (Syncrolift), which enabled Tandanor to place large ships in dry dock and move them to workstations, thus providing the largest and very latest ship repair capacity in the region.

It was the center of one of the first and worst privatizations of the early 1990s, which ended up in a scandal, followed by bankruptcy and renationalization in 2007, with 10% of the shares held by employees.

Along with the construction of the Syncrolift, the Navy partnered in a mega-project, with a German shipyard which sold to Argenina a fleet of submarines and a highly modern shipyard that was supposed to build more of them, though this never actually happened.

This magnificent shipyard, under the authority of the Ministry of Defense was originally named Astillero Ministro Domecq García and then renamed Almirante

Storni. Of all shipbuilding facilities in the country, Storni is the best, with the highest potential, yet also the most wasted asset of all (Fig. 6).

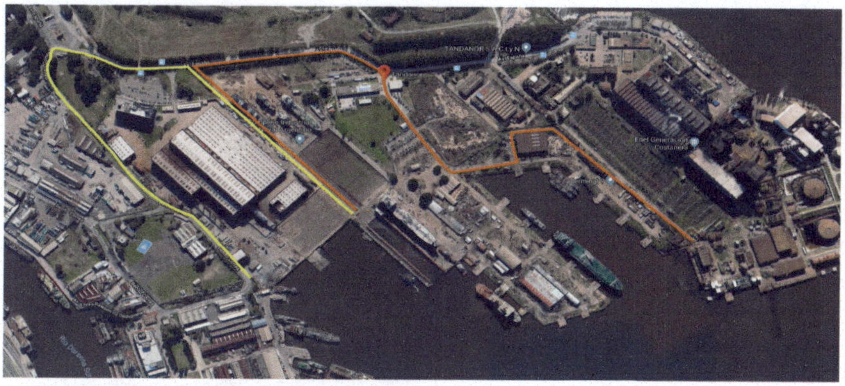

Fig. 6. Tandanor + Storni (CINAR) aerial view

The aerial picture shows on the left (yellow line) the STORNI (ex Domecq García submarines yard) huge facilities.

The orange line marks the perimeter of TANDANOR, the largest repair facility in the region with a Syncrolift and ample wharfs.

The CINAR complex is located at a premium real estate area in downtown Buenos Aires city, only blocks away from the rich Puerto Madero zone and the Pink House (Government Headquarters).

5.3 Large Private Shipyards

Promising state policies of the mid 1960s fostered the modernization and enlargement of traditional shipyards from the district of Tigre and the Riachuelo in Buenos Aires province [1].

Astarsa: This was the first to appear, continuing the historical Hansen and Pucini (1927) in the district of Tigre, already a player in the modern shipbuilding industry since 1937. It built some 35 ships of considerable size, totaling 210,000 DWT. Apart from ships up to 30,000 DWT, it built locomotives and road equipment. Worth mentioning are the merchant ships of the *Río Limay* series, the bulk carriers of 11,500 DWT, the research vessel *Puerto Deseado*, as well as Ro-Ro and containerships.

Principe, Menghui y Penco: In the area of the Riachuelo – Avellaneda, partners Principe and Menghi associated with Penco to form the second large argentine private shipyard in 1967, which built some ten large ships, mainly for the Navy, the polar ship ARA *Bahía Paraíso* standing out among them. It also built a series of lash type barges for an American shipowner, and several river vessels for the regional market.

Alianza: A few blocks from Principe, in the early 1970s, naval architect and marine engineer Héctor Rodríguez Zubieta established the region's most efficient and state-of-the-art shipyard, Alianza. This yard built over thirty large ships, 750,000 DWT on aggregate, and raised the Argentine shipbuilding industry to the highest productivity standards in history (Fig. 7).

Fig. 7. View of Alianza

Alianza's sales to Chile were the first exports of large ships for the national industry, and a series of bulkcarriers to Poland were the largest exports. Altogether, it exported over 300 MMusd, representing 37% of its production.

It also delivered dredge *Capitán Nuñez*, reefers of the *Glaciar* series, *Freedom* ships and oil tankers.

Led by Alianza, the three large private shipyards taken together built 10% of the total of units, invoicing 24% of the value, and their clients procured 71% of the National Fund of the Merchant Marine (FNMM, in Spanish), conceived to finance and subsidize national shipbuilding.

5.4 Middle Sized Shipyards

The authors studied [2] the performance of some 54 middle-sized private shipyards from 1937 to our days, which on aggregate delivered 81% (1,043) of the units, generating 36% (2,109 MM dollars) of total sales, at an average of four vessels per shipyard per year, mainly supplying markets of fishing, river transportation, tourism and port services, with vessels that had an average value of 2 MM dollars per unit.

Today, there are 17 mid size yards along the maritime and river coastline of the provinces of Corrientes (1), Santa Fe (2), Buenos Aires (12), Chubut (1) and Santa Cruz (1). They are all SMEs combining shipbuilding and repairs.

Fifty-nine percent of the value of ship exports was created by this private industrial sector, and it provided 66% of sales to national private shipowners. Its market share to supply government needs was minimal. Middle-sized private shipyards were mainly financed with private and export funds, using only 19% of those available in the public funding system of FNMM [1].

Bibliographic References of this Chapter are presented in Chapter 'Argentine Shipbuilding Industry 100 Years (1937–2036): Future'.

Argentine Shipbuilding Industry 100 Years (1937–2036)

Policies

Raúl Rafael Podetti and Raúl Eugenio Podetti(✉)

Asociación Argentina de Ingeniería Naval – AAIN, Tucumán 1438, piso 7 (704) - CABA, C1050AAD Buenos Aires, Argentina
podettiraul@gmail.com

Abstract. This work seeks to be a contribution and a respectful but firm demand to those in charge of governmental policies, which are required so that the ships argentinians need are built efficiently in national yards. Excluded from this respect are those officials, civilian and military, who have not fulfilled their obligation to guard Argentine work and who, on the contrary, have facilitated the shady dealings associated with the tax free mass importation of ships, particularly second-hand ones, which could have built in local yards at competitive prices.

Keywords: Shipbuilding · Marine industrial policy in Argentina · Shipyards

1 Introduction

Argentine shipbuilding industry has been immersed in a far too long depression stage due to the combination of the worst industrial policies.

This is the second chapter of a paper on shipbuilding policies in Argentina. Having described the Industry in a previous section, this one summarizes the study and discussion of cause and effect relationships between state policies and development, as presented in the book *Argentine Shipbuilding Industry, 100 years (1936–2037)* published in 2018 by the authors of this paper.

2 Policies

Effective actions, rather than discursive speeches, make policies that indicate what rulers really think and intend to see happen.

Ministers, following presidential guidelines, regulate, bid and contract for a certain thing to occur, allegedly in the nation's best interest. Legislators, following their convictions or the instructions of their political leaders, draft laws that rule and orient reality in one direction or another. This "given direction" towards which they tilt the balance marks the intention of a public policy and reveals whom it really benefits.

It is thus worth looking into the impact of state actions (or omissions) through the government's executive and legislative deeds. This enables finding out the relations

V. A. J. E. Carreño Moreno et al. (Eds.): CIDIN 2019/COPINAVAL 2019, pp. 11–20, 2020.
https://doi.org/10.1007/978-3-030-35963-8_2

among state policies and development, the latter being measured through concrete notions, such as employment, tax revenue and foreign reserves. To do this, the authors examine sectorial policies and their effects, the international context, financing, government roles and, finally, six specific naval cases of recent times, in each case assessing the impact caused by the policies.

Some nations travel the hard path to attain shipbuilding development, and after many years of striving only a few reach a noteworthy level. And still, many countries reach sufficient capacity to at least meet the domestic demand concerning fishing, defense, port services, and coastal trade. There is barely a record of nations which have gained such strength only to subsequently waste it. Argentina is a tragic case.

A recent study carried out by the OECD reveals the importance assigned by governments to the shipbuilding industry, concluding that 90% assigned great significance to its employment generation as well as to the increase in industrial and technological capacity.

After analyzing over twenty cases of shipbuilding nations on four continents, embracing over 97% of world output, virtually all of them have in common a resolute state involvement as promoter of the industry's development, through financing schemes, protection on the domestic market, and support for an increasing export-oriented competitiveness.

Among the reasons for state support are: the concept of strategic industry for defense, security in domestic trade, easy access to international markets, demand for diverse qualified jobs, protection of foreign reserves, and generation of technological development.

Although all these reasons are valid for Argentina, two are worth considering, and they both come from Brazil, namely, governmental policies endorsing the shipbuilding industry as a means to distribute natural wealth and as a way to add value to other economic activities.

Natural Wealth: In 2003, having found great offshore oil wealth, Brazil's President defined a vital state policy summarized as follows [4]:

> This oil does not belong to Petrobras, nor to the oil-field workers, much less to the government. It belongs to all the Brazilians. If the benefits of this natural wealth are to reach the highest number of people, ships and offshore platforms for their exploitation must be built in Brazil.

His speech didn't come to nothing. Brazil applied active policies regarding financing, support and supervision, so that the shipbuilding industry might play a role as competitively as possible. The outcome was an exponential industrial and employment growth; in only twelve years, employment rocketed from 2,000 to 80,000 jobs in the shipyards, and was three times higher when including concurring industries.

Argentina is blessed with extensive waterways and abundant aquatic life but policies promote massive imports of tax free second hand vessels.

Added Value: In a conversation with the author, a Brazilian government officer explained the reason of support to shipbuilding industry, helding up a soybean and stating that:

At the end of the day, what Brazil and Argentina produce the most is this – soybeans. Almost without added value involved. But from the farm's gate to the end consumer market, the highest value added is freight by ship. Therefore, to add this extra share of value to our main production, we must build the ships to transport soybeans. It is simple. We do it, you do not.

In Argentina, at least we should build all barges and pusher to carry the soybeans to the shipping port; tugs and port service vessels to enable huge bulk carriers to maneuver; dredges and bouy tenders to help them take their load from our ports. It is simple.

3 Global Context

No correlation was found between countries' shipbuilding industries and their overseas merchant marines orders although local markets for fishing, defense, port service and coastal traffic are almost always reserved for national shipyards.

When analyzing the relative importance in each economy, it is noticed that, except for Korea and Croatia – whose shipbuilding industries reach 2% of the GDP – in the rest it is below 0.5%, while Argentina's is barely 0.007% – 38 times lower than in 1981 (0.26% GDP) (Figs. 1 and 2).

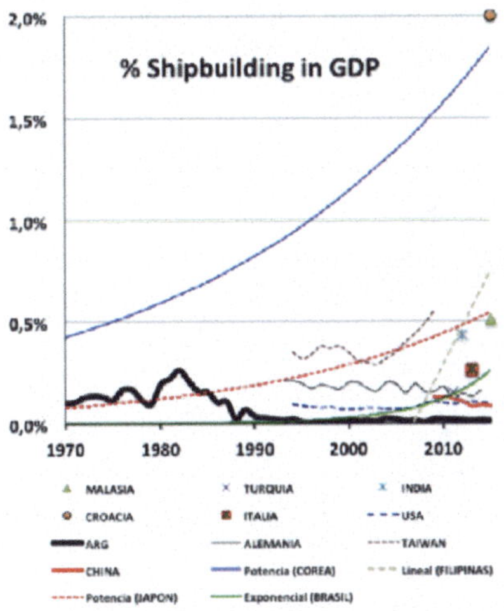

Fig. 1. % Shipbuilding in GDP

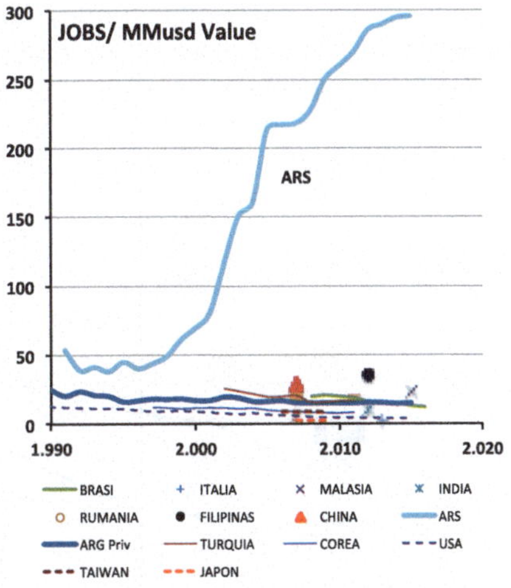

Fig. 2. Shipyard Jobs/MMUSd

To compare competitiveness, apart from the countries' labor costs, productivity was measured according to the amount of jobs required per production unit. ARS stands out, with a productivity over fifteen times worse than the rest of the world and local yards.

3.1 Naval World Map

It is proposed a new map displaying Wealth (GDP per capita) vs. Development (in shipbuilding) coordinates. A series of countries are charted, eventually forming the "continents" responding to given industrial models.

The countries in the Technological model (Germany, Holland, US, UK, etc.) are those of greatest wealth and development, selling marine equipment with high added value; Mega-producers (China, Korea and Japan) concentrate the highest volume, dominating the market in cost and capacity.

This right hand side hemisphere of the new map also embraces the Developed countries (Brazil, India, Portugal, Australia, Iran, etc.) which meet their own needs. Some crosses over to the group of ship Exporters, among them Romania, Taiwan, Ukraine, Croatia, Malaysia, Vietnam, Indonesia, etc.

Most of the Latin American countries are located in the Developing continent, where they strive to become self-sufficient in their marine needs.

The Importers (left) hemisphere has two continents. To the north, the more affluent countries, import new vessels. To the south, the poorer, less developed countries, lacking shipbuilding capacity (Africa, Haiti, Bolivia), get second-hand ships discarded by other nations. These countries receive them readily, because it's assumed that this is what they deserve: their structural poverty, government corruption or negative state policies keep them immersed in this "Scrap" model. Argentina has been forced, by bad policies into this undeserved world corner.

It is shows a navigation course followed by Argentina. In 1967 crossed the producers continent and reached the group of ship exporters. Since 1982 it was steered towards Underdevelopment were it is beached since then (Fig. 3).

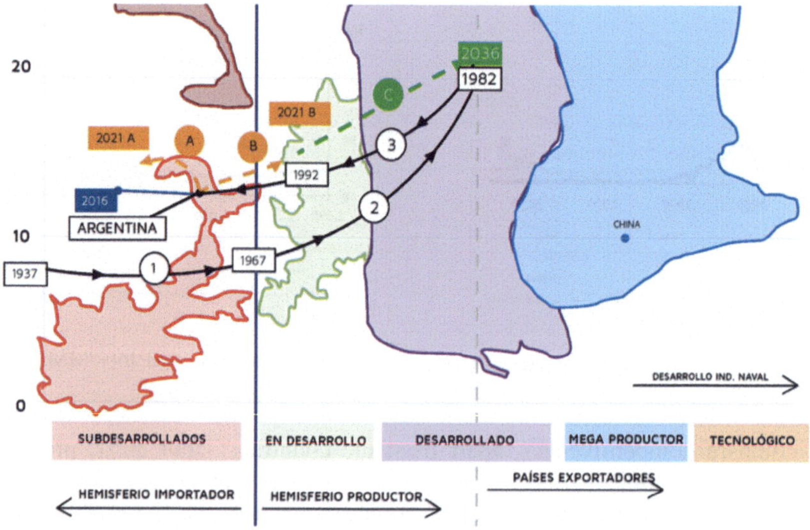

Fig. 3. Argentine navigation course un naval world map

3.2 Marine Financing

Without proper marine financing systems, there is no chance for a shipbuilding industry to function properly.

Given this understanding, all states that are determined to attain this development conceive mechanisms to ensure that shipbuilding projects will not lack financial flow. Sometimes the government intervenes by contributing any necessary collateral, so as to smoothen the access of shipyards and shipowners to the capital market; at other times, the state directly funds sound operations with promotional schemes.

State intervention is almost a constant, and moreover, it is one of the most common ways to provide incentives, subsidies and competitiveness enhancement to its maritime industries in the global context.

In Argentina, most ships have been financed by private funds (42%) or by export funds (27%), representing 17 and 26%, respectively, of the total industrial value. Public financing of shipbuilding came primarily from the National Merchant Marine Fund (FNMM, in Spanish), which provided barely nine percent of the built value. Other public financing sources, such as the BIRA, BIND, BND, contributed a mere one percent extra. Using all these forms of public financing, 14% of the units were built.

Several private owners (Astramar 17%; Ciamar 14%; Maruba 13%; Del Bene 9%; Esso 5%) used the FNMM for new constructions, almost exclusively at the large shipyards (Alianza, 48%; Astarsa, 17%; Príncipe, 11%; AFNE, 10%) [1, 2].

The US policy (guarantees provision) contributed an average of 600 MM dollars/year to the system, while Brazil went through two distinct stages. In the 1970s, Brazil and Argentina converged in setting in motion their Merchant Marine Funds, yet on a very different scale: Brazil (direct funding) assigned 400 MM dollars/year, while Argentina barely 40 MM dollars/year [4].

Brazil second phase started at the year 2000 and for more than fifteen years thereafter, contributed an annual average of 900 MM dollars to finance (and activate) the marine industry. Also evident is the link between ship financing and job generation in these countries. The same analysis shows that the complete demise of marine financing in Argentina since the mid-1980s is one of the main causes of the lack of employment in local shipyards.

There are other financing schemes involving less state intervention, as is the case with foreign marine industry groups. In these cases, the external group contributes any necessary imported equipment, technology and full financing, while the local shipyard provides labor, local goods and services, thus giving rise to a win-win situation among marine industrial players. In Argentina, SANYM used this financing system to build a series of containerships as well as inland pushers and barges (Fig. 4).

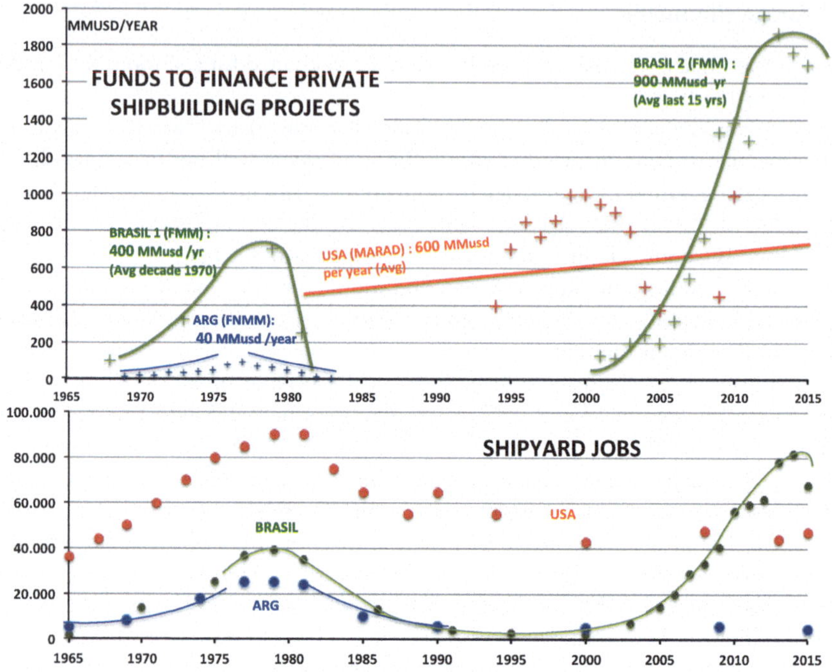

Fig. 4. Ship financing and employment in US, Brazil and Argentina

4 The State Roles

4.1 The Regulator State

Regulating the activity is done through Congress, in its legislative function, and through the presidency by way of the management exerted by the cabinet ministries related to the marine sphere.

For many years now, a marine regulatory framework has been under consideration, and at the end of 2017, Congress granted approval to legislative bills on the Merchant Marine and Shipbuilding Industry. These laws, very poor, became even worse by a governmental veto of the most critical articles (marine financing).

4.2 The Shipowner State

Argentine state played an active role as overseas carrier over the 1940–1980 period, with its height in the 1970s with the creation of the merchant marine company (ELMA) and the enactment of the Cargo Reservation Law [1]. The state also played this role through the YPF oil carrier fleet, until it gave up this function (mid 1980s) as a result of new rules governing the international market and in national politics and economy.

But there were other "shipowning" roles which the state logically kept for itself, for they are vested upon it and cannot be delegated, as is defense (ARA), security (PNA), waterways maintenance (SSPyVN) and fisheries research (INIDEP). But authors found that:

- PNA and INIDEP have avoided providing work for Argentines, always seeking to import, in ways usually contrary to the national interest.
- For the last 38 years, ARA has failed to buy any significant ships in the country, and has fallen to the worst degree of obsolescence (+40 years)

Only 21% of PNA + ARA fleet was locally built, 50% was imported second-hand and at present all the new vessels are planned to be imported.

4.3 The Shipyard State

Internationally, state-owned shipyards peaked from the 1950s to the early 1970s (ARS 1953; Tandanor 1971), on account of the political and economic views of the times. Towards the close of the century, those massive shipyards began to reduce their size or to become privately owned, to fit the guidelines of the new world economic policies. The European post-communist nations resolved to cease underwriting the significant losses of the state shipyards, which proved to be overstaffed, unproductive and unable to compete.

A group of state owned shipyard cases was analyzed, involving twelve countries and over 230,000 jobs, which after the reorganizing process downsized their staff to 30% and increased production, which reveals the large unproductive overstaffing inherent in state-owned shipyards.

The Australian Maritime Complex (AMC) is a successful model to consider.

The largest industrial defense machine, which is the US, entrusts private shipyards with the building of military ships. It is also the case of Germany, the UK, Israel, Portugal, or the former Soviet bloc. Noteworthy is the case of India, whose significant military contracts render their state shipyards appealing to private investors on the stock market.

The "Social Value – Economic Result" matrix – developed by Price Waterhouse Coopers [5]-yields, on one hand, a successful group of state shipyards which generate real social value without economic loss-making, and on the other, a cluster of national state shipyards which have shifted from a strategic position of Social Value generation with low losses to the worst quadrant, with a very arguable value contribution and high structural economic losses (Fig. 5).

4.4 LatAm's State Owned Shipyards

In the case of Latam countries on the Pacific, where private construction shipyards are a minority, special situations arise. Their navies actively control the shipyards, which also are highly market dominant.

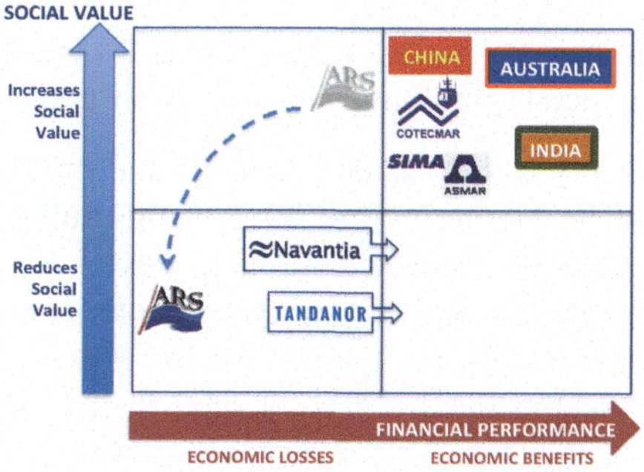

Fig. 5. Social value vs financial performance matrix

They stay profitable, avoiding overstaffing and unnecessary expenses, and they are awarded naval contracts at fairly competitive values, generating development, employment, tax revenue and saving foreign reserves. Political and military powers are in genuine tune with the state shipyards. The opposite happens in Argentina since 1980s.

When comparing the performance of Argentine state yards with similar ones throughout the region, it is found that in Argentina overstaffing is significant, a very high percentage is in administrative tasks, with no flexibility and a very low production level. Thus, productivity (Employment/MMUsd) is much worse, and structural losses have remained unresolved for years. This is partly due to a very poor industrial state policy in general, in this case further worsened by low confidence of the government in its own yards [2, 6–8] (Fig. 6).

AVERAGE DATA OF LAST 2 YEARS (20014-6)

	COTECMAR	ASMAR	SIMA	ARS	TANDANOR
SALES	80	160	92	14	20
GROSS PROFIT	2,8	6	3,5	-170	-30
PROFIT / SALES	3,5%	3,8%	3,8%		
SALES /GDP	0,020%	0,080%	0,030%	0,002%	0,004%
JOBS	1600	2900	1700	3500	1100
JOBS / MMUSD	20	18	18	250	55
PROD. JOBS/TOTAL	74%	78%	80%	30%	80%
FIX JOBS / TOTAL	40%	78%	76%	99%	65%

Fig. 6. Performance of regional state owned yards

4.5 Naval Cases

A number of cases are mentioned, where sectorial state policies have doomed the Argentine shipbuilding industry to an undeserved regression.

The Navy (ARA) Case: Only in the brief 1970–1982 period were a few ships contracted in Argentina, but 70% of the military fleet was imported. Moreover, 75% of these imports were second-hand vessels averaging over 17 years of age. While the world's navies are the primary promoters of national shipyards, in the case of Argentina this hasn't been the case at all for a long time. It's been almost forty years since a military ship has been contracted in the country, even though it had a fit and competitive industry. Regional navies have done the opposite with very good results.

The Coast Guard (PNA) Case: Few countries with shipbuilding capacity fail to supply their own Coast Guards, with much lower requirements than warships. This is why these vessels are nearly always built by local shipyards. But this is not the case of PNA that has never contracted a medium-sized new ship in Argentina. Instead, it imported plenty of them. From 1978 to 1982, PNA imported over 36 units, 100% of which could have been built in Argentina. PNA is still doing the same today.

The Waterway Case: In the 1990s began a huge development of the river push transportation system, requiring a large number of barges and towboats that Argentine shipyards were more than able to build at competitive values. But a governmental policy prevented this, instead prompting the tax-free import of thousands of second-hand barges and towboats from the Mississippi. Due to bad policies combined with corruption, Argentina has only managed to obtain 10% of this ¨brown water¨ market valued at 3,500 MM dollars. Its industrial capacity and competitiveness would have enabled access to over 70%.

The Fishing Case: As of the 1970s, Argentina used the wealth of the national fisheries as a bargaining chip to negotiate other priorities (external financing, foreign policy, and more), favoring foreign ships with fishing permits, to the detriment of the fishing vessels built in the country, for which the catch quotas "were no longer sufficient." Of the 185,000 m3 hold capacity of the Argentine fishing fleet (485 ships), 85% was imported (91% were second hand ships whose mean age exceeded 36 years). In twenty-five years, this state policy merely enabled Argentina to get a 13% share of a fishing-ship market of 1,700 MMusd, despite its capacity to take all of it. By 2017, the importation of fishing ships is still being enforced under the local policies.

The INIDEP Case: Funded by the IDB, Argentina's National Fisheries Research and Development Institute (INIDEP, in Spanish) offered two tenders for the purchase of fishery research vessels in 2015/2016. These were ideal ships for local yards who were banned to participate. A yard form Vigo, surprisingly was the only bidder. The tender was clearly oriented from the outset towards this Spanish shipyard.

The ENARSA Case: In 2016, argentine government put out a tender for the tugboat service of the LNG vessels for the state-owned ENARSA energy company. But the successful bidder was a group with no available vessels so that it partnered with the world's largest tug company, which itself had no ships available in the country, as the bid would have required. Finally, the government approved the import of twelve old tugboats generationg the reaction of the whole marine sector, denouncing the multiple irregularities and the flagrant inconvenience for the country.

The River Patrol Boats Case: In January 2017, it was announced that PNA contracted in Israel, for 49 MM dollars, four gunships to use in river patrolling. This was done directly and secretly, failing to consider much better offers from the national industry. It was denounced that Argentina was paying twice as much as Nigeria for the same vessels from the same yard in Israel; that the number of boats were insufficient to cover the huge area; that they were designed for the sea rather than for river use and that their excessive armament posed a great peril to coastline populations. More and better river patrol boats, designed and built in Argentina, could have been bought with the same investment.

The OPV Case: In 2016/2017, the Navy moved to directly purchase four ships from a French state owned shipyard, denying the Argentine yards any chance to compete. The foreign shipyard favored, a partner of Odebrecht, is part of Brazil's Lava Jato scandal, and it ranks top in corruption.

Authors acted as consultant for President Macri on this decision, strongly recommending to avoid the import of the vessels and to reorient the purchase to the local industry. In order to answer the argentine presidential question regarding the competitiveness of the french offer of 110 MMusd/unit, the author researched hundreds of OPV orders for a 20 years period and found that the french offer was very overpriced (70%). The report also offered a scheme to build equivalent vessels in Argentina in less time and better financial terms and demonstrated that 92% of OPVS for latin american nations with lower shipbuilding capacity have been ordered locally. In mid-2017, President Macri ordered to cancel the purchase, but in 2018 french pressure came back to life in even worse conditions for Argentina and the worse navy contract of all times was signed.

Impact of policies is summarized in a table for each of the naval cases presented. In turn, this analysis is divided by accountability level between the "Menem + Kirchner" and Macri administrations (Fig. 7).

	LOSSES ATTAINED			
CASES	JOBS	TAXES	FRG.EXCHG	VALUE
		MMusd	MMusd	MMusd
FISHING	32.611	316	1.042	1.488
WATERWAYS	50.472	476	310	2.095
SUBTOTAL 1991-2015	**83.083**	**792**	**1.352**	**3.583**
MENEM-KIRCHNER ADMINIST.	*3.323*	*32*	*54*	*MMusd/year*
FISHING	4.560	46	150	174
WATERWAYS	14.646	138	90	606
INIDEP	575	9	24	34
ENARSA	1.212	3	34	70
RIVER PATROL	572	11	22	49
OPV	2.464	27	221	308
SUBTOTAL 2016-2021	**24.029**	**234**	**541**	**1.241**
MACRI ADMINISTRATION	*4.005*	*39*	*90*	*MMusd/year*
GRAND TOTAL 1991-2021	**107.112**	**1.026**	**1.893**	**4.824**

Fig. 7. Losses in naval cases due to bad industrial policies

Bibliographic References of this Chapter are presented in Chapter 'Argentine Shipbuilding Industry 100 Years (1937–2036): Future'.

Argentine Shipbuilding Industry 100 Years (1937–2036)

Future

Raúl Rafael Podetti and Raúl Eugenio Podetti(✉)

Asociación Argentina de Ingeniería Naval – AAIN, Tucumán 1438, piso 7 (704) - CABA, C1050AAD Buenos Aires, Argentina
podettiraul@gmail.com

Abstract. Authors add up to a century devoted to the shipbuilding industry, in which they have pursued their career as naval architects and marine engineers.

This third chapter is about the Future of shipbuilding industry in Argentina. It presents a series of proposals for this future to be positive. Hard information is presented and explained in a soft manner.

It discusses the main matters to be solved (demand, financing and role of state) and develops a model of industial reorganization with a modern government role, providing underutilized industrial assets for the private sector to hire it to build vessels for the Navy and other demanding sectors.

An economic projection is made for five markets measuring their combined impact on Production, Jobs, Taxes, Foreign Exchange and Financing requirements.

Keywords: Shipbuilding · Marine industrial policy in Argentina · Shipyards

1 Introduction

Previous chapters have explained the shipyards performance and policiies affecting argentine shipbuilding industry between 1937 and 2017.

This third section -Future- from 2017 onwards, is aimed at projecting the following twenty years, rounding out one century as the temporal scope of the three chapters paper.

In a spirit of realistic optimism, this section presents conceptual and concrete proposals, developing scenarios and possible projected results.

2 Future

2.1 Matters to Be Solved: Demand, Financing, and Role of the State

Demand from the Argentine shipyards has been blocked for years due to the excessively generous terms offered for the import of second-hand vessels, thus rendering fair competition imposible. This is particularly serious in the case of the fishing, port

V. A. J. E. Carreño Moreno et al. (Eds.): CIDIN 2019/COPINAVAL 2019, pp. 21–29, 2020.
https://doi.org/10.1007/978-3-030-35963-8_3

activities, and river transportation markets which, in the rest of the world, are normally supplied by the shipyards of the country concerned. Normalizing the situation in the fishing and ports industrial activity is very simple and has a significant immediate impact; doing so with river transportation system is only slightly more complex.

The volume of governmental demand may immediately become very substantial if it is reoriented towards the national competive yards, as it should.

Marine Financing for Argentina's commercial shipbuilding is one of the world's simplest cases to solve, because it mainly refers to standard vessels of relatively low unit value. The ship mortgage instrument, leasing, collateral systems, and the participation of overseas supplier groups are highly helpful elements which may combine with the essential governmental involvement. Upon solving the irregular situation of the import of second-hand vessels, financing becomes the primary instrument of industrial recovery. For government use, the solution is also simple, by means of the Multilateral Credit Agencies (IDB, WB and others), Country to Country loans, or the financing schemes of countries supplying imported equipment required to build the vessels.

The Role of the State, in turn, has three distinct aspects: The "Regulator State" still has to write proper legislative rules for the development of the shipbuilding industry and the merchant marine. The new laws (2017) are extremely poor when not negative. Still, through its role as administrator, the state may cause a genuine shift towards positive state policies. The "Shipowning State," although lessened from what it used to be up to the 1980s, has a great potential to generate work in the Argentine industry. The "Shipyard State" must reconsider its main capacities so that these no longer create deficits (200 MM dollars annually) and add genuine social value. The most complicated case is that of ARS, due to the heavy structural deficit caused by its numerous staff members not related to production. Tandanor should easily attain financial balance and Storni iss the most modern and spectacular facilities for ship construction, but with an extremely low exploitation level. Its reorientation may be one of the cornerstones to buttress national industrial offerings under a modern concept of "maritime cluster" combining public and private strengths.

2.2 CINAR 2.0

This is a new version of the Argentine Naval and Industrial Complex (CINAR, in Spanich), which intends to redefine it as "second-tier shipyard" in order to provide it with a new "operational system" based on a steady and sound complementary relationship with private activity (see Australian Maritime Complex). Significant structural changes are proposed, with positive impact on its profitability and the potential to turn it into a driver of development for the local shipbuilding industry.

Created in 2010, CINAR combines two key assets (Tandanor and Storni). When comparing their strength and weaknesses with those of private industry clear opportunities for a complementary relationship are found.

It is necessary to expose and highlight the significance of certain key aspects, such as Governance, Joint Ventures (UTE, in Spanish), and Importing Packages + Local Financing Schemes (PIFLO, in Spanish).

Governance: CINAR should become functionally related to the Ministry of Production, which has the closest affinity with industrial development matters, productive investments, project financing and export promotion.

Joint Venture Systems: These schemes are proposed between CINAR and private shipyards on a "project basis". Private part should be the primary effective manager of the operations, hiring staff and running the entrepreneurial risk. CINAR would contribute the use of partial facilities during the agreed term, without increasing staff.

PIFLO – Imported Packages + Financing: This system is ideal for large projects, where equipment imports are a substantial component. Full project financing shall be provided by the foreign vendor of equipment package (Fig. 1).

The trajectory of CINAR's predecessors is shown in the PWC's "Strategic Matrix ", having lost much social value from 1978 to the present.

A recovery path is projected for CINAR 2.0 model, to get profitable and to increase social value for the Argentinians.

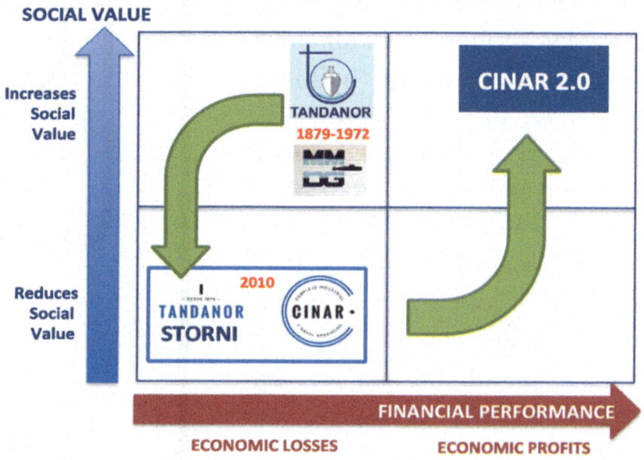

Fig. 1. CINAR in social value-financials matrix

2.3 OPV @ CINAR 2.0

The direct overpriced purchase of the OPV fleet should be avoided and an international tender set in place. The tender conditions for the Navy's most significant project in forty years would be vital, for they define a state policy. The tender conditions would be the most concrete expression of the development model for Defense Production upheld by the government regarding the national shipbuilding sector.

Does it make any difference for Argentina to buy ships abroad or to let the Argentinians to build them in a competitive manner? Assuming, as is indeed the case, that quality, price, terms and collaterals are equivalent, no one should doubt that the national option is much better. If so, why not propose this "preferred model" as cornerstone of the tender?

We might, for instance, put the matter out to international public bid, stipulating as a condition that the ships be built in Argentina. Nobody prevents Argentina from doing the best for the argentinians. There is no international treaty forcing this mistake.

Fortunately, we have this alternative at hand, unlike other countries which are relatively poor or underdeveloped in industrial terms and have no other choice than to import the ships they need.

For the sake of comparison, considering all expenses, spare parts and additional requirements raised by the Navy, the final value is assumed to be over 400 MM dollars for all four ships; in the national building alternative, the local content is 46%, with a 30% total labor incidence and a mean labor hour value of 15 dollars (2017).

The local alternative generates 4,000 jobs over 4 years. Besides, there is a tax revenue of 55 MM dollars, and foreign currency savings of around 184 MMusd.

Considering the impact on tax revenue, it might be concluded that a 16% higher value of the national alternative would mean economic equivalence. In fact, this percentage will be much higher still, because it does not consider the positive effect of employment generation, foreign currency savings and general technological development resulting from the fact that a whole industrial sector rises again.

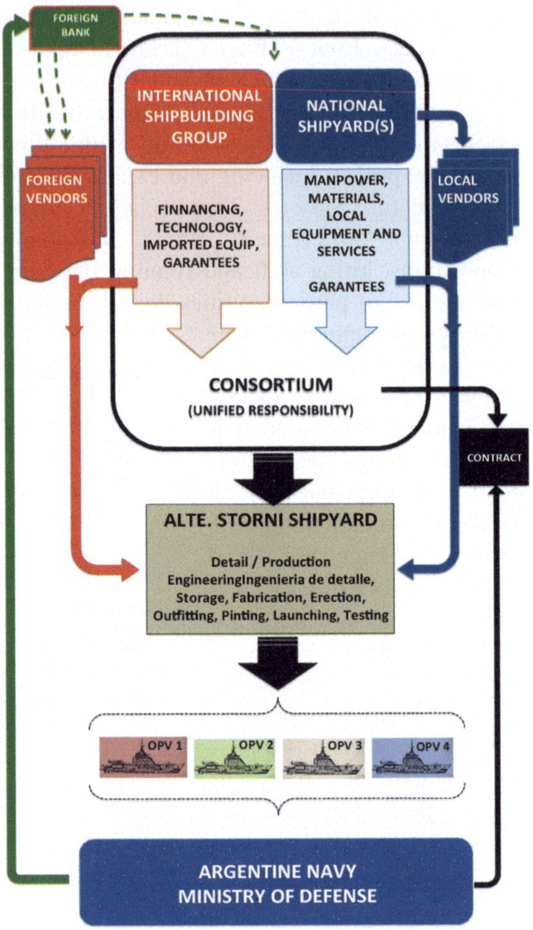

Fig. 2. OPV@CINAR 2.O model

Authors have proposed President Macri a scheme (see Fig. 2) based on CINAR 2.0 model so that building OPVs domestically, in addition to being a success in itself, may lay the foundations for a replicable and sustainable model for the future development of the Argentine shipbuilding industry.

3 2036 Projection

It is presented a future scenario for the Argentine shipbuilding industry, feasible only insofar as the already discussed institutional hindrances are removed.

3.1 Economic Projection

To make this analysis, it is considered the local market demand as a whole and the five market sorces separately. In each case, the volume of global demand is projected up to the year 2036. It follows the estimate of impact in Taxes, Foreign Exchange and jobs (Fig. 3).

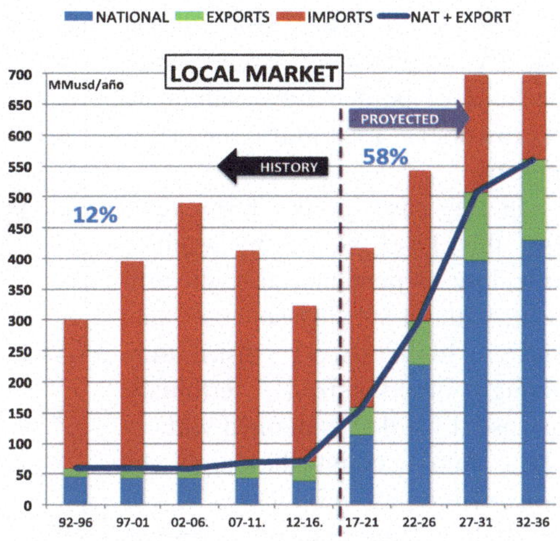

Fig. 3. Argentine market projection to 2036

3.2 Tax Revenue

The current local content (63%) is projected to increase to 82% on account of higher demand for local ship parts suppliers. By applying a mean tax rate of 30% on the net value of national production, this results in a total tax collection of 1,715 MM dollars, averaging resource generation of 86 MM dollars annually, which is 22% of total value (Fig. 4).

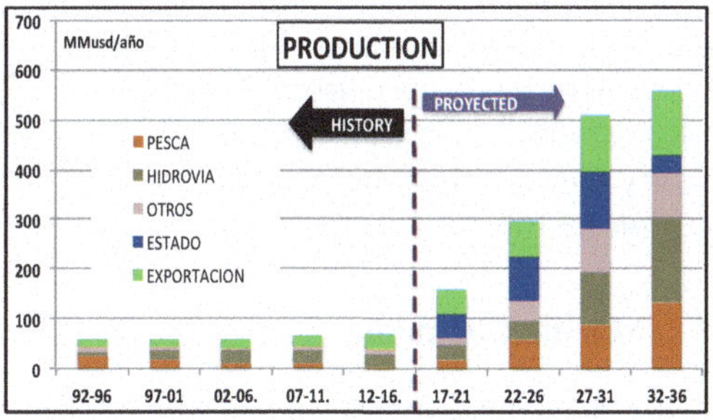

Fig. 4. Ship production projection to 2036

3.3 Foreign Exchange

Because of the replacement of ship imports by others to be built nationally, and due to higher exports, a trade surplus would be reached by 2036 (Fig. 5).

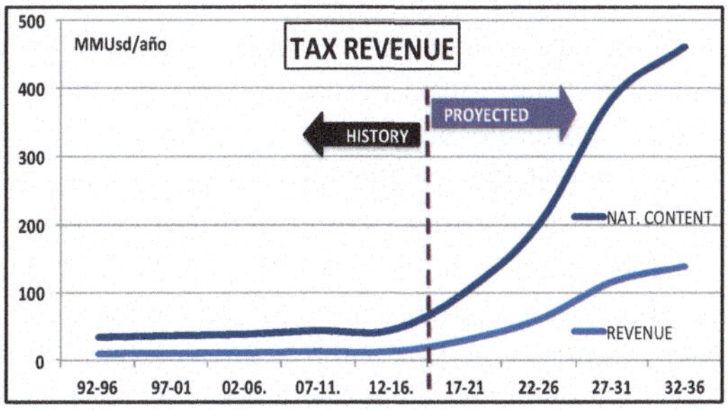

Fig. 5. Tax revenue projection to 2036

3.4 Employment

For the employment projection, the applied parameters are those previously analyzed of 14.5 direct production jobs for each million dollars in sales; 2.8 times the direct employment figure is taken to obtain the total employment in shipyards -including repairs-; and 2.5 is taken as multiplier to consider, also, concurring industries.

Thus, in 2036, total shipbuilding industry employment would reach the level of 60,000 jobs, similar to its historical maximum in 1978 (Fig. 6).

In this case, unlike forty years ago, this level would be attained with low dependence on government, without subsidy schemes, and with a higher work distribution among middle-sized terminals, because it would be attained without taking into account the large shipyards of the past (Fig. 7).

High employment growth indexes are the product of today's very low activity level, and are modest when compared with those recently attained in Brazil. In that case, in only twelve years, shipyard employment has leaped from 2,000 to 80,000 jobs, with a growth factor of forty times, while in Argentina, the projected factor is set at eight.

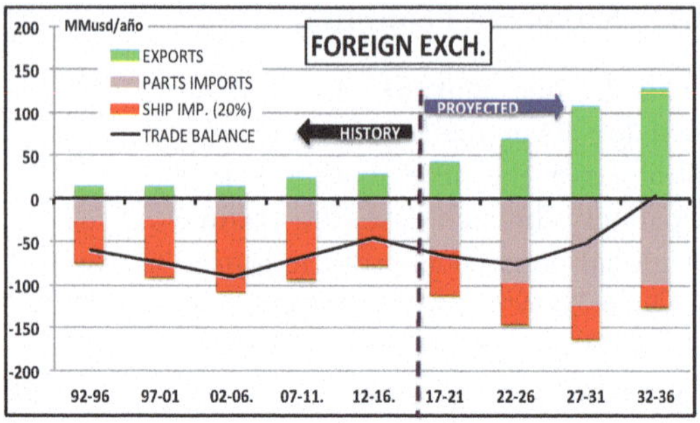

Fig. 6. Foreign exchange in projection to 2036

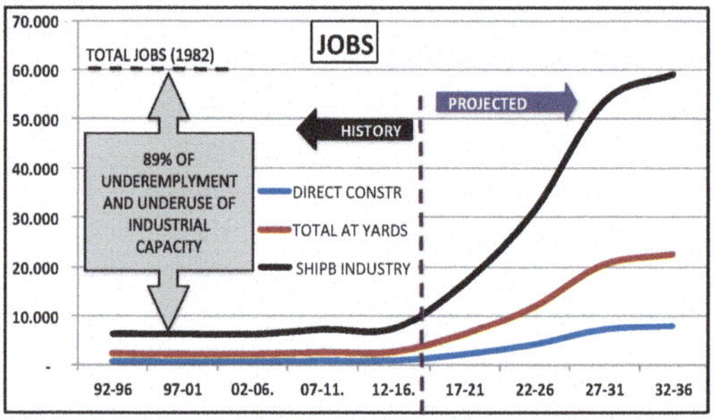

Fig. 7. Employment in projection to 2036

3.5 Marine Financing

A ship financing scheme is proposed for the private sector that encourages the quick generation of contracts, since the paying-back period starts at twenty years and is reduced to ten years in 2027, while the coverage percentage diminishes from 90 to 60% in the same time-span, with collateral and interest suitable to a promotional system.

The annual limit of 75MM dollars for 15 years is considered enough for the first period but larger funding would be required in the future.

In general, an initial period is projected with very significant growth, followed by a period of moderate but steady development.

The main driver of the initial recovery are state purchases (primarily, the OPVs), followed by the fishing market and that of ships for port services.

In the second phase, state purchases start to lose prevalence, because the state's most urgent initial needs have already been met, thus ending up at 19%. This second period should also witness the positive effects of the increase in the river-traffic market demand, expected to be the slowest to react. Penetrating the export market should also start to yield increasing results.

The combination of both effects is projected to compensate by far for the reduction in the governmental market (Fig. 8).

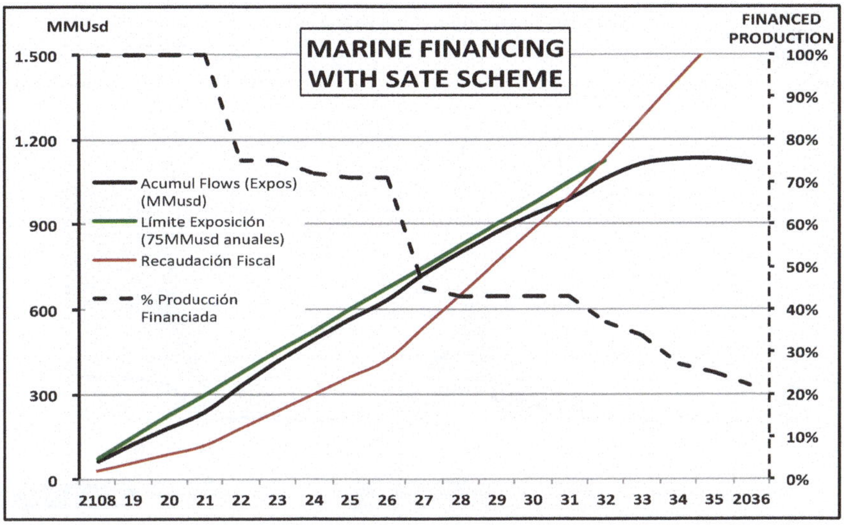

Fig. 8. Marine financing projection to 2036

4 A Century in Perspective

In conclusion, the proposal is to see the century as a whole, the eighty years of history and the twenty years projected ahead, from the point of view of produced Value and generated Employment. Because, at the end of the day, it all comes to this: adding value that generates decent work, and viceversa.

Added to the examined historical stages (Preparation, Growth, Collapse and Underdevelopment), two future stages are proposed: Recovery and Consolidation. This is a way of distinguishing the first coming years, possibly marked by significant development, from the years further into the future, when industry will have already reached a more stable position (Fig. 9).

In the graphic that shows the Value (left axis), the projection might seem too optimistic compared even to the best years of the historical period. But the Consolidation period reaches merely 0.07% of GDP versus the historical max of 0.28%, fourfold higher. Furthermore, for instance, the projected growth indexes are five times lower than the ones yielded by the Brazilian shipyards over the last decade. Employment (measured along the right vertical axis) might reach the same historical maximum as 1978 (60,000 jobs).

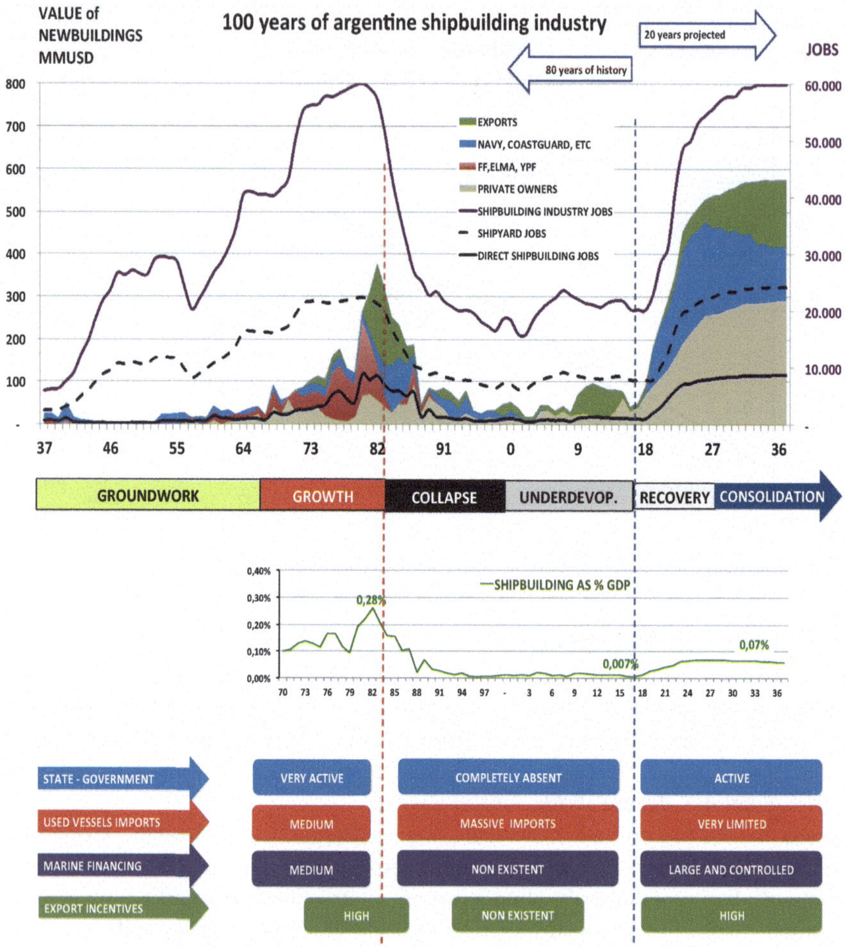

Fig. 9. One century of Argentine shipbuilding (1937–2036)

5 New Opportunities

New technologies will have an impact on construction, designs and processes. But in order to keep pace, Argentina must invest in ongoing education and technological updating.

The environmental aspects shall have an increasing impact, generating ship replacements and ship transformation opportunities (e.g., double hulls, clean propulsion) to adjust them to the new requirements.

New and existing inland waterways and port systems shall require increasing efficiency. Local engineering and industry should provide it.

Rivers and seas will be increasingly valuable to exploit their fishing and mineral resources, or for clean energy, such as tidal or wind power – like the huge offshore windmill fields that already exist in the northern seas. This poses new demands to our marine industry and confers greater significance on the recent extension of the Argentine Sea, which shall require more OPV.

In the (near) future, the demand for ships and Antarctic logistic services will be increasingly important. Argentina must seize the opportunity provided by its geographic position. The industrial experience gained with the refurbishing of the icebreaker ARA *Almirante Irízar* and, soon, with the polar ship construction, should be a valid background if we are to develop in that specialty niche: the polar shipbuilding industry.

But best opportunities will arrive with the most expected change which is the one involving the government's sectorial policies.

References

1. Gonzalez Climent, A.: Historia de la Industria Naval Argentina. AFNE, Buenos Aires (1972)
2. FINA: Federación de la Industria Naval Argentina. Archivos (2016)
3. Observatorio de Empleo y Dinámica Empresarial -SSPTyEL- Ministerio de Trabajo, Empleo y Seguridad Social, República Argentina (2016)
4. SINAVAL: Sindicato Nacional da Indústria da Construção e Reparação Naval. A Indústria da Naval e o Desenvolvimento Brasileiro, 22 August 2012. Disponible en www.sinaval.org.br/docs/IndNaval-DesBrasil-2011.pdf
5. Pricewaterhouse: State-Owned Enterprises Catalysts for public value creation (2015)
6. ASMAR: Astilleros y Maestranzas de la Armada. Memorias anuales y estados financieros.
7. COTECMAR: Corporación de Ciencia y Tecnología para el Desarrollo de la Industria Naval Marítima y Fluvial, Colombia. Memorias y Balances.
8. SIMA: Servicios Industriales de la Marina. SIMA PERU. Memorias anuales

Future-Proofing Navies for Uncertainty

Geoffrey Till(✉)

US Naval War College, Newport, RI, USA
Geoffrey.Till.UK@usnwc.edu

Abstract. Naval planners face a future rendered uncertain by competing priorities that result from competing visions of the international future at sea. To this must be added radical technological and legal change. Managing the consequent problems will require a holistic and considered maritime approach, effective selling of the sea, strategic patience, clarity in operational priorities and cost-effective acquisition and fleet management systems.

Keywords: Navies · Future-proofing · Management

Perhaps the least surprising thing about the maritime situation in which naval planners find themselves these days is that they are constantly surprised. Taking everything into consideration, when considering the future, we are all, in the biblical phrase 'looking through a glass darkly.'[1] As Clausewitz warned us 'war is the realm of uncertainty; three quarters of the factors on which action in war is based are wrapped in a fog of greater or lesser uncertainty.'[2] In peace time too, the 'black swans' and 'pink flamingos' abound (pink flamingos being things people know might happen but do not want to think about - like, in Colombia's case, a renewal of conflict with FARC'.[3]) For all that, naval planners try to construct a future for themselves in which they reconcile the predictable (deduced often from projections forwards of the past and present) and the unpredictable 'black swans' the seemingly impossible that apparently appear from nowhere.

So how are navies to manage 'a surprising future'? The first step seems to be picking out the issues in the main elements of that future - the structure of international politics and their country's place in it, the law of the sea and the impact of new technology.

1 Competing Visions of the Maritime Future

As far as the international scene is concerned, two paradigms seem on offer. The first is post-Westphalian and system-centred. It is about multilateral cooperation to defend the sea-based trading system between states in which common problems such as economic recession, climate change, food, energy and water shortages, pandemics and

[1] I Corinthians, 13:12.

[2] von Clausewitz (1984), p. 101.

[3] Hoffman (2015).

G. Till—The opinions expressed in this article are my own and should not be taken necessarily to represent those of the US, UK or any other government or institution.

V. A. J. E. Carreño Moreno et al. (Eds.): CIDIN 2019/COPINAVAL 2019, pp. 30–42, 2020.
https://doi.org/10.1007/978-3-030-35963-8_4

international terrorism and other such threats to international peace and stability would be addressed and managed cooperatively. The second paradigm by contrast reverts to a Westphalian world in which traditional 19th and 20th Century balance of power conceptions will be expanded to draw in the rising powers of the Asia-Pacific region, particularly China, a possibly revitalised Russia and other major in the Middle East, South America and, perhaps, Africa. Within a Westphalian system dominated by considerations of 'who gets what when and how', the main actors, essentially, compete with each other, while perhaps conceding the need for compromise and stable balances of power.

Naval planners need to assess their country's likely place in the system, its strategic requirements, and critically the future direction of the economy and the budgetary resources likely to be available. In consequence navies across the world have adopted a mix of two rather different sets of roles and missions which are increasingly being presented in the form of publicly available doctrinal statements intended to explain and justify naval expenditure and activity and to act as a headmark for training and acquisition planning. One mission set, roughly in alignment with the post-Westphalian paradigm identified earlier, would comprise cooperative and collaborative maritime strategies that derive from the need to protect the international sea-based trading system against a spectrum of threats ranging from the depredations of such maritime criminals as pirates and human traffickers at one end of the scale to inter-state war at the other. These strategies are often based on the need to react to distant crises lest their effects come home. The ability to do so depends on 'the capacity to maneouvre', namely the ability to do what their leaders think must be done in order to achieve desired objectives. This requires 'sea control' - but a newer version of this traditional concept which is less about one nation or group of nations commanding or dominating the sea and more about making it safe and fee for legitimate use by all peoples.

This kind of sea control creates the condition for the conduct of expeditionary and humanitarian operations of one sort or another, where self-contained sea-based forces are able to operate for long periods away from home, bringing 'good order from the sea' against instability ashore. Humanitarian assistance and disaster response (HADR) becomes a high priority task. Both expeditionary and humanitarian missions encourage the development of the concept of 'sea-basing' – the ability to conduct sustained operations with only a minimum military footprint ashore. 'Maritime Security Operations' against disorder at sea, where the emphasis is on collaboration with others against common threats like piracy, are likewise given emphasis. This in turn requires the habits of cooperation that derive from collaborative naval diplomacy operations which are designed to improve the political and technical ability of the navies of different nations to work together in common cause. Usually this collaborative approach will imply a contributory approach since none of these problems are solvable by a single nation, however powerful, acting alone because the challenges are diverse and apparently infinite while the resources available are often decidedly finite. In the same way, the collaborative approach is often distinguished by the development of shared procurement in arms acquisition and the maintenance of niche capabilities in the expectation that allies and partners will fill the resultant gaps.

This collaborative and cooperative approach is in strong contrast to traditional concepts of maritime strategy which are state rather than system based and therefore in line with the second paradigm. Here the focus is on the behaviour of other states and their navies and there is often a preoccupation with peer competition with other navies. Accordingly, there will be an emphasis on the preparation of high-intensity warfare capabilities such as ASW, AAW and surface warfare which often do not really make sense in campaigns against drugs smugglers or pirates. Sea control here, *does* mean an aspiration for the ability to dominate the sea, to set the rules, as much as resources will allow. It facilitates the old-fashioned projection of military power against conventional military forces ashore. Maritime security in this case is shaped by the desire to guard sovereignty, and to protect maritime resources against foreigners. Naval diplomacy is more competitive, about showing the flag, about the compellence and deterrence of other states. Independence of national decision is hopefully secured though the maintenance of as balanced a fleet structure and defence industrial base as national resources will permit, in order to reduce reliance on other states as much as possible.

Making planning choices is further complicated by the renewed visibility of 'hybrid' forms of conflict associated with Russia's General Sergei Gerasimov back in the 1990s, in which nation states conscious of their numerical inferiority against a more powerful adversary or generally concerned about the costs of outright conventional war adopt the asymmetric style of insurgents or cooperate with them as proxies. Alternatively the Tamil Sea Tigers fighting the Sri Lankan Navy, or the Houthi rebels of the Yemen take on some of the characteristics of a nation-state and fight accordingly. Such approaches endanger the concept of any kind of 'rules-based order', and pose seemingly intractable problems for conventional military forces to respond to. The challenges posed by these muddy conflicts require the 'comprehensive approach' outlined in NATO by which the military contributes to an overall policy involving social, economic and legal lines of development as well. Having successfully participated in grueling counter-insurgency campaigns, both Peru and Colombia are well aware of the planning complexities this brings.[4]

Accordingly most major navies and very many small and medium ones too will tend to adopt a mixture of these two notional alternative sets of maritime strategies, rather than exclusive versions of either and accept the likelihood of styles of conflict that are in a grey area halfway between the two.

1.1 Changing Concepts of Maritime Law

The UN Convention on the Law of the Sea has transformed our notions of the sea space, provided many countries with much greater areas of responsibility, and in many cases much more to argue about when it comes to claimed jurisdiction. Buttressed by existing customary law, developing treaty law and international convention, UNCLOS

[4] Koven (2016).

affects navies more and more, both in what they do and where they do it. The capacity to manage the law of the sea and to turn it to strategic advantage has now become so important it has given rise to the concept of 'lawfare' and almost becomes a constituent of seapower in its own right.[5]

This problem is aggravated by the fact that treaty law is a matter of negotiation and compromise and the results often leave scope for diverse interpretation and implementation. Events in the South China Sea demonstrate all too obviously that claimant states vary considerably in terms of what they regard as their areas of responsibility and in what maritime rights these areas entitle them to. Establishing those areas and enforcing those rights has therefore become an increasingly important task for many of the world's navies.

1.2 Transformative Technology

Planners are also beset with bewildering visions of the future transformational effect of new technology. In the nuclear age, there is nothing new about this but several developments seems to demand attention, none of which are totally new but which have nonetheless markedly developed in recent times. The exponential rate of development in digital and wireless technology, and the incipient revolution that many expect to derive from Artificial Intelligence for example hold the promise, (or the threat !) of revolutionising nearly all aspects of human life and of military operations.[6] Accordingly, most militaries now take an acute interest in cyber operations and the general battle for what might be called the 'decision space.' of one sort or another. The deployment of unmanned systems in naval operations is another area of technical advance since a whole variety of unmanned and relatively cheap maritime autonomous systems (MAS) or 'drones' can now be deployed under, on and over the surface of the sea to perform a whole variety of reconnaissance, strike and protective functions. In themselves such developments will not change the purpose and objectives of navies although they will of course provide yet another capability that navies have to achieve within the limits of their resources. They also explain the heightened interest in developing strategies for effective innovation.[7]

Many of these developments do however come together to shape the technological balance between offence and the defence and so may determine the crucial operational issue of access. Arguably, the biggest current operational issue facing the world's navies is the rise of land and sea based capabilities for denial, which puts at risk maritime access by even the strongest naval forces. Some of these pose such military-technical and politico-strategic challenges to even the most capable of naval powers as to raise doubts about the extent to which they can go on defending their national interests (and perhaps the international system more widely) through a forwards presence and the capacity to project limited power ashore. Recently modern technology

[5] A point developed in the 4th edition of my Till (2018), pp. 138–141.

[6] 5G (2019).

[7] Kamara (2018).

in the varied shape of mines, small fast combatants, coastal diesel submarines and shore based airpower and missile when allied with hypersonic speeds, autonomous systems and cyber attacks have persuaded many that forward maritime operations against countries with developed military capabilities have become too hazardous. This could have a transformative effect on the role and even the utility of navies for all but the most limited and local of purposes. Other, though, argue that such capabilities of sea denial (or Anti-Access Area denial as it has become widely known) could be countered by the adaptation of the very technologies that currently seem to threaten it.

If A2/AD is thought of as a *systemic* threat, then the response has also to be based on a systems approach. Included within this systemic response are emerging enhanced missile defence for ships and bases, cyber offensives and counter-space operations against enemy ocean surveillance systems, directed energy weapons, air and missile attacks on his air bases and surveillance systems, offensive mining, enhanced mine counter measures and anti-submarine capabilities. A complementary response will be the development of 'enhanced long-range strike capabilities' as a means of staying outside the strike range of A2/AD forces while preserving the capacity to project power ashore.[8]

Much of this comes together in the concept of network enabled operations popularised in the 1990s by US Secretary of Defense Donald Rumsfeld. This envisaged much tighter and faster communication of intelligence, surveillance and targeting information around the dispersed components of a fleet or task force, and their associated air and land equivalents that would provide decisive tactical and operational advantage over a force not so well endowed. Concentration of force, in short, would be digital not physical and much less subject to attack. Inevitably, this concept has leached into the current preoccupation with cyber operations which are designed to attack exactly those communications. Tactically, a cyber attack could perhaps take down the operating system of a warship such that its sensors, weapons and propulsion systems would not work properly; operationally it could have the same effect on a much bigger scale against an adversary's task force or fleet. This was the scenario for the initial defeat of the US Navy by the Chinese in Peter Singer's *Ghost Fleet.*[9]

The point is that defending oneself against a cyber attack while being able to mount one oneself has become another capability that navies have to aspire to. But, because technology has always historically provided responses and answers to the problems it poses, decisive operational advantage is only likely to come to the side which somehow manages to achieve a decisive technological lead that its adversary is unaware of or unable to counter in the time available. Doing their best to ensure that this does not happen will therefore continue to be an important priority for the world's navies, just as it always was.

[8] Gunzinger and Dougherty (2011).

[9] Singer and Cole (2015).

2 So How is the Maritime Future to Be Managed?

Managing such a complex and demanding future seems to require a capacity to respond in an integrated fashion to a whole range of economic, social, environmental, political and military developments which work together or against each other to shape the world's future.[10] Confronting this range of challenges and opportunities has to be in the words of Australia's Chief of Navy, 'a national enterprise' handled holistically.[11] A well-rounded maritime policy comprising six key components seems called for.

2.1 Sustaining a Maritime Approach

From the top of the hierarchy of government to the bottom, maritime policy needs to be decided and strategy set. China's President Xi provides us with a contemporary example of just such a maritime vision encouraged from the top. At the Third Plenum of the 18th Party Congress in November 2013 he announced the intention to turn China into a maritime power, an aspiration which built on the expectations of his predecessor and which seems supported by a whole variety of maritime and naval enterprises. China's subsequent *2015 Military Strategy* is unequivocal:

> The traditional mentality that land outweighs sea must be abandoned, and great importance has to be attached to managing the seas and oceans and protecting maritime rights and interests. It is necessary for China to develop a modern maritime military force structure commensurate with its national security and development interests, safeguard its national sovereignty and maritime rights and interests, protect the security of strategic SLOCs and overseas interests, and participate in international maritime cooperation as to provide support for building itself into a maritime power.[12]

Colombia's *Politica Nacional del Oceano y de los Espacios Costeros* of October 2013, provides an interesting alternative conceptualisation of a country's 'ocean policy.' While statements of a 'vision' - a grand overview of all of a country's maritime objectives - have now become relatively common as awareness of the importance of the maritime domain has risen, Colombia's version is less usual in identifying both the requirements of each maritime policy objective and the courses of action necessary to achieve them. This would seem usefully to apply the 'ends, ways and means' approach to maritime policy making and by entrusting the task of monitoring performance to the body that produced it in the first (namely the Ocean Commission of Colombia, chaired by the Vice President of the Republic) will hopefully translate aspiration into product.

[10] One of the best and most convenient reviews of all this are the Global Strategic Trends reports produced annually by the UK Ministry of Defence's Developments, Concepts and Doctrine Centre available at www.gov.uk/mod/dcdc.

[11] Barrett (2017), p. 57.

[12] China's Military Strategy (2015). Available at http://end.mod.gov.cn/Database/WhitePapers/index.htm, p. 4.

Its success, however, remains to be seen[13] varied as such statements of maritime intent are they all at least provide something of a headmark, an overall objective, that others can aim at.

Such a 'champion' might well need to advance a coherent maritime, rather than merely naval, vision that is capable of persuasively integrating the interests of the country's major stakeholders, especially but not exclusively the maritime ones. An integrated maritime infrastructure in terms of shipping, ship-building, fisheries, off-shore oil and gas interests and so forth, help sustain naval power by embracing it in the maritime not only as the kind of constituents of seapower that Mahan discussed, but also providing a degree of protection in bad times. Russian experience suggests that if a navy is not closely linked to the wider maritime scene, then its fortunes will over-depend on the transient caprices of the country's leaders.

Accordingly countries like Peru and Colombia which are clearly intent on developing an indigenous defence industrial base, through organisations like SIMA and COTECMAR, are indeed the 'cornerstone of naval rejuvenation.'[14] Establishing a Defence Industrial base (DIB) thus becomes a way of insuring against an unsettled future. Through long experience in catering for the needs of the armed services during the conflict with FARC, COTECMAR has already developed experience in producing vessels for riverine operations and is in a good position to export some of these products abroad.[15] This involves the acquisition of technological transfer through constructive partnership through foreign firms like Spain's Navantia, Germany's Fassmer and France's Thales and DCNS, the accumulation of more sophisticated ship maintenance, refurbishment design and construction skills, the completion of more ambitious projects like SIMA's 7300 ton multipurpose LPD based on Indonesia's Makassar class (which itself owed much to its Korean predecessor) and finally in the case of COTECMAR a 1700 ton OPV[16] the sale of these products to other countries such as the latter's sale of a landing ship to Honduras.[17] By such means these two countries will acquire not only an important constituent of naval power but also a significant stakeholder in the maritime cause that in the future will help to protect it.

President Xi is in a position of great authority in the Chinese system, for example once chairing both the Leading Small Group on maritime affairs and the Central National Security Commission and so able to act as the 'champion' of the maritime interest. Able to replace retiring members in the system with people of his own choice, his future authority is likely to increase still further, a development bound to help him secure his maritime objectives.[18] By contrast President Jokowi, trying to do the same in Indonesia, will need to work hard to secure his own position in order to push through his maritime agenda. The demands of generating the maritime resilience needed to cope with an uncertain future are such, that top-down 'shove' is probably critical.

---

[13] Politica Nacional del Oceano y de los Espacios Costeros (2013).

[14] Torremans (2015).

[15] Oliver (2015).

[16] Ing (2015).

[17] Sanchez (2016).

[18] Char (2016).

2.2 Effective Selling of the Sea

Seapower does not have a face, an address or a phone number. It is easy for it to drop out of sight and out of mind in the public's more pressing day-to-day preoccupations. Accordingly, constantly emphasising its importance to the peace and prosperity of the country by a variety of mean seems necessary. Nearly all sailors and those interested in their ways believe their own particular country to be afflicted with sea-blindness, even those with manifest maritime interests like Australia.[19]

In such cases, there needs to be investment in a campaign of persuasion of the ordinary citizen and civil society generally.[20] Devices tried out include ventilating the obvious connections between shipping and shopping in the media, establishing maritime research centres, incorporating maritime interests in educational syllabi, holding maritime days of one sort or another, or more dramatic public gestures such as President Jokowi holding a cabinet meeting on a ship in disputed waters around Natuna island or publicly blowing up intruding foreign fishing boats. Whatever the methods adopted, the notion is of a specific campaign, a conscious and resourced exercise in strategic communication.

2.3 Strategic Patience for the Longer Term

Whether maritime power grows naturally and organically from the bottom, or has to be consciously cultivated from above, there are, finally and very evidently, no 'quick and easy fixes' in the demanding task of growing or maintaining maritime power. It takes on average sixteen years to grow a Commanding Officer of an offshore boat.[21] Generating maritime power requires major effort on the part of every member of the maritime community over a long period and perhaps above all, continuing clarity of vision at every level.

2.4 The Identification of Operational Priorities

Future proofing the navy of course requires first-rate planning that focuses on deciding overall policy objectives and the allocation of appropriate resources. In many countries the approach is conveniently encapsulated in the ends-ways-and means process for deciding policy objectives and setting strategies at the operational and tactical levels to achieve them. A country's operational priorities depends on its assessment of its domestic and international environment.

In their struggles against the continent's last two left-wing insurgencies, FARC and Shining Path, the navies of Colombia and Peru have both developed considerable expertise in riverine and coastal COIN operations. Both are now actively considering their 'post conflict scenario,'[22] which requires the regeneration of capabilities for other missions, such as conventional war-fighting roles of the sort previously discussed under

---

[19] DoD (2016).

[20] Narula (2017).

[21] White (2015).

[22] Guevara (2015).

the 'competitive paradigm' described above and the humanitarian disaster operations at a distance and peace keeping operations that feature in its 'cooperative' alternative. This is a major challenge for the ARC since it involves both a major shift in focus and requires resetting the balance between a variety of 'home' and 'away' options.

The ARC of course still needs to maintain order within the country and to sustain the hard-won peace. Hence the need for new generation vessels for riverine operations and the opportunity to capitalise on this experience through the export of equipment and training missions abroad.[23] Hence also the need for OPVs capable of more extensive counter-narcotics operations in the Caribbean and elsewhere. In a potentially volatile area there is also deemed to be a need for a deterrent capacity near to home. Nor can the possibility of the world's wider tensions making themselves felt in the region be easily dismissed any more now than they could be in Cold War days. But at the same time the ARC needs to expand its horizons out to a wider world, and in fact has embarked on this course already, with its counter-piracy operations off Somalia, operations off the Panama canal, RIMPAC, and even an Antarctic expedition in early 2015.[24]

While there seems little sign of the authorities in the country seeking a 'peace dividend' after the closure of the war against FARC[25], the Navy nonetheless faces a real challenge in balancing all these competing priorities, since the country has limited access to the option pursued by many other medium navies, of specialising in just a few roles while looking for strong support in the others from collective security arrangements like NATO. For Colombia this does not appear to be an option, although it does of course have major foreign partners.

2.5 Effective Defence Acquisition

The acquisition of naval *materiel* is intrinsically difficult since both the lead times normally required to produce sophisticated naval weapons, sensors and platforms and their probable service life are likely to be very long. The requirement is to provide governments with choice when facing the demands of an uncertain future. The highly varied and operationally effective career of the UK's *Invincible* class of light carrier – ordered in 1973, and only phased out in 2015-suggests that large, flexible naval platforms in principle can be a useful means of delivering such choice.[26]

However, one recent study has with considerable justice called defence acquisition one of the most demanding forms of human activity.[27] It necessarily results in a future-oriented procurement strategy that tends to suffer from the unpredictability of the future economic, budgetary and strategic environment, during the platform's gestation period and very possibly during its service life too. The fact that defence inflation continually and significantly outstrips conventional inflation resulting in an annual increase

[23] Guevara (2015).

[24] Oliver (2017).

[25] Wasserbly (2015).

[26] Childs (2009).

[27] Hambleton et al. (2013).

according to some estimates of ten per cent since 1945 poses a particularly vexatious problem.[28] The long gestation periods of complex and ambitious projects such as the UK's *Queen Elizabeth* (ordered 1993 and initial commissioning 2020) make them particularly vulnerable. Typically, this will result in constant delays, cost increases and iterative tinkering with the original specification - and eventually in the failure or chronic delay of the programme in ways which mean that the navy tends to acquire new *materiel* in a piecemeal, opportunistic way rather than as part of an overall cohesive plan. It may well produce capability gaps and short-falls likely to the navy's capacity to perform its present roles, let alone its future ones. Instead, a navy will benefit from a real partnership with industry 'in a fully integrated approach to naval capability planning, development, delivery and sustainment.'[29] For this planned and 'continuous build' may well be necessary.

Growing the Smart Customer. An effective partnership with industry requires that the navy produces 'smart customers' who can also represent naval interests in the maritime commercial world, throughout the governmental hierarchy and amongst a wider audience both at home and abroad. This is in effect an investment in a navy's future rather than in its present and as such often has to take a lower priority than it deserves, especially in times of resource constraint. Typically this will require periodic campaigns to address the deficiencies that have been allowed to develop in the past. China is currently conducting just such a campaign in its major reform of its massive Professional and Military Education and training system. Excessive cut-backs in its personnel or suspensions in its training and PME programmes in the name of short-term economies are likely to prove expensive in the long run.

2.6 Effective Fleet Management

Navies clearly need to make the most of what they've got in terms of fleet assets as well as people. Increased budgetary pressure has increased interest in 'organisational arrangements that maximise decision efficiency, weapons effectiveness and the attainment of the strategic goal.'[30] Recently such arrangements have focussed on the effects needed rather than on the various platforms delivering them. 'This approach maximises the commonalities that derive from integrated systems, and facilitates the flexibility and agility that are key to contemporary and prospective seapower.'

Organisationally, this requires two things, first effective force design and second effective fleet operation. Relating to the first, there has developed something of a false dichotomy between strategically-led approaches where force design is based on the characteristics of real world, if putative, adversaries with a geographic setting, strategic objectives and a way of war on the one hand and capability based planning on the other. The latter is a consequence of an inability to identify enemies and has been described as inherently technology-based, treating adversaries in effect as a collection of weapons systems operating across a flat and featureless battlespace, an approach

28 Kirkpatrick (1997), p. 11.

29 Barrett, op cit, p. 60.

30 Barrett, op cit, p. 55.

intended to deliver capabilities that may nonetheless come in handy for unknown contingencies.[31] 'No matter how enmeshed a commander is in the elaboration of his own thoughts,' Churchill warns us however, 'it is sometimes necessary to take the enemy into account.' Both approaches have their advantages and their disadvantages; a judicious mix of the two would seem the best way forward.

As far as fleet design is concerned, Frank Hoffman has argued that like all force planning it should be based on five principles[32]:

- Force design and posture must embrace uncertainty – we have to plan in the expectation of surprise
- Force design must be strategically driven- we need to define priorities for the long term
- Force design must be risk-informed – we need to insure against unknown but intelligent adversaries
- Force design must emphasize versatility over adaptability – our fleet needs to be able to do everything all the time, not simply shift from one task to another
- Force design must ensure a degree of balance as a hedge against uncertainty – we need a fleet without major capability gaps, including in research, training and PME.

Of course, these are observations about desirable habits of mind in fleet planners rather than specific recommendations but they do capture just how difficult the task is and how it needs to be taken very seriously indeed. They reinforce the idea that navies need an organisational culture and a hierarchical structure that encourages innovation because radical change now seems to be a fact of life.

When it comes to the maintenance and operation of the fleet things are hardly less complicated. Maintaining capability is essentially an anticipatory process; clearly it is no use waiting for systems to break down or wear out before seeking their replacement – or, as is often done, disguising this reality by unrealistically extending their operational life until, hopefully, the new equipment comes in. Moreover, most of a project's outcome costs in terms of maintenance and operation are set in its first design phase. The introduction during the early stages of the Cold War of the US Planning, Programming, Budgeting, and Execution (PPBE) process was intended as a through-life way of avoiding such problems. It has been constantly refined ever since, with some success. The US CNO in 2014 released some interesting figures to the effect that the US Navy at the time had 104 ships operational, out of a fleet total of 285. This is about the same number as it used to maintain twenty years ago when the US Navy was around 400 ships.[33] Through its 'Optimised Fleet Response Plan' the US Navy, anxious about the gap between resources and commitments, still tries to refine the balance between deployment and maintenance cycles to ensure as much sustainable time-at-sea from each platform as possible.[34] Other countries in their own ways seek to do the same.

[31] Pietrucha (2015).

[32] Hoffman (2015).

[33] Greenert (2014).

[34] Cavas (2016).

3 Conclusion

This article has focused less on what change might be and more on how naval decision-makers might need to manage it. The future remains uncertain, arguably to a degree worse than it was before since, to repeat the phrase, we are all 'looking through a glass darkly.' Nonetheless it is a necessary task and not one that can simply be put onto the 'too difficult' shelf. To an extent planners are responsible for their own futures and given that for all current distractions the 21st Century is likely to prove to be the maritime century so often proclaimed in the past, countries that are more successful in the difficult business of 'future-proofing' their fleets will be much better placed to face its challenges than those that do not.

References

Barrett, T.: The Navy and the Nation: Australia's Maritime Power in the Twenty-first Century. Melbourne University Press, Melbourne (2017)

Cavas, C.P.: Carrier Bush Under the Gun to Deploy on Schedule. Defence News, Washington (2016)

Char, J.: Reclaiming the party's control of the gun: bringing civilian authority back in China's civil-military relations. J. Strateg. Stud. **39**(5–6), 608–636 (2016)

Childs, N.: The Age of Invincible: The Ship that Defined the Royal Navy. Pen and Sword Maritime, Barnsley (2009)

von Clausewitz, C.: On War (trans Michael Howard and Peter Paret). Princeton University Press, Princeton (1984)

Department of Defence: 2016 Defence White paper. Department of Defence, Canberra (2016)

Government of Colombia: Politica Nacional del Oceano y de los Espacios Costeros. Government of Colombia, Bogota (2013)

Greenert, A.J.: Remarks at Center for Strategic and International Studies, Washington, 19 May 2014

Guevara, I.: Colombia reshuffles military for 'post-conflict scenario.' Jane's Defence Weekly, London, 8 August 2015

Guevara, I.: Colombian marines train other river forces. Jane's Defence Weekly London, 19 May 2015

Gunzinger, M., Dougherty, C.: Outside-In: Operating from Range to Defeat Iran's Anti-Access Area-Denial Threats. CBSA, Washington (2011)

Hambleton, K., Holder, I., Kirkpatrick, D.: Ten chronic challenges in UK Defence acquisition. Defence Stud. **13**(3), 361–371 (2013)

Hoffman, F.: Black Swans and Pink Flamingoes: Five Principles for Force Design. War on the Rocks, Washington, 19 August 2015

Ing, D.: Navantia to sign Colombia deal. Jane's Defence Weekly, London, 24 March 2015

Kamara, H.M.: 'Military Transformation' Joint force Quarterly, no. 91, 4th Quarter 2018

Kirkpatrick, D.: Rising costs, falling budgets and their implications for defence policy. Econ. Aff. **17**(4), 10–14 (1997)

Koven, B.S.: Emulating US counterinsurgency doctrine: barriers for developing country forces, evidence from Peru. J. Strateg. Stud. **39**(5–6), 878–898 (2016)

Narula, K.: Building India's Maritime Resurgence Through Citizen Engagement. SeaGull, Pune (2017)

Oliver, D.: Colombia is Winning the War. European Defence Review, June/July 2015
Oliver, D.: Boldly going where no Latin American Navy has gone before. Warships International Fleet Review, London, May 2017
Pietrucha, C.M.W.: Capability-based planning and the death of military strategy. In: Proceedings of the USNI, Annapolis, MD, 5 August 2015
Sanchez, A.: COTECMAR delivers two new Colombian vessels. Jane's Defence Weekly, London, 27 July 2016
Singer, P., Cole, A.: Ghost Fleet: A novel of the Next World War. Houghton, Mifflin, Harcourt, New York (2015)
State Council Information Office of the People's Republic of China China's Military Strategy. State Council Information Office of the People's Republic of China, Beijing (2015)
Torremans, G.: Building the Future. Warships International Fleet Review, London, March 2015
Till, G.: Seapower: A Guide for the 21st Century, 4th edn. Routledge, London (2018)
5G: What's Next, Wall Street Journal, 27 February 2019
UK Ministry of Defence: Global Strategic Trends 2019. Ministry of Defence, London (2019)
Wasserbly, D.: Colombia likely to maintain military end strength. Jane's Defence Weekly, London, 19 Nov 2015
White, J.: The Rise and Fall of the Sierra Leone Navy and the UK's role in its downfall. 201 The Corbett Centre for Maritime Policy Studies, Corbett Paper No 17, Shrivenham, UK, May 2015

Why Should Fishing Research Vessels (FRV) Be Certified? Their Technological Role in the Current Challenge to Reduce the Underwater Signature of Commercial and Navy Ships

Publio Beltrán Palomo(✉)

TSI-Técnicas y Servicios de Ingeniería, S.L., Avda. PIO XII, 44, 28016 Madrid, Spain
publiobp@tsisl.es

Abstract. During the last decade, most of the maritime countries in the world have been upgrading their corresponding FRV Fleets. In most cases, these fleets are made up of old fishing vessels updated with electronic equipment and echosounders to develop the fishery and oceanographic tasks entrusted. The age of these vessels, their low performance in terms of the underwater acoustic signature and high level of interferences with the electronic equipment are main reasons for the renewing of the fleet.

Biological community claims and the latest public claims against the negative impacts of submarine noise pollution on marine fauna have encouraged policymakers, international organizations and country administrations to develop regulatory frameworks, resulting in a new challenge for the maritime engineering and shipbuilding. This challenge is being dealt through the knowledge gained by dynamic structural design and outfitting equipment integration on board by engineering companies such as TSI, which has been successfully applying its Noise & Vibration Control methodologies in the designing and shipbuilding of the most modern FRV.

For too long time there was no regulation of Underwater Radiated Noise, but finally ICES Report Nº 209/1995 was the most important milestone in civil shipbuilding in terms of URN introducing limits for FRVs with two objectives: (i) avoiding the disturbance of the targeted fish species to be studied as well as (ii) reducing the echosounders interferences. Until then, the reducing of the vessel noise was only important for war vessels and submarines.

As not all of the new FRV were build taking ICES 209/1995 into account, it is highly recommendable to assure that the tag of Silent Vessel is obtained by means of experimental data issued by third party and its requirements should be totally defined and transparent. Otherwise the credibility of all noise signature reduction effort would be seriously mined, what would lead to delays in solving a serious men's health and environmental problem.

This study tries to present an overview of proper methodology, lessons learnt and recommendations to the stakeholders involved in vessel noise control.

Keywords: URN · Noise · Vibration · Ships · Design · Silent

V. A. J. E. Carreño Moreno et al. (Eds.): CIDIN 2019/COPINAVAL 2019, pp. 43–60, 2020.
https://doi.org/10.1007/978-3-030-35963-8_5

1 Introduction

At first sight, the title of this paper might look a marketing slogan. This is not the intention of the author, but quite the contrary. Therefore, the author will justify technically the reasons of this controversial question, as well as the advantages of the formalism of certification. The author also wants to highlight that this paper is a short and modified version of a large paper presented in the 6th INTERNATIONAL SHIP DESIGN & NAVAL ENGINEERING CONGRESS held in Cartagena de Indias (Colombia) from the 13th to the 15th of March 2019.

Old FRVs (Fishing Research Vessels), many of them still in operation, were former fishing vessels improved and updated solely by means of the installation of electronic equipment and echo-sounders without any preventive action to control noise and vibration on board and much less the underwater acoustic signature of them. The age of the vessels and the low performance of the electronic equipment caused by the interferences from the underwater background noise of these ships are two main leitmotifs for a technological upgrade.

The released of ICES (International Council for the Exploration of the Sea) Report Nº 209 in 1995, was a milestone in Civil Shipbuilding, establishing limits for the URN emitted by FRVs. Based on more than five decades of experience, it addresses two essential targets of any ship dedicated to perform fishery and geological research. Firstly, the disturbance of the targeted fish species subject of the observations and secondly the performance of the sophisticated and expensive electronic equipment minimizing the interferences in their echo-sounders and transducers.

Consequently, all vessels denominated with the label "Fishery and Oceanographic Research Vessel" should be designed and build trying to achieve this distinction. In fact, Noise, Vibration and URN requirements are nowadays included in the majority of Technical Specifications of experienced shipowners of FRVs. Additionally, these Specifications require contractually the presence of a Noise and Vibration Consultant as a help to the Shipyard with application of appropriate Vibration and Noise control methodology. That methodology concerns not only to the shipyard itself but also to the suppliers of main machinery and propeller. It includes the definition of the corresponding acceptance criteria as well as monitoring of the whole process from the earliest stages of the design, during the building stage and until the official sea trials.

It must be highlighted that current ICES CRR209 is only applicable to a short segment of the marine market (Fishing Research Vessels and Oceanographic Research Vessels), what implies that only few shipyards and suppliers are able to be included in the state-of the art of the underwater acoustic signature management. Regardless to the results obtained, it is obvious that these stakeholders are in the forefront and with greater technical capability than their competitors to face up the new Regulatory Framework that will become mandatory by laws related to the Health Protection of the Seamen and the Environment as well as protection of the Marine Fauna in its natural habitat in the forthcoming years.

FRV and other typologies of vessels were object of the specific research carried out in the framework of the European R&D Projects "BESST", "SILENV" and "AQUO" [1], funded by the European Commission in FP7 program. All of these projects have addressed the characterisation and the reduction of the underwater acoustic signature of all type of commercial ships. Thus, to address the definition of the guidelines to reduce the underwater signature of the commercial ships and achieved relevant findings, all these aforementioned projects have used the segment of the FRV as reference. This findings and lessons learnt were focused on the design, retrofits and improvements in FRVs, particularly in the machinery. The improvements in machinery, as one of the main URN sources, can be extrapolated to all type of commercial ships.

Additionally, it is important to highlight that it has produced a small technological revolution in Noise & Vibrations Control Techniques, including FAT (Factory Acceptance Tests) as a critical stage in quality control of the vessel, as well as a software development for measurement procedures of URN.

In the field of regulatory frame, the Shipowners will face important challenges and it is recommended to be prepared for them. The mandatory nature of the coming Regulations and Directives (e.g. European Directive 2008/56/EU) will result in specific requirements that will imply changes in design and construction, becoming one of the most outstanding and difficult challenges ever addressed by the civil shipbuilders and the auxiliary industry. The majority of the stakeholders of the marine sector (i.e. main suppliers of machinery and propellers, hydrodynamic research centres, suppliers of insulation material, etc.) are already developing new products and systems to deal with this increasingly demanding and growing market. Considering that this Directive and other non-European are already implemented in some countries, sooner than later they will be extended to the rest of maritime areas as it can be seen by the movements of IMO-International Maritime Organization, being Canada the best example in protection of its marine fauna. Furthermore, it also obeys to the social claim regarding the Protection of the Health of the Workers against noise and vibrations exposures in all workplaces including all type of ships independent of their tonnage and dimensions. The "exceptions" of some type of ships included in the old IMO Noise Code A468 (XII) and maintained in the new one MSC 337(91), represent an obsolete regulation, because for instance, it contradicts the current Regulations and European Directives as well as the recommendations of the ILO (International Labour Organization) [2].

On the other hand, the Biological Community has identified the anthropogenic noise from shipping as one of the major contributors to a downgrading of the environmental status of the oceans during the last five decades with an increase of approximately 19 dB during the period 1950–2007 [3], corresponding to a rate of increase of almost 4 dB per decade. If no mitigation measures are implemented in the forthcoming years, it is expected that this growth of noise rate will get worse considering the marine traffic constant growth. The protection of the Marine Fauna is another social claim that the Marine Sector cannot continue to ignore.

Based on the extended author´s experience in the design and construction of seven (7) FRVs, it can be ensured that reducing URN by minimizing the vibrational energy on board contributes in addition to (i) significant reductions of the vibration and noise levels on board; and (ii) reduction of weight of noise insulation materials, implying a reduction of draft and consequently an improved propulsion efficiency of ships. For all executed FRV's projects, the vibration levels in all spaces and structure were below 1 mm/s-rms and the noise levels in all the accommodation spaces, which only depend on the quality of the HVAC system, achieved values in compliance with the BV Comfort Class Notation COMF-N 2 [4].

According to these proven and consistent experiences provided by the FRV projects, it seems feasible to integrate the Noise and Vibration comfort class notations of most of the Classification Societies into a potential unique class notation or certification regarding to URN. In this case, the control of the HVAC and exhaust system as well as the attenuation of the airborne noise in the partitions will enable to guaranty the compliance with the comfort class notations. It would be well appreciated by the shipowners if the Classification Societies would move in this sense.

Another critical point of concern is the several and different URN measurement procedures. There are procedures of different nature of ANSI/ASA, DNV-GL, BV, RINA, LR and ABS, as well as others as STANAG and NATO used in the navy's facilities and military vessels. It is needed harmonization of them to avoid measurement uncertainties and be able to correlate and compare data regardless of the measurement procedure followed.

2 The Context

2.1 Trend Overview and Regulatory Scenario

In the last two decades the different policies of the European Commission (EC) within the framework of the Green Policy have been focused on combating and reducing the current environmental impact of all type of vessels that make up the current European Fleet. Other international institutions such as IMO and ILO are moving in the same direction. To fulfil this main target and engage companies and regulatory bodies, specific R&D projects [1] have been promoted and launched by the EC. A common conclusion brought out from all these projects is that the N&V-Full Signature (N&V-FS) of the vessel is required to assess properly its environmental impact. This key indicator consists of (i) the noise and vibrations (N&V) on board, (ii) the airborne noise radiated to the harbour (NRH) and (iii) the underwater radiated noise (URN), being URN the greatest challenge to be faced by the Civil Shipbuilding Industry, because if a vessel has a low URN emission, N&V and NRH will be automatically low.

Within this regulatory trend, European Union is considering different Directives such as 2003/10/EC (to protect the health of all the seamen without exceptions against the noise exposure levels) and Directive 2006/56/EC (to avoid the annoying acoustical impact of the ships in the surrounded residential areas of the harbours) among others, which have been released and extrapolated to EU Member States as mandatory local laws.

The results made by all players involved, including the author, during decades of efforts in the abatement of the noise in the marine sector, are finally tangible. Nowadays the URN from commercial shipping has definitively been admitted by Public Institutions and Regulatory Bodies, as an anthropogenic pollution that may have adverse impact on the marine environment and so have brought short and long-term negative consequences on marine life, especially marine mammals. Although, different international environmental conventions and directives have tried, for many years, to address the different noise polluters, the continuous noise from commercial shipping remained out of the scopes until 2008.

The Marine Strategy Framework Directive (MSFD) [5] issued in 2008, recognized that all efforts put in preserving the biodiversity of the EU marine environment had to include the anthropogenic underwater noise pollution. This directive aims at defining and reaching a Good Environmental Status (GES) in all EU waters by 2020.

Shipping is admitted to be one of the largest contributors to the low frequency ambient noise in the oceans. The issue of URN was first raised at International Maritime Organization (IMO) in 2004. The Marine Environment Protection Committee (MEPC) on Noise Impact on Marine Life has finalized the Non-Mandatory "Guidelines for the Reduction of underwater noise from commercial shipping to address adverse impacts on Marine Life" [6] with its approval on April 2014. Despite of the non-mandatory character of these guidelines, they include in the chapter 5 (Predicting underwater noise) different engineering and simulation tools that may assist to this task. Among others and according to author's experience, it is worth mentioning the following ones: (i) using of CFDs for improvement of the wake field of the propeller as well as predicting propeller cavitation, and other phenomena produced by hydrodynamic noise such as bubble sweep-down circulation along the hull; and (ii) Finite Element Analysis (FEA), Boundary Element Method (BEM) and Statistical Energy Analysis (SEA), for the simulation of the noise transmitted from the machinery to the water through the structure of the ship.

2.2 Methodological and Technical Standards for Measurement and Assessment of the URN of the Ships

It is important to mention that there is still today a lack of agreement between main stakeholders about methodological and technical standards for the measurement of the URN of ships and their corresponding assessment. It cannot be forgotten that the characterization of each ship as a source of noise radiated to water and its harmonization is essential from the point of view of the noise control engineering [7]. Currently, the methodologies and measurements procedures for civilian ships could be classified into two different groups: (1) those coming from international standardisation bodies, mainly addressed to the noise measurement procedures without any noise limits for assessment; and (2) those released by the Classification Societies as URN Class Notations with their own noise limits and measurement procedure.

An essential parameter is the transmission loss (TL), used to calculate the radiated noise at 1 m at both sides of the ship from the noise signal recorded at the closest distance between the target ship, according to the defined test speeds and the hydrophones hanging line. The real transmission loss depends on several parameters and it is usually estimated. It represents a current uncertainty that must be harmonized, because if it is not used the same TL, obviously the URN values obtained for the same vessel and recorded noise signal would be different.

Below, it is summarized the several measurements procedures and methodologies regarding URN measuring:

ANSI/ASA-2009

The ANSI/ASA standard "Quantities and Procedures for Description and Measurements of Underwater Sound from Ships" [8] was released in September 2009 by the American National Standard organization. This standard is applicable to measure the levels of underwater radiated noise in the far field and deep waters for all sizes of vessels operating in the market. The sound pressure levels are acquired in thirds of octaves and the energy bands are determined for all frequencies in the range of interest.

ISO-17208-1

The ISO standard "Underwater Acoustics-Quantities and procedures for description and measurement of underwater sound from ships – Part 1" [9] was published by International Standard Organization (ISO) in 2016. This document specifies a new regulation to calculate the emission of ship radiated noise and the levels of underwater noise in the oceans. This approach can be used for multiple purposes as per example the demonstration of vessels' compliance with contract requirements, the evaluation of the ship acoustic signature, the periodic assessment of URN, and the development of research and development activities to prove the high levels of noise emitted by the shipping activity.

The following assumptions for the measurement of the ship underwater radiated noise are made for this standard. First, this method can only be used for ships transiting at speeds lower than 50 knots (25,7 m/s). Second, this method does not exclude the possible influence of sound reflections and absorption effects which can affect significantly to the propagation of the acoustic noise in the marine environment. Thirdly, this standard is only suitable for the measurement of deep waters and therefore, it is not an appropriated tool for the measurement of URN in shallow waters. As a result, this methodology is not recommended for a wide number of sea environments as the Adriatic Sea, Baltic Sea, North Sea and other shallow water seas for the rest of the world which can be seen in Fig. 1. Thus, it is of vital importance to develop an effective procedure to measure the URN levels in all locations of the sea as per example rivers, seaports and shallow waters.

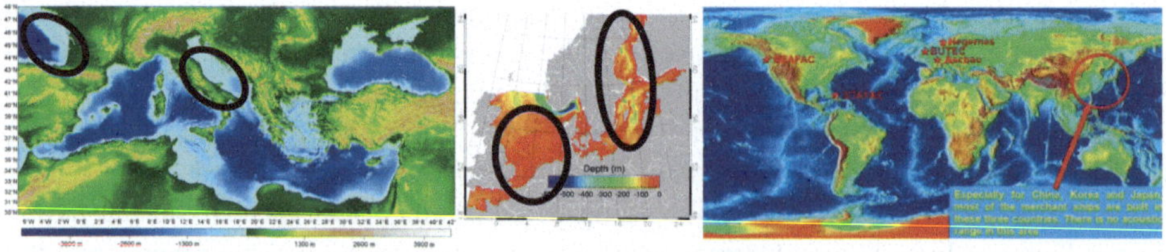

Fig. 1. Limitations of the ANSI/ASA and ISO-17208-1 standards for the measurement of URN in shallow waters of Europe and Asia. (Source: China Ship Scientific Research Centre)

ICES CRR N°209

It represents the URN regulation of Fishing Research Vessels (FRV) in the framework of civil ships and it was released in 1995 by ICES (International Council of Exploitation of the Sea) [10]. Based on more than 40 years in biological research of this institution, it was stated the recommended URN target limits detailed in the Fig. 2 from 10 Hz to 100 kHz. These limits are expressed in term of 1/3 octave band spectrum levels referred at 1 Hz, at 1 m of both hull sides and at an operational speed of 11 knots. These limits, with two different slopes for two specific frequency ranges, were defined to ensure the optimal performance by avoiding the disturbance of the target fish species subject to observation (limit in the frequency range from 10 Hz to 1 kHz) as well as the interferences in the echo-sounders (from 1 kHz to 100 kHz).

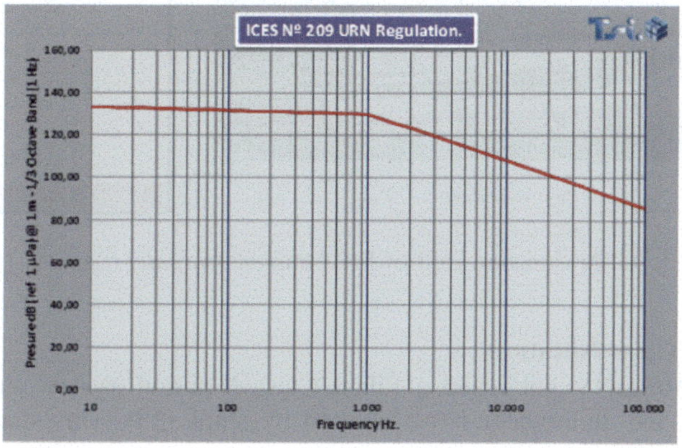

Fig. 2. ICES N° 209 Regulation. URN target limits for FRVs (Source: TSI)

The positive effects of its implementation in the segment of FRVs have been the following ones: (i) inclusion of a specific clause of strict compliance with the URN limits in the contracts with the Shipyard by shipowners of all the modern FRV vessels; (ii) implementation of significant changes and improvements in main propulsion systems; and, (iii) overcome the lack of available experimental data, emphasized and identified by the author long time ago [7, 11, 12] and many other researchers is starting to be covered, allowing the improvement of prediction methods and measurement capabilities.

Considering the concern of harmonization and transparency, the author wants to warn that there are some ships of recent construction that are wrongly denominated as "*Oceanographic ship*" because they are not really. It has been identified a set of conceptual mistakes in their design, inadequate selection of their main machinery or its assembly on the structure, making not possible their compliance with the URN limits of the ICES in the low-frequency range (10 Hz to 1 kHz). For the transparency of the market, it is essential that all FRV should have a corresponding certificate issued by an "*independent company*", which guarantees that it strictly complies with the criteria defined by ICES. As an example of this certificate, Fig. 3 shows an URN signature card of FRV *Ramón Margalef*, showing the thresholds and the URN emitted in all frequencies.

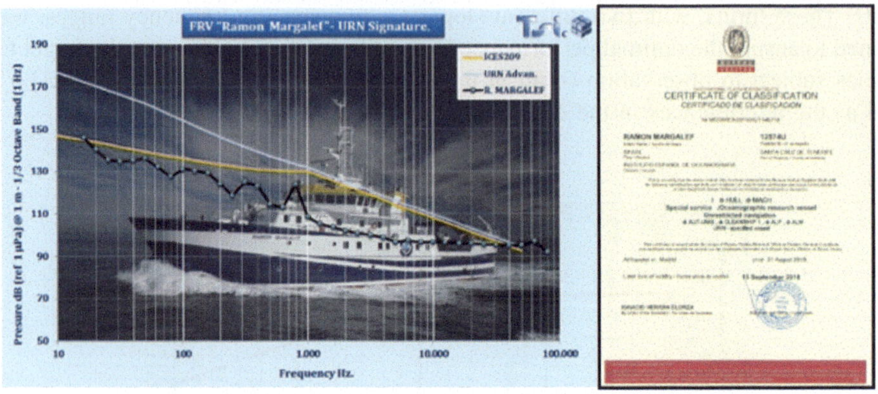

Fig. 3. URN signature card of FRV *Ramón Margalef* (Source: TSI)

DNV-SILENT Class Notation

On January 2010, DNV published a Silent Class Notation as new standard for the measurement of the underwater noise emitted by ships [13]. This standard specifies targeted limits and associated measurement procedures for shallow waters. Depending of the compliance with the targeted limits the Classification Society provides different type of Notations: "A", "*S*", "*F*" and "*SILENT-R*", being the last one for vessel engaged in research or other noise critical operations. The URN limits for this last category "R" are the same required by ICES CRR Nº209.

Regarding to the measurement procedure the following points are highlighted: (i) only one hydrophone is employed, see the Fig. 4; (ii) According to this Class Notation, DNV-GL decides the speed/s at which the vessel shall be tested, based on the vessel operational speed profile for typical hydro-acoustics operations preliminary submitted to it; (iii) Only two runs (one for each side of the vessels) are required; (iv) TL parameter (Transmission Loss) is calculated according to the following formula: $TL = 18\ Log_{10} DCPA$ (DCPA stands for Distance of Closest Point of Approach) and a correction of minus 5 dB must be included due to surface reflections; (v) And finally, another important issue is the "data window length" used.

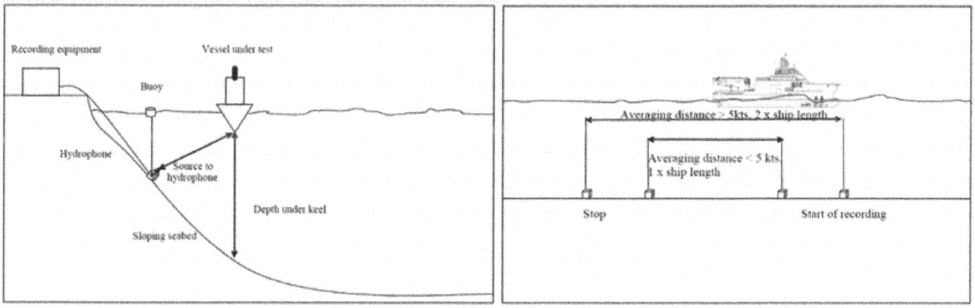

Fig. 4. DNV-GL SILENT URN measurement procedure (Source: DNV-GL)

Regarding (ii), it is important to keep in mind that speed of the ship becomes essential for the suitable dynamic-acoustic design of the silent vessels, and it should be defined and agreed at the beginning of the project and the Classification Society should not decide on it or change it at a convenience. For instance, ICES CRR209 defines its contractual noise limits at the speed of 11 knots, driving the entire dynamic-acoustic design of the ship to comply with this operational condition, which is established at the signing of the contract.

Additionally, this Classification Society proposed an alternative way to obtain noise radiated by ships through a making a hole in the aft part of the hull of the ship, close to the propeller, to measuring noise with a hydrophone or a pressure sensor. Despite this proposal was presented and discussed in technical meetings of experts promoted by European shipyards as an alternative to reduce time and cost of URN measurement, it was rejected by the majority of their attendees.

BV-NR 614 URN Notation

Bureau Veritas (BV) issued in October 2014 its first URN Notation, BV NR614 [14] along with a proposal of a European URN measurement standard [15] & [16] in the framework of AQUO project after a collaboration with TSI S.L. in this R&D project [1]. AQUO project was focused on underwater noise emitted by vessels, and its main objectives were to develop guidelines and recommendations to reduce shipping noise to achieve a quieter ocean and to provide support to policy makers, as well as to propose design solutions and mitigation measures to manage URN emission of ships to reduce its impact on the marine fauna. BV NR614 URN Notation was developed considering the URN database managed by TSI in SILENV project and AQUO project, bringing realistic, consistent and reasonable justification of the down-selected limits and measurement procedure.

This notation aims to control and limit the environmental impact on marine fauna of all self-propelled vessels providing an URN measurement and compliance assessment procedure according to its specific URN vessel requirements, considering a minimum set of three runs for each side of the vessel and a set of speeds of the vessel according to its own operational modes. For further details regarding measurement and post-processing procedure, see the reference [14].

While ANSI/ASA is a URN measurement procedure that is focused mainly on deep waters (with limitation of its application in some areas of Europe) and DNV-SILENT measurement procedure is for shallow waters, BV NR 614 is the only URN measurement procedure that considers both deep waters and shallow waters, covering the widest range of measurements without any restriction.

Other URN Notation of the Classification Societies: RINA, LR and ABS

The rest of the Classification Societies (CCSS): RINA (Registro Italiano Navale), LR (Lloyd's Register of Shipping) and ABS (American Bureau of Shipping), released recently their corresponding URN voluntary Class Notations: Rina Dolphin on 1st July 2014, LR on February 2018 and ABS on July 2018. Most of these class notations have adopted the corresponding URN measurements procedures defined recently by ISO-17208-1 and ANSI/ASA.

Considerations of the Methodologies Comparison and the Importance of Consistent URN Data

The comparative analysis among the measurement procedures of the different CCSS reveals that while all of them, with the exception of DNV-GL, are moving toward the "harmonization", adopting the same approach for TL (Transmission Loss), TL = 20 Log10 DCPA recommended by ANSI/ASA and ISO-17208-1. It is rather curious that DNV-GL, without any technical or experimentally supported data, has adopted the formula TL = 18 Log10 DCPA instead of the other one used by the others. The possible reason is that for a value of the DCPA around 100 m, measurement distance range recommended by the Standards, DNV-GL, will always obtain a value for the underwater acoustic signature characterization of the ship (as a spherical source), *4 dB below* with respect to the values provided by the others procedures for the same ship. And that figure is without taking into account the additional *reduction of 5 dB* justified as a correction due to the reflection of the noise on the seabed.

Accordingly, the same ship would present a completely different picture about its URN signature depending on how it is measured and how is the data analyzed. The *repeatability of the test and comparability of results*, which is one of the main objectives of the Standards, is definitively impractical and unfeasible. In order to support this controversial statement and with the sole purpose to contribute to solve this *black spot* of measurement heterogeneity, Fig. 5 presents the underwater acoustic signatures of an FRV obtained with different post-processing methodologies. The input data have been the same in all cases and as it has been shown, the results vary depending on the methodology.

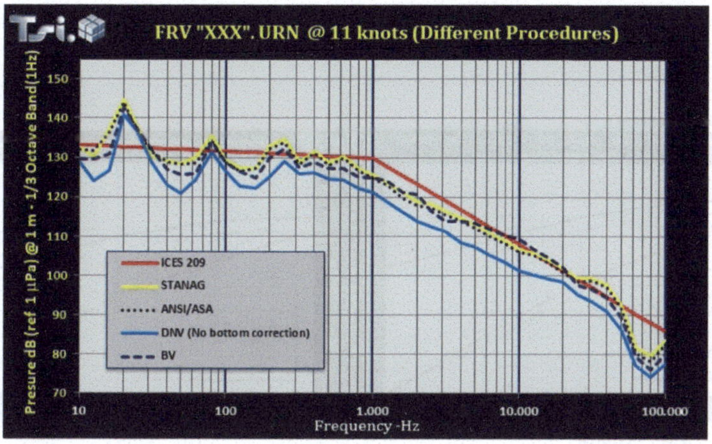

Fig. 5. Comparison of URN results depending on post-processing methodology (Source: TSI)

It is clear that the availability of consistent experimental URN data is of crucial importance as "starting point" in the challenge to reduce the URN by the ships. Fortunately, this issue is more and more consolidated, considering the movement of the marine sector and the different publications released lately related to URN experimental data analysis [17] and [18] as well as ECHO program (Port of Vancouver, Canada) and R&D projects [1] such as AQUO, BESST, SILENV and SONIC, in which were also performed URN measurements of different vessels making possible to extend vessel URN database. Recently, it can be seen on the marine sector that some shipowners and equipment suppliers demand voluntarily URN measurements in order to prepare themselves better for the future environmental limits.

2.3 URN Signature of a Ship Depending on the Classification Society Selected: Consensus as a Need of the Sector

As it has been highlighted above by the author that it is needed an agreement to reduce the market confusion regarding URN emitted by vessels, in order to be efficient in the common commitment of its reduction. The risk of confusing the market is evident, and it should be solved by all the stakeholders, specially by the Classification Societies as *independent parties* of the marine sector. It is confusing and far from the seriousness that the same type of ship can receive the certification of *silent vessel* or not depending on the Classification Society chosen. And on top of that, when such methodology or limits of acceptance have been defined without any technical support that endorses them. These issues, sooner than later, would be perceived by the marine sector and society worried about the pollution of the oceans, as a commercial strategy instead of a serious commitment to the challenge of reducing URN.

One of the main Classification Societies has released to the market a set of very flexible noise limits in all the frequency ranges where the noise induced by the machinery and propeller is always located, even when the propeller is cavitating. Once again, it will be very easy to understand what will be the *most convenient choice* from

Shipowners and Shipyards' point of view to avoid problems during the building and the certification obtaining.

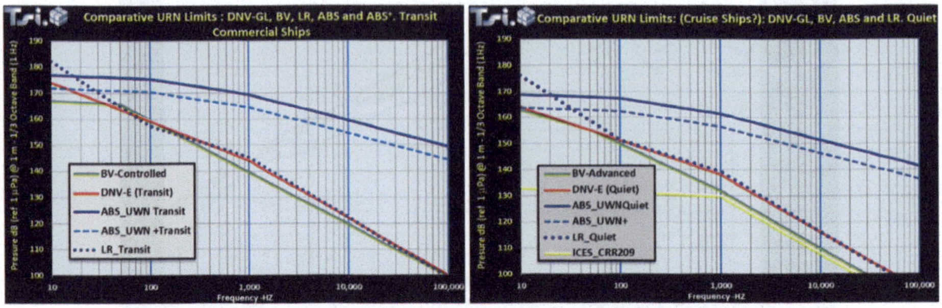

(a) Transit (b) Quiet

Fig. 6. CCSS URN limits and ICES209 comparison (Source: TSI)

As a comparative analysis, in Fig. 6 is presented the URN limits required by several classification societies: BV, DNV-GL, LR and ABS, for *"Transit"* and *"Quiet"* operational conditions and for two types of ships: commercial vessels and cruise ships. The curve of limits required by ICES CRR209 for the FRV's has also been included. This type of vessel is named as "R" by all the Classification Societies that maintain for them the same criteria adopted in 1995 by ICES CRR209.

As it can be seen in Fig. 6(a), while the limits required by BV, LR and DNV-GL for the same commercial ships in Transit condition are very close for the selected frequency range (10 Hz–1 kHz), the ABS's limits are much less restrictive than the others, with deviations that varies from +18 dB (at 100 Hz) and +25 dB (at 1 kHz) to +40 dB (at 10 kHz). In the case of cruise ships or commercial ships in Quiet condition the comparative trend is almost the same. BV, LR and DNV-GL require noise limits very close while ABS's limits are above those. In this case the deviations are around as follow: +16 dB (at 100 Hz), +22 dB (at 1 kHz) and +41 dB (at 10 kHz).

Secondly, In the Fig. 6(b), for comparative analysis has also been included the noise limits curve required by ICES CRR209 for the FRV's. As it can be seen, the noise limits required to any FRV are located, at least, ***−36 dB***, in the frequency range 10 Hz–1 kHz, and ***−43 dB***, at 10 kHz, below the noise limits required by the least restrictive Classification Society, for any cruise ship or commercial ship in Quiet condition. It should be reminded that FRV sector is a small segment of civil ships that must respond, as it is already doing, to the most restrictive limits of URN. The marine sector must find the right guidelines to reduce the URN by all types of ships through technical advances and innovations applied to FRV technology in the last two decades.

Curiously, all classification societies have achieved a *harmonization agreement* for the FRV not only regarding to the noise limits in all frequency ranges (10–100 kHz), but also related to the operational speed conditions under which the experimental tests for assessment must be performed. Thus, the dynamic and acoustical design of these ships should be performed in the way to meet the noise limits at 11 knots, what is well known since the beginning of each project. On the contrary, the *ambiguity* of the term Transit and Quiet from some of the URN notations, regarding to the speed (power rate) in which experimental tests are performed should be clarified from the engineering point of view.

As a summary, while the FRVs as civil ships, must face the strictest URN limits at the defined speed of 11 knots (also adopted by all Classification Societies), the rest of commercial vessels are able to select the laxest Classification Society as the most convenient one to obtain the formalism of certification. Another inconsistency is related to the operational speed of the URN measurements for commercial vessels, since it is not defined (unlike FRV sector) or is a decision of the Classification Society, creating possible suspicions and mistrust on the methodology followed.

In order to clarify this existing *black spot* abovementioned for all the ships that have obtained the certification of silent ship, a set of parameters should be publicly accessible to avoid the lack of transparency in the marine market: corresponding ID card (underwater acoustic signature) of the vessel, tests speed and information related to the measurement procedure used. At least, this clarification should be addressed for ships which have the possibility to access to economic incentives coming from public money for the prevention of fraud or abuse.

2.4 Technical Topics Related to the Main URN Sources, Definitively Clarified

There is some light at the end of the tunnel. The good news is that it has been developed a *harmonized format* to present URN Spectra adopted for most of the professionals of the URN sector, presenting the pressure (ref 1 μpa) at 1 m of the hull and 1/3 Octave Band (1 Hz) against frequency (Hz) in the range of 10 Hz to 100 kHz. In Fig. 7 is shown an example of this representation of the URN signature:

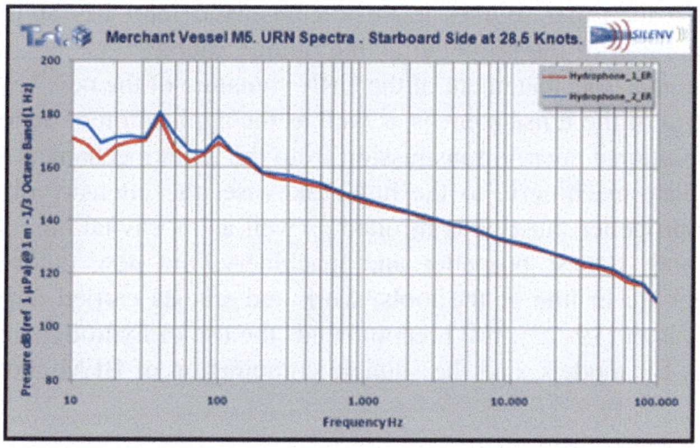

Fig. 7. Typical URN 1/3 octave band spectra of a commercial vessel at 28.5 knots (Source: TSI)

In parallel, since a long time ago, an intense debate was opened about which were the main sources that contribute to URN in a vessel. Traditionally, the propeller was considered as the principal root cause of the URN signature of the vessels, without any experimental support. This kind of assertions were made based on outdated information, in which the propeller appeared as one of the main root causes of vibrations and noise on board. To bring light into this matter, in the framework of the SILENV project, a set of on-site measurements activities were carried out to obtain the URN signature with satisfactory frequency accuracy and resolution (0,5 Hz) in order to identify in the energy spectrum the contribution of each one of the main URN sources (machinery and propeller). Figure 8 presents the spectrum in narrow band corresponding to a ROPAX ship as an example of the results obtained.

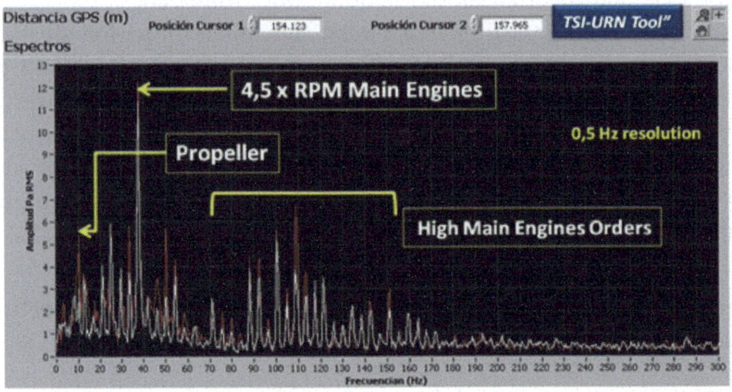

Fig. 8. Typical URN 1/3 octave band spectra for merchant vessel at 28,5 knots (Source: TSI)

Figure 8 shows that the narrow band spectrum analysis allows identifying with enough accuracy the contribution of each main source, both machinery and propeller, and depending of each case, the global noise radiated by each ship. This sort of analysis demonstrates that the other sources must be considered for URN and neglecting the traditional assumption of the propeller as the unique source.

According to that, the abatement of the URN signature of the new vessels requires a preventive control of the noise sources as well as the application of actions focused on reducing the vibrational energy transmission from the main mechanical sources, as the main and auxiliary machinery, to the hull. Likewise, the intensity of the excitation induced by the propeller must be controlled as well as its cavitation.

To tackle both issues: propeller and machinery, the new guidelines provided recently by IMO are in line of the tools, work and studies carried out by the author throughout his career [19–23]. IMO recommends the use of hydrodynamic simulations based on CFDs, FE models, and the suitable combination of BEM and SEA models.

3 Towards Harmonised Strategy of Reducing the URN of Ships

The shipping noise and its impact on the marine fauna and environment is a global issue that cannot be reduced by acting, only, in one or another maritime region. Thus, it requires global and harmonized efforts from all the stakeholders and countries around the world. In this sense, Canada's initiative deserves especial mention. It is the results of the efforts made by Vancouver Fraser Port Authority who in 2014 started the ECHO (Establishing Cetacean Habitat and Observation) program to better understand and manage the impacts of shipping noise in the endangered resident population of Southern Resident Killer Whales (SRKW) located in their natural habitat in the Salish Sea (British Columbia). In a clear vision of the maritime market, the policy applied by Vancouver Port had consisted in the reduction of the port's taxes for all ships that could prove that they are implementing mitigation measures or changing its operational profiles to reduce their URN emissions.

Once again, as public money is involved, the transparency and homogenization of measurement procedures and limit values are essential. In the next paragraphs are summarized a set of proposals addressing several corrective actions that would contribute to clarify the current confused scenario:

- **Harmonization of the Regulatory Framework by the different States.** Even though it is not an easy task for the policy-makers, the regulatory framework should be harmonized, applying different mechanisms such as specific requirements stated by IMO as well as the different States, Ports and the Flag administrations.
- **Harmonization of the URN Measurement Procedures.** Significant efforts have been carried out by ANSI/ASA and ISO in this sense, particularly related to the URN measurement procedure in deep waters. But due to water depth limitations observed in the proximity of several very active shipbuilding areas, additional efforts should be applied by both regulatory bodies to the measurement procedure in shallow waters. In this concern, the author suggests the convenience to review the "European URN Measurement Procedure" for both, "deep" and "shallow" waters and its corresponding uncertainties assessment. The homogenization is also strongly required in terms of propagation (TL) model and seabed effects model for comparative analysis and practical engineering applications. These topics were addressed, during 3 years of research in the framework of the AQUO project, by a wide variety of experts including shipyards, Classification Societies, universities, research centers, towing tanks, marine biologist and engineering consultants.
- **Harmonization of the URN Limits of the different Classification Societies (CCSS).** The harmonization of URN limits for the different types of ships is an absolute "must" in order to avoid acoustic signature discrepancies for the same ship depending of the classification society chosen, that might drive to confusion of the label of a vessel as *silent* or *noisy*. A desire for harmonization is a *permanent clamor* in all technical meetings promoted by FINCANTIERI (one of the most active worldwide shipyards specialized in the construction of cruise vessels), and with the participation of all relevant Classification Societies and engineering experts. As a result of these technical meetings, BV and RINA are moving forward

to the harmonization of their corresponding URN limits and corrections. On the other side, the double role assumed by DNV-GL, acting as URN consultant and also as a certification entity with its own URN limits, does not seem to be compatible and consistent with the transparency demands of the market.

- **Management of the URN: Old Fleet and new commercial ships. URN Identification Card.** It is challenging to control and manage the environmental impact of a huge fleet of existing ships with an average age above 22 years. Conservative actions might be carried out to reduce the URN of the fleet focusing on operational mitigation measures such as navigating far away from protected areas and reducing their speeds. In both cases, *"how far away"* and *"how much reduce the speed"* are technically undefined. These types of actions, which have significant impact on the exploitation of the ships, will drive to shipowners to include technical changes in their future vessels to reach future limits of URN. In the meantime, the required distance and speed should be defined based on real underwater signature of each ship, by means of an URN Identification Card (URN_ID) that would include the URN levels measured at different operational speeds by certified independent companies. Additionally, for the FRV sector, the URN_ID would avoid the wrong denomination of "Fishery and Oceanographic Research Vessel" to vessels that do not meet the URN limits required by ICES CRR209.

4 Conclusions

As it has been shown in Sect. 2.3, the URN levels emitted by a standard commercial ship built on 2014, at some frequencies and at transit speed achieve values that exceed around 60 dB the corresponding values for any FRV that meets ICES CRR209 limits. For those not familiar with decibels (dB), it means that URN produced by modern commercial vessel is almost one thousand times higher than the URN corresponding to a FRV (see Fig. 3, FRV Ramon Margalef). Therefore, it can be said that the most modern FRVs built according to ICES 209 requirement can be considered as technological references for the Civil Shipbuilding Industry as well as for the Naval Shipbuilding Industry.

The main reasons for such a different underwater acoustical behavior are related to important differences in the conceptual design, the careful selection of critical noise sources (especially machinery, propeller and other sources of vibrations and noise) and the applying of the most efficient vibrations and noise control tools in all the stages of the project as well as the final sea trial to obtain URN signature to ensure that all actions adopted were the proper ones.

Below are described the conclusions achieved:

- FRV sector is a small segment of civil ships that must fulfill the most restrictive limits of URN, accepted and consensual by all the Classification Societies. Consequently, FRVs have become a *technological reference* and a clear example of the technical advances and innovations that should be applied to the rest of the Civil Shipbuilding Industry to minimize the vessels impact in the oceans. Of course, other

factors such as cost/benefit ratios and fuel efficiency should be considered and properly weighted.
- The validation of a vessel as FRV, and thus as *technological reference,* should be performed by a third independent party, issuing the corresponding official certification including its URN_ID.
- The harmonization of the URN measurement procedures and URN limits are inescapable to avoid confusion of the marine sector and increase the competitiveness of all the players involved in this market. Only in such case, it will be possible to guarantee that effective innovations and technical solutions will be useful to other potential shipowners and will avoid replication of mistakes of FRVs that in fact do not fulfill the ICES CRR209 limits.
- The tools and studies recommended by IMO to analyze the effect of noise sources on the URN of vessels are in line with the current studies of some engineering companies with broad expertise in noise and vibration control, however, further research is needed to perfectly understand the behavior of them and its implications on the global vessel URN.

There are available effective and proven tools at disposition to the Maritime Sector and the Shipbuilding Industry to tackle the challenge of the reduction of the URN by ships. The challenge is how to motivate their implementation and how to make compatible the economic development with the sustainability of the marine environment. The process will take time, thus, no more delays in actions are affordable. Only willingness and decision-making are needed. The future generations will value our responsibility.

References

1. BESST-Breakthrough in European Ships and Shipbuilding Technologies, SILENV-Ships oriented Innovative soLutions to rEduce Noise & Vibration and AQUO-Achievement Quiet Oceans, projects founded by the EC within the 7° F.P (2012)
2. Beltrán, P., Salinas, R., Moreno, A.: The new IMO noise code: a lost technical opportunity. Irreversible and high cost consequences for fishermen and other seamen that will continue being deaf. In: ICSV (2014)
3. Frisk, G.V.: Noiseconomics: the relationship between ambient noise levels in the sea and global economic trends. Sci. Rep. **2**, 437 (2012)
4. Bureau Veritas. Comfort Class Notation
5. Directive 2008/56/EC of the European Parliament and of the Council of 17 June 2008 establishing a framework for community action in the field of marine environmental policy (Marine Strategy Framework Directive)
6. IMO: Guidelines for the reduction of underwater noise from commercial shipping to address adverse impacts on marine life. MEPC.1/Circ.833, 7 April 2014
7. Beltran, P., Diaz, J.I., et al.: Achievement of the new underwater radiated noise requirements by the Spanish shipbuilding industry. The FRV "Ramon Margalef". In: Proceedings of the 11th European Conference on Underwater Acoustics (2012)
8. ANSI/ASA S12.64-2009/Part 1: Quantities and Procedures for Description and Measurement of Underwater Sound from Ship – Part 1: General Requirements

9. ISO-17208-1:2016: Underwater acoustics - Quantities and procedures for description and measurement of underwater sound from ships - Part 1: Requirements for precision measurements in deep water used for comparison purposes
10. ICES CRR N° 209 Regulation (1995)
11. Palomo, P.B.: The most recent noise & vibration assessment of the European fleet, within the framework of the "SILENV Project". Ship Sci. Technol. **7**(14), 43–66 (2014)
12. Palomo, P.B.: Reduction of underwater radiated noise from ships: new shipbuilding challenge. The vessels "Ramón Margalef" and "Ángeles Alvariño" as technological references of how to build silent vessels. Ship Sci. Technol. **8**(15), 45–60 (2014)
13. DNV-Silent Class Notation. Part 6, Chapter 21, January 2010
14. Bureau Veritas, Rule Note NR614 DT R00E, Underwater Radiated Noise (URN). Bureau Veritas, October 2014
15. AQUO European R&D Project, deliverable D3.1, European URN Standard Measurement Method, April 2014
16. Rodríguez, A.M.: New european underwater noise measurement procedure standard developed in AQUO project. In: OCEANS 2015, Genoa. IEEE, May 2015
17. McKennaa, M.F., et al.: Underwater radiated noise from modern commercial ships. J. Acoust. Soc. Am. **131**(1), 92–103 (2012)
18. Gassmann, M., Wiggins, S.M., Hildebrand, J.A.: Deep-water measurements of container ship radiated noise signatures and directionality. J. Acoust. Soc. Am. **142**(3), 1563–1574 (2017)
19. Beltrán, P.: Noise vibration on ship prediction: basic engineering tool to fulfil the current requirements of comfort and quality. Naval Engineering Magazine (2001)
20. Beltrán, P., et al.: Silent fishing vessels. A milestone for small and big Spanish shipyards. Lessons learnt. Naval Engineering Magazine (2006)
21. Beltrán, P.: The Fishing Research Vessel "Miguel Oliver". Naval Engineering Magazine (2007)
22. Beltrán, P.: Oceanographic Ship "Miguel Oliver": The excellence on Noise and Vibration on board fulfilling ICES-Nª209. First Prize 47th Congress of Naval Engineering and Maritime Industry (2008)
23. Beltrán, P., Tesorero, M.A.: Spanish Shipbuilding ready to meet the new and demanding Environmental Requirements: New Challenges and Business Opportunities. Ingeniería Naval 887. Silver Price 49th Congress of Naval Engineering and Maritime Industry (2010)

Risk Based Design with Goal Based Standards and Safety Level Approach. An Example Based on a Spanish Fishing Vessel Fleet

M. J. Núñez-Sánchez(✉) and Luis Pérez-Rojas

Doctor Ingeniero Naval y Oceánico, Universidad Politécnica de Madrid, Avenida de la Memoria s/n, Madrid, Spain
nunez.miguel@gmail.com

Abstract. Due to the quick advance in technology there is already a need to develop designs considering risk from a formal point of view to escape from prescriptive regulatory approaches and allow flexibility. In order to do this, there is a benefit in providing the current regulations with functional requirements with the additional benefit to set the frame to develop a structure providing rules for rules. It is proposed that a structure with goals, functional requirements and open and shared high-level risk models are developed at the level of the Administration that can be later complemented with low level risk models. In order to put this into practice a complete example of goal-based standard embedded in risk models was developed leading to more rational, cost effective and explicit regulations.

Keywords: Alternative design · Risk · Goal Based Standards · Safety level approach

1 Introduction

Naval engineering and ship construction has undergone many technological changes in the last few years. There are new areas in which the progress has been noticeable such as new materials, more and better sensors, the development of autonomous systems combining hardware and software, more capacity to analyse data, optimized propulsion systems and a considerable improvement in communications systems. In particular the large amount of information that may be obtained throughout ships equipment, systems and incidents may be used for allowing flexibility in the design. All of the above, if combined with competitive costs of manufacturing, will increase the trend to incorporate technology in the designs.

With this it is foreseeable that ship-owners will demand more personalised and specialized designs with lower operational costs and it is also expected that the design process will be even more important so that shipyards may incorporate more technology with less interferences in the design process.

In order to be more competitive and be able to provide the client what is needed it is necessary to take these decisions at pre-contractual level and this is why it is of the utmost importance to gather as much knowledge as possible from the very beginning.

V. A. J. E. Carreño Moreno et al. (Eds.): CIDIN 2019/COPINAVAL 2019, pp. 61–78, 2020.
https://doi.org/10.1007/978-3-030-35963-8_6

However, from the regulatory point of view the novel engineering solutions need to fit in the classic regulations, therefore one of the key aspects will be being able to break ground in technical, legal or financial matters and this is also why risk management will become more important in the future [1].

The approximation to risk in the maritime industry started at the beginning of the 60s within the subdivision and damaged stability in a probabilistic framework and was later applied in the offshore industry in areas such as fire protection [2].

This approach found resistance in the sector due to a conservative attitude and the regulatory framework. Ship owners, design departments, companies, shipyards and administrators had the compliance of prescriptive regulations as a limitation to general arrangements, on board installation of equipment and ship dimensioning as it can be seen in Fig. 1, which clearly shows that the better designs may be away from the optimal. However, when compared with the traditional approach most ships are still designed following in their entirety prescriptive regulations, meaning that there is a barrier for the design. This kind of resistance is still found today *inter alia* in the oil tanker industry [3].

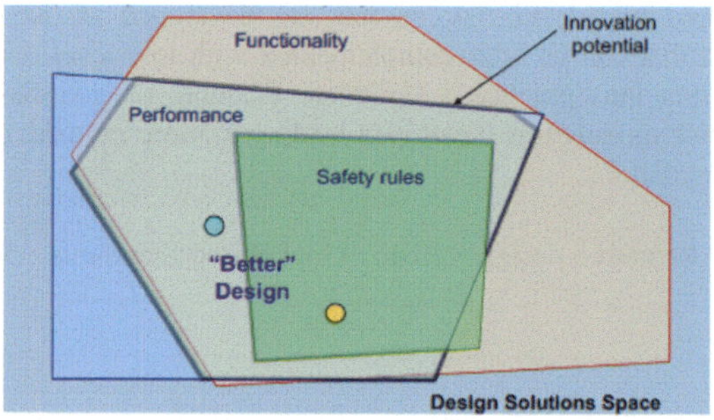

Fig. 1. Optimal design solution out of the regulatory framework [4].

Furthermore, at the level of the International Maritime Organization (IMO) issues such as ships' stability, structural fire protection and life-saving appliances have been studied in an isolated manner, without considering the ship as a whole and as a reactive response to accidents, meaning that the consequences of an accident or a series of accidents are triggers to amend or add new regulations.

Nevertheless, the IMO worked in providing a legal framework for alternative designs adopting SOLAS Chapter 2-2 regulation 17, on alternative design and arrangements to comply fire safety with engineering analysis. This solution was later adapted to other parts of SOLAS related to life-saving appliances and arrangements, machinery, electricity and automation or for gas fuelled ships under the International Code for Safety of Ships Using Gases or Other Low Flash Point Fuels (IGF Code), mainly on LNG. Alternative design has been used in the cruise industry for ad-hoc regulations due to the complexity of the process of having alternative design approved, but one of the fundamental problems when incorporating risk design relates to the need to provide transparency to the approval process to be able to be continuously subject to revisions, both internally and externally. In this regard this paper will explore how to provide transparency for these designs considering risk methods.

In other engineering fields risk methods have become more frequent for decision taking to reach what may be called "safety objectives". There are some principles to follow such as the use of a consistent measure of safety using a formal procedure to allow its quantification, models able to measure functionality and expected performance and to find threshold values considering that:

- the measurement of safety will also need to be integrated in the design process to be able to compensate safety and other design factors taking into consideration the necessary performance, costs during the "lifetime" as this will provide a decision tool for final decision making and optimization. This means that it is necessary to find the main hazards and risk factors related to the accident categories to be able to apply formal safety tools.
- with regards to functionality it is necessary to take into account the level of calculations to be able to deal with all safety aspects and the potential effects in designs and the expected performance. This means the need to use parametric models that would allow checking how the changes in ship design compensate with security aspects when they overlap.
- regarding the threshold values these are one of the keys for the approval of risk-based designs. In this regard in order to guarantee an equivalent safety level the new designs will need to be assessed in a quantitative manner, meaning a quantitative risk assessment (QRA) estimating probability, consequences and expressed in individual or societal risk parameters.

With all of the above if quantitate risk assessment can test possible engineering solutions and risk models demonstrate that the solutions are superior, without compromising the ship's efficiency or performance, existing designs may be improved and be approved with an equal or higher safety level something that cannot be approved in the current regulatory context except for ad-hoc regulations.

There are some difficulties in these processes:

- first, the only reference that is currently provided is the prescriptive regulation, with its implicit safety level, and all innovative solutions would have to be as safe as them. One of the main weaknesses of this process is that the current regulations may well have been developed through methodologies that cannot be sufficiently justified or be far away from the optimal solution and with no consistent safety level, and
- second, the development of risk models for a particular ship is expensive and requires considerable engineering effort. Hence these tools would not be accessible to the whole sector and usually need proper risk-based objectives.

Although there is not a simple policy to solve these problems, there are different stages the risks tools that may be used during risk design: Hazard identification (HAZID); risk analysis decision making; document preparation of risk-based design and finally the approval given by the Administration that considers that the safety level required by the Administration has been met.

In conclusion it would be very important that the Administration can contribute to this process to avoid ambiguities. On the other side there is a need to develop regulations with a basis and familiarize the sector with risk. With this the future might provide the following possibilities [2]:

- *"partial risk-based designs with a safety level for some functional requirements that can be increased to a function of the ship;*
- *partial risk-based designs with a safety level met by function of the ship which encompasses several functional requirements; and*
- *full risk design".*

In spite of the fact that risk design might not be able to be currently used for low added value ships or ships that are normally customized there should be a need offer this possibility in the future. Therefore, it was decided to work in this direction so that Maritime Administrations may offer a basic layer of functional requirements that will also contribute to develop long lasting regulations and contribute to disseminate and improve the lessons learn in maritime accident investigation.

This paper intends to demonstrate that the use of risk-based design is feasible when data is available and suitably processed and also beneficial to provide clarity in regulation development.

2 Tools for Risk-Based Design and Rule Development

Taking into account the objective of the paper indicated above while looking for a solution at the level of a Maritime Administration it was considered unavoidable to take into account IMO trends and methodologies as these offer the potential for allowing alignment in the future.

At MSC 66, IMO agreed to a new proactive approach to develop regulations before disasters occur. Following the above the IMO also developed some tools and concepts for risk-based methodologies in order to develop regulations which may well live with flexibility in the design. These tools are Formal Safety Assessment, Safety Level Approach and Goal Based Standards.

2.1 Formal Safety Assessment

FSA is a tool with some steps as outlined in Fig. 2 [5] which evaluates new regulations and helps to compare proposed changes with existing standards, enabling a balance to be drawn between the various technical and operational issues, including the human element, and between safety and costs. Parts of these tools, in particular the definition of goals systems and operations, HAZID, and scenario definition (cause analysis – consequence analysis –risk control) are used in risk design. In essence the methodology consists in carrying out a risk assessment and considers the costs and benefits using different techniques.

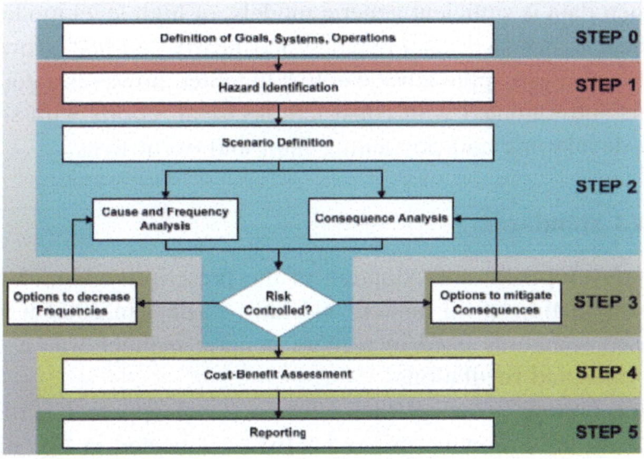

Fig. 2. FSA step approach [5]

In relation to the development of regulations FSA defines recommendations, known as risk control options, which should be presented to the decision-makers in an auditable and traceable manner. These recommendations are based upon:

- the comparison and ranking of all hazards and their underlying causes;
- the comparison and ranking of risk control options as a function of associated costs and benefits; and
- the identification of those risk control options which keep risks as low as reasonably practicable.

The FSA approach has been extensively used with a relative success for the main ship types, with models sometimes focused on scenarios with fault and event trees, relating causes to consequences, and this has helped IMO Members States to consider the necessity of new or amended regulations.

Notwithstanding the above, this risk-based approach has some challenges, such as:

- the quality and quantity of the data collected in order to support monitoring and development of safety regulations;
- the integration of risk-based methodologies and the latest analysis techniques into the safety regulatory framework to provide a sound scientific and practicable basis for the development of future safety regulations; and
- the know-how required to use these tools versus the traditional approach to propose new rules or amend new rules with justifications that are not required to contain detailed documented rationales, basis for assumptions, description of uncertainties or sensitivity analysis.

The main obstacle is data as ship owners and the sector is normally reluctant to provide information related to near misses or faults that could feed into event trees to enrich them. This is why FSAs are disputed as these could provide opposite results depending on the dataset used.

However, when data is sufficient generic models, or high level models, can be used and whenever there is not such data detailed and nearly "ad-hoc", low level models, need to be developed. Both approaches can live together, however a good basis lays in high level models substantiated with consistent accident reports with similar structure that could allow developing and populating fault and event trees.

2.2 Goal Based Standards

Even before the development and adoption of the prescriptive Goal Based Standards (GBS) for the construction of oil tankers and bulk carriers in 2007 it was considered that the goal-based standards used in a more generic manner was a solution in the development of rules and regulations.

GBS is a "top-bottom" concept that offers a tiered approach, "rules to develop rules", working with the following principles [6] as indicated in Fig. 3:

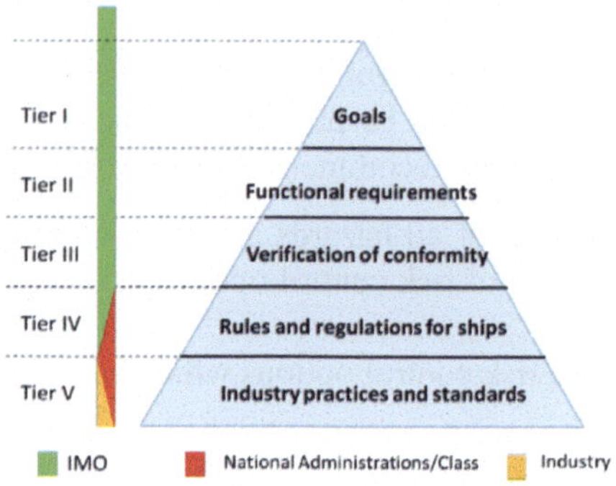

Fig. 3. Goal Based Standard tiered structure

- tier I- Goal, which is a high-level objective to be met that should address an issue of concern;
- tier II- Functional Requirements, which provide the criteria to be complied in order to meet the goals and are developed after the goals and considering the relevant hazards;
- tier III- Verification of Conformity, which provides a transparent instrument necessary for monitoring and verifying that the associated rules and regulation for ships conform the goals and functional requirements;
- tier IV- Rules and regulations for ships, which are the detailed requirements (developed by IMO, a national administration, a Recognized Organization or a classification society) and applied by them that need to meet the goals and functional requirements; and

- tier V- Industry practices and standards, developed as a consequence that may be referenced in the rules and regulations.

In connection with alternative design the idea to establish functional requirements was used in the latest revision of Chapter SOLAS II-2 on Construction - fire protection, fire detection and fire extinction. Sometime later, the idea to use some parts of this structure, through a Goal Based approach, in particular the use of the concepts of Tier I and II exclusively, has been applied in the development of IMO instruments such as the above mentioned IGF Code and the Polar Code:

- with the aim to provide more clarity and, in addition;
- to be able to provide a possible flexibility in order to comply with these functional requirements by means of risk management tools in alternative designs; but
- without a pre-established agreed criterion in their definition; and
- following a "bottom-up" approach that makes the functional requirement self-limiting.

Hence GBS tiers I and II introduce objectives and functional requirements that may allow providing clarity in the necessary functions in ship design similarly to what was stablished in alternative designs and arrangements, however these will need to be prepared for their quantification.

2.3 Safety Level Approach

The concept of safety level approach (SLA) refers to the structured application of risk-based methodologies to reach an explicit safety level or to verify compliance with the rules. The aim is to have quantitative and rational safety levels to be able to be used and provide a way to measure safety in the ship concept and the human element in the IMO rule making process. This safety level should be able to be adjusted when needed *inter alia* based on cost criterion to implement measures; i.e. a higher safety level would allow more costly measures.

This approach needs the development of quantitative or qualitative safety levels and processes to be used for achieving a practicable safety level, or an implicit safety level such as that in the "as low as reasonable as practicable" (ALARP) principle [7] with F-N curves which takes into consideration the number of accidents with N equal to 1 or more fatalities and the frequency as per Fig. 4. These curves are a measure of societal risk [8] and exclude time exposure. With this Administrations or the IMO could revise and adjust the safety level as needed when this is not sufficient, or it is overestimated by setting a curve in an area ranging from negligible to intolerable.

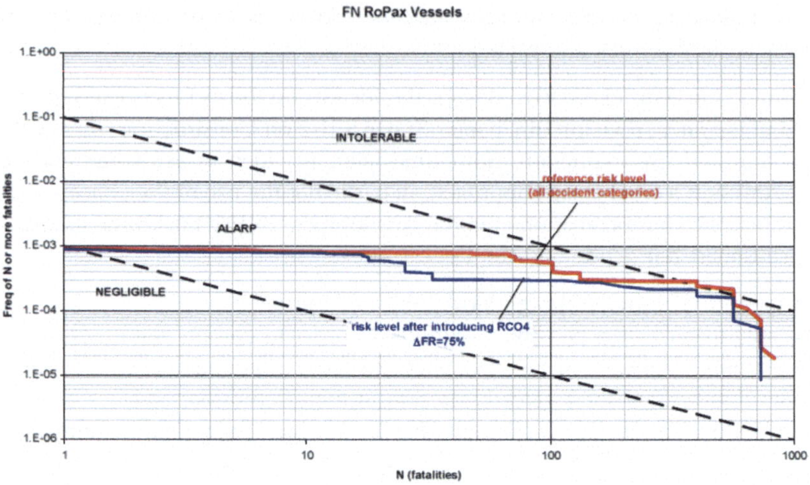

Fig. 4. FSA Ropax. Change on societal risk after introducing an improvement in evacuation which consider that the risk is lowered 75%. This measure didn't pass the cost benefit analysis when examined in an FSA due to a cost superior to 3 M$ [7]

The safety level would assist to determine whether the costs of measures can be increased, taking into account the historical data or the ALARP curves (by setting the current safety level as acceptable).

2.4 Extension of Risk-Based Approach. GBS-SLA

When considering the above indicated tools a complete regime based in risk can be implemented both for risk design and for developing regulations. Combining the safety level with the GBS and the SLA the concept of Goal Based Standards and Safety Level Approach (GBS-SLA) may be established, but in order to provide a quantitative risk basis there is a need to incorporate formal safety assessment. This combination, with suitable models, will also allow designers to overcome prescriptive regulations.

The way to achieve this is by introducing safety levels that will have to fit in the previously mentioned ALARP curves to manage the risks. If the safety levels are adjustable risks could be better handled depending on the context. As an alternative the safety level based on cost criteria could be assessed due to the fact that the ALARP curves provide a statistical reality but may consider accidents with heterogeneous regulatory schemes.

In this particular case it will be considered very advantageous to develop high risk models with the aim to be made open to the sector, to be able to input data, sharing them with the industry, and to be able to implement the FSA in the Maritime Administration. The whole process may be seen in Fig. 5. This can lead to develop very collaborative approach between designers, accident investigators and Administrations, creating a very positive synergy.

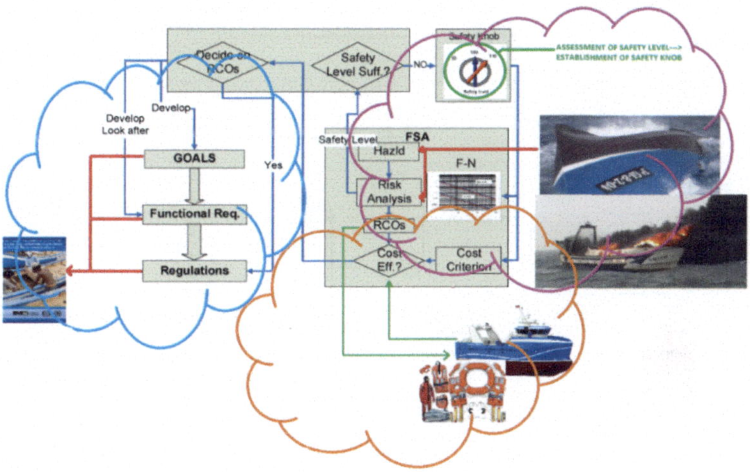

Fig. 5. Combinación of risk assesments, FSA, GBS y SLA [8]

Once this has been developed the rest of the GBS tiers would be simpler and may need that the verification for conformity.

In relation with the use of the FSA in the Maritime sector it also needs to be considered whether the risk analysis and the cost benefit analysis needs to be compared with other industries. This can be easily done for an individual country.

In considering the whole integration indicated in Fig. 4 it was identified that functional requirements require to incorporate the hazards and risk factors that may arise with priorities and trying to eliminate ambiguity. In addition, functional requirements would have to be based on the ship functions and for each of the functions an acceptance level would have to be defined with criteria for acceptance. These criteria may be availability, reliability and maintenance.

3 Practical Application

In order to put the above methodology into practice it was decided to apply it in an area with sufficient data to be able to build high level models in an Administration. In this regard the maritime subsector with large data available for a ship type but also with the highest rate of fatalities and a considerable number of accidents is fishing.

In this regard it was decided to adapt a database of fishing vessel 975 accidents happened during the years 2000 to 2014 and calculate costs of repairs, cost of replacement and loss of income. Based on the available information the models considered sequence in the fault tree before the accident happened and events leading to consequences in an event tree after the accident has occurred with conditional probability and considering uncertainty as indicated in Fig. 6, and each accident with enough information was adjusted to the shape of bow-tree analysis.

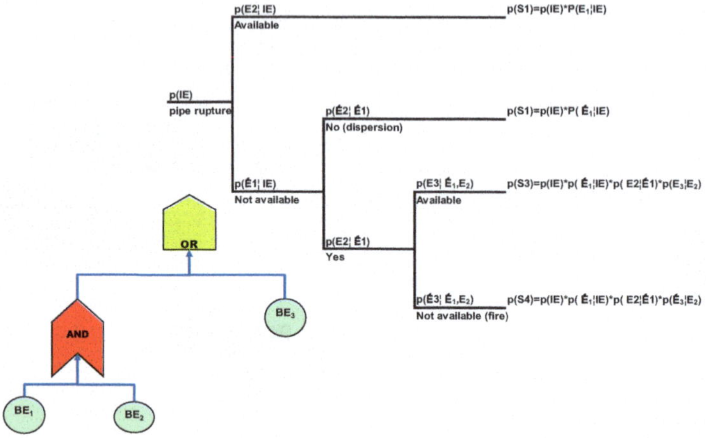

Fig. 6. Uncertainty in high level risk models stemming from frequency.

Before developing the models some of the data would be discarded with the use of sensitivity analysis which implied a limitation in the desired granularity, meaning detailed studies on fishing vessel subtypes or length ranges, cannot be done as the sample needs to be statistically significant.

Five Bow tie high level risk models were developed and considered for the following types of accidents: collision, fire-explosion, grounding, foundering and list considering the risk factors and hazards, mainly in the fault trees. One example is provided in Fig. 7.

When models were prepared the dataset was introduced and the probabilities derived from the models checked against the frequencies, in order to check how the models adjusted. This was carried out by means of ALARP curves. An example of how all the models worked in combination in calculating probability can be seen in Fig. 8 where the blue line indicates the frequency of N or more fatalities and the red line the accumulated probability. It can also be seen in the Fig. 7 that the models worked well in predicting probability, however the granularity is lost, by including a vast number of fishing vessels between 12 and 24 m L. This can be problematic whenever risk control options are introduced without using low level risk models.

After this step it would be possible to introduce potential risk control options in the models, following the flow of an FSA to develop prescriptive regulations, to assist the industry in introducing risk designs or to justify exemptions or equivalences in new designs.

In a similar manner, instead of introducing risk control options as barrier to prevent the escalation of a fault or event, it was considered that functional requirements could be introduced in the models. Therefore, the models could help the purpose of developing new prescriptive regulations based on the GBS.

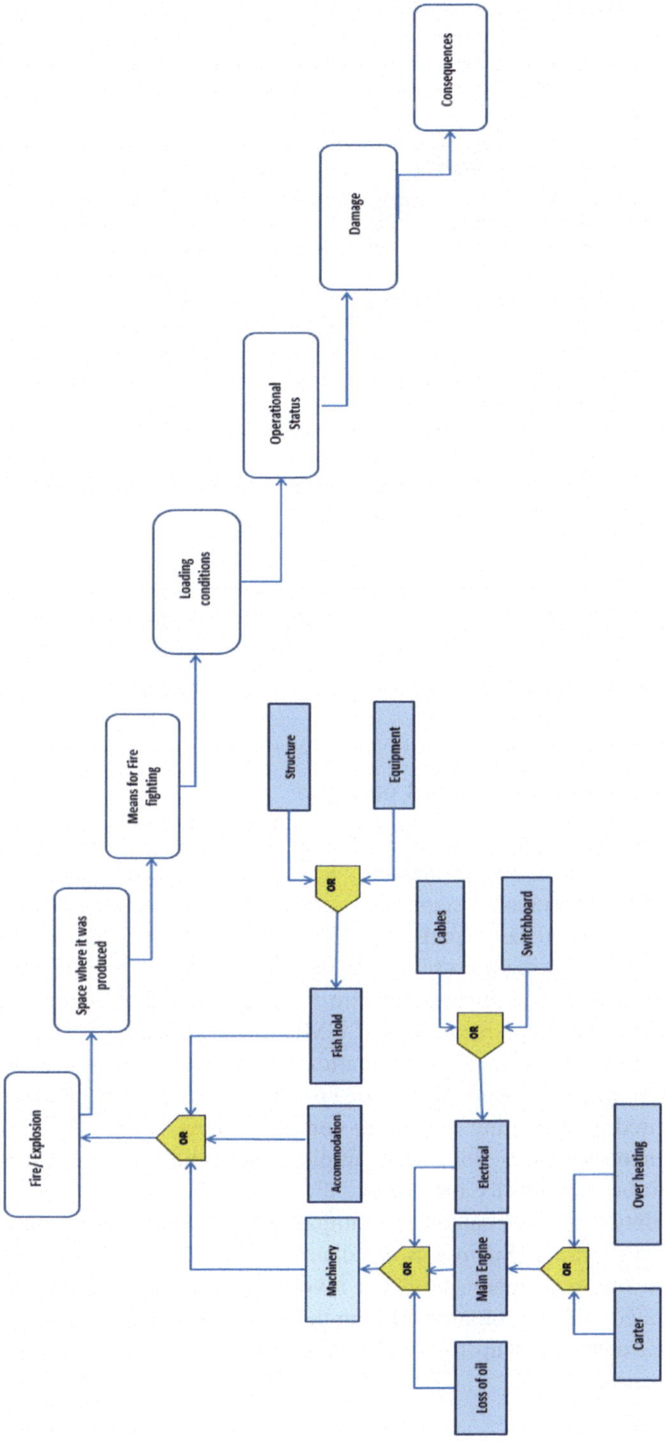

Fig. 7. Risk model for fishing vessels less than 24 m L for fire and explosion.

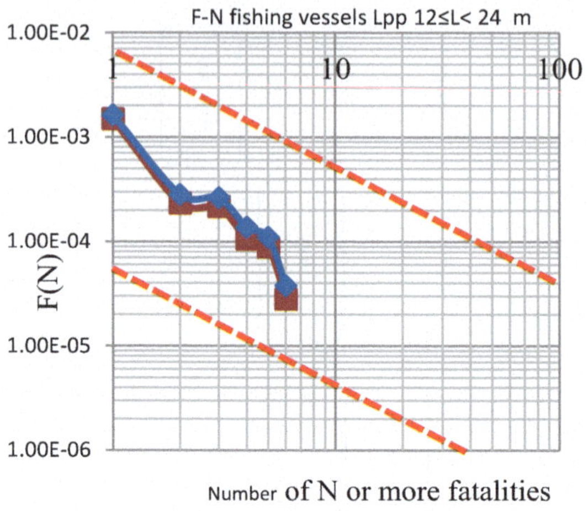

Fig. 8. Societal risks with ALARP areas showing the probability extracted from all models compared with the frequency in the accident database.

Hence taking into account the GBS and considering how difficult is to develop prescriptive regulations for these types of ships with the aim to develop a long-lasting framework and be able to properly develop and justify amendments it was decided that two steps were needed: establish the current safety level and develop GBS goals and then draft the functional requirements.

3.1 Safety Level Calculation Based on Costs

It was also considered that the risk models could well serve to provide a safety level approach. In order to assess the current safety level it was decided to calculate the costs and benefits of the implemented regulations in the Spanish national Administration during the period 2007–2010, in particular "The Code of Safety for fishing vessels" adopted in 2007 [9] consisting on a set of regulations introduced requirements for new and existing ships related to construction, fire-fighting and safety equipment, and the regulations related to the implementation of GMDSS for all fishing vessels [10], with equipment depending on the range of operation of the fishing vessel. It needs to be indicated that these measures were accelerated due to a process of regularization of the fleet which required 1300 fishing vessels to comply as new vessel. This would mean that the requirements for new ships were implemented much faster, and the impact should be able to be more easily measured.

Hence the safety level was calculated comparing the value of preventing a fatality (VPF), which for the case of Spain was calculated as 4.5 million € taking into account the GPD and the life expectancy, with the invested gross cost of averting a fatality (GCAF) as indicated in Eq. 1, obtained diving the cost of a measure by the reduction in risk. This factor f determines up to which extent the threshold value of preventing a fatality VPF was used.

$$GCAF = \frac{\Delta C}{\Delta R} \quad (1)$$

The effectiveness of the above measures was inferred considering the incremental risk during the time which took into account the reduction of the fleet and considering that the measures were set at 100% effectiveness, meaning full contribution of the measures to the reduction of fatalities, which allowed a conservative value of the factor. This simplification is due to the complexity to calculate the effectiveness of the whole package and based on the fact that the risks reductions in the period 2008–2013, 2008–2014 and 2008–2015 increased the f values from 1 to 1,1 for fishing vessels less than 12 m in length between perpendiculars (L) and from 0.5 to 0.8 for fishing vessels on or over 12 m L as indicated in Fig. 9, though these figures tend to stabilise with time.

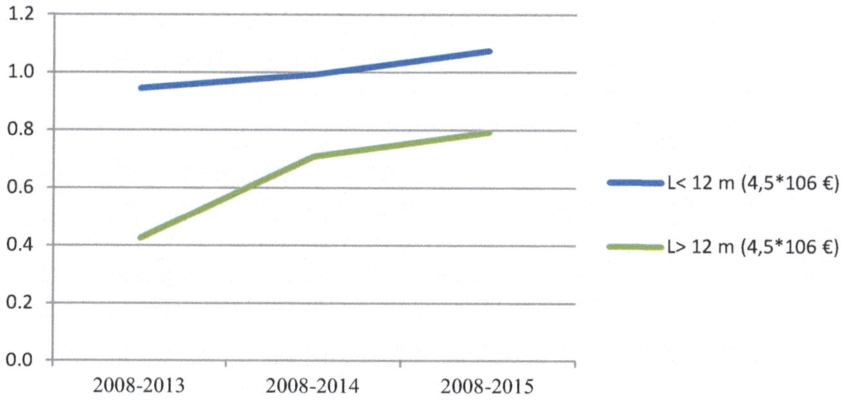

Fig. 9. Variation of f factor depending on the sample taken [8]

Alternatively to this cost estimation values the simplification that could be have been done was to consider the current historical values in the curve as the current safety level and work in this level to keep it ALARP which would mean that the gross cost of the proposed risk control options (new regulations) can go up to the VPF and this can be changed when the F-N curves are in the intolerable region.

The next step was to develop an umbrella, in particular functional requirements, to justify the regulations encompassing the current applicable national safety regulations with the future.

3.2 Development of Goals and Functional Requirements

Due to the fact that the functional requirements would assist in flexible design as well as to consider potential regulations it was decided that these requirements could act in the models considering the implementation of a functional requirement through a preventive measure at fault tree level or mitigating measure at event tree level.

Overarching goals and functional requirements for the safety of fishing vessels following tiers I and II of GBS as indicated in Fig. 2 needed to be developed.

Functional requirements were built up with priorities trying to eliminate ambiguity. In addition to this it was decided that functional requirements would have to be based on the ship functions, for each of the functions an acceptance level would have to be defined with criteria for acceptance. The criteria for availability, reliability and maintenance was be discarded at this time in order not to overcomplicate the requirements as this would require further examination of standards such as those of equipment.

The current works in NATO [11] were used to develop functional requirements with the three following elements: a description, performance requirements/rationale and justification.

The rationale and description were based on the analysis of the current regulations that could also address the hazards. The justification was developed to address the risks and risk factors indicated in Table 1, thus making the functional requirements quantifiable.

Table 1. Risks hazards and risk factors considered in the models and thereafter in the functional requirements

Risks	R	Hazards	P	Risk Factor	FR
Foundering	R1	Water ingress	P1	Structural failure	FR1
List	R2	Bad weather	P2	Failure of closing appliances	FR2
Collision	R3	Loss of weather tightness or water tightness	P3	Failure mooring and anchoring	FR3
Grounding	R4′	Loss of stability	P4	Failure fuel oil retention (includes gas systems for galley)	FR4
Water ingress	R4′′			Failure oil retention	FR5
Fire/explosion	R5			Steering gear failure	FR6
				Overheating	FR7
				Loss of fuel oil supply	FR8
				Human error	FR9
				Failure electrical equipment	FR10
				Loss of electrical supply	FR11
				Navigation error	FR12
				Fatigue	FR13
				Give way	FR14
				Cargo Shift	FR15
				Machinery failure	FR16
				Bilge system failure	FR17
				Carter explosion	FR18
				Failure of navigation equipment	FR19

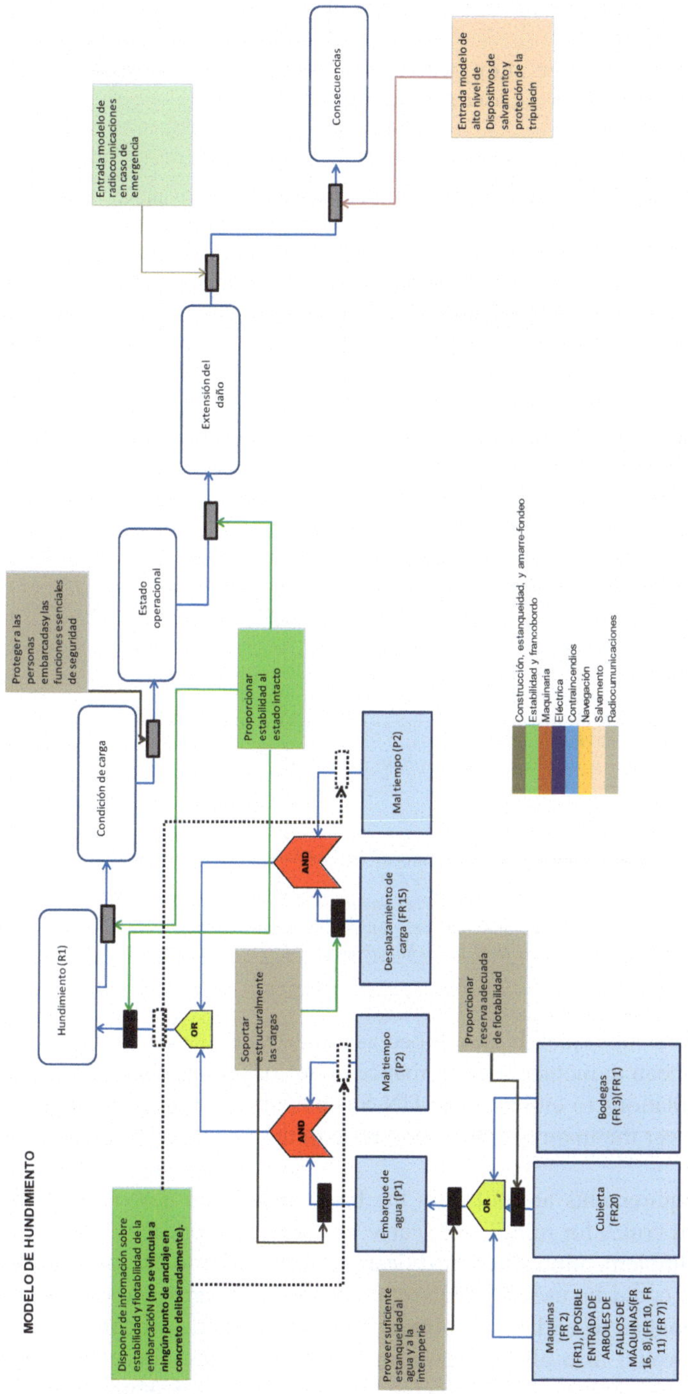

Fig. 10. Foundering model with functional requirements interacting as barriers in a similar manner to RCOs or alternative measures

Fifty functional requirements were developed and connected to the high-level models developed during the risk assessment in step 2 of the FSA as shown in Fig. 10 for foundering [8].

The functional requirements for ship structure and buoyancy are indicated in Table 2 below containing risks, risk factors and hazards.

Table 2. Goal and Functional requirements for structure and buoyancy.

<u>Goal</u>
"To design, build and maintain structure, closing appliances and mooring and/or anchoring equipment to serve the purpose of navigation and fishing offering protection for the crew against the weather conditions and sea state assisting in the economical sustainability of the sector by minimizing the risk of ship loss and assisting in the social sustainability by reducing the risk of loss of lives"

<u>Functional Requirement 1</u>
<u>1.Description and expected results</u>

- *Provide sufficient water tightness and weather tightness*
  - *In all operational conditions that may be foreseen*
  - *Withstand structural loads*
- *For all operational conditions that may be foreseen*
- *Allow personnel on board to carry out their tasks in a safe manner*
- *Protect personnel on board and the essential safety functions*
  - *Against emergencies and accidents*
  - *Providing a safe location on board*
  - *Until hazards have been dissipated*
- *To provide for a safe storage of fuel*
- *For the safe functioning of the engines*
  - *To be able to available and being used on board*
- *Provide accommodation to the personnel*
  - *Suitable for the crew*
  - *In accordance with the size and operability of the ship*
- *Provide means for mooring and anchoring*
  - *In all foreseeable conditions*

<u>2.Justification:</u>

- *Foundering (R1) or list (R2) due to embarkation of water (P1), bad weather (P2) due to failure of the ships structure (FR1) and failure of closing appliances (FR2)*
- *Collision (R3), grounding (R4') due to failure of anchoring and mooring equipment (FR3)*
- *Fire or explosion (R5) due to failure in fuel tanks and fuel systems (FR4) or oil leakage (FR3)*
- *Collision (R3), grounding and water ingress (R4) due to human error (FR9) related to fatigue (FR13).*

Qualitative goals were subsequently developed as indicated in the above table and were incorporated as a mean to include a performance indicator for sustainability in the future allowing the regulations to embrace the UN Sustainable development Goals of 2030 (SDG2030), however the main quantification rests in the functional requirements themselves.

These functional requirements are not only the basis to develop new regulations and therefore considered "rules for rules", applicable to a very heterogeneous fleet, but also the means to start implementing risk-based design [4]. In order to do so the high-level models developed could be used by the sector (mainly shipyards and designers) and enriched by stakeholders to be later combined with low level risk models. These low-level risk models would also allow measuring the effectiveness, instead of just applying a qualitative expert judgement. Following this approach safety would not only be in the hands of the regulatory bodies when they establish the prescriptive regulations but also in the hands of the whole sector and would allow consistency and alternative design. This layer of functional requirements are also the basis for updating regulations.

The high-level risk models were used to benchmark a complete FSA incorporating the safety level estimated. The result was satisfactory due to the fact that the effectiveness of the measure was introduced by qualitative assessment with experts during a workshop and the functional requirements in the models found consistent.

4 Conclusions

Risk based design has been used in Maritime applications in a limited manner as it is more complex than the traditional prescriptive ship design, however it provides considerable benefits in terms of flexibility. On the contrary prescriptive ship design will be a limitation in the coming future.

Taking into account the rapid technological changes it is essential that designers and shipyards familiarize more and more with these tools to be able to provide innovative design. This will improve competition and allow exploring new possibilities considering ships in a holistic manner.

In risk-based design the technical solutions will depend upon the proper identification of risks and the means to revise these designs so that they are acceptable at Administration level. If there is not flexibility by maritime administrations, the innovation will be restrained.

The final acceptance and approval by the Administration will depend upon how compliance has to be demonstrated. In any case the approval of risk-based designs will need to be transparent, predictable and consistent.

One of the main difficulties for risk-based design lies in the establishment of the minimum safety level required and the definition of quantitative functional requirements with the use of tools such as FSA, to be able to develop new regulations or alternative to allow designers to construct novel designs. The use of quantitative functional requirements allows combining all this and expanding into sustainability through the goals.

There are different obstacles for the acceptance and implementation of risk designs and the amount of work needed to obtain their approval will have to be reduced to the minimum to be able to ensure effectiveness. A way to provide this is through the GBS-SLA framework. GBS and SLA approach in ship design will complement alternative designs if open high-level models that may be used by everyone.

It is important to define goals and functional requirements to allow risk-based design. If these goals and functional requirements can also be used for developing prescriptive requirements the consistency will be ensured.

In order to be able to merge risk-based designs under alternative designs regulations and under goal-based standards there is a need to connect the performance of the safety criteria required by alternative design with the expected results required by the GBS functional requirements. In order to do the models with suitable barriers (functional requirements) that can be later enriched by the sector and be complemented with low level risk models to assess the effectiveness of the measures offer considerable potential.

Taking into consideration the practical example in a fleet of approximately 10000 (2000 data) to 8000 (2014 data) fishing vessels it has been possible to carry out an initial approach to GBS-SLA and FSA. By mining the database of accidents to adapt them to the models and carrying out calculations the impact of the current regulations was be measured with the potential to be adjusted in the future, with a safety level based on costs, and also calculate the impact in terms of cost and benefits of new potential measures, which will comply with the functional requirements and also to link them with sustainability through the goals.

Acknowledgements. Thanks to Pitchanat Khomsupphawit for the continuous support.

Disclaimer. The content of this paper does not necessarily reflect the official opinion of the European Maritime Safety Agency. Responsibility for the information and views expressed in this paper lie entirely with the authors.

References

1. Núñez, M.: La construcción de buques basada en objetivos y el enfoque del nivel de seguridad en el entorno normativo. Una nueva oportunidad para la industria marítima española. Ingeniería Naval (943), 77–83 (2016). ISSN 0020-1073
2. Núñez, M.: Towards a new SOLAS Convention: a transformation of the ship safety regulatory framework Transactions RINA. Int. J. Maritime Eng. **158**(Part A2), A139–A146 (2016). ISSN 1479-8751
3. Intertanko, GBS Goal-Based Standards. www.intertanko.com/upload/ISTEC-5-6GBS-Little.pdf. Accessed 12 Dec 2018
4. Papanikolaou, A., et al.: Risk-Based Ship Design Methods, Tools and Applications. Springer, Heidelberg (2009). https://doi.org/10.1007/978-3-540-89042-3
5. IMO, MSC-MEPC.2/Circ.12/Rev2 Revised guidelines for formal safety assessment (FSA) for use in the IMO rule-making process
6. IMO. MSC.1/Circ. 1394/Rev.1 Generic Guidelines for Developing Imo Goal-Based Standards
7. Kristiansen, S.: Maritime transportation. Safety Management and Risk Analysis. Capítulo 2 y 12. Elsevier (2005)
8. Núñez, M.: Establecimiento y evaluación de un nivel de seguridad de la flota pesquera nacional de menos de 24 m de eslora (2017). Chapters 11 and 12
9. Gobierno de España, Ministerio de Fomento, RD 543/2007 de 27 de abril, por el que se determinan las normas de seguridad y de prevención de la contaminación a cumplir por los buques pesqueros menores de 24 metros de eslora (L). «BOE» núm. 131, de 1 de junio de 2007, páginas 23806 a 23891, 86 p.
10. Gobierno de España, Ministerio de Fomento, RD 1185/2006, de 16 de octubre, por el que se aprueba el Reglamento por el que se regulan las radiocomunicaciones marítimas a bordo de los buques civiles españoles. BOE. núm. 261, de 1 de noviembre de 2006, páginas 37906 a 37944, 39 p.
11. NATO Standardization Office (NSO) © NATO/OTAN NATO Standard, ANEP-77, Naval Ship Code Edition F Version 1, August 201

eMaster: Ship's Longitudinal Resistance

J. Freiria[1(✉)] and E. Philippi[2(✉)]

[1] Facultad de Ingeniería, Montevideo, Uruguay
jfreiria@fing.edu.uy
[2] Arquitectura Naval Asociados, Montevideo, Uruguay
Enrique.philippi@arquitecturanaval.net

Abstract. The conditions of loading in the Maritime Transport are becoming increasingly fundamental and demanding, as the ships increase their size and the efforts begin to travel through the limits of the properties of the materials.

These conditions became critical and require a very demanding control, as loading and unloading operations can collaborate in the collapse of the resistant structure, either by a bending or excessive cutting effort.

In same way of improving the assessment of the stability conditions of the vessel, this work aims to show the new computational development that Arquitectura Naval Asociados is integrating to its platform, which will allow an integral approach by Operators, Ship Owners, Maritime Authorities and Training Institutions as well.

Keywords: Ship longitudinal resistance · Naval software · Bending moment

1 Introduction

For the International Maritime Organization, shipbuilding safety is a priority. This position is consolidated by the adoption of the rules called Goal-Based Ship Construction Standards, which among other points emphasizes in the Rules and Regulations for the design and construction of ships, which must be applied by the Maritime Authorities or Representative Organizations (Nitta et al. 1992).

The application of this regulation has led to a progressive decrease in the loss of vessels, above 300 annual in the period 1966–1985, to less than 120 in 2010, after having been approved in its 23 Session in November 2003 as part of its Strategic Plan, and after having been put into mandatory use for Bulk Carriers and Oil Tankers, through Resolution MSC.290 (87) (IMO 2010; IMO 2015; Lupascu and Chirica 2015).

A systematic evaluation of the losses of bulk carriers and tankers led to the conviction that the main faults were associated with a weak structure under a system of distributed loads, difficult navigation conditions and high tensions in relation to fatigue efforts.

This leads to the conclusion about the importance of longitudinal resistance assessment during the complete life cycle of the vessel, from the design stage, through the construction phase, ending with a systematic control during the entire operational life of the vessel.

V. A. J. E. Carreño Moreno et al. (Eds.): CIDIN 2019/COPINAVAL 2019, pp. 79–88, 2020.
https://doi.org/10.1007/978-3-030-35963-8_7

Special emphasis was given to the case of passenger ships, including those of the Ro - Ro type. In these case, the connections of structural members, large exceptional openings, lack of pillars between roofs, type of vehicles entered, makes the deformations between decks cannot be neglected, making it necessary a deeper study that will include efforts of fatigue and buckling.

Taking care of that, the standards indicate the use of structure evaluations using finite element analysis based on complete ship models or structural details (Bureau Veritas 2018).

2 Estimation at Design Stages

Traditional design procedures in longitudinal structure assessment involves development of weight distribution for light ship, full and intermediate cargo conditions, still water and wave buoyancy distribution (Fig. 1), in order to calculate maximum bending moment and shear forces.

The values defined by this procedure was then compared with stresses supported by structural member as designed, including a security factor.

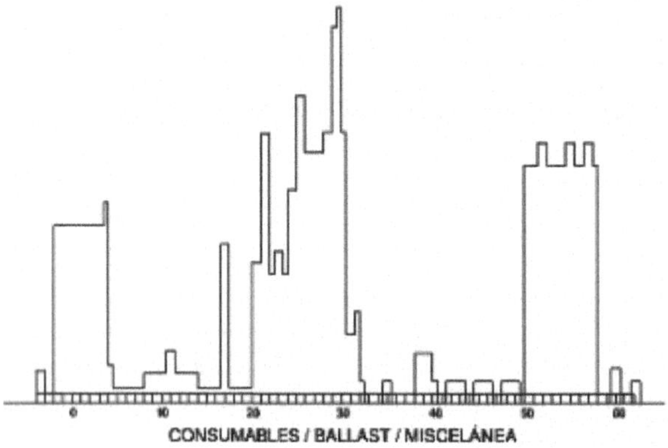

Fig. 1. Weight distribution, loaded condition

The information was sent to the Classification Society to check everything was according to norms before subscribe formal approbation.

Derived formulae for calculating shear force and bending moment arise from traditional beam resistance analysis:

$$p' = \frac{dS}{dx} = \frac{d^2M}{dx^2} \quad (1)$$

Where p' represents the net load by longitudinal unit, S the shear force and M the bending moment along the longitudinal positions of the beam.

In a graphical development, it can be say that shear distribution could be drawn with net load integration and so on moment from shear distribution, as is shown on Figs. 2 and 3.

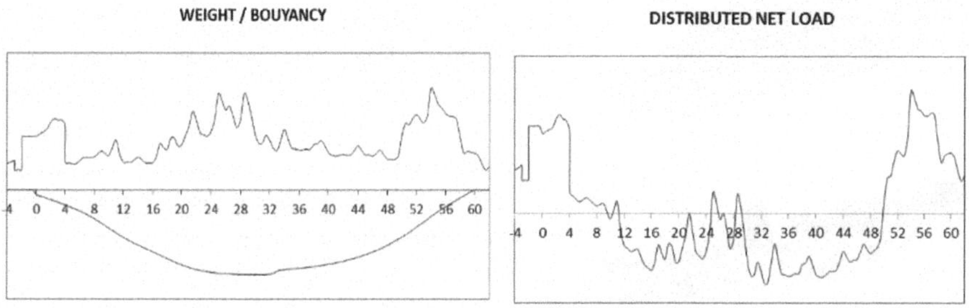

Fig. 2. Individual and net distributed forces

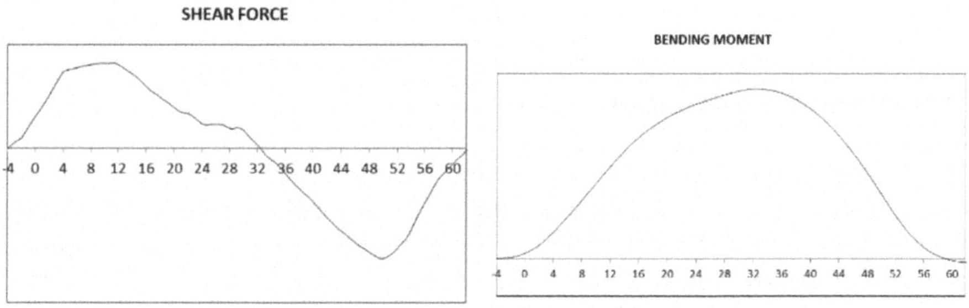

Fig. 3. Shear force and bending moment integration

In terms of obtaining a fine analysis of stress distribution, mathematical models were tested and used to determine the stresses derived from bending and shear moment (Bayraktarkatal et al. 1996).

Closer in time, an excellent approach to collapse analysis is found in Finite Element Methods (FEM) applied to non-linear theories such as the modified RIKS method using three-dimensional models.

This methodology shows impressive comparative results in relation to experimental data with differences in the order of 2%, allowing extending the calculation of the ultimate value of longitudinal resistance through this type of analysis in the design phase (Van-Vu 2015).

Once the stress mapping is ready all over the ship's shell and structure for critical conditions, dimensions could be definitive defined or modified so the net will absorb stresses within a certain grade of security.

Finally hull designer is prepared to dimension the complete set of frames, web frames, transverse and longitudinal bulkheads and shell plate which leads to longitudinal functions of actual section modulus and plate thickness, parameters to compare with normalized data from Classification formulae (Fig. 4).

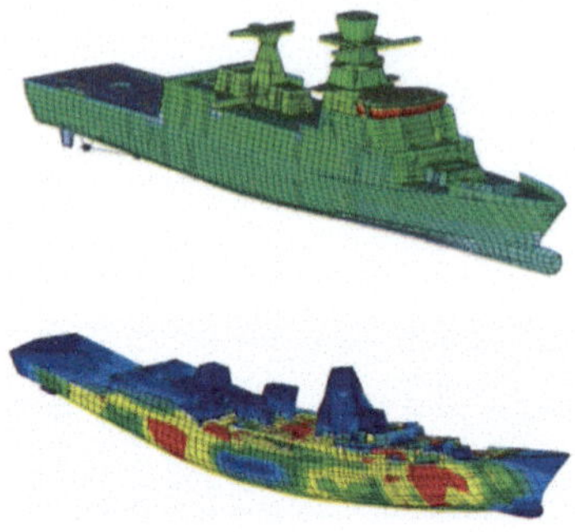

Fig. 4. FEA analysis for determining hull stress distribution (Granneman 2009)

3 Normalized Calculations

The loads distributed along its length, (static and dynamic) affect the longitudinal resistance of the ship. They are represented by a set of loads constituted by a bending moment, a shear force and a torsional moment. The international regulations define the lower limits of the inertial modules of the marine structures taking into account the bending moments estimated for calm waters and over the waves.

The static loads are derived from net difference between hydrostatic and distributed weights components, and are fully determined in design stages for the light ship condition, remaining variable loads that must be calculated and added case to case.

The dynamic loads are induced by the effect of the waves (break and fall), and for their determination mathematical models have been defined, which are considered in the different Rules of the Classification Societies. These rules had be standardized by the IACS (International Association of Classification Societies) and were universally accepted by their associates.

However, the forces acting on the hull are uncertain to the extent that sea conditions vary as well as loading conditions due to intermediate loading/unloading operations (Nitta et al. 1992; Det Norske Veritas 2008).

The loads induced by the waves are estimate by IACS according to the following formulations (IACS 2010):

Shear stress in conditions of hogging and sagging:

$$Fw(+) = +0.30.F_1.C.L.B.(Cb+0.7) \tag{2}$$

$$Fw(-) = -0.30.F_2.C.L.B.(Cb+0.7) \tag{3}$$

Where *Fw* (+) y *Fw* (−) are shear efforts on hogging and sagging conditions, which depends on principal dimensions[1] (*L*, *B* and *Cb*) and longitudinal distribution factors (*F1* y *F2*)

Bending Moment in conditions of hogging and sagging:

$$Mw(+) = +0.19.C_1.C_2.L^2.B.Cb \tag{4}$$

$$\mathrm{Mw}(-) = -0.11.\mathrm{C}_1.\mathrm{C}_2.\mathrm{L}^2.\mathrm{B}.(\mathrm{Cb}+0.7) \tag{5}$$

Where *Mw (+)* y *Mw (−)* are estimated bending moments for hogging and sagging conditions, which depends on principal dimensions[2] (*L*, *B* and *Cb*) and longitudinal distribution factor (*C2*).

These values must be sufficiently broad to prevent extraordinary situations, including fine-tuned safety coefficients associated with different parameters.

The formulations make it possible to determine the admissible values of section modules for each of the frames and/or web frames that make up the ship's structure. In the same way, side shell or longitudinal bulkheads thickness in way of longitudinal position are find.

As it was relate before, the information that must be submit by the owner must include a complete set of section module and side shell/longitudinal bulkheads thickness must be supplied, so that the real elements may have an equivalent module or thickness or a higher ones.

4 Loading Computer System Definitions

IACS and associated members define a loading computer system (LCS) as an on board digital computer system, a hardware and software combination which permits stability and strength verification for any cargo on a single ship.

Both hardware and software module are subjected to specific requirements and controls as an assembly, procedures that might be stated under a Quality System. QS procedures may satisfy either requirements of a recognized Standardization Organization of some National Accreditation Agency or a specific Class Society through assessment (IMO 2006a, b, c; IMO 2007; ISO 2006).

Some of them include responsibilities, system documentation, configuration management and competent staff (IACS 2006).

Particularly, strength and stability software module are built for performing stability and strength calculations to determine, at specified points, that in any load or ballast conditions, actual values will not exceed the specified permissible ones, eventually integrating some alarm to show cases witch no fulfill limiting requirements.

In the same way, it is built to perform and compare stability calculations with stability standard requirements in any load or ballast condition.

[1] *C* is a function of *L*.

[2] C_1 is a function of *L*.

Integrated system may also combined the computer base system with tank or load sensors, which are interconnect in order to allow real time operation.

Approval of the software means that the Class Society or National Agency approves hardware and software for installation on board a specific vessel. The approval is based on a review and acceptance of design, calculations method, and verification of stored data and test conditions for the specific vessel.

The approval is a confirmation that the software is able to give correct results provided the user's input is correct.

It may be remarked that an approved loading instrument do not replace an approved loading manual authorized by Maritime Authority, but concerns an on-board computer system capable of assessing operational conditions not included for stability and strength verification.

5 eMaster® Technical Description

eMaster® is a loading computer system which, among stability assessment done by proper module (Freiria 2017), is also used to perform still water and seagoing bending moments and shear forces calculations, and compare all over the ship stress distribution against normalized values and specified permissible values.

Shear forces and bending moments are being calculate for any loading condition at actual still water or wave flotation line as result of the static equilibrium of intact ship condition, integrating fixed light ship weight condition and specific dead weight by means of a composed matrix along longitudinal step - by - step calculation.

Detailed and proper loading distribution for tanks, load and miscellanea is use for arithmetical superposition in order to approach total weight, while hydrostatics and wave approximation is used for buoyancy development, with automatic procedure to correct longitudinal misalignment for gravity and buoyancy centers.

As it was state before, first and second integration of the approximated loading distribution give still water shear force and bending moment values at any point along the ship's length, and this is done in a finite element logical using matrix calculations.

Results from module calculations include actual stress values compared versus allowable shear forces and bending moments limits either in harbour or seagoing condition as far as applicable.

The program is written in Java language. Class encapsulation mechanism of Java provides integration of functions with relevant data, which guaranties that correct data are assigned to functions that operate on it. As a result the calculations are very safe and particular data structures together with calculation methods can be altered without many interventions to the code. Also communication between user interface and calculation is made client-server, so changes in any of the elements don't imply changes in another one.

Software is designed to developed personalized modules for ship officers, shore – based cargo planners and ship management companies for all types of vessels. Special use is planned for Training Centers adding a proper integrated dynamic 3D interface.

Personalization required specific documents as Lines Plan, Body Plan and/or Hull Offsets, General Arrangement Plan, Tank Sounding, Shell Expansion, Mass Distribution, etc. Typical components of a mass distribution are steel weight, equipment and accommodation (resulting in the light ship weight), bunkering, water ballast, and cargo.

5.1 Selection of Load Cases

IACS guideline for global strength analyses is used for normalized loading conditions, as Maximum Still Water Bending Moment (Maximum displacement at scantling draught with maximum permissible vertical hogging still water bending moment - Max SWBM) and Minimum Still Water Bending Moment (Maximum displacement at scantling draught with minimum possible vertical hogging/maximum sagging still water bending moment - Min SWBM).

Different conditions but Max SWBM or Min SWBM for intermediate conditions will be set by operator, who will be allowed to enter dead weight values including drilling or volumes of liquids in consumables tanks, the load distributed in holds, and the inclusion of items that they are not covered in the general conditions.

eMaster® may be operated as a passive system requiring manual data entries, but it is possible to perform it as an active system when sensors reading are available.

5.2 Menu and Basic Operation

This application is a personalized software that shows a menu of commands with basic commands of user interaction. The operation begins by opening a loaded project (Fig. 5), in this case the ship M/V Gaucho, for which the necessary surveys have been made and all data have been previously loaded, showing for example general identification, dimensions, condition of the light ship, etc.

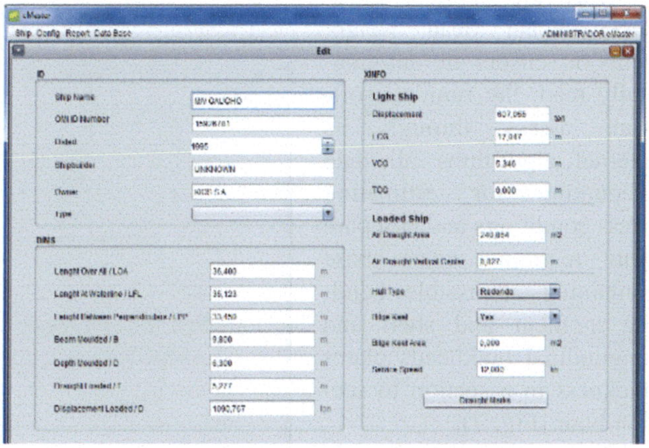

Fig. 5. Open new project window

Then in the menu Report lightweight distribution, full load distribution and bending moment and shear force limits (Fig. 6).

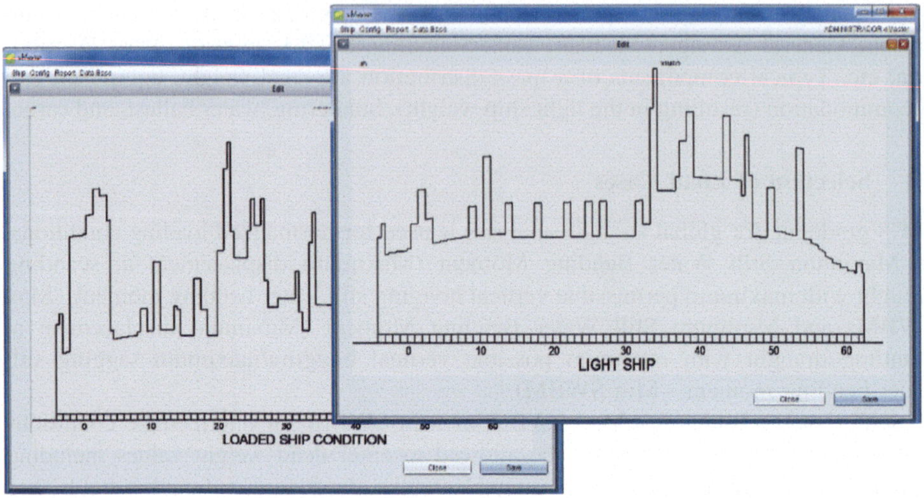

Fig. 6. Weight distribution window

The operation manual is simple and straightforward, containing full descriptions and instructions, as general description of the application use, a copy of the type approval certificate, installation procedures, etc.

Once the operator introduces actual dead weight items or sensors are automatically read, the running software assess through its internal algorithms all data accounting for estimating stress conditions and compare with resistance data base containing admissible bending moment and shell and longitudinal bulkheads plates thickness in a section to section base (Fig. 7).

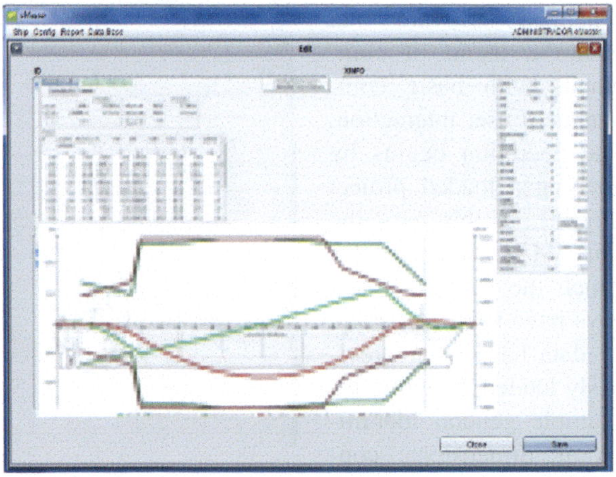

Fig. 7. Bending moment and shear force comparison with ship parameters

Inertia and section module for each one of complete set of structural members as hull steel planks thickness is introduced carefully as a data base from data supply from original project or elaborated by trained officers, in order to be used as comparative figures from proceeded results. Spreadsheet or graphic presentation may be available.

Warnings are displayed when some load operation generates over stress on a particular station, or when it is particularly high, so Officer in charge could prevent and put into force correct action to relieve tensions.

5.3 Longitudinal Resistance Module Validation

To become a confidence and powerful tool for seaman it has to probe that functioning and algorithm represents what is mean to represent.

In this way, verification and validation actions has to be taken; try out test ships whit known properties and behavior will be done, some hard work is expected to do with verification in order to be safe.

On the other hand, adjusting code to conform basic norms within allowable uncertainties is also necessary for Class validation, before software would be in the market (Det Norske Veritas 2006; IACS 1997).

5.4 Data Entries for Strength Module

For customized ship, following data shall be pre charged to configure structural stress characterization, in order to check actual values for a given condition. This data will be a enclose folder for most program operators:

Holding both stability and strength calculus, light ship weight and its center of gravity position, hydrostatic, stability arms and cargo/consumable tanks data; specific for longitudinal resistance calculations, actual distribution of structural sections strength will be part of the customized module for application set up.

Input data might be entered by authorized operator, and covers all item that identify ship condition, as mass cargo distribution, individual tank condition, passenger and crew, etc.

Particularly regarding longitudinal resistance, once application calculates still water shear forces SF and bending moments BM all over the ship's length, immediate stress data will be shown, indicating oversized stresses as an alarm indication if corresponding.

6 Conclusions

The results obtained through the eMaster® longitudinal resistance application are being refined to resolve some technical details; first data obtained for prototype vessel are accounting highly positive. Final settings and validation are been developed.

Additionally to the first developed stability module, this module complete the objectives of sharing a fully loading computing system with easy and friendly management and high speed processing.

It is not found an enterprise or project at the regional level tending to cover this sector with similar products. Internationally, some software references appears that offer similar features to eMaster.

With a modern implementation in terms of control by remote servers and automatic communication with volumetric sensors in tanks, it is understood that a differential value is it regional characteristics development, and the introduction into the training system through a simulation system in real time.

References

Nitta, A., Arai, H., Magaino, A.: Basis of IACS unified longitudinal strength standard. Mar. Struct. **5**(1), 1–21 (1992)

Bayraktarkatal, E., Ünsan, Y.: An approximate method for the calculation of shear force and bending moment loading in tankers and bulk carriers. In: Proceedings of the First International Conference on Marin Industry MARIND 1996, vol. III, p. 117 (1996)

Bureau Veritas: Guidance Note for Structural Assessment of Passenger Ships and Ro-ro Passenger Ships, Guidance Note Ni 640 (2018)

Det Norske Veritas: Rules for Classification of Ships; Hull Structural Design Ships with Length 100 Metres and Above (2008)

Det Norske Veritas: Approval and certification of the software of loading computer systems; Class Guide Line, DNVGL-CG-0053 (2016)

Freiria, J., Philippi, E.: eMaster: herramientas para la prevención y predicción de estabilidad de buques. XXV COPINAVAL, Panamá (2017)

Grannemann, F., Bertram, V.: Design of high-performance ships using simulations. Ship Sci. Technol. **3**(6) (2009). ISSN 1909–8642

Van-Vu, H.: Prediction the ultimate longitudinal strength of intact ship by finite element method. Int. J. Mech. Eng. Appl. (Special Issue: Transportation Engineering Technology) **3**(1–3), 18–23 (2015)

IACS: S11 Longitudinal Strength Standard. International Association of Classification Societies and International Association of Classification Societies (2010). http://www.iacs.org.uk/publications/unified-requirements/ur-s/ur-s11-rev8-cln/

IACS: On Board Use and Application of Computer. International Asociation of Classification Societies. IACS, Londres (2006)

IACS: Recommendations on Loading Instrument. International Asociation of Classification Societies. IACS, Londres (1997)

International Maritime Organization: International Goal - Based Ship Construction Standards for Bulk Carriers and Oil Tankers, Focus on Imo, January 2015

International Maritime Organization: Adoption of the International Goal-Based Ship Construction Standards for Bulk Carriers and Oil Tankers, International Resolution MSC.287 (87). adopted on 20 May 2010

International Maritime Organization: MSC 836, Recommendation on loading instruments. IMO, Londres (2006a)

International Maritime Organization: MSC 854, Guidelines for ship-board loading and stability computer programs. IMO, Londres (2006b)

International Maritime Organization: MSC 891, Guidelines for the on-board use and application of computers. IMO, Londres (2006c)

International Maritime Organization: MSC 1229 Guidelines for the approval of stability instrument. IMO, Londres (2007)

International Organization for Standardization: ISO 16155:2006 Ships and marine technology - Computer applications - Shipboard loading instruments. ISO, Londres (2006)

Lupascu, D., Chirica, I.: Assessment of seagoing ships longitudinal strength in the context of international rules, important factor for safe operation. World J. Eng. Technol. **3**, 291–310 (2015)

Analysis of Bow Shapes Proposed on a High-Speed Displacement Hull

Franklin J. Domínguez Ruiz[1,2(✉)], Lino Andrés García Jaramillo[1], and Patrick Roger Townsend Valencia[3]

[1] TECNAVIN S.A., Guayaquil, Ecuador
jdominguez@tecnavin.com, lagarcia@espol.edu.ec
[2] Universidade da Coruña - UDC, A Coruña, Spain
[3] ESPOL, Guayaquil, Ecuador
ptownsen@espol.edu.ec

Abstract. This study uses the computer fluid dynamic (CFD) software to compare the resistance between six different bow shapes for a high-speed displacement hull – type NPL. To begin, the original forms of a conventional shape hull are compared between the resistance results from tank test measurements with the CFD results. Then, six bow shapes are also proposed and analyzed, including a bulbous bow, wave piercing bow, axe bow, modified axe bow, and reverse piercing bow; these shapes are sized according to analytical and experimental recommendations. At final on this research, a new hull bow shape is proposed: by adding a ballast bulb to the axe bow to improve ship performance. The CFD resistance calculation is performed for calm water, while for the sea condition is considered added resistance estimation, calculated from seakeeping analysis. A predictive resistance analysis has been performed to show ship behavior and compared with the CFD results. Finally, the seaworthiness analyses of the models are calculated using strip theory to compare the ship comfort and performance of the six studied models.

Keywords: CFD · High speed ship · Bow · Optimization

1 Introduction

Reducing ship CO2 emissions, fuel consumption, and operating costs have gained importance over the last few decades. Currently, hull form optimization is a significant factor in improving ship performance and seaworthiness, and has been studied extensively for example: Takahei (1960) and Takahei (1961) introduced a bulbous bow in the hull to realize a wave less hull form. Keuning et al. (2002) performed towing tank test for the axe bow concept founding improving to the seakeeping behavior and operability compared to a conventional bow hull. In this paper a conventional hull form has been studied using a tank test with a model of the ship, obtaining model results including resistance and propulsion tests that can be transferred to a full-scale hull according to ITTC recommendations.

V. A. J. E. Carreño Moreno et al. (Eds.): CIDIN 2019/COPINAVAL 2019, pp. 89–103, 2020.
https://doi.org/10.1007/978-3-030-35963-8_8

Computer fluid dynamics (CFD) is the numerical modeling of fluids that allows the interactions of liquids, gases, and solids. Initially, tank test results of a conventional bow hull are compared with the CFD software results for validation. The expected results for this study will allow a determination of which of the studied bow shapes shows the best performance with regard to hull resistance, dynamic pressure, and free surface elevation as well as heave and pitch for six different models.

2 Analysis of Hull Bow Shapes

The study compare bow shapes developed in previous studies that where tested independently, as Wigley (1936), Hirota et al. (2005) and Keuning et al. (2002). Five different bow shapes shown in Figs. 2, 3, 4, 5 and 6 are proposed for a comparison with conventional hull

Model 1. The first model is a conventional high-speed displacement hull, type NPL. It measures 45 m in length and has an optimized hull form in basin tests. The conventional hull shape is shown in Fig. 1 and its characteristics are given in Table 1.

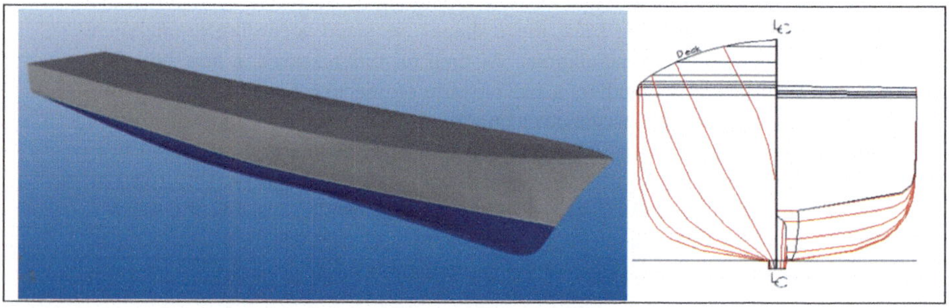

Fig. 1. Conventional optimized hull shape – Model 1

Table 1. Model 1, principal dimensions

Main characteristics		
Overall length	45.0	m
Beam	7.0	m
Depth	4.0	m
Draft	1.85	m

Model 2. The second model is the same as Model 1, with the addition of a bulbous bow of the Nabla type (see Kyriazis 1996). The bulb lies below the waterline and protrudes by 1 m. The Model 2 hull shape is shown in Fig. 2.

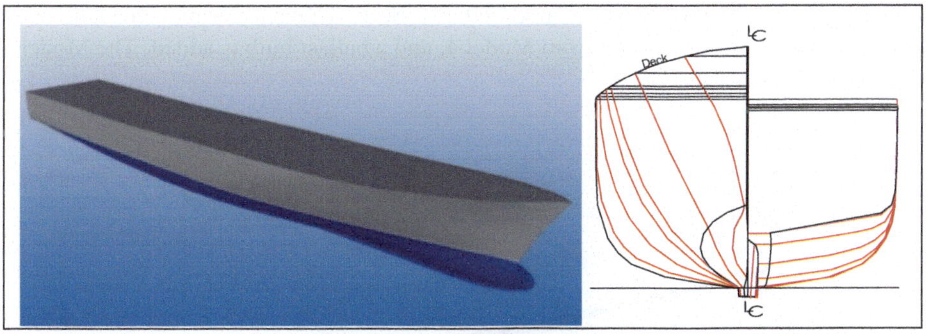

Fig. 2. Bulbous bow – Model 2

Model 3. The third model follows the dimensions of Model 1 and adds a wave piercing bow as well as a new fine body form just above the waterline. Its body protrudes 2 m from its original shape. The Model 3 hull shape is shown in Fig. 3.

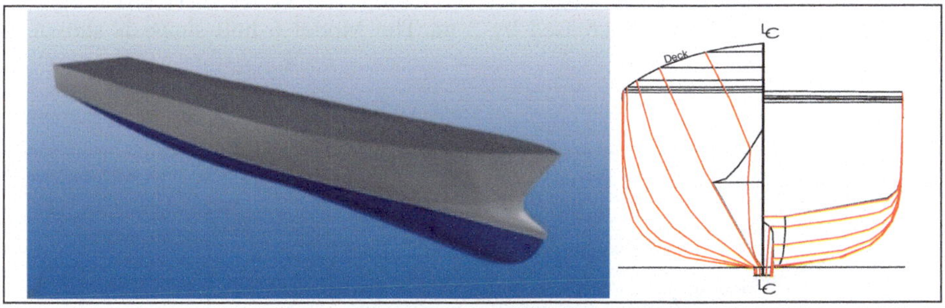

Fig. 3. Added wave piercing – Model 3

Model 4. The fourth model, the axe bow, has a completely vertical wave piercing bow, which increases the ship's overall length by 6 m; it also has a very fine bow shape without flare. The Model 4 hull shape is shown in Fig. 4.

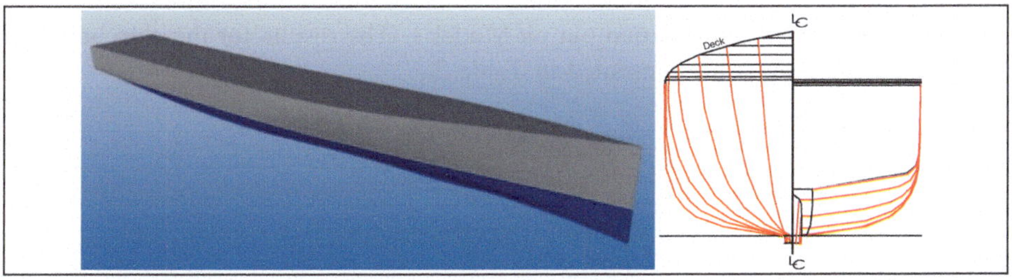

Fig. 4. Axe bow – Model 4

Model 5. The fifth model has a modified axe bow. This means that the hull body below the waterline is modified from Model 4, and a ballast bulb is added. The Model 5 hull shape is shown in Fig. 5.

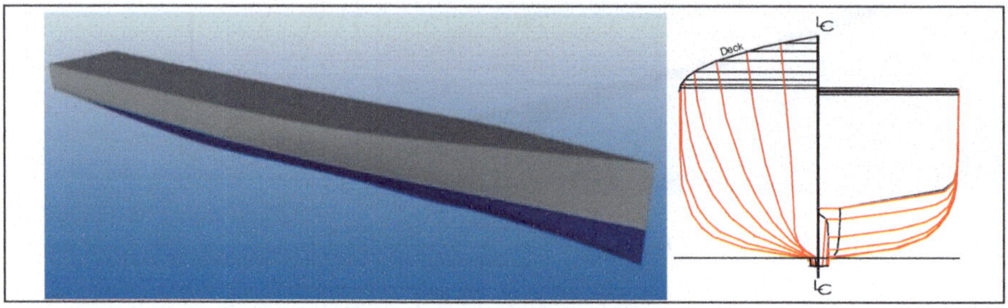

Fig. 5. Modified axe bow – Model 5

Model 6. The final model has an inverted bow. This increases the ship's length below the waterline, obtaining a finer bow, a bow flare, and a bulb at the lower forward point. The ship's overall length is increased by 3 m. The Model 6 hull shape is shown in Fig. 6.

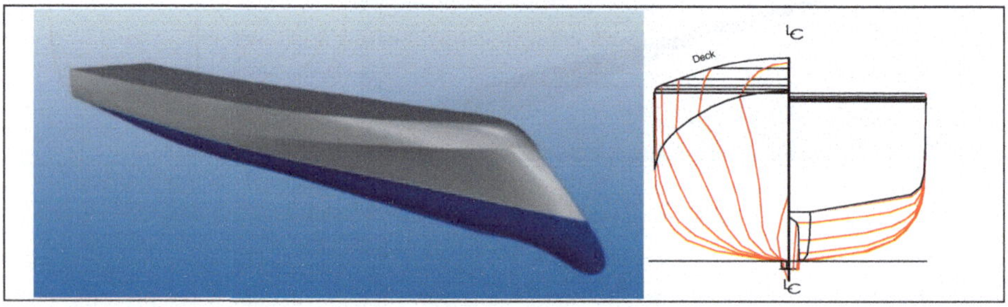

Fig. 6. Inverted bow – Model 6

3 Tank Test Results

A tank test experiment was carried out on Model 1. The results for the ship bare hull resistance in calm water are presented in Table 2.

Table 2. Tank test results for conventionally shaped - Model 1

Ship speed	Resistance	Froude	Heave	Pitch
[kn]	[kN]	Fnv	[m]	[°]
14	41.6	0.8919	−0.12	−0.11
16	65.6	1.0193	−0.17	0.04
17	84.2	1.083	−0.2	0.27
18	104.2	1.1467	−0.22	0.54
19	124.1	1.2104	−0.23	0.84

4 CFD Test Condition

The six ship models detailed in Sect. 2 were tested under the same load conditions and without considering weight increase or center of gravity modification. The results are shown in Table 3.

Table 3. CFD test conditions

Speed	19	Kt
Displacement	294	Ton
Hull rugosity	0.15	Mm
Initial trim	0	Degrees
Sea state	–	Calm water

5 CFD Results (Calm Water)

Each model's results for resistance, heave and pitch performance were obtained from the CFD software and are shown in Table 4. The results are compared with the conventional hull shape in Table 5.

Table 4. CFD results at 19 knots

Model	Resistance	Heave	Pitch
	N	M	degrees
1. Conventional	124486	−0.210	−0.943
2. Bulbous bow	119986	−0.216	−0.864
3. Wave piercing	112310	−0.210	−0.845
4. Axe-standard	82286	−0.172	−0.478
5. Axe-bulb	80606	−0.172	−0.449
6. Inverted bow	96324	−0.209	−0.506

Table 5. CFD results compared to the conventional Model 1

Model	Resistance	Heave	Pitch
	Reduction	Difference	
2. Bulbous bow	4%	−3.1%	8.4%
3. Wave piercing	10%	−0.2%	10.3%
4. Axe-standard	34%	17.9%	49.3%
5. Axe-bulb	35%	18.0%	52.3%
6. Inverted bow	23%	0.6%	46.3%

5.1 Free Surface Results (Calm Water)

Free surface is the contact surface between water and air. Free surface results demonstrate the waves generated by the ship in calm water. The free surface elevations in the wave patterns generated by the respective models are shown in Fig. 7; blue shows wave depression while magenta shows wave crest.

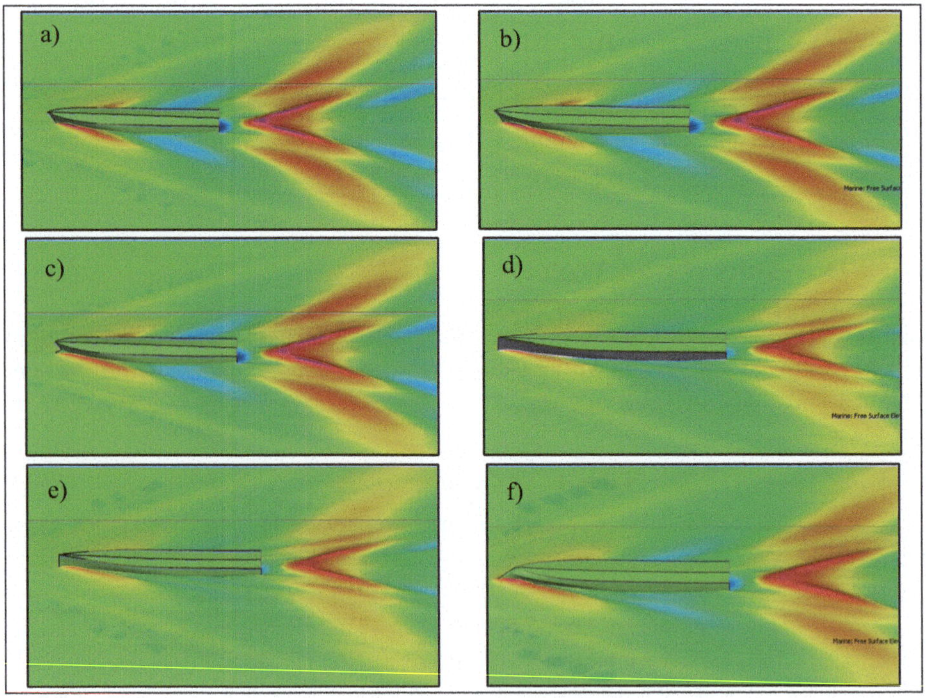

Fig. 7. Free surface elevation: (a) model 1, (b) model 2, (c) model 3, (d) model 4, (e) model 5, and (f) model 6

5.2 Bottom Dynamic Pressure Results

Bottom dynamic pressure is the pressure generated by the water due to ship movement. The results of the bottom dynamic pressures are shown in Fig. 8: (a) model 1, (b) model 2, (c) model 3, (d) model 4, (e) model 5 and (f) model 6.

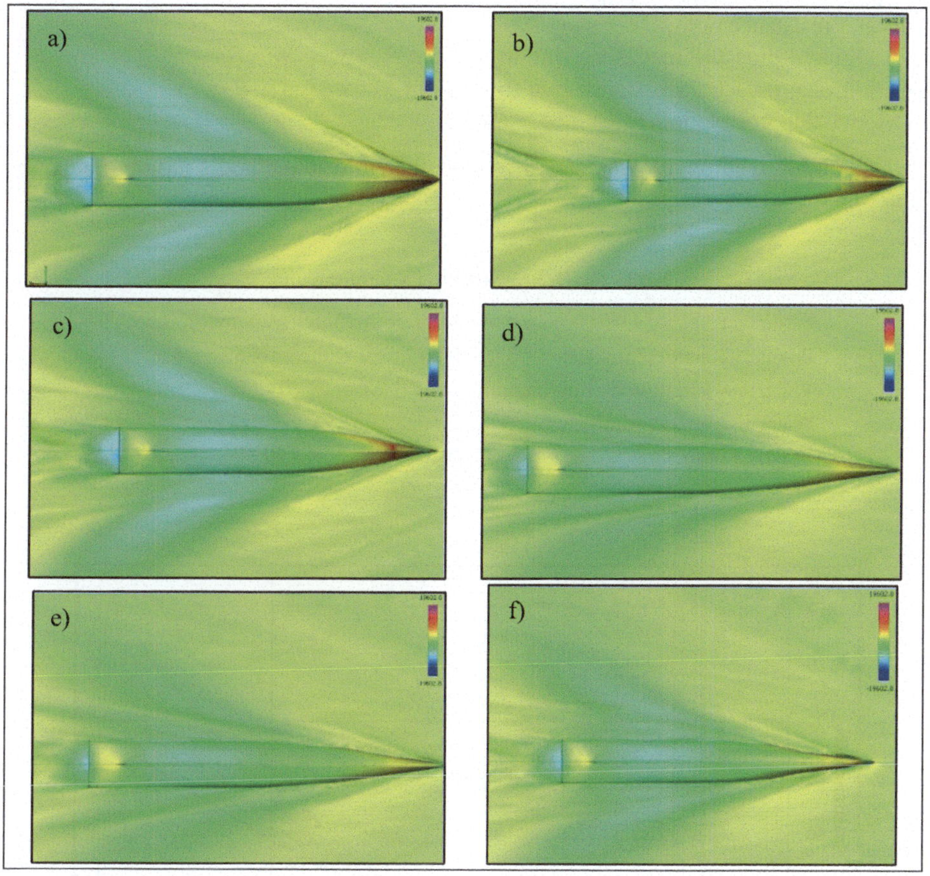

Fig. 8. Bottom dynamic pressure: (a) model 1, (b) model 2, (c) model 3, (d) model 4, (e) model 5, and (f) model 6.

6 Bare Hull Resistance Prediction Results

The preliminary resistance prediction analysis was determined with the Maxsurf Resistance software using two different methods, namely: Holtrop regression method (Holtrop 1984) and the Slender body method at different speeds in calm water. The tank test was performed only for the conventional hull, Model 1, thus, a regression resistance calculation was performed to provide a comparison with the CFD results.

Figure 9 and Table 7 show the resistance prediction using the Holtrop method (Holtrop 1982, 1984), which is based on the test results for certain types of ships.

The resistance calculated with the Holtrop method includes:

1. Frictional resistance according to the ITTC formulation – R_F
2. Form factor of the hull – $1 + k_1$
3. Appendage resistance – R_{APP}
4. Wave resistance – R_W
5. Additional pressure resistance due to a bulbous bow – R_B
6. Additional pressure resistance due to transom immersion – R_{TR}
7. Model-ship correlation resistance – R_A

The proposed ship hull characteristics meet the Holtrop method's requirements; see Table 6.

Table 6. Holtrop method's requirements

Parameter	Minimum value	Maximum value
Prismatic coefficient - CP	0.55	0.85
Waterline length/waterline breadth	3.90	14.90
Waterline breadth/draft	2.10	4.00
Froude number	0.10	0.80

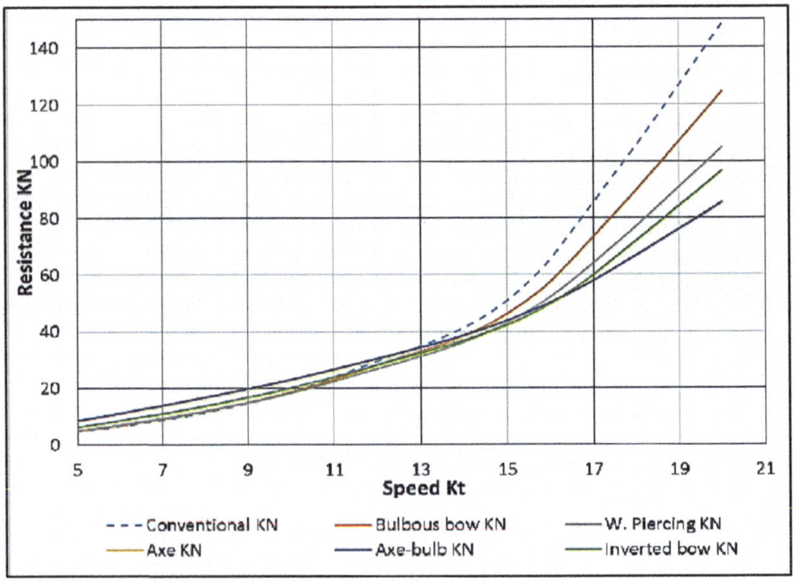

Fig. 9. Holtrop resistance prediction

Table 7. Holtrop resistance prediction at 19 knots

Model	Resistance KN	Resistance reduction
1. Conventional	126.9	–
2. Bulbous bow	107.1	16%
3. Wave piercing	90.8	28%
4. Axe-standard	76.1	40%
5. Axe-bulb	76	40%
6. Inverted bow	84.2	34%

Table 8. Slender body resistance prediction at 19 knots

Model	Resistance KN	Resistance reduction
1. Conventional	125.3	–
2. Bulbous bow	120.1	4%
3. Wave piercing	118.2	6%
4. Axe-standard	96.2	23%
5. Axe-bulb	95.3	24%
6. Inverted bow	103.5	17%

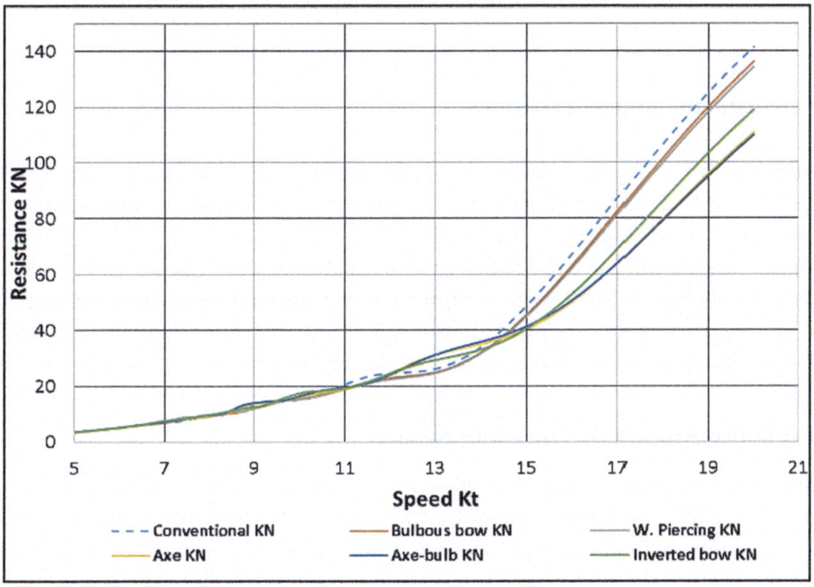

Fig. 10. Slender body theory resistance prediction

Figure 10 and Table 8 show each ship's resistance prediction using the slender body theory method. The Maxsurf Resistance software producer Bentley Systems (2013), uses a Michell paper (1898) based approach to compute the wave resistance of a port/starboard symmetrical mono hull to calculate the total resistance.

7 Hydrostatics

As the hull shape has been modified for each model, the hydrostatic conditions have changed especially the buoyancy longitudinal center (from Aft Perp.) and the water line length; the displacement was considered 294 ton and the trim of 0°. The results are shown in Table 9.

Table 9. Hydrostatic parameters for 294 ton displacement

Model	Buoyancy length	Waterline length	Draft
	m	M	m
1. Conventional	−2.08	43.90	1.845
2. Bulbous bow	−1.61	43.92	1.832
3. Wave piercing	−1.54	45.91	1.830
4. Axe-standard	1.06	52.74	1.682
5. Axe-bulb	1.17	52.74	1.681
6. Inverted bow	−0.07	48.33	1.739

8 Preliminary Seaworthiness

Seaworthiness is an important parameter to be analyzed due to that greater sea added resistance and accelerations decrease ship performance, comfort, and safety at sea. A number of previous studies (Ghassemi et al. (2015); Keuning and Pinkster (1995); Salvesen (1978); Faltinsen (1990) have investigated seakeeping behavior. The Maxsurf Motions software is a preliminary seakeeping analysis program using a linear strip theory method that is based on the work of Salvesen (1978). It is used to calculate in preliminary way the coupled heave and pitch response of the vessel and uncoupled roll motion.

8.1 Response Amplitude Operator Results

The Response Amplitude Operator (RAO), also referred to as a transfer function, describes how the response of a vessel varies with wave frequency. These RAOs are normally non-dimensional through the wave height or wave slope (Bentley Systems, Maxsurf Motions 2013). The RAO calculations in this paper were made using the Maxsurf Motions software using the linear strip theory method.

Waves at sea increase resistance due to ship motions. Figures 11 and 12 and Table 9 show the response amplitude operators (RAO), including added resistance from waves for the six models at sea state 5 (wave amplitude 3.26 m) at 19 knots and with head waves. Salvesen (1978) and Prpić-Oršić et al. (2008) present computation methods for added resistance from waves (Table 10).

Table 10. Added resistance

Model	Added resistance (KN/m^2)
1. Conventional	83
2. Bulbous bow	86
3. Wave piercing	82.5
4. Axe-standard	72
5. Axe-bulb	73
6. Inverted bow	84

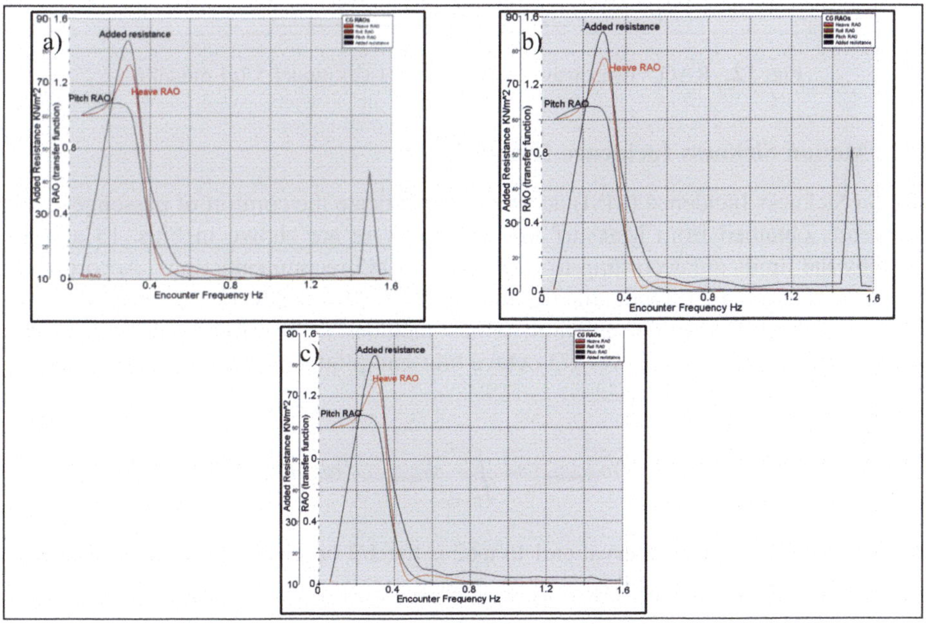

Fig. 11. RAO - head waves – (a) model 1 (b) model 2 (c) model 3

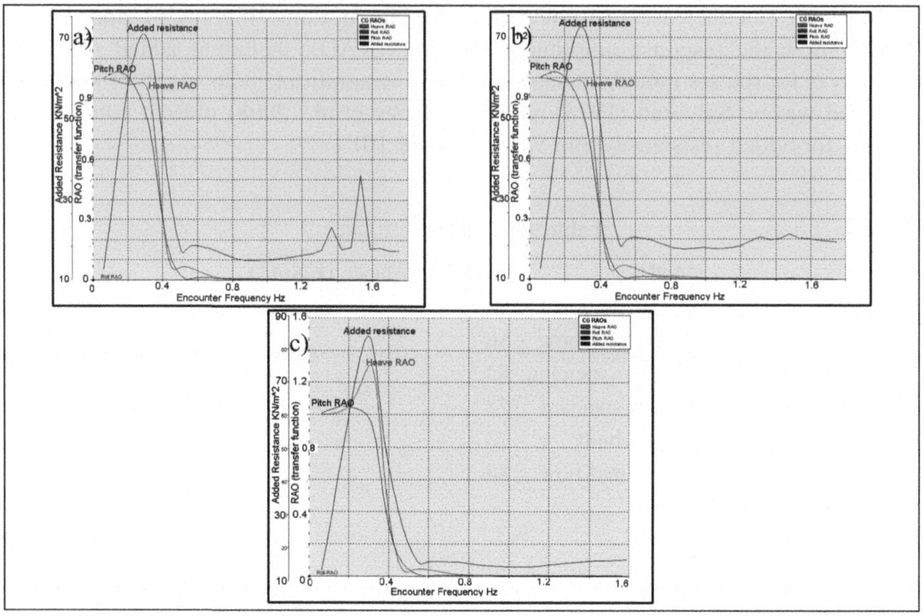

Fig. 12. RAO - head waves – (a) model 4 (b) model 5 (c) model 6

8.2 Motion Sickness Incidence Results

Motion Sickness Incidence (MSI) accelerations indicate the comfort of passengers, and the results obtained from Maxsurf Motions software are shown in Figs. 13 and 14, whereby the limits are according to ISO 2631/3 1985 recommendations. According to the Maxsurf Motions software, the calculation of the MSI is developed according to the McCauley et al. (1976) formulation, which includes an exposure time, while MSI accelerations are integrated over 1/3 octave bins and plotted against acceleration limits. The following equation is used:

$$MSI(\omega_{ecentre}) = \int_{\omega_{e1}}^{\omega_{e2}} S_{vertaccel}(\omega_e) d\omega_e$$

Where the frequency interval ωe1 to ωe2 is the 1/3 octave range centered about the ωe center and Svert accel is the absolute vertical acceleration. Bentley Systems (2013).

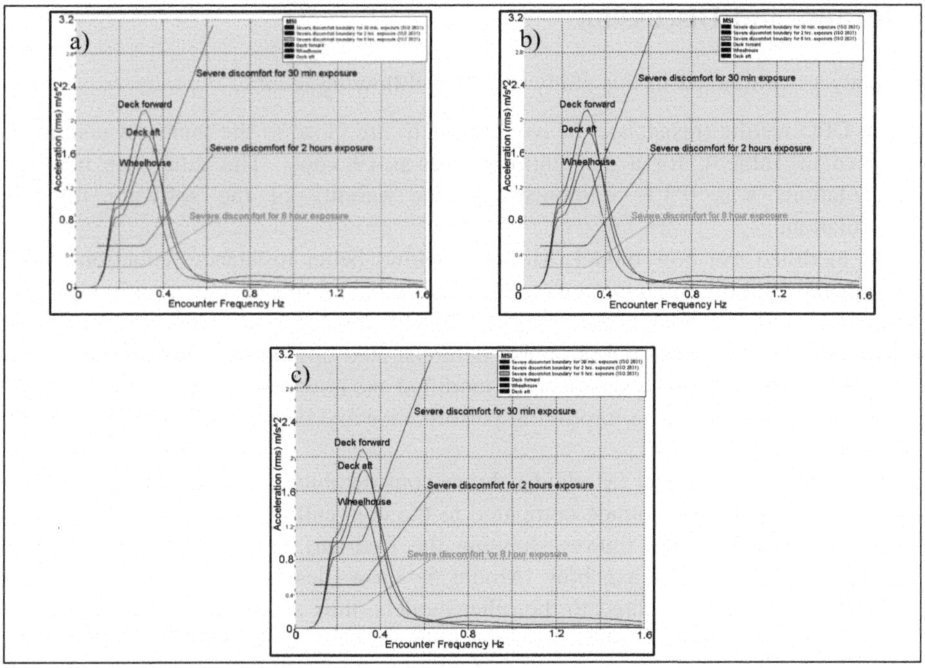

Fig. 13. MSI head waves – (a) model 1 (b) model 2 (c) model 3

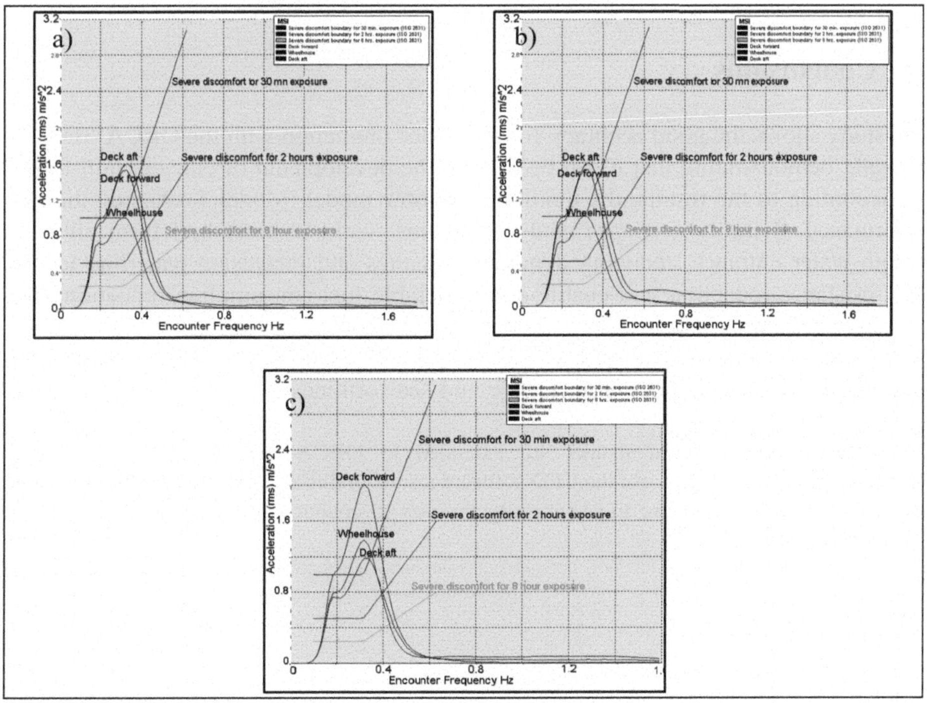

Fig. 14. MSI head waves – (a) model 4 (b) model 5 (c) model 6

9 Results and Discussion

The analyses conducted in this study offer the following results.

- The CFD results (resistance, heave and pitch) are close to the tank test results, as shown in Table 4, for a conventional hull at 19 knots. The difference between calculations was 0.3%, demonstrating the validity of the software for this calculation.
- The modified axe bow model (Model 5) shows better resistance reduction (35%) compared with the conventional models. This model result the better option because improves ship efficiency due to its fine and slender bow, while an improved reduction in MSI acceleration is also demonstrated upgrading ship comfort.
- Regarding wave added resistance, the modified axe bow model (Model 5) shows the smallest added resistance improving performance and decreasing fuel consumption at sea.
- The inverted bow model (Model 6) shows considerable resistance reduction (23%) and a slight comfort upgrade compared to the conventional model (Model 1).
- As can be seen in the figures showing the free surface elevation and dynamic pressure, the modified axe bow (Model 5) shows less wave elevation and less dynamic pressure compared to the other models, thus avoiding loss of energy.
- The resistance reduction shown on the bulbous bow model (Model 2) and wave piercing model (Model 3), are 4% and 10%, respectively. These resistance reduction results are less than the other models studied.

10 Conclusions

This study shows the good accuracy for resistance prediction with the ORCA-3D CFD software, demonstrating that it can be used in the development of efficient hull forms.

According to the results, the modified axe bow model (Model 5) showed the best performance of the bow shapes analyzed in this work; its finer hull form allows a smooth water entrance, reducing dynamic pressures and increasing waterline length. This resistance reduction will enable a considerable fuel consumption reduction, thus reducing CO2 emissions.

It is recommended that further analysis should examine aft shape modification possibilities to increase ship performance and seaworthiness.

Acknowledgements. Special thanks are extended to ORCA 3D MARINE CFD AND SIMERICS for permission to use the CFD software results in this analysis. Special thanks also to ESPOL for permission to use the University licenses to conduct this study according to the investigation programs.

References

Bentley Systems: Maxsurf motions user manual, version 20 (2013)

Bentley Systems: Maxsurf resistance user manual, version 20 (2013)

Faltinsen, O.M.: Sea Loads on Ships and Offshore Structures. Cambridge University Press, Cambridge (1990)

Ghassemi, H., Majdfar, S., Gill, V.: Calculations of the heave and pitch RAO's for three different ship's hull forms. J. Ocean Mech. Aerospace Sci. Eng. **22** (2015)

Hirota, K.: Development of bow shape to reduce the added resistance due to waves and verification on full scale measurement. In: International Conference on Marine Research and Transportation, The Island of ISCHIA, Italy (2005)

Holtrop, J., Mennen, G.G.J.: An approximate power prediction method. Int. Shipbuilding Progress **29**, 166–170 (1982)

Holtrop, J.: A statistical re-analysis of resistance and propulsion data. Int. Shipbuilding Progress **31**, 363 (1984)

Keuning, J.A., Pinkster, J.: Optimization of the seakeeping behavior of a fast monohull. In: Fast 1995 Conference (1995)

Keuning, J.A., Pinkster, J., van Walree, F.: Further investigation into the hydrodynamic performance of the AXE bow concept. In: Proceedings of the 6th Symposium on High Speed Marine Vehicles, Castello di Baia, Italy, pp. II 25–II 38 (2002)

Kyriazis, G.: Bulbous bow design optimization for fast ships. M.Sc. thesis, Massachusetts Institute of Technology (1996)

Michell, J.H.: The wave resistance of a ship. Philos. Mag. **5**(45), 106–123 (1898)

McCauley, M.E., Royal, J.W., Wylie, C.D., O'Hanlon, J.F., Mackie, R.R.: Motion sickness incidence: exploratory studies of habituation, pitch and roll and the refinement of a mathematical model. Human Factors Research Inc., Technical Report 1733-2, 1976. (USN ONR contract N00014-73-C-0040)

Prpić-Oršić, J., Nabergoj, R., Trincas, G.: The methods of added resistance estimation for ships in a seaway. Pula, Croatia, Symposium Sorta (2008)

Salvesen, N.: Added resistance of ships in waves. J. Hydronaut. **12**(1), 24–34 (1978)

Takahei, T.: A study on the waveless bow Part I. J. Soc. N. A. Japan **108**, 53–61 (1960)

Takahei, T.: A study on the waveless bow Part II. J. Soc. N. A. Japan **109**, 73–85 (1961)

Wigley, W.C.S.: The Theory of the Bulbous Bow and Its Practical Application. Newcastle (1936)

Harmonization in the Multimodal Transport of Chemical Substances and Mixtures

José Ángel Fraguela-Formoso(✉), Asunción López-Arranz, Almudena Filgueira-Vizoso, and José de Troya-Calatayud

University of A Coruña, A Coruña, Spain
jafraguelaformoso@gmail.com, {alarranz, almudena.filgueira.vizoso, joaquin.troya}@udc.es

Abstract. Multimodal or intermodal transport is carried out by road, rail, airplane and ship and the goods are transported in different forms: bulk, packages, containers, cisterns, various packages. A special type of transported products is dangerous goods which, because of their danger to people, goods and the environment, require a specific legal framework relating to their packaging, identification, storage, handling and type of transport. Furthermore, the majority of dangerous goods are restricted by airway and traffic between continents is carried out by sea. The different legislative frameworks in the world represent an added problem for all those involved in the logistics chain and users of these products, which deserve to be analysed and revised from a current perspective. The international harmonization of the legal framework for classification and labelling, registration, evaluation, authorization, restriction and transport of chemical substances represents undoubted advantages for manufacturers, importers, transporters and users of these products.

Keywords: Ship · Transport · Chemical substances

1 Introduction

At international level, both the movement of goods using two or more modes of transport (multimodal transport) and the movement of goods in the same unit or vehicle, successively using two or more modes of transport without handling the goods in the changes of mode (intermodal transport) are usual.

In these transports, a Multimodal Transport Operation is used. In this modality, the responsible for the load makes a single contract dealing with all the logistics between origin and destination. It has several advantages: the multimodal operator is responsible for the operation; the multimodal operator is the only interlocutor until the arrival of the goods at the destination; a single Multimodal Transport Document is needed; greater agility in the departure of the load from the ports.

The multimodal or intermodal transport is carried out by road, rail, airplane and ship and the goods are transported in different forms: bulk, packages, containers, cisterns, various packages.

A special type of transported products is dangerous goods which, because of their danger to people, goods and the environment, require a specific legal framework

V. A. J. E. Carreño Moreno et al. (Eds.): CIDIN 2019/COPINAVAL 2019, pp. 104–112, 2020.
https://doi.org/10.1007/978-3-030-35963-8_9

relating to their packaging, identification, storage, handling and type of transport. This represents an added problem for all those involved in the logistics chain and users of these products.

The international harmonization of the legal framework regarding classification and labelling, registration, assessment, authorization, restriction and transport of chemical substances and preparations represents undoubted advantages for manufacturers, importers, transporters and users of these products.

2 The Legal Framework for the Transport of Dangerous Goods

The Economic and Social Council (ECOSOC) is the United Nations central platform for reflection, debate and innovative thinking on sustainable development. The Recommendations on the Transport of Dangerous Goods [11], published for the first time in 1956, are intended to serve as reference for governments and international organizations when legislating on each of the modes of dangerous goods transport, excluding river and maritime transport, liquefied gases in bulk, chemical products in bulk and petroleum products, which are regulated by the International Maritime Organization (IMO).

These recommendations are intended to achieve international harmonization in the field of dangerous goods transport, contributing to the control of possible damage to people, goods and the environment, favouring and simplifying all the procedures for the handling and control of multimodal and intermodal transport. So as to achieve a greater effectiveness, they are revised every two years in order to adapt them to the evolution of the technique and the modes of transport.

Their content has been divided into seven chapters and appendices and it classifies dangerous goods into nine classes. Each dangerous goods transported is assigned a UN (United Nations) number, name and description, class and division, secondary risk, UN packaging group, special provisions, limited and excluded quantities per packaging, packing instructions, packaging special provisions, transport instructions for portable tanks and bulk containers and special provisions for portable tanks and bulk containers. In addition, it has a labelling system (Fig. 1).

These recommendations have served as reference for the elaboration of regulations on each of the modes of dangerous goods transport.

The maritime transport of dangerous goods is regulated by the International Maritime Dangerous Goods Code (IMDG) of the International Maritime Organization (IMO 2016) [10], a United Nations specialized organization responsible for the navigation safety and the prevention of marine pollution from ships. It is mandatory for all countries which are part of SOLAS (Safety of Life at Sea) Convention and updated every two years. It extends the provisions of SOLAS and MARPOL (International Convention for the Prevention of Pollution from Ships) Conventions.

IMDG follows the Recommendations on the Transport of Dangerous Goods - Model Regulations, using the same classification and labelling, the same UN numbers and official transport designations and the multimodal dangerous goods form. The list of dangerous goods, special provisions and exceptions of IMDG is more complete.

Fig. 1. Labelling system

Each substance, matter or article is classified and has specific provisions about its packaging, labelling and stowing on ships. In its supplement, it has a guide on emergency intervention procedures for ships carrying dangerous goods and another guide on first aid in order to use them in the case of accidents related to dangerous goods. Dangerous goods are assigned to their corresponding UN numbers and shipping names based on their risk classification and their composition. The shipping name allows to find quickly: the IMDG page on which this dangerous good appears; its UN number (4 digits); the class it belongs to; the packaging group; the secondary risk label; the number of the Emergency Procedures Card and the number of the First-Aid Guide.

Road transport is regulated by the European Agreement concerning the International Carriage of Dangerous Goods by Road (ADR 1957) [12]. It was done in Geneva on 30 September 1957 under the auspices of the United Nations Economic Commission for Europe (UNECE). It was approved in 1968 and it is regularly reviewed by working group WP.15, within the United Nations. The latest amended version will enter into force on 2015.

Rail transport is controlled by the Regulation Concerning the International Carriage of Dangerous Goods by Rail (RID 1999) [8]. It is carried out by the Intergovernmental Organization for International Carriage by Rail (OTIF), located in Bern, of which 45 countries of Europe, North Africa and the Middle East are part. It is appears in the Appendix C of the Convention concerning International Carriage by Rail (COTIF), concluded, as amended, in Vilnius (Lithuania) on 3 June 1999.

The European Agreement concerning the International Carriage of Dangerous Goods by Inland Waterways (AND 2000) [13], is intended to facilitate the transport, guarantee the safety of dangerous goods transport and prevent the pollution derived.

The International Civil Aviation Organization (ICAO), a United Nations specialized organization, located in Montreal, publishes the Technical Instructions for the Safe Transport of Dangerous Goods by Air (ICAO 2015) [9], used by the International Air Transport Association (IATA) and applied by all the IATA member operators.

Fig. 2. International regulations on the transport of dangerous goods

These regulations of the different modes of dangerous goods transport (Fig. 2) are very similar and the tendency is towards a total harmonization. Meanwhile, in order to make multimodal transport easier, an agreement has been reached on the mutual recognition of the documentation and the labels of the products transported.

The Directive 2008/68/EC of the European Parliament and of the Council of 24 September 2008 and later versions, on the Inland Transport of Dangerous Goods [1], Official (EU Directive 2008, 2009, 2010 and 2011) [1–5], are applied to the dangerous goods transport by road, rail or inland waterway among Member States or within them, including loading and unloading activities, the transfer among the modes of transport and stops required by transport circumstances. It is not applied to the transport of dangerous goods by seagoing ships on maritime waterways forming part of inland waterways, by ferries only crossing an inland waterway or a port or wholly performed within a delimited perimeter.

3 The Legal Framework for Classification and Labelling of Chemical Substances

The Committee of Experts on the Transport of Dangerous Goods of the Economic and Social Council of the United Nations sponsors the Global Harmonized System of Classification and Labelling of Chemicals (GHS) [14]. It was approved in December 2002 and it is reviewed and updated every two years. The objectives are:

- Improving the level of protection of the health and the environment during the handling, transport and use of these products.
- Providing a legal framework for all countries.
- Reducing trials and evaluations of chemical substances.

- Harmonizing the different systems of classification and labelling of chemical products present in the world.
- Facilitating the international trade.
- Establishing globally harmonized criteria and elements to classify and report the hazards of chemical products.

The Regulation of the European Parliament and of the Council, on Classification, Labelling and Packaging of Substances and Mixtures (CLP) (EU Labelling 2008) [6], amended by various subsequent regulations in order to adapt it to the technical and scientific progress, applies the classification criteria and GHS labelling standards in the European Union and the provisions of REACH (EU REACH 2006) on the inventory of classifications and labelling.

This Regulation increases the protection of the users' health and the environment and contributes to improve the free circulation of chemical substances and mixtures of two or more chemical substances.

For the classification of chemical substances and their mixtures, three types of hazards, twenty-eight hazard classes and seventy-nine hazard categories are established. Physical hazards (16 classes and 45 categories); health hazards (10 classes and 28 categories); hazards to the environment (2 classes and 6 categories). The annexes include the criteria for the classification and labelling of dangerous substances and mixtures, the list of hazard statements, the list of precautionary statements, the pictograms for each hazard class and the classification and labelling lists harmonized at Community level.

CLP introduces substantial changes in the identification system, danger pictograms, words of warning, hazard statements, precautionary statements and additional information (Fig. 3).

The Registration, Evaluation, Authorisation and Restriction of Chemicals (EU REACH 2006) [7] constitutes an integrated system of standards for chemical companies and its fundamental objective is to ensure a high level of protection of the health of the workers and the users of chemical substances and preparations and to protect the environment in a sustainable development system. Another purpose is the free movement of these products and the promotion of competitiveness and innovation.

After the approval of REACH, the European Chemicals Agency (ECHA) is created to manage the technical, scientific and administrative aspects of this regulation. It will manage its central database, in which all substances manufactured or marketed in the European Union will be compulsorily registered. On this purpose, manufacturers or importers must provide the Agency with a technical dossier containing information about the properties, use and precautions for use of each of the chemical substances. It also develops the technical guides for the implementation of REACH by manufacturers, importers and competent authorities.

The key elements of REACH system are the registration of chemical substances and preparations in a central database managed by ECHA, the assessment which allows ECHA to verify that the industry fulfils its obligations and avoids unnecessary tests on vertebrate animals, the authorization of the substance and the restriction of the manufacture, commercialisation and use of chemical substances and preparations which pose unacceptable risks to the health of users and the environment.

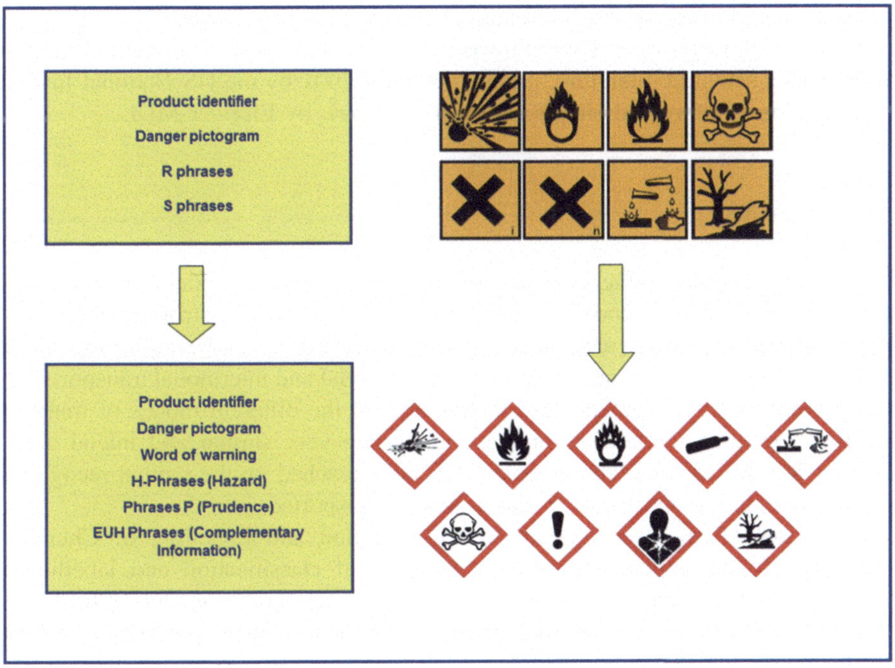

Fig. 3. CLP Regulation. Changes in identification, pictograms and phrases

The harmonized regulation on classification and labelling of substances and mixtures recommends the update of the International Chemical Safety Cards (ICSC), which were developed on the initiative of the International Program on Chemical Safety (IPCS), being a joint work of the International Labour Organization (ILO) and the World Health Organization (WHO), with the cooperation of The European Commission (EC). One of their main objectives is the assessment of the risks to the health and the environment caused by chemical substances. Although their follow-up is not mandatory, they are an important tool for workers and entrepreneurs. They are cards in which all possible international identification numbers of a chemical substance appear:

Number EC. Number of the European Union.
Number CAS. Number of the Chemical Abstracts Service, a division of the American Society of Chemistry.
Number EINECS. Number of the European Inventory of Existing Chemical Substances. It is a registration number given to each chemical commercially available in the European Union between 1 January 1971 and 18 September 1981.
Number ELINCS. Number of the European List of Notified Chemical Substances. As of 19 September 1981, for all new substances entering the European market.
Number UN (United Nations). Four-digit identification number of the materials or articles, taken from the United Nations Standard Regulation.
Number ICSC. Project Identification Number of the International Chemical Safety Cards, initiative of the International Program on Chemical Safety (IPCS).

Number ICSC. Number of the International Chemical Safety Cards.
Number RTECS. Registry of Toxic Effects of Chemical Substances, compiled from the available open scientific literature, maintained until 2001 by the US National Institute for Occupational Safety and Health (NIOSH) and now by Elsevier MDL.

4 Conclusions

The Recommendations on the Transport of Dangerous Goods - Model Regulations aim to harmonize the different modes of dangerous goods transport, controlling the possible damages to people, goods and the environment, favouring and simplifying all the procedures of handling and managing of the multimodal and intermodal transports. The different regulations, agreements and instructions of the different modes of transport, developed according to these recommendations, are very similar and intend a total harmonization. Meanwhile, an agreement has been reached on the mutual recognition of the documentation and labels of the products transported.

The Global Harmonized System of Classification and Labelling of Chemicals (GHS) tends to harmonize the different systems of classification and labelling of chemical products present in the world, improving the level of protection of health and the environment during the handling, transport and use of these products and facilitating the international trade.

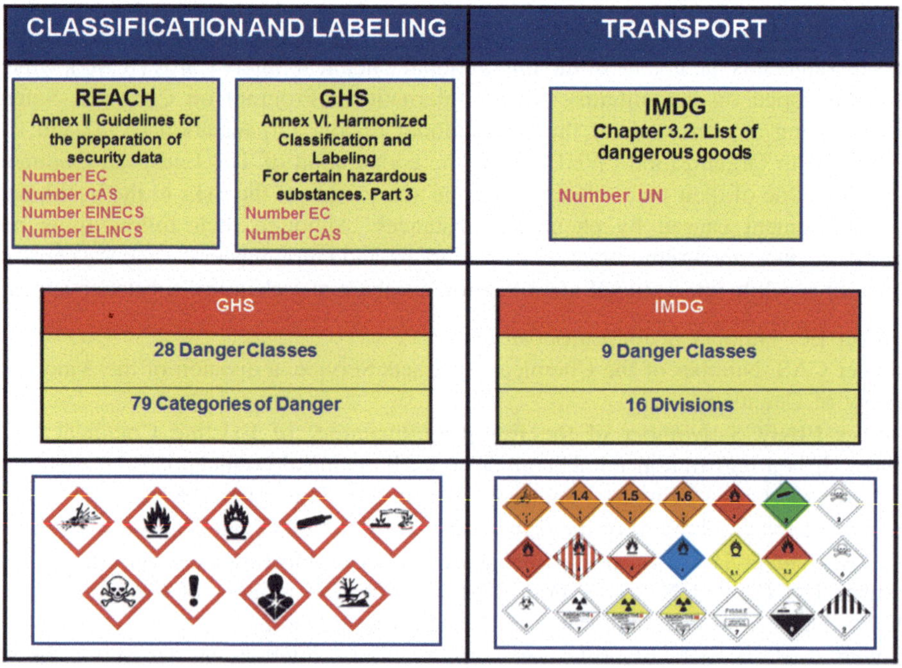

Fig. 4. Differences between Transport systems and Classification and Labelling systems

The Regulation on Classification, Labelling and Packaging of Substances and Mixtures (CLP) applies the classification criteria and GHS labelling standards in the European Union and the provisions of REACH on the inventory of classifications and labelling, increasing the protection of users' health and the environment and improving the free movement of these products.

The Registration, Evaluation, Authorisation and Restriction of Chemicals (REACH) intends to ensure a high level of protection of the health of the workers and the users of these products and the protection of the environment, as well as their free movement and the promotion of competitiveness and innovation.

The European Chemicals Agency manages the technical, scientific and administrative aspects of REACH and registers all these substances which are manufactured or marketed in the European Union.

There is a lot of work to be done to harmonize the transport systems and the classification, labelling and packaging systems of chemical substances and mixtures, which is essential to facilitate the multimodal and intermodal transport of substances and the free movement of chemical substances and their mixtures (Fig. 4).

References

1. European Union: Directive 2008/68/EC of the European Parliament and of the Council of 24 September 2008 on the Inland Transport of Dangerous Goods, Official Journal of the European Union L 260/13, pp. 13–59, 30 September 2008
2. European Union: Directive 2009/240/EC, of the Commission, of 4 March 2009, authorising Member States to adopt certain derogations pursuant to Directive 2008/68/EC of the European Parliament and of the Council on the Inland Transport of Dangerous Goods, Official Journal of the European Union L 71/23, pp. 23–58, 17 September 2009
3. European Union: European Union, Directive 2010/187/EU, of the Commission, of 25 March 2010 authorising Member States to adopt certain derogations pursuant to Directive 2008/68/EC of the European Parliament and of the Council on the Inland Transport of Dangerous Goods (notified under document C(2010) 1610), Official Journal of the European Union L83, pp. 24–58, 30 March 2010
4. European Union: Directive 2010/61/EU, of the Commission, of 2 September 2010 adapting for the first time the Annexes to Directive 2008/68/EC of the European Parliament and of the Council on the Inland Transport of Dangerous Goods to scientific and technical progress Text with EEA relevance, Official Journal of the European Union L233, pp. 27–28, 03 September 2010
5. European Union: Directive 2011/26/EU, of the Commission, of 14 January 2011 authorising Member States to adopt certain derogations pursuant to Directive 2008/68/EC of the European Parliament and of the Council on the Inland Transport of Dangerous Goods (notified under document C(2010) 9724), Journal of the European Union L13, pp. 64–93, 18 January 2011
6. European Union: Regulation (EC) 1272/2008 of the European Parliament and of the Council, of 16 December 2008, on Classification, Labelling and packaging of substances and mixtures, amending and repealing Directives 67/548/EEC and 1999/45/EC, and amending Regulation (EC) 1907/2006, Journal of the European Union L 353, pp. 1–1355, 31 December 2008

7. European Union: Regulation (EC) 1907/2006 of the European Parliament and of the Council of 18 December 2006 concerning the Registration, Evaluation, Authorisation and Restriction of Chemicals (REACH), establishing a European Chemicals Agency, amending Directive 1999/45/EC and repealing Council Regulation (EEC) No 793/93 and Commission Regulation (EC) No 1488/94 as well as Council Directive 76/769/EEC and Commission Directives 91/155/EEC, 93/67/EEC, 93/105/EC and 2000/21/EC (Text with EEA relevance), Journal of the European Union L 396, pp. 1–852, 30 December 2006
8. Intergovernmental Organisation for International Carriage by Rail: Convention Concerning International Carriage by Rail (COTIF). Appendix C- Regulations concerning the International Carriage of Dangerous Goods by Rail (RID). 2017 edition of RID in force from 1 January 2017
9. International Civil Aviation Organization (ICAO): Technical Instructions for the Safe Transport of Dangerous Goods by Air. 2015–2016 Edition, Addendum NO. 2/Corrigendum No. 2, The attached addendum/corrigendum should be incorporated into the 2015–2016 Edition of the Technical Instructions (Doc 9284) (2016)
10. International Maritime Organization (IMO): International Maritime Dangerous Goods Code (IMDG) (2016)
11. United Nations: Economic and Social Council (ECOSOC). Recommendations on the Transport of Dangerous Goods-Model Regulations, 19th edn. (2015)
12. United Nations: European Agreement concerning the International Carriage of Dangerous Goods by Road (ADR), New York and Geneva (2016). ISBN 978-92-1-139156-5
13. United Nations: European Agreement concerning the International Carriage of Dangerous Goods by Inland Waterway (AND), New York and Geneva (2017). ISBN 978-92-1-139157-2
14. United Nations: Work of the Committee of Experts on the Transport of Dangerous Goods and on the Globally Harmonized System of Classification and Labelling of Chemicals (GHS) (2002)

The Potential for the Growth of Maritime Transport in Brazil: Focus on Cabotage/Short Sea Shipping

Delmo Alves De Moura[1(✉)] and Rui Carlos Botter[2]

[1] Federal University of ABC, Santo André, São Paulo, Brazil
delmo.moura@ufabc.edu.br
[2] University of São Paulo, São Paulo, Brazil
rcbotter@usp.br

Abstract. Brazilian transport matrix has a strong predominance of road transport, approximately 62%. The country has a territorial extension of a continent, with a navigable sea coast of 8.5 thousand kilometers, with 37 public ports, 39 fluvial ports which serve cargo ships that dock in the most significant public ports in the country, and more than 164 terminals for private use (TUPs). The majority of the Brazilian population, approximately 80%, which corresponds to 156 million inhabitants, inhabit up to 200 km from the coast, which would justify even more the use of maritime transport and leave the modal road for small distances and door-to-door deliveries.

The focus of this work is analyzing the products that can be transported by cabotage/short sea shipping between Northern and Southern regions of Brazil and contribute to a more sustainable transport matrix, besides helping to reduce logistics costs, promoting the competitiveness between modals and increasing the use of maritime transport. The study develops a detailed analysis of the current situation of maritime cabotage/short sea shipping, its characteristics, and its logistical bottlenecks. It studies how to use this mode of transport to improve the national logistics of goods movement throughout Brazil.

The use of sustainable transport, with a lower logistic cost, is a challenge that Brazil must face and implant in its transport matrix, and cabotage/short sea shipping is a perfect way to integrate all modes transportation.

Keywords: Cabotage/short sea shipping · Maritime transportation · Sustainable transportation

1 Introduction

1.1 Brazilian Coast

The Brazilian coast is 8,500 km navigable, has 37 organized seaports, 156 Private Use Terminals (TUPs), 25 Transhipment Stations of Charges (ETCs) and 2 Port Facilities of Tourism (IPT), besides some river ports which serve coastal vessels (in Porto Alegre, Manaus, and Belém, for example). Cabotage/Short Sea Shipping transport and port

V. A. J. E. Carreño Moreno et al. (Eds.): CIDIN 2019/COPINAVAL 2019, pp. 113–123, 2020.
https://doi.org/10.1007/978-3-030-35963-8_10

operations are strongly linked. Figure 1 shows the relation of these organized ports and Fig. 2 shows the list of Private Use Terminals.

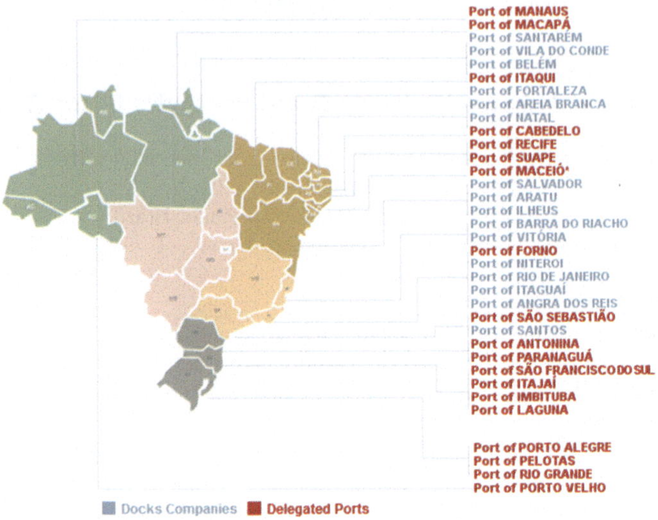

Fig. 1. Organized ports

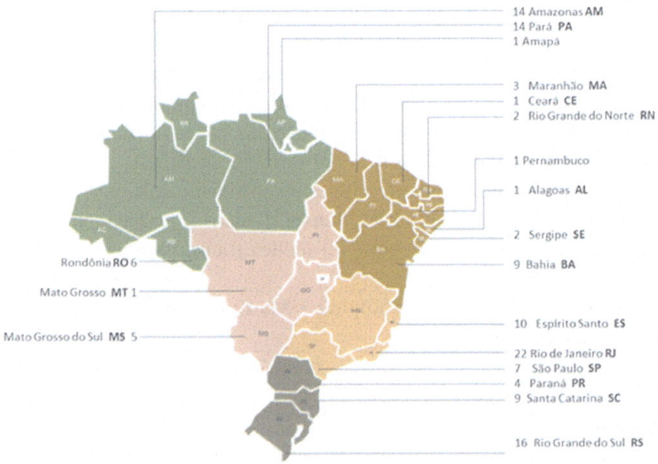

Fig. 2. Private Use Terminals (TUPs)

The maritime sector is subject to the rules and regulations of the Ports and Coasts Directorate of the Brazilian Navy, the National Water Transportation Agency (ANTAQ) and the National Petroleum Agency (ANP).

ANTAQ, the main regulatory body, was created under Law 10.233/2001, which provides for the restructuring of waterway and land transportation, establishes the National Council for the Integration of Transport Policies, the National Land Transport Agency, the National Transportation and the National Department of Transport Infrastructure. That Law establishes that ANTAQ is responsible for regulating, supervising and supervising the waterway sector (the port, fluvial, lake, crossing, maritime support, cabotage/short sea shipping, and long-haul ship navigations are part of its scope), the transport of dangerous and unique cargo waterways, organized ports and private port terminals and the exploitation of federal waterway infrastructure. Figure 3 shows the waterway transport system in Brazil.

Fig. 3. Cabotage/short sea shipping, long course, inland navigation and maritime support.

The private sector operates in cabotage/short sea shipping with 40 companies, through authorization provided by ANTAQ [1]. The estimated fleet is 153 vessels. The largest companies are Petrobras/Transpetro, Aliança, Norsul, Elcano, Log-in and Mercosur. The shipping company Norsul has 24 boats, being a specialist in the transportation of bulk, general cargo and barge ships in cabotage in Brazil. This company has dedicated lines for its clients in the transportation of wood, pulp, and steel. Elcano operates in the transport of bulk carriers, oil tankers, and liquefied petroleum gases - LPG [2].

Regarding the transportation of cargo with regular services, there is the Alliance (BR-Marítima), Log-in and Mercosul Line, which carry the door-to-door transport.

These companies that offer regular lines operate in cabotage//short sea shipping in Brazil and extend the service to Argentina and Uruguay. The Maestra shipping company operates in the container transportation sector and offers regular cabotage//short sea shipping services Transpetro, (a Petrobras subsidiary in the area of transportation), works in the transportation of petroleum and its by-products, natural gas, ethanol, and biofuels in Brazil. Brazil has enormous potential to grow in the maritime transport sector, through cabotage//short sea shipping and integrate this modal with the road, rail, waterway, and pipeline and minimize the country's dependence on the road modal, which currently corresponds to 61.1% of the matrix of Brazilian transport [2].

2 Methodology

Thorugh the point of view of its primary objectives and procedures, it is considered exploratory research, involving a bibliographical survey about the theme. The research instruments of this work were raising the scientific articles about the study and evaluating its essential points that contributed to this study. Besides, there were visits to navigation companies, that operate in the sector of cabotage//short sea shipping in Brazil. The advantage of this methodology is to be able to understand a system theoretically and have the opportunity to verify its main characteristics in loco.

3 Features of Transport System in Brazil

There is intense competition in the road modal for the transportation of cargo in Brazil, compared to the maritime modal. Most of the cargo in Brazilian transportation is by road (approximately 61.1%). Nowadays, the transport policy favors road transport to the detriment of maritime cabotage/short sea shipping. As a characteristic of this modal one can cite the higher speed, depending on the route and distance covered, greater offer and frequency. The railroad model corresponds to approximately 20.7% of freight transport, waterway mode with 13.6%, a pipeline with 4.2% and airway with 0.4%.

The maritime mode of cabotage/short sea shipping has lost competitiveness in freight due to the development of the road, through the construction and maintenance of highways, the technological development of vehicles and fuel subsidies, given by the Federal Government in the 50s and 90s.

A single transport contract characterizes the multimodal transport operation. Two or more transport modes are used, from the origin to the destination. The performance is under the sole responsibility of a Multimodal Transportation Operator (MTO). This operator is a company contracted to carry out the multimodal transport of the cargo from the origin to the destination, either by own means or through third parties. In addition to the transportation itself, the Multimodal Freight Transport comprises the collection, consolidation, cargo handling and storage services, deconsolidation and delivery, in short, all the essential steps to the complete execution of the task.

Operating in the potential markets requires frequency and reliability of the services provided, minimization of intermodal costs, a team vision, control of operations, integrated systems, own or outsourced intermodal structure, intermodal terminals and integrated decision making [3].

Figure 4 shows the waterway movement, in millions of tons, by segment: Inland navigation, cabotage/short sea shipping, and long haul, between the years 2010 to 2016.

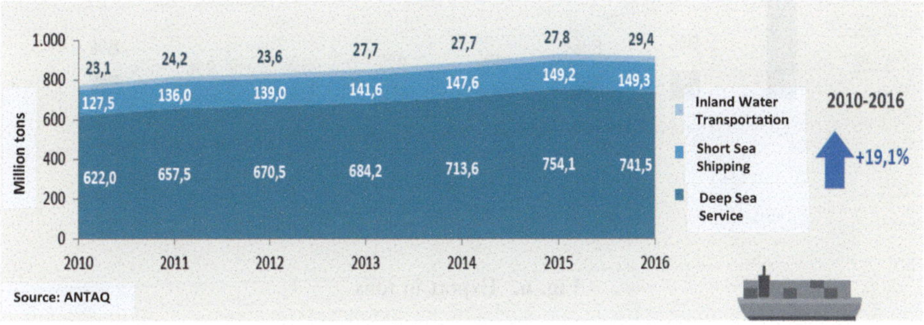

Fig. 4. Waterway movement, in millions of tons.

3.1 Main Products Transported in Cabotage/Short Sea Shipping

Figure 5 shows the import and export sectors of Brazil, from 2010 to 2017.

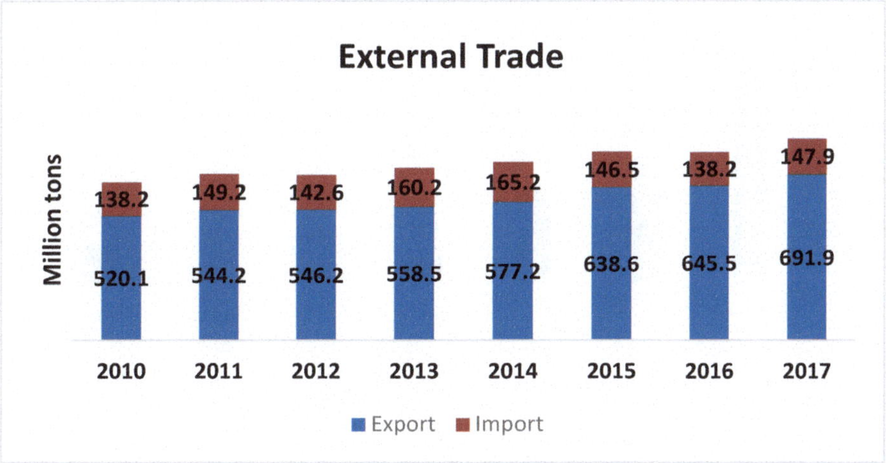

Fig. 5. Foreign trade in Brazil

Figure 6 shows the main products exported by Brazil over the last years, which require water transport.

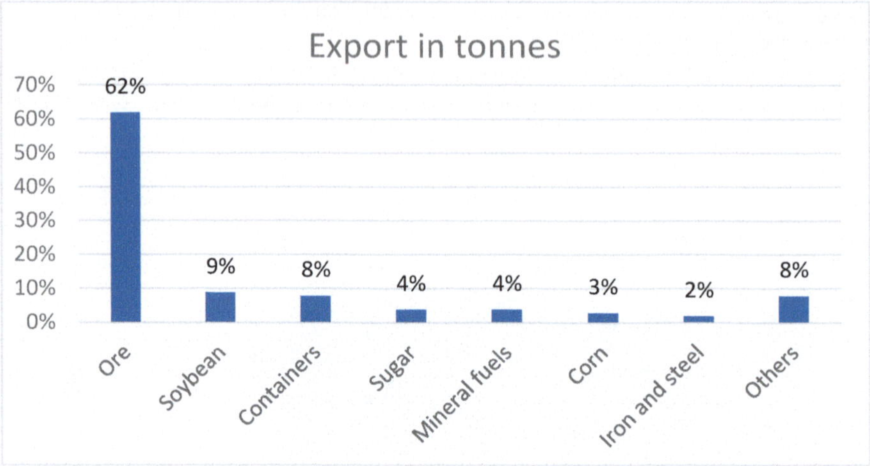

Fig. 6. Export in tons

Figure 7 shows the main products imported by Brazil over the last few years, which require water transport.

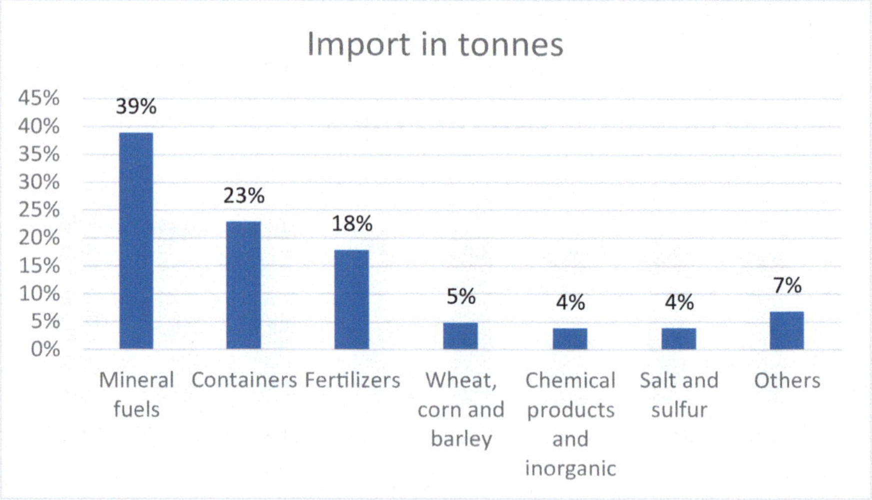

Fig. 7. Imports in tons

Figure 8 shows the percentage of the total number of containers handled in the country, in the main ports and private terminals. The locations are: Port of Santos (São Paulo - SP), Private Use Terminal of Porto Nave (Santa Catarina - SC), Paranaguá (PR), Rio Grande do Sul State, RS, Itapoá Private Use Terminal (North of Santa

Catarina - SC), Embraport Private Port Terminal (Porto de Santos - São Paulo - SP), Porto de Rio de Janeiro (RJ) and Port of Salvador (Bahia - BA).

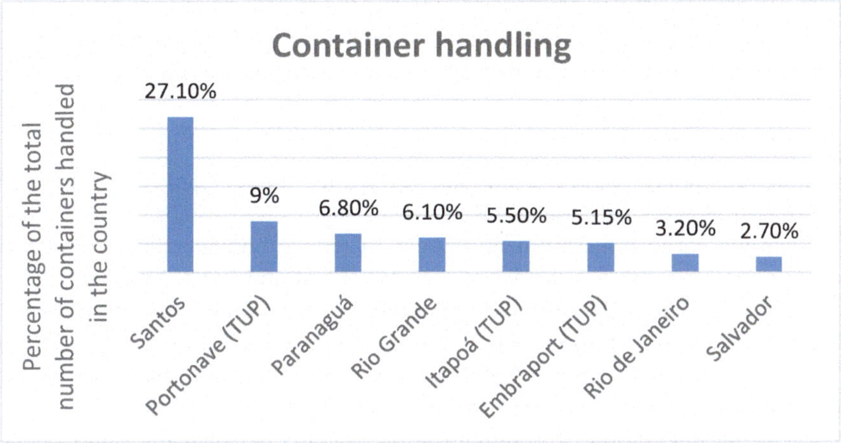

Fig. 8. Percentage of the total number of containers handled in the country

Figure 9 shows the growth in cabotage/short sea shipping transport in Brazil, over the years 2010 to 2016.

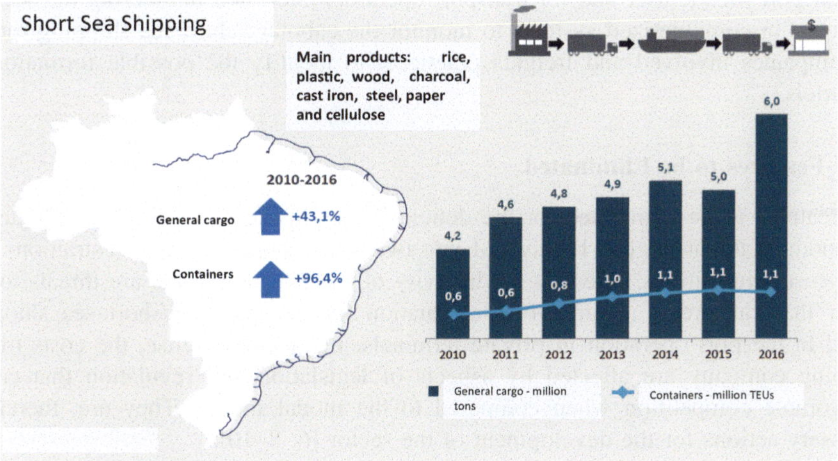

Fig. 9. Growth in cabotage/short sea shipping transport in Brazil.

4 Factors that Can Leverage the Shipping Industry

Regarding shipping, an essential factor for the user market of the system refers to aspects of multimodality or intermodality that Brazilian cabotage/short sea shipping needs to meet efficiently and effectively. If, on the one hand, some deficiencies are inherent to the activity - less speed, lack of versatility - on the other hand, they are barriers to entry for the sector that needs to be integrated and coordinated with different modes [4–7].

They are, therefore, necessary actions for the development of the sector:

- To think of cabotage/short sea shipping always in an integrated way to the various modes, and as a door-to-door service;
- Regulate the cargo transshipment process to reduce bureaucratic procedures and facilitate the emergence of feeder service for Brazil and Latin America;
- Invest in information technology to integrate modalities, facilitate the tracking and clearance of cargo and allow optimized decision making by the cargo transporter;
- Invest in mechanisms to monitor the market, with the purpose of enabling the preparation of forecasts and the identification of investment potential;
- Invest in the integration of communication systems between the Latin American ports aiming at the Latin American feeder service and a possible integrated market in the medium term;
- Intervene, as a regulatory agency, in the mechanism for releasing MMF resources and related accounts, with the objective of protecting the national fleet, guaranteeing the supply of vessels in the future and the technological advancement of the fleet for the transportation of containers;
- Carry out institutional campaigns on the advantages of cabotage/short sea shipping transport for the user and for transport operators interested in entering the sector;
- Invest in computerized systems to monitor the cabotage/short sea shipping routes, companies involved and freights practiced, to identify the possible formation of cartels.

4.1 Features to be Eliminated

The features to be eliminated, or the deficient ones in the sector which the external environment threatens, can be divided into two main categories: port restrictions and legal costs impositions. The low productivity of the Brazilian ports are threats to the sector that can create an unfavorable situation for the cabotage/short sea shipping vessel in the port operation in private terminals. In the same sense, the costs of the shipping company are affected by aspects of legislation and regulation that create unfavorable competition when compared to the modal model. They are, therefore, necessary actions for the development of the sector [6, 8–10]:

- Ensure, as a regulatory agency, that coastal shipping vessels do not impair port service, due to congestion or the least economic interest in their service;
- Act to reduce the number of port taxes applied in cabotage/short sea shipping, to reduce bureaucracy and rationalize the taxation of the sector;

- Ensure, as a regulatory agency, the isonomy in the amount paid by fuels for coastal vessels;
- Regulate the clearance of the cabotage/short sea shipping cargo to expedite the port operation;
- To monitor supply and demand in the shipbuilding market: shipyards, orders under construction, portfolio orders, the age of the fleet, the forecast of renewal, resources available in the FMM, among others. Publicize the market situation and act to ensure the supply of yards for the renewal of the national fleet in the long term;
- Invest in information technology to monitor the Brazilian port productivity.

Features to be potentialized

The elements to be explored are economic, social and environmental, boosted by the situation experienced worldwide pressure by the less use of fossil fuels. Nowadays, there are few explorations of these elements as a means of advertising the sector. They are, therefore, actions to be examined:

- Act within the FMM to accelerate the release of resources and protect the national fleet;
- Invest in institutional advertisements on the advantages of coastal shipping regarding lower emissions of pollutants and greater road safety.

4.2 The Main Obstacles for Cabotagem/Short Sea Shipping are

Bureaucratic

- The documents required for cabotage/short sea shipping are comparable to long-haul freight. In all, one only vessel needs, at least, 44 records, and another nine may be necessary depending on the type of cargo. This system/bureaucratic structure ends up burdening the sector and making it less attractive to the transport market [6].
- High labor cost, due to low supply and restriction of hiring and training of the maritime sector employees.
- A high tax burden directly related to the provision of the cabotage/short sea shipping service and also associated with the cargo transported by this modal.

Operational

- Low port infrastructure, with a shortage of berths and backyards.
- Lack of adequate equipment for handling cargo.

Investment

- Regarding obtaining funding through the FMM, the process is long and time consuming for construction and maintenance of boats.
- The volume of investment in infrastructure is low.

Top Proposed Solutions

Main Solutions Proposals

- Reduce bureaucracy and simplify processes and volumes of documents requested for cabotage/short sea shipping transport and obtaining financing.
- Increase the supply of port infrastructure, aiming to expand the terminals, berths, and backyards.

- Implement hub ports and feeder ports to promote long-haul and cabotage/short sea shipping navigation.
- Increase the number of skilled labor in the market, technical and official, which would require increasing the number of entities that form these professionals.
- Promote the development of the shipbuilding sector, through economic mechanisms, promote fleet renewal, reduce bureaucratic access to credit lines and reduce the number of procedures for access to the Merchant Marine Fund (FMM) resource.

The Merchant Navy Fund (FMM) is an accounting fund from the collection of AFRMM (Additional Freight for Merchant Marine Renewal) to provide resources for the development of merchant shipping and the maritime industry. The following banks operate it: BNDES, Banco do Brasil, Caixa Econômica Federal, Bando da Amazônia and Banco do Nordeste.

Advantages of the FMM: Lower interest rates and exemption of some contributions. Disadvantages of FMM:

- The financing only occurs for vessels produced in the national shipyards.
- The process is very bureaucratic and time-consuming.

The request for funding from FMM resources must be carried out by the companies concerned. One needs to complete 20 steps, and this can take 25 months on average.

5 Conclusion

Brazil can mirror the international model, such as that implemented in the European Union, to develop its potential for cabotage transport, as it relates exclusively to the integration of road, rail and waterway modes. The European Union uses these modes in an integrated way and exploits the potential of each one, aiming to eliminate or reduce obstacles along the entire transport chain and optimizing the role of each modal within a sustainable transport corridor. In this way, Brazil needs to put into practice public and private policies that encourage the use of maritime transport along its coast, integrating with other modalities and reducing bureaucracy and optimizing the transportation system to:

- Increase the frequency of cabotage routes to attract more cargo to this transportation system;
- Increase the speed of delivery of merchandise and greater versatility in the service offered to the market;
- Improve coordination and synchronization of existing manners;
- Have effective data transfer in all stages of the cargo transportation process, involving all the necessary modalities between the origin and destination of products;
- Have a feeder service organized for Brazil;
- Improve the port congestion system to promote cabotage transport further;
- Have well-structured transshipment areas to carry out the operations with the other modes in the integrated transport system;

- Eliminate bureaucracy for the cabotage cargo in the customs area, as well as the costs related to this transport operation and
- Invest in information technology to integrate modalities, facilitate tracking and clearance of the cargo and allow optimized decision making by the cargo transporter.

References

1. ANTAQ: Agência Nacional de Transportes Aquaviários. Centro de Informação em Transporte Aquaviário (2018)
2. Fonseca, R.O.: A navegação de cabotagem de carga no Brasil. The cabotage in Brazil. Mercator **14**(1), 21–46 (2015). Revista de Geografia da Universidade Federal do Ceará, Fortaleza
3. Kuznetsov, A.L., Kirichenko, A.V.: Methodological problems of modern transportation logistics. Int. J. Mar. Navig. Saf. Sea Transp. **12**, 611–616 (2018)
4. Castells, M., Usabiaga, J.J., Martínez, F.: Road and maritime transport environmental performance: short sea shipping vs road transport. J. Marit. Res. **9**(3), 45–53 (2012)
5. Kontovas, C.A.: The Green Ship Routing and Scheduling Problem (GSRSP): a conceptual approach. Transp. Res. Part D Transp. Environ. **31**, 61–69 (2014)
6. Moura, D.A., Botter, R.C.: Challenges for implementation of the green corridor to Brazil. In: Maritime Transportation and Harvesting of Sea Resources. International Congress of the International Maritime Association of the Mediterranean, Lisbon, Portugal, October 2017
7. Çetin, İ.B., Akgül, E.F., Koçak, E.: Competitiveness of Turkish coaster merchant fleet: a qualitative analysis by short sea shipping perspective. Int. J. Mar. Navig. Saf. Sea Transp. **12**, 389–396 (2018)
8. Lee, T.-C., Lam, J.S.L., Lee, P.T.W.: Asian economic integration and maritime CO_2 emissions. Transp. Res. Part D **43**, 226–237 (2016)
9. Qu, Y., Bektas, T., Bennell, J.: Sustainability SI: multimode multicommodity network design model for intermodal freight transportation with transfer and emissions costs. Netw. Spat. Econ. **16**, 303–329 (2016)
10. Zis, T., Psaraftis, H.N.: The implications of the new sulfur limits on the European Ro-Ro sector. Transp. Res. Part D **52**, 185–201 (2017)

Modeling of the Learning Curve Corresponding to the Transit Through the New Expanded Panama Canal

Luis Manuel Carral Couce[1(✉)], Javier Tarrío-Saavedra[2], Aladar Rodríguez-Díaz[3], and Salvador Naya[2]

[1] Department of Naval and Industrial Engineering, Mixed Engineering Group, Escuela Politécnica Superior, Universidade da Coruña, A Coruña, Spain
lcarral@udc.es
[2] Department of Mathematics, MODES and CITIC Groups, Escuela Politécnica Superior, Universidade da Coruña, A Coruña, Spain
[3] Universidad Marítima Internacional de Panamá, Panama City, Panama

Abstract. The pilotage in the Panama Canal (PC) is unique due to the importance it acquires; in fact, the pilot relieves the captain in his/her duties during the transit. This exceptional nature has meant that prior to the commissioning of the expanded canal; a small group of experienced pilots of the existing Canal was enabled for the transit of Neopanamax ships through a planned and innovative learning.

Once the Canal is extended in service, routine learning begins, so that the learning curve in performing these maneuvers represents the growing skill acquired by the pilots and the organization.

The learning effect is quantifiable. Indeed, the aim of this work is to study and model the pilot learning curve corresponding to the maneuvers of the total time in transit through the Canal, including the navigation routes and locks of Cocolí and Agua Clara. Exploratory analysis, control charts and multivariate linear models have been applied in order to identify and model learning patterns. The application of these techniques has been demonstrated that there is actually a learning effect of linear nature on PC transits.

Keywords: Expanded Panamá Canal · Learning effect · Learning curve · Time in transit · Control charts · Regression

1 Introduction

The aim of this work is to identify and model the learning patterns corresponding to the process of vessel transit across the new expanded Panama Canal (PC). This learning pattern also known as learning curve should be related to the increasing experience of pilots and the continuous improvement of PC methods and installations. Thus, in this section, both Panama Canal organization (mainly the PC pilotage) and the learning curves background are briefly introduced.

V. A. J. E. Carreño Moreno et al. (Eds.): CIDIN 2019/COPINAVAL 2019, pp. 124–133, 2020.
https://doi.org/10.1007/978-3-030-35963-8_11

1.1 Learning Curves

The spent time, even the cost or the hours required to develop a specific work can be described in a graphical way by estimating learning curves [1]. These curves are obtained by monitoring the time corresponding to the developing of a specific product or the mean time concerning to the production of a number of products [1, 2]. The estimation of the latter is straightforward, consisting on dividing the total time corresponding to the production of the group of production units by the number of units that have been produced [3]. It is also important to stress that statistical theory is usually applied to this domain. Namely, the parametric regression model fitting (linear or nonlinear) is often applied to estimate the learning curves of a specific process [1, 4–6]. In addition, at a lesser extent, once the model that explains the process is determined, the application of simulation studies has been also applied [7–9]. Concerning the mathematical model for learning curves, those can be applied to retrospective data [2] to explain the past performance, and also can be estimated to make predictions [3]. The learning curves are applied to many different domains. Among others, there are many interesting works concerning the building construction [5, 6], and the performance of infrastructure projects [1, 9]. The most used regression model for learning curves estimation is the linear model (or Boeing/Wright) [3], often applied after logarithmic transformation of the variables [7]. Hinze and Olbina [3] also discuss the application of nonlinear models, but the differences in accuracy with respect to the linear model application are not significant. This fact support the application of linear regression models in this work.

1.2 Pilotage Activity in the Expanded Panama Canal

In the maritime navigation phase, in the field of Transportation, a good is no longer produced, but just a service is provided. The navigation maneuvers of vessels during are activities significantly different from others corresponding to industrial sectors, because there are many procedures related to each mile navigated, each one different and unrepeatable. However, the case of transit through a canal such as the PC is similar to a production activity that provide units from a succession of repetitive processes (all the maneuvers corresponding to the navigation through the sea lanes, locks, and Culebra Cut and Gatun Lake) [10–14]. Under this assumption, the estimation of the learning curve of the whole of the pilotage activity can be feasible and applicable.

The so-called Expanded Canal is made up of the access routes and the Locks complex corresponding to Cocolí and Agua Clara [10, 13]. In the Canal, traffic control services are needed through a Maritime Traffic Control Center (CCTM) (Rodriguez et al. [18]), which has been reinforced by the incorporation of the expanded channel. In addition, the pilots training is provided entirely in the Canal facilities, mainly through the Canal Maritime Simulation and Development Center (SIDMAR), which is certified by the Panama Maritime Authority under the schemes of the International Agreement on Training Standards, Certification and Guards for Seafarers, 1978 (STCW Convention).

2 Methodology

Once opened on June 26, 2016, the third set of locks began to work. Balboa Maritime Traffic Control Center monitored the maneuvers of the ships that transit through the expanded Canal. Consequently, the Panama Canal Authority (PCA) could conduct a census on the time-vessels spent in the Canal. 200 transits took place during the observation period. Taking into account that the vast majority of vessels correspond to the types of Container, Liquefied Petroleum Gas (LPG) and Liquid Natural Gas (LNG), these types were studied, and tankers and Ro-ro's were omitted. The resulting data set consists of different critical variables during the Canal transit process.

An exhaustive statistical analysis was performed and shown in the following section to identify the most influential factors and quantitative variables in transit time, to model these dependence relationships, to estimate and control the time variability, and to evaluate the pilot learning effect over transit time.

3 Results and Discussion

In this work, statistical process control tools such as quality control charts have been used to identify possible learning patterns in the time in transit, the critical to quality (CTQ) variable for overall transit across the Canal. In any case, prior to control charts and regression modelling is necessary to identify those influencing factors on the time in transit.

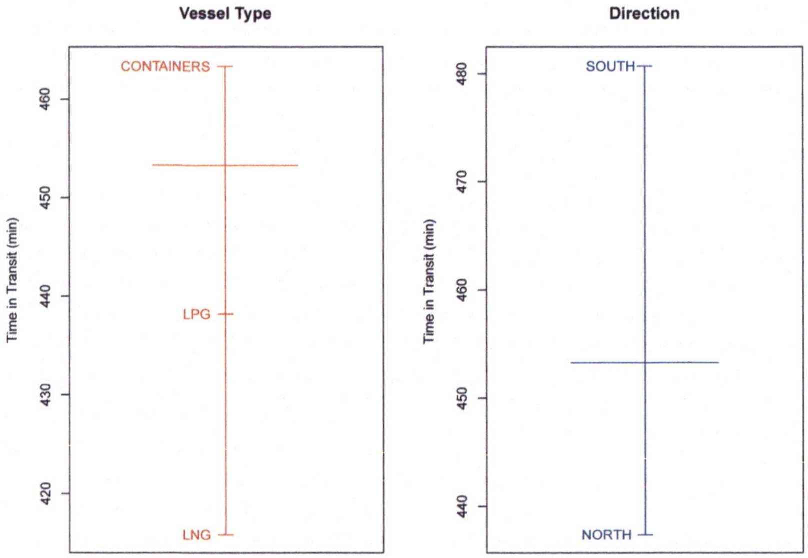

Fig. 1. Main effects of the vessel type and the direction over the time in transit trough the PC.

Thus, ANOVA analysis [15] is performed, including the ANOVA table with F test and factor main effect estimation, to evaluate if the time in transit is different depending on the transit direction (from North to South –South level– or from South to North –North level), the type of vessel and, in addition, the interaction between the two above mentioned factors. The Table 1 shows that the direction and type of vessel significantly affect to the time in transit (p-values < 0.05). The main effects of the latter factors are shown in Fig. 1. The transit duration of containers is about 40 min greater than the corresponding to the LNG vessels, and 20 min greater than the transit time of LPG vessels. These results, in accordance with Carral et al. [10–12], are due to the special conditions of transit corresponding to LNG and LPG vessels in addition to their different dimensions and load. It is important to note that the duration of transits from Atlantic to Pacific (South) are 40 min greater than those with North direction.

Table 1. ANOVA table with F test and the corresponding p-values for Direction, Type of vessel and their interaction.

	Sum of squares	Degrees of freedom	F statistic	p-value of F test
Direction	169211	1	39.474	0.000000001124
Vessel type	101612	2	11.852	0.000010963409
Direction and vessel type interaction	3986	2	0.465	0.6286
Residuals	1328856	310		

Afterwards, Shewhart control charts for individual measurements [15] have been implemented for the CTQ variable to identify possible learning patterns. Considering that the time in transit is different depending on the direction, control charts have been applied separately for each level of direction (North and South). In addition, Cumulative sum control charts (CUSUM) are applied to identify small changes (smaller than two standard deviations with respect to the process mean) [15]. CUSUM chart shows the cumulative sums of the differences with respect to the process mean, both those positive and negative. The yellow and red points indicates that the process have changed with respect to the calibration sample (observations obtained in 2016) from which the upper and lower natural control limits are calculated. The yellow ones are corresponding to runs (more than six points at one side of the central line) and the red ones indicate that the point is out of the control limits. The charts of Fig. 2 shown that a learning pattern has been identified for the time in transit corresponding to North direction, whereas no learning curve is observed for the South direction (Fig. 3), by observing the monitored sample (new data) corresponding to the measures obtained during 2017. It is interesting to note that those detected patters for North direction are identified in the negative cumulative sums of the CUSUM chart, i.e. the time in transit is decreasing with respect to the time passed from the expanded PC opening. The assignable causes, concerning to the special operation maneuvers, and currents, among others, for each direction will be studied in further studies. This slightly gradual change in the transit process can be observed in a more proper way by means the CUSUM chart.

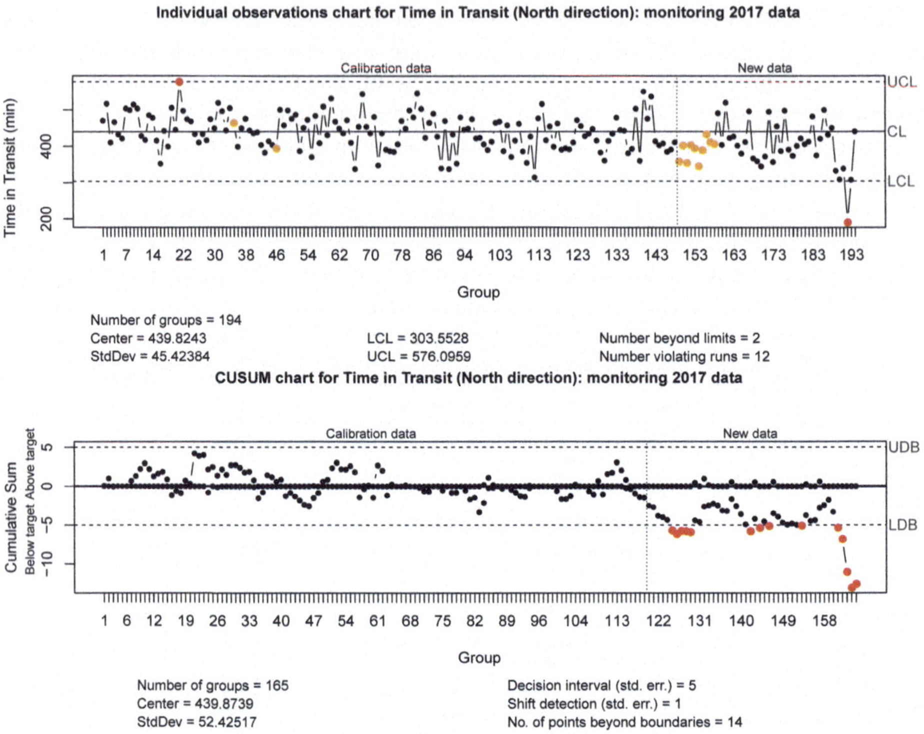

Fig. 2. Control chart for individual measurements (on the top) and CUSUM control chart (below) corresponding to the time in transit corresponding to vessels going from the Pacific to the Atlantic. The control limits are estimated assuming normal distribution from the calibration sample composed of the measurements obtained in 2016. The monitor sample (new data) accounts for the measurements obtained in 2017; they are not considered to estimate the limits.

The next step is to model the relationship between the time in transit critical to quality response and its influencing predictors. Taking into account the nature of the effects of the predictors over the response, a multivariate regression model is applied [15]. The ANOVA showed that the vessel direction and the type of vessel have to be included in the model. In addition, the vessel dimensions are potential influencing predictors, thus the vessel length overall (LOA) is also included. Finally, the time passed from the expanded PC opening is included taking into account that the aim of this work is to model the PC learning effect. The estimation of the model is

$$\widehat{y} = 356.4 - 0.086x_1 + 0.097x_2 + 45.43x_3 - 5.952x_4 - 3.315x_5, \text{ with } R^2 = 0.32,$$

whereby $\widehat{y}$ are the estimates of the time in transit, x_1 the time from the PC opening, x_2 the vessel LOA, x_3 the dichotomous variable defined by South direction, x_4 indicates that the type of vessel is LNG, and x_5 accounts for vessel type equal to LPG. The parameters shown in the latter expression are significant different from zero as shown in Table 2 (except x_4).

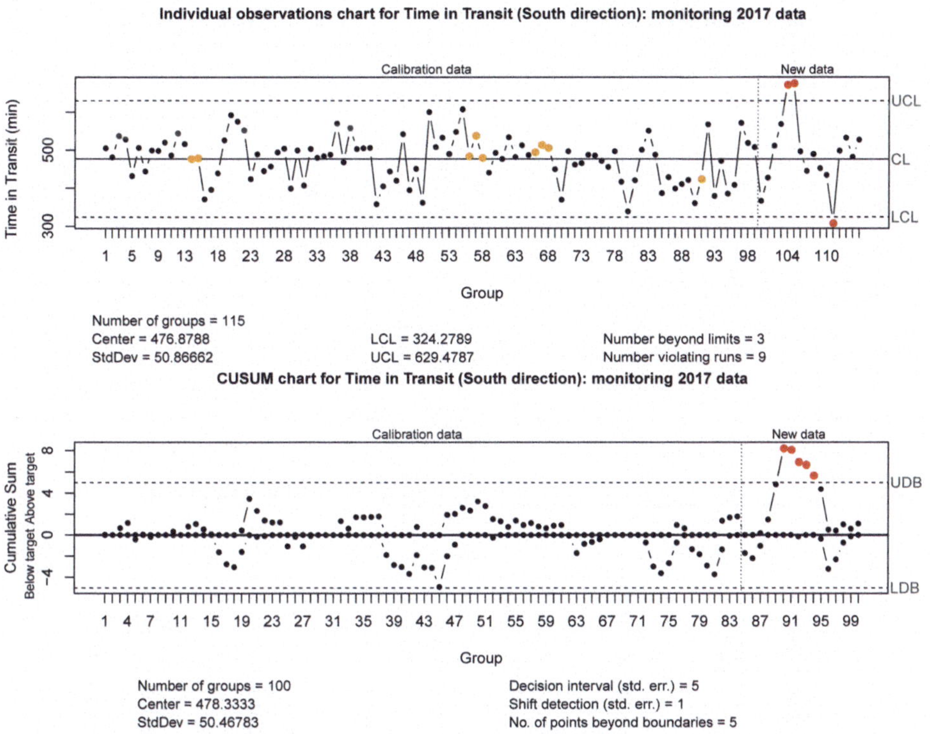

Fig. 3. Control chart for individual measurements (on the top) and CUSUM control chart (below) corresponding to the time in transit corresponding to vessels going from the Atlantic to the Pacific (South). The control limits are estimated assuming normal distribution from the calibration sample composed of the measurements obtained in 2016. The monitor sample (new data) accounts for the measurements obtained in 2017; they are not considered to estimate the limits.

Table 2. Estimates of the multivariate linear model parameters for predicting the time in transit. Signification analysis based on t test is also included.

	Estimates	Standard error	t value	p-value of t test
(Intercept)	356.4	44.58	7.995	2.98e−14
Direction[SOUTH]	45.43	5.842	7.776	1.27e−13
Vessel type[LNG]	−5.952	11.95	−4.980	1.09e−06
Vessel type[LPG]	−3.315	14.19	−0.234	0.8154
Time from the expanded PC opening	−0.086	0.019	−4.631	5.47e−06
LOA	0.097	0.042	2.274	0.0237

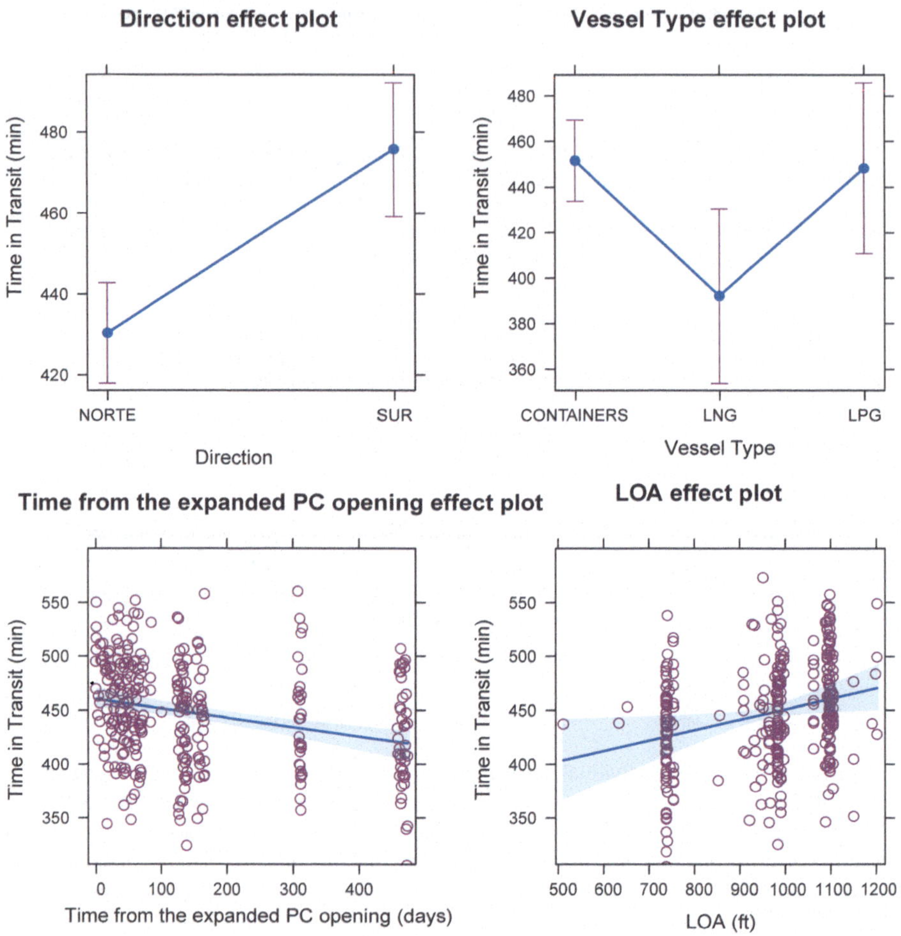

Fig. 4. Partial effects (corresponding to the multivariate regression model) of the direction, vessel type, time passed from the expanded PC opening, and LOA over the time in transit response variable. Scheffé confidence bands for partial effects and model partial residuals are included.

The explained variability of the model accounts for the 32% of the overall. The no explained variability may be related to other variables that are not measured in this study such as the meteorological variables, the currents, the different ability and skills of the PC pilots [10, 13], among others. In any case, the aim of this work is to model the learning curve and this has been achieved using this model. Namely, for each additional day passed from the opening of the expanded PC, the time in transit has been reduced in 0.086 min, at a constant rate. Thus, the learning plans of PC Authority have provided positive results. This linear effect can be observed in the Fig. 4, in addition to the effects of LOA (the time in transit increases 0.097 min for each addition ft), type of vessel and direction. Scheffé confidence bands [16] and confidence intervals are included for each effect. Moreover, the partial residuals for each effect are also included to indicate the overall variability. Finally, the contribution of the learning effect of the explained variability for the time in transit is about the 20% as shown by the R^2 contribution study (see Fig. 5) [17]. Taking into account the continuous learning plans implemented for the PC Authority and the increasing experience of the pilots and remaining personal, a continuous decrement of time in transit is expected.

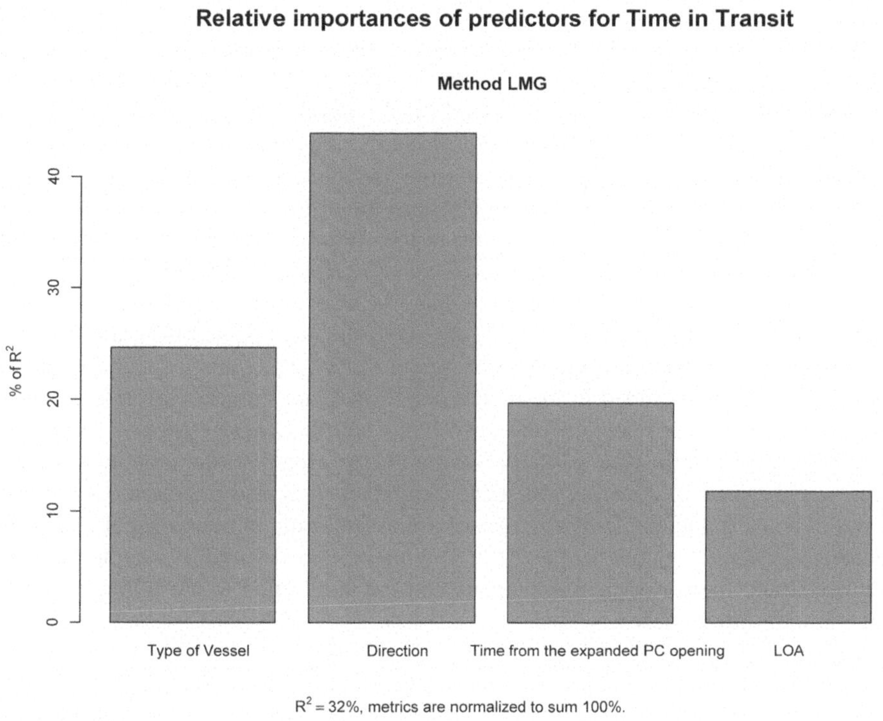

Fig. 5. Relative importance of predictors for the time in transit response in terms of percentage of determination coefficient, R^2. The metric LMG has been implemented.

4 Conclusions

In this work, the learning curves have been introduced and their estimation in the framework of PC pilotage has been considered feasible, adequate, useful and informative. This can be done taking into account the PC pilotage process is a special case of the field of transportation where the final service/product is the result of a repetitive maneuvers.

In this study, the variable that defines the quality of a transit is the time in transit of each vessel. Thus, statistical quality control techniques have been applied to the vessel time in transit (composed of the lockage time, and the time expended to cross the Gatun lake and Culebra Cut) in order to identify possible learning patterns, and, further, to estimate statistical regression models that best explain them.

Prior to the estimation of learning curves using regression analysis, an ANOVA study have been performed in order to identify of which factor variables the time in transit depends. As a result, we have identified that the vessel type and the transit direction through the PC significantly affect to the time of transit. Consequently, these variables have to be taken into account before applying techniques to detect patterns and also included in the regression modelling of the time in transit.

The learning patterns have been identified by using individual measurements and CUSUM control charts. The time in transit slightly decreases following a gradual pattern if the measurements observed in 2017 (monitoring sample) are compared with those obtained in 2016, when the expanded canal was opened. Those patterns are more clearly identified by using CUSUM control charts, thus the change of the process is small (less than two standard deviation from the process mean) and gradual. It is important to note that a learning pattern has been identified when the transit direction is from South to North, i.e. from the Pacific to the Atlantic. Nevertheless, there is not a clear learning curve when the transit form North to South is studied.

Multivariate linear regression models have been estimated to explain the time in transit variation with respect to time (expressed as the days passed from the Expanded Canal opening). In addition, the type of vessel (LNG, LPG and containers) and the transit direction are included in the model due to they are influencing variables on the time in transit response feature. Consequently the learning curve estimates corresponds to the linear effect of the time passed from the PC opening over the time in transit. The effect over the time in transit is statistical significant. It is also important to stress that the multivariate model explains the 32% of the overall time in transit variability, where the learning effect accounts for the 20%. The no explained variability could be related to seasonal and meteorological assignable causes, in addition to the different skills of pilots, among other causes.

Acknowledgments. This work has been founded by MINECO grants MTM2014-52876-R and MTM2017-82724-R, and by the Xunta de Galicia (Grupos de Referencia Competitiva ED431C-2016-015 and Centro Singular de Investigación de Galicia ED431G/01), all of them through the ERDF. Also in part financed by the 1-FCN-007 Project of the International Maritime University of Panama (UMIP).

References

1. Farghal, S.H., Everett, J.G.: Learning curves: accuracy in predicting future performance. J. Constr. Eng. Manag. 41–45 (1997). https://doi.org/10.1061/(ASCE)0733-9364(1997)123:1(41)
2. Jarkas, A., Horner, M.: Revisiting the applicability of learning curve theory to formwork labour productivity. Constr. Manag. Econ. **29**(5), 483–493 (2011)
3. Hinze, J., Olbina, S.: Empirical analysis of the learning curve principle in prestressed concrete piles. J. Constr. Eng. Manag. 425–431 (2009). https://doi.org/10.1061/(ASCE)CO.1943-7862.0000004

4. Thomas, H.R., Mathews, C.T., Ward, J.G.: Learning curve models of construction productivity. J. Constr. Eng. Manag. 245–258 (1986). https://doi.org/10.1061/(ASCE)0733-9364(1986)112:2(245)
5. Couto, J.P., Teixeira, J.C.: Using linear model for learning curve effect on highrise floor construction. Constr. Manag. Econ. **23**(4), 355–364 (2005)
6. Pellegrino, R., et al.: Construction of multi-storey concrete structures in Italy: patterns of productivity and learning curves. Constr. Manag. Econ. **30**(2), 103–115 (2012)
7. Hijazi, A.M., AbouRizk, S.M., Halpin, D.W.: Modeling and simulating learning development in construction. J. Constr. Eng. Manag. 685–700 (1992). https://doi.org/10.1061/(ASCE)0733-9364(1992)118:4(685)
8. Lutz, J.D., Halpin, D.W., Wilson, J.R.: Simulation of learning development in repetitive construction. J. Constr. Eng. Manag. **120**(4), 753–773 (1994)
9. Naresh, A.L., Jahren, C.T.: Learning outcomes from construction simulation modeling. Civ. Eng. Syst. **16**(2), 129–144 (1999)
10. Carral, L., Tarrío-Saavedra, J., Naya, S., Bogle, J., Sabonge, R.: Effect of inaugurating the third set of locks in the Panama Canal on vessel size, manoeuvring and lockage time. J. Navig. **70**(6), 1205–1223 (2017)
11. Carral Couce, L., Tarrio-Saavedra, J., Castro-Santos, L., Lamas-Galdo, I., Sabonge, R.: Effects of the expanded Panama Canal on vessel size and seaborne transport. Promet-Traffic Transp. **30**(2), 241–251 (2018)
12. Carral, L., Tarrio-Saavedra, J., Crespo-Pereira, D., Fernandez-Campoamor, M., Sabonge, R.: Analysis of the competitive position of the expanded panama canal in the shipment of LNG between the atlantic and pacific. Proc. Inst. Mech. Eng. Part M: J. Eng. Maritime Environ. (2018). https://doi.org/10.1177/1475090218812242
13. Carral, L., Tarrio-Saavedra, J., Alvarez-Feal, C., Naya, S., Sabonge, R.: Statistical methodology to determine the transit time required for Neopanamax vessels in the Panama canal. J. Mar. Sci. Technol. (2019). https://doi.org/10.1007/s00773-019-00650-3
14. Carral, L., Fernández-Garrido, C., Vega, A., Sabonge, R.: Importance of the Panama canal in the reduction of CO2 emissions from maritime transport. Int. J. Sustain. Transp. (2019). https://doi.org/10.1080/15568318.2019.1632994
15. Montgomery, D.C.: Introduction to Statistical Quality Control. Wiley, New York (2009)
16. Draper, N.R., Smith, H.: Applied Regression Analysis, vol. 326. Wiley, Hoboken (2014)
17. Carral, L., Rodriguez-Guerreiro, M., Tarrio-Saavedra, J., Alvarez-Feal, J.C., Fraguela Formoso, J.: Social interest in developing a green modular artificial reef structure in concrete for the ecosystems of the Galician rias. J. Clean. Prod. **172**, 1881–1898 (2018)
18. Rodriguez, M.: El esquema de practicaje en el canal de panama. monografía—seminario internacional: Canal de panamá: El reto de la ampliación de la vía interoceánica o cómo diseñar, construir y gestionar una de las obras de ingeniería más grandes del mundo (2011). https://www.fundacionareces.tv/watch/canalde_panama?as=4f0ca68694a05fc458070000. Accessed May 2019

Hydrodynamic Analysis of Inland Self-propelled Vessel for Cargo Transportation in the Magdalena River

Oscar D. Acosta Lopera[1(✉)], Kazuo Nishimoto[1(✉)], and Jairo H. Cabrera Tovar[2(✉)]

[1] Numerical Offshore Tank, Polytechnic School, The University of São Paulo, Avenida Professor Mello Morais 2231, 05508-030 São Paulo, SP, Brazil
acosta.lopera@alumni.usp.br, knishimo@usp.br

[2] The Technological of Bolívar University, Parque Industrial y Tecnológico Carlos Vélez Pombo, Km 1 Vía Turbaco, Cartagena, Bolívar, Colombia
jcabrera@utb.edu.co

Abstract. A hydrodynamic analysis is realized to predict the ship resistance of an inland vessel for cargo transport the Magdalena River lower course. The analysis is applied numerically by solving the 3D Reynolds-Averaging Navier-Stokes (RANS) equation together with the standard k-ε model for turbulent effects, allowing the simulation of the viscous and pressure effects around the hull in a towing tank as long as the wall bottom and the side effects in the shallow water navigation. The Finite Volume method is applied for the discretization of flow domain. The Volume of Fluid (VOF) is used for the capture of the free surface, in which the formation of the waves is generated during the navigation of the inland vessel.

In this paper is presented a methodology that allows the calculation of the ship resistance, where are validated the experimental results obtained by Technological Research Institute of the State of São Paulo (IPT, acronym in Portuguese), along with the numerical simulation partial results and the exiting empirical methods for shallow waters. Finally, is presented the conclusions and the future work to improve this study.

Keywords: CFD · Free surface flow · FVM · Inland vessel · *k*-ε turbulence model · Magdalena river · RANS · Resistance · Restricted waterways · VOF

1 Introduction

Since the European conquest of the Americas, the Magdalena river has been the principal fluvial artery in Colombia. During the colonization period, the river served as a single route; and during the independence period, the patriotic armies used the river to dominate the Spanish colony. This event was described in the historical novel *The General in His Labyrinth* by Gabriel Garcia Marquez. The fluvial activity was of great importance in 19th century; nevertheless, in 20th century, the commercial activities in the river, as far as the transport of the commercial goods is concerned, started to decline due to other means of transport as railway and road [1].

V. A. J. E. Carreño Moreno et al. (Eds.): CIDIN 2019/COPINAVAL 2019, pp. 134–152, 2020.
https://doi.org/10.1007/978-3-030-35963-8_12

Aiming to the increase in the activities of the river, the Colombian Constitution of 1991 created the Magdalena Grand River Corporation (Cormagdalena, acronym in Spanish) and since then, the fluvial transport has been recovering due to projects developed or being developed that enhance the navigability of the river through the construction of fluvial ports, dredging works and the maintenance of the river. Nowadays, the cargoes are transported in convoys. In 2017, 3.67 millions tons of a large variety goods, including hydrocarbons and dry cargoes such a coal and cement, were mobilized by the river, indicating an increase of 68.5% compared to the previous year [2].

The calculation of the resistance is based on Froude hypothesis, and it is considered that the resistance in this case is mainly composed by frictional and residual resistance. ITTC [3] established an equation for the frictional resistance coefficient. Guldhammer and Harvald [4] used diagrams for the calculation of the residuary resistance coefficient; Georgakaki and Sorenson [5] demonstrated that this variable can be extrapolated providing reasonable results.

Considering the effects of shallow waters, Molland et al. [6] explained the wave generating phenomena that would affect the resistance calculation; Latorre et al. [7] presented the empirical methods for the ship resistance analysis applied to European and American vessels; ITTC [8] considered some typical parameters to be considered in the definition of the level of waterway confinement in shallow water; Pompée [9] reviewed many empirical methods to determine the hull resistance depending on the type of vessel, as well as, the physical aspects of a river (small, medium and large, confined, semi-confined and laterally unrestricted water-ways).

In case of large rivers, three methods are applicable: Schlichting [10], Landweber [11] and Lackenby [12], while in case of resistance calculation for vessel intended to operate in medium rivers, Karpov [13] interpreters the phenomena of shallow water with effective velocities and Artjushkov [14] improves this work correcting wall effects. For small rivers, the calculation of the inland vessel resistance is associated with the existence of a limited velocity in restricted waterways that cannot exceed, where could be causes by a steep ship resistance rise.

The numerical calculation of the hydrodynamic flow around the vessel conducted during this study was based on the applicable theories in order to obtain reliable results. From the conservation of mass to the conservation of momentum, the Navier-Stokes (NS) equations are the basis of the description of fluid motion. Euler [15] initiated these descriptions for incompressible fluids and non-friction flows, Navier [16] analyzed the friction effects for viscous fluids, and Stokes [17] improved this work completing the solution. Reynolds [18] introduced the time-averaging of the flow for turbulent effects, defining the decomposition of the mean and fluctuating parts of a variable. This is applied to the NS equations obtaining the Reynolds-Averaging Navier-Stokes (RANS) equations. Launder and Spalding [19] were the first to introduce the standard k-ε turbulence model. This model was refined and is called RNG (Re-Normalization Group methods) k-ε model and is was developed by Yakhot et al. [20]. Later, Shih et al. [21] improved the turbulence model and is called Realizable k-ε. Wilcox [22] published another turbulent model denominated standard k-ω. To improve k-ω model, Menter [23] developed the baseline (BSL, also called BSL k-ω model) which was later refined for the transport of the turbulence shear stress and is called Shear Stress Transport

(SST) k-ω model. Noh and Woodward [24] and Hirt and Nichols [25] developed the Volume of Fluid (VOF) method to track and locate the free-surface, based on Marker and Cell (MAC) method. The interpolation of the convection term is used in the transport equation applied to Finite Volume Method (FVM). Courant, Isaacson and Rees [26], and independently, Gentry et al. [27], Barakat et al. [28] and Runchal et al. [29] introduced the first-order upwind scheme. Another type of interpolation, the second-upwind scheme, was initiated by Warming and Beam [30], and Hodge et al. [31] for finite difference discretization. Later, it was implemented for finite volume by Tamamidis and Assanis [32] as an explicit transient scheme, and by Thompson and Wilkers [33] as a steady state implicit version. Patankar and Spalding [34] introduced the iterative method called Semi-Implicit Method for Pressure Linked Equations (SIMPLE) algorithm for the calculation of pressure-velocity couple.

Celik et al. [35] described the Grid Convergence Index (GCI), used in the estimation and report of uncertainty results in CFD applications; Ji et al. [36] studied the prediction of the relationship between the geometrical and the kinematic parameters of a convoy and the amplitude of the waves generated by the vessel in a restricted waterway. In this paper, the calculation of the grid size in x direction is conducted and the transversal wave length λ is estimated its value being divided in 10 points obtaining, therefore, the grid size; Liu et al. [37] evaluated the inland vessel resistance in confined waters, analyzing the squat effect. The computational domain, the mesh configurations and the time step are adopted in this paper.

The objective of this work is the study of the resistance of a 2700 TDW self-propelled inland vessel by numerical simulation. The numerical results are compared to the existing empirical formulas and the experimental test results being, therefore, validated.

The outline of this paper is as follows: The ship geometry, as well as the physical properties, are presented in Sect. 2; the governing equations for the numerical calculation, including the turbulence model, are presented in Sect. 3; the empirical models used in this study are mentioned in Sect. 4; the methodology is detailed in Sect. 5 and the results are presented in Sect. 6. Finally, the conclusions and the future work are given in Sect. 7.

2 Ship Geometry

The 2700 TDW inland vessel model was tested in the towing tank at the Technological Research Institute (ITP, acronym in Portuguese) [38] and the results are considered in this work. The geometry of the hull and the physical characteristics are described in Fig. 1 and Table 1.

The model was tested under 4 conditions. For the numerical computation, the most critical condition (highest draft) is used in the comparison of the experimental results. The scaled model (scale factor equal to 20) was towed at 6 different constant velocities. The width of the tank was 3.5 m and the depth of the water was 30 cm.

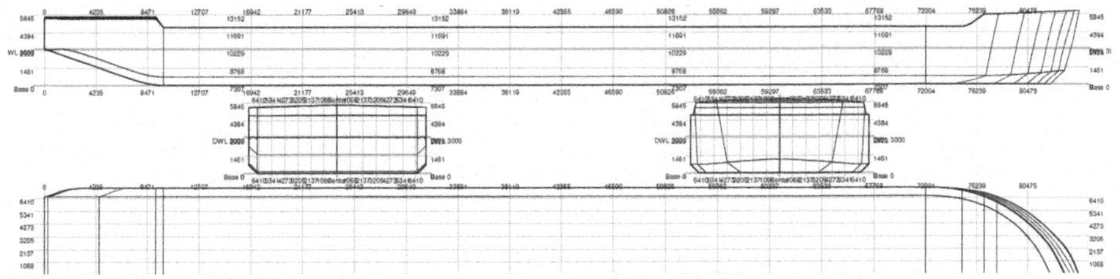

Fig. 1. Lines-plan of the self-propelled inland vessel.

Table 1. Main particulars of the 2700 DWT inland vessel.

Particulars	Full scale	Model scale
Length between perpendiculars L_{pp} (m)	83.86	4.193
Breadth B (m)	14.50	0.725
Design draft T (m)	3.20	0.160
Wet surface area S_{ws} (m^2)	1564.40	3.911
Displacement Λ (m^3)	3560.00	0.445

3 Numerical Model

3.1 Governing Equations

For describing the flow motion, the governing equations for the mass and momentum conservation are the RANS equations. Using the Einstein notation, the equations are

$$\frac{\partial \rho}{\partial t} + \frac{\partial}{\partial x_j}(\rho U_i) = 0, \tag{1}$$

$$\begin{aligned}\frac{\partial}{\partial t}(\rho U_i) + \frac{\partial}{\partial x_j}(\rho U_i U_j) &= -\frac{\partial P}{\partial x_i} + \frac{\partial}{\partial x_j}\left[\mu\left(\frac{\partial U_i}{\partial x_j} + \frac{\partial U_j}{\partial x_i} - \frac{2}{3}\delta_{ij}\frac{\partial U_i}{\partial x_i}\right)\right] \\ &\quad + \frac{\partial}{\partial x_j}\left(-\rho\overline{u_i' u_j'}\right),\end{aligned} \tag{2}$$

$$\rho = \sum_{n=1}^{2} \alpha_n \rho_n, \; \mu = \sum_{n=1}^{2} \alpha_n \mu_n, \tag{3}$$

where ρ is the density of the fluid, t is the time, $x_i (i = 1, 2, 3)$ are Cartesian coordinates, $U_i (i = 1, 2, 3)$ are the velocity components, P is the pressure, μ is the dynamic viscosity, δ_{ij} is the Kronecker delta, u_i' is the fluctuating velocity and $-\rho\overline{u_i' u_j'}$ is the Reynolds stress tensor that is expressed by

$$-\rho\overline{u_i' u_j'} = \mu_t\left(\frac{\partial U_i}{\partial x_j} + \frac{\partial U_j}{\partial x_i}\right) - \frac{2}{3}\left(\rho k + \mu_t \frac{\partial U_i}{\partial x_i}\right)\delta_{ij}, \tag{4}$$

where μ_t is the dynamic turbulent viscosity and k is the Turbulent Kinetic Energy (TKE). In this study, the standard k-ε model is adopted and is given by

$$\frac{\partial}{\partial t}(\rho k) + \frac{\partial}{\partial x_i}(\rho k U_i) = \frac{\partial}{\partial x_j}\left[\left(\mu + \frac{\mu_t}{\sigma_k}\right)\frac{\partial k}{\partial x_j}\right] + P_k - \rho\varepsilon + S_k, \tag{5}$$

$$\frac{\partial}{\partial t}(\rho\varepsilon) + \frac{\partial}{\partial x_i}(\rho\varepsilon U_i) = \frac{\partial}{\partial x_j}\left[\left(\mu + \frac{\mu_t}{\sigma_\varepsilon}\right)\frac{\partial \varepsilon}{\partial x_j}\right] + C_{1\varepsilon}\frac{\varepsilon}{k}P_k - C_{2\varepsilon}\rho\frac{\varepsilon^2}{k} + S_\varepsilon, \tag{6}$$

where P_k represents the generation of TKE due to the mean velocity gradient, S_k and S_ε are user-defined source terms. P_k and μ_t are given by

$$\mu_t = \rho C_\mu \frac{k^2}{\varepsilon}, \tag{7}$$

$$P_k = -\rho\overline{u_i' u_j'}\frac{\partial U_i}{\partial x_i}. \tag{8}$$

The model constants $C_{1\varepsilon}$, $C_{2\varepsilon}$, C_μ, σ_k and σ_ε have the following default values: $C_{1\varepsilon} = 1.44$, $C_{2\varepsilon} = 1.92$, $C_\mu = 0.09$, $\sigma_k = 1.0$ and $\sigma_\varepsilon = 1.3$.

3.2 Numerical Methods

In this study, the flow simulation - in way of the RANS equations' resolution for specific points on the mesh - was realized with the aid of the CFD (Computational Fluid Dynamics) software STAR-CCM+ [39].

The numerical method of flow properties' analysis, used in this study, is the FVM in which the flow domain is discretized into control volumes (CV). First-order temporal discretization scheme (Euler implicit form) is used for the transient term. For the convection term, second-order upwind scheme is performed. The diffusion term is based on Central differentiating.

For capturing the free surface (between water and air) the VOF method is applied. Tracking between phases is described by the solution of the continuity equation. For nth phases, the equation is expressed by

$$\frac{\partial \alpha_n}{\partial t} + \frac{\partial}{\partial x_j}(U_i \alpha_n) = 0, \tag{9}$$

where α is the volume fraction, represented as

$$\sum_{n=1}^{2} \alpha_n = 1. \tag{10}$$

For resistance calculation, the time step could be determined according to the ITTC guidelines [40]; however, in this study, a smaller time step is used in accordance with Liu et al. [37] aiming to a more reliable result

$$\Delta t < 0.002 \frac{L_{pp}}{\upsilon}, \tag{11}$$

where υ is the velocity of the ship.

The SIMPLE algorithm is applied to complement the numerical procedure that is described in Fig. 2 introducing the concept of iteration process on the calculation of pressure and velocity fields.

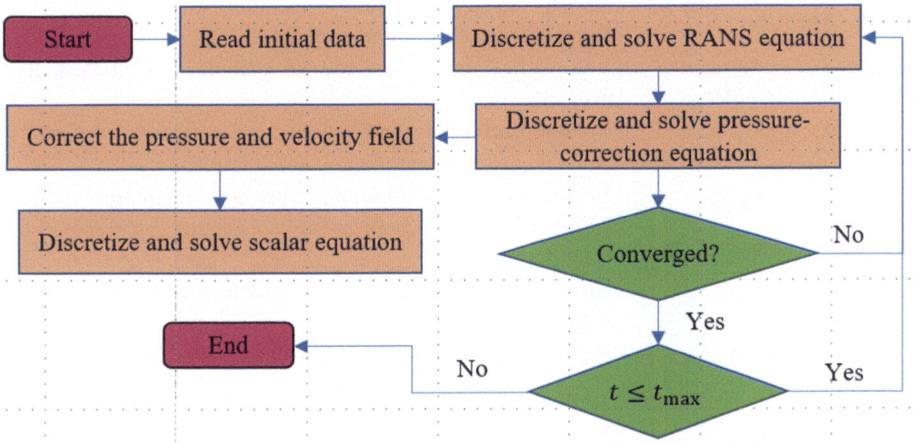

Fig. 2. Main flow chart in the STAR-CCM+ solver.

3.3 Grid Generation and Computational Setup

The computational domain and the grid mesh are shown in Fig. 3. This domain is composed by 7 boundaries: inlet plane, outlet plane, top plane, tank bottom, tank side walls and hull surface. To reduce computational cost, half of the domain is modeled both hull and tank are symmetrical in x (longitudinal) axis and the hull centerline coincides with the tank's centerline, indicating a symmetry plane configuration. The domain extends over $0.98L_{pp}$ from bow to the inlet plane, $1.98L_{pp}$ from the stern to the outlet plane and $0.19L_{pp}$ from the free surface to the top plane.

Fig. 3. Overview of the computational domain.

The boundary conditions are given in Table 2. In order to remove fluctuations of waves on inlet and outlet plane, a numerical damping wave is configured with a damping length of 9 m (Fig. 4). No-slip condition is applied on the bottom plane, tank's walls, and the hull. In the symmetry plane, the shear stress is zero.

Table 2. Boundary conditions properties in *STAR-CCM+*.

Boundary	Condition	Properties
Inlet	Velocity inlet	Normal velocity with volume fraction of water and air. Damping wave reflections avoided
Outlet	Pressure outlet	Volume fraction of water and air. Damping wave reflections avoided
Wall banks	Wall	Volume fraction of water and air
Top	Velocity inlet	No-slip condition and motion: x-velocity
Bottom	Wall	No-slip condition and motion: x-velocity. Prism layer mesh with $y+=30$
Symmetry	Symmetry	Default
Vessel hull	Wall	No-slip condition and smooth wall (default). Prism layer mesh with $y+=1$
Vessel stern and deck	Wall	No-slip condition and smooth wall (default). No prism layers

The importance of the mesh in the numerical simulation is considerably high. In fact, depending on the mesh adopted, a higher degree of detail can be achieved in the most critical zones where the gradients variation is high. Considering this fact in this study, five (5) zones of the volume domain were refined. These zones are located around the hull and at the free surface interface in order to obtain adequate resolution at these areas and, therefore, adequate viscous flow simulation and analysis of the complex flow around the hull. It is worth mentioning that the mesh in the rest of the zones (non-critical zones) is coarser in order to avoid unnecessary expenditure of computation time. The mesh created in this study, which is similar to those presented by Liu et al. [37], is composed of both the coarser zones and the refined zones, as mentioned before, and is shown in Fig. 5. For the definition of the cells size in x and y axes at the free surface region, the wavelength λ generated by the inland vessel is calculated using the following equation as [6]

$$\upsilon_{\text{model}}^2 = \frac{g\lambda}{2\pi}\tanh\left(\frac{2\pi h}{\lambda}\right) \tag{12}$$

where h is the depth of the water. Ji et al. [36] recommend using 10 points per length of the transverse waves. The cell size for both axes (x and y) is 2.125E-2 m, because the velocity reference is 0.576 m/s.

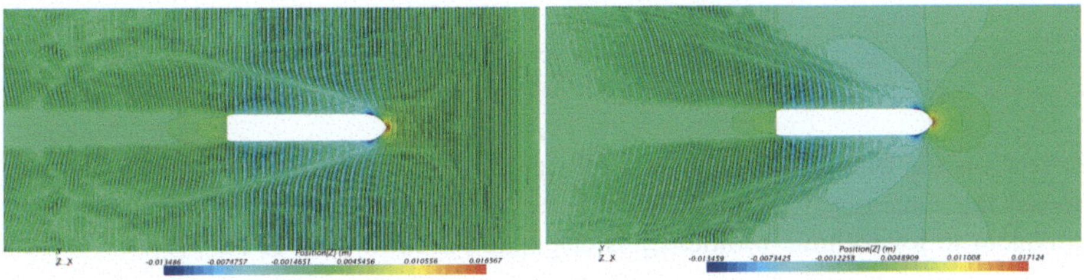

Fig. 4. Numerical damping on *STAR-CCM+*, where no numerical damping affects the true waves generated by the inland vessel (left), unlike numerical damping with length of 9 m applied only at inlet and outlet boundary conditions (right).

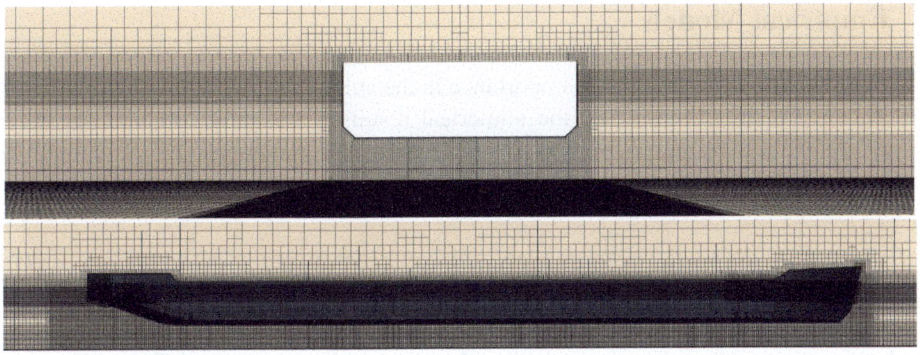

Fig. 5. Grid structure around the vessel. Top, midship section; bottom, longitudinal section at symmetry.

The boundary layer is important because the near-wall flow solution allows the determination of forces and flow features that depend on the velocity gradients [39]. The near-wall tangential velocity is related to the wall shear stress t_w

$$u_\tau = \sqrt{\frac{t_w}{\rho}}, \tag{13}$$

and the dimensionless wall distance y^+ and velocity u^+ are expressed by the following equations

$$y^+ = \frac{yu_\tau}{\nu} \text{ and } u^+ = \frac{u}{u_\tau} \tag{14}$$

where y is the normal distance of the first thickness of the cell mesh away from the wall and ν is the kinematic viscosity. The value of y is calculated according to ITTC [40] and is determined as

$$y = \frac{y^+ L_{pp}}{Re_L \sqrt{\frac{C_f}{2}}} \tag{15}$$

where C_f is the frictional resistance coefficient, Re_L is the Reynolds number. The user chooses the value of y^+. For finite mesh, $y^+ \leq 1$; and for coarse mesh, $30 \leq y^+ \leq 100$, equivalent to logarithmic profile. C_f is calculated according to ITTC [41]

$$C_f = \frac{0.075}{[\log_{10}(Re_L) - 2]^2} \tag{16}$$

4 Empirical Models

The empirical models are of great importance in the ship resistance analysis because the experimental results, as well as, the numerical results are usually compared to these models for validation purposes. The Froude hypothesis is composed by frictional R_f and residual R_r resistance. That is

$$R_t = R_f + R_r = \frac{1}{2}\rho S_{ws} \upsilon^2 C_t = \frac{1}{2}\rho S_{ws} \upsilon^2 (C_f + C_r) \tag{17}$$

where S_{ws} is the wetted area of the hull; C_f, C_r and C_t are the frictional, residual and total resistance coefficients, respectively. The frictional resistance coefficient is calculated using the Eq. 16, and the residual resistance is determined according to the method of Guldhammer and Harvald [4]. The total resistance for open water can be calculated using the Holtrop's method [42, 43]. Pompée [9] classified the calculation of the resistance depending of the physical conditions of the river, as well the type of vessel. As far as the physical conditions of the river are concerned, the rivers are categorized into large rivers and middle rivers. In case of large river, the methods of Schlichting [10], Landweber [11] and Lackenby [12] are used; whereas in case of middle river, the methods of Karpov [13] and Artjushkov [14] are applied.

To estimate the effect of shallow water in large rivers, Schlichting [10] used the velocity estimation, where he only took into account the reduction of the water depth without considering the increasing influence of the banks (walls) in the towing tanks. Landweber [11] analyzed the Schlichting's method, proposing the substitution of the water depth by the hydraulic radius. Lackenby [12] improved the Schlichting's method proposing a semi-empirical formula as a complement.

For middle rivers, Karpov [13] used two velocity estimations calculated by diagrams that he created. Artjushkov [14] included the restricted width effect adding a third velocity correction. Georgakaki and Sorenson [5] demonstrated that the velocities estimation and the variables of width effect could be extrapolated.

5 Methodology

In this study the market in which the vessel will operate, the physical restrictions of the river and the hull data of existing inland vessels were considered. Once the hull design was chosen, the resistance of the inland vessel was calculated by three ways. The first being the resistance calculation by means of the experimental test carried out in the towing tank on ITP; the second being the empirical procedure (Fig. 6) considering the parameters and the third being the numerical procedure (Fig. 7) in which the mesh was generated with the design of the geometry volume domain. Later, initial physical conditions were configured. Before running the numerical solution, time step and maximum simulation time were determined; a practice that is necessary in order to obtain reliable results avoiding unnecessary time expenditure. After that, the numerical solutions were analyzed and compared to the empirical results and the experimental results. The above are described in Fig. 8.

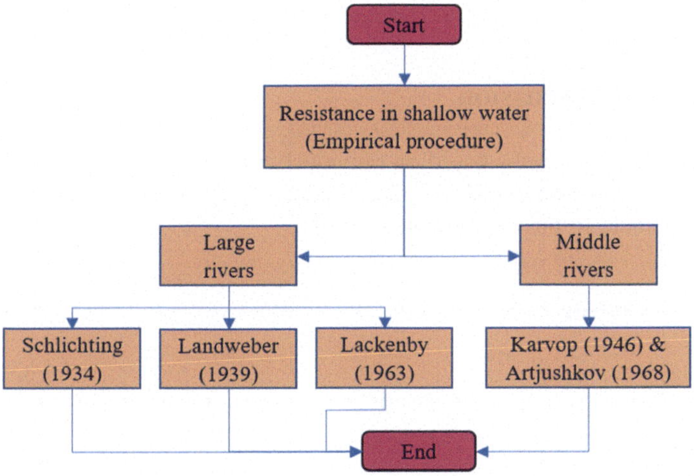

Fig. 6. Empirical procedure.

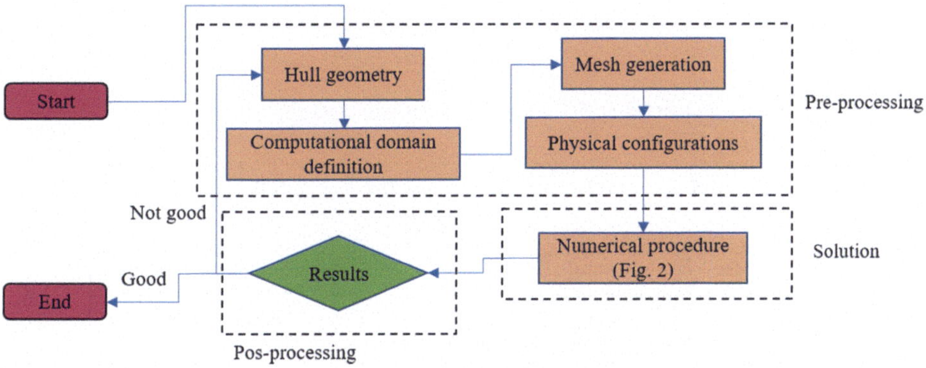

Fig. 7. Numerical procedure.

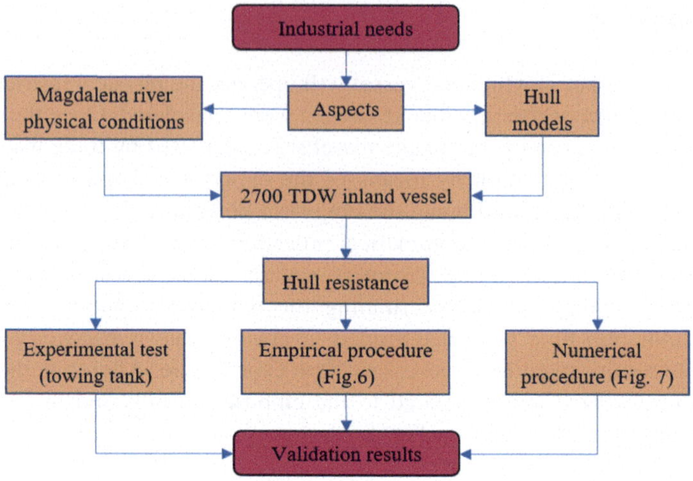

Fig. 8. General methodology.

6 Results and Discussion

The comparison of the results based on previously mentioned methods is illustrated in Fig. 9, where it can be observed that the numerical and empirical methods for middle rivers are proximate to the experimental results.

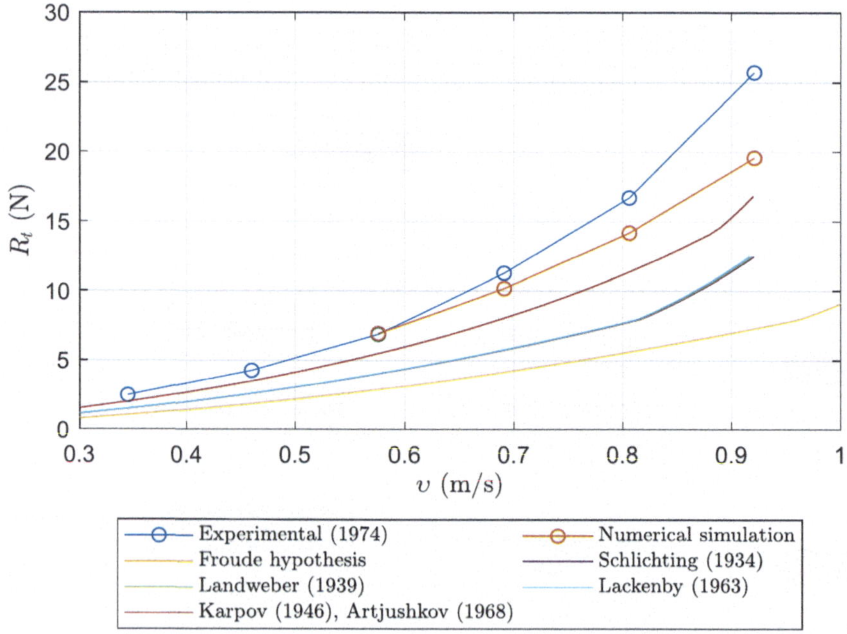

Fig. 9. Comparison of methods with 2700 TDW inland vessel.

The average loss of the velocity using empirical methods applicable to large rivers (Schlichting, Landweber and Lackenby) is 15% approximately and the resistance values obtained applying these methods do not offer an accurate prediction of the vessel resistance, as can be seen in Fig. 9. In case of large rivers, the values of the residuary resistance are higher than the values of the frictional resistance and the residuary resistance coefficient shall be used instead of the wave-making resistance.

The values of the inland vessel resistance using empirical methods for middle rivers are in better agreement with the experimental results. The velocity estimations determined by Karpov's method are higher than the velocities in deep water (maximum of 12.5% approximately). For length Froude number values up to 0.15 the residuary resistance coefficient is equal to 2.44, and for length Froude number equal to or higher than 0.15 this coefficient varies until 2.473. Raven [44] affirms that the most frequently used empirical methods of shallow water resistance estimation for inland vessels have a very weak theoretical and empirical basis. He recommends the development of new prediction methods that correct separately the components of total resistance [45].

The results of the numerical method are more accurate at the beginning, but as the velocity becomes higher the discrepancies appear, as can be observed in Table 3. The relative error that represents these discrepancies be-tween the approximation values and the exact values is given by the following equation

$$E_r = \frac{F_m - F_{CFD}}{F_m} \tag{18}$$

where F_m is the resistance of the experimental model and F_{CFD} is the resistance calculated numerically.

Table 3. Comparison of the numerical and empirical method with the experimental results.

υ (m/s)	R_t (N)			E_r (%)	
	Experimental	CFD	Empirical	CFD	Empirical
0.576	6.89	6.96	5.59	−1.02	18.87
0.691	12.29	10.20	8.00	9.65	29.14
0.806	16.71	14.20	11.52	15.02	31.06
0.921	25.72	19.60	16.06	23.80	37.59

Figure 10 shows the numerical results of the hull resistance for each velocity and, in this figure, it can be observed that there is a similarity in the behavior (in terms of resistance versus time for different velocities). At the beginning, the values are higher and at approximately 100 s the resistance values are stabilized. As can be seen in Fig. 11, the same behavior is observed with y^+ in the wall bottom and hull vessel, because in *STAR-CCM+* the default *Two-Layer All y+ Wall Treatment* is used.

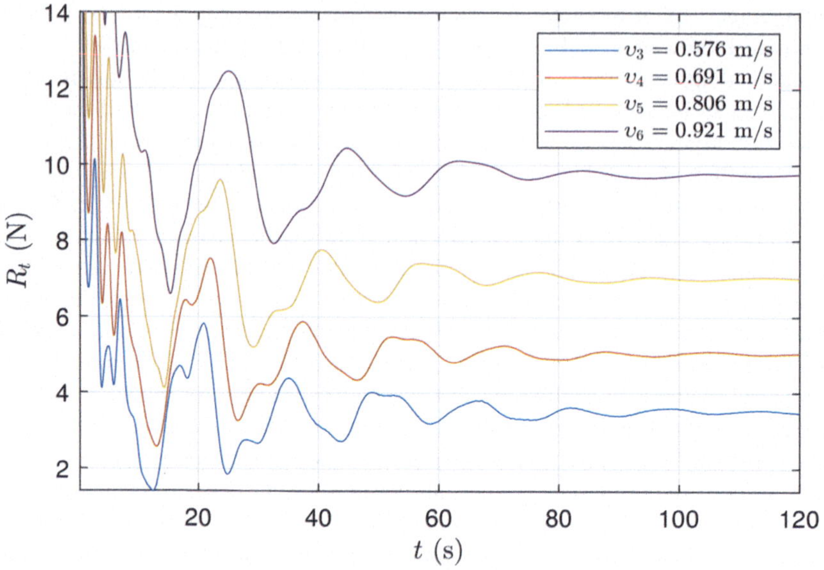

Fig. 10. Results of the total resistance of 2700 TDW inland vessel calculated numerally, where the values are half of the hull.

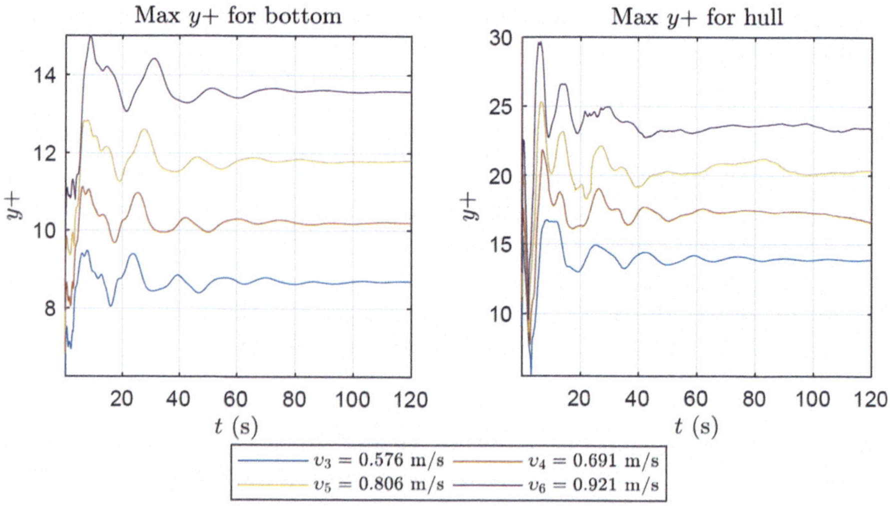

Fig. 11. Max $y+$ for tank bottom and hull inland vessel.

The wave pattern generated by the inland vessel's hull for each velocity can be visualized in Fig. 12, showing the Kelvin waves system. The presence of the walls causes reflection of the waves. For lower velocity, the minimum waves height is −0.015 m, and the maximum is 0.011 m. In case of velocities increase, wave height also increases by a minimum of −0.05 and a maximum of 0.024 m.

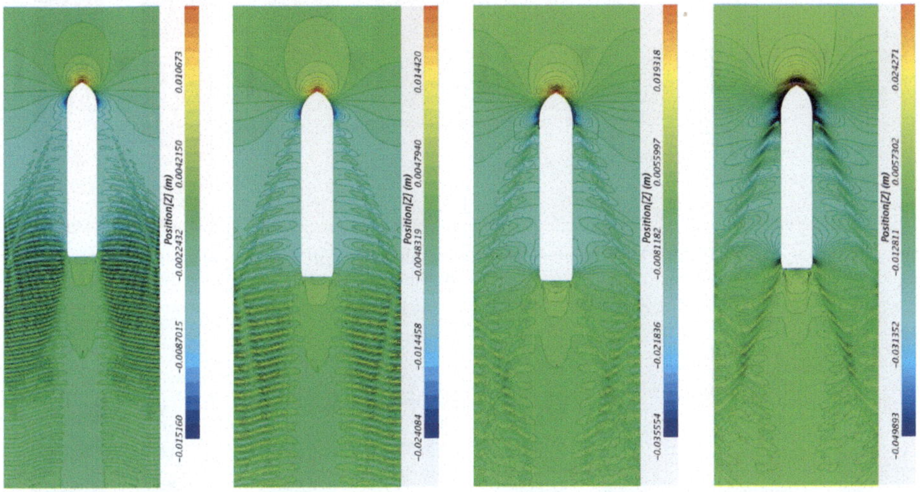

(a) $v = 0.576$ (b) $v = 0.691$ (c) $v = 0.806$ (d) $v = 0.921$

Fig. 12. Wave pattern generated by 2700 TDW self-propelled inland vessel.

At different velocities, the water that flows between the hull and wall bottom is faster, but the pressure is low according to the Bernoulli's principle. Another observation is the distribution of the velocities at bow and stern, where the values are zero. These observations are illustrated in Figs. 13 and 14.

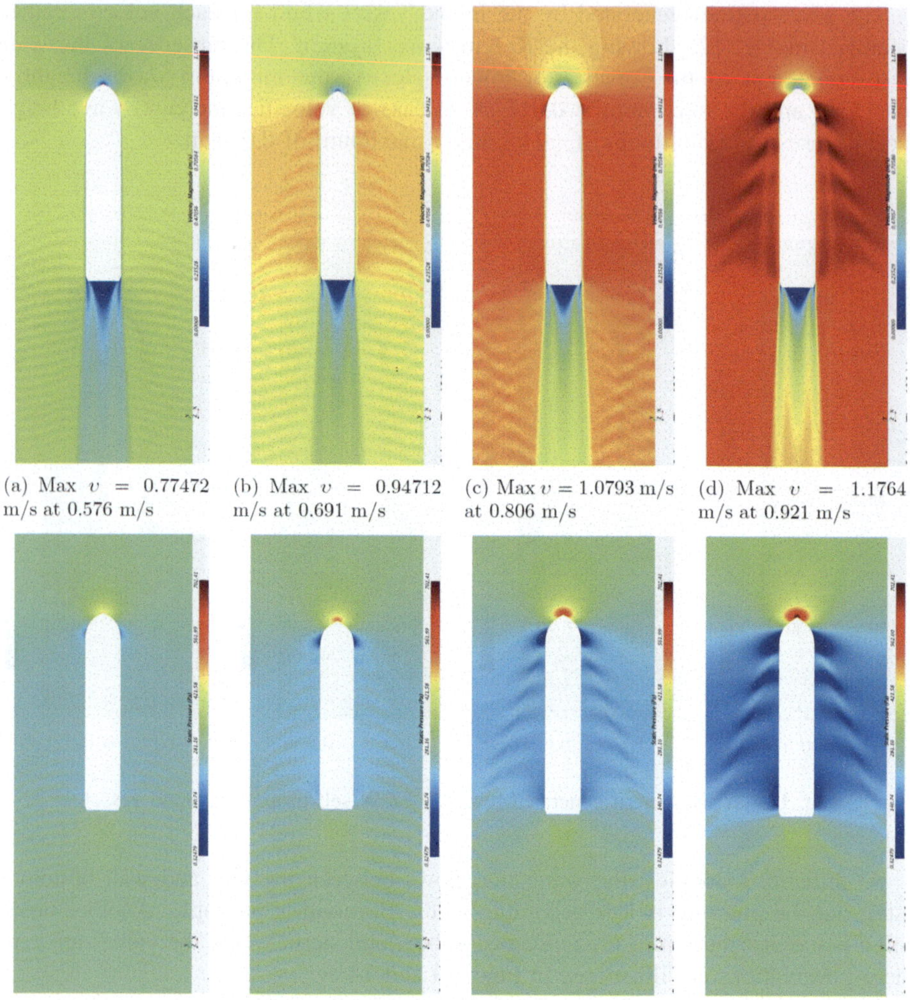

(a) Max v = 0.77472 m/s at 0.576 m/s
(b) Max v = 0.94712 m/s at 0.691 m/s
(c) Max v = 1.0793 m/s at 0.806 m/s
(d) Max v = 1.1764 m/s at 0.921 m/s
(e) Min p = 144.63 Pa, Max p = 455.77 Pa at 0.576 m/s
(f) Min p = 58.274 Pa, Max p = 525.56 Pa at 0.691 m/s
(g) Min p = 0.47628 Pa, Max p = 607.50 Pa at 0.806 m/s
(h) Min p = 0.32479 Pa, Max p = 702.41 Pa at 0.921 m/s

Fig. 13. Velocity magnitude (top) and static pressure (bottom) of 2700 TDW self-propelled inland vessel at different velocities (top view).

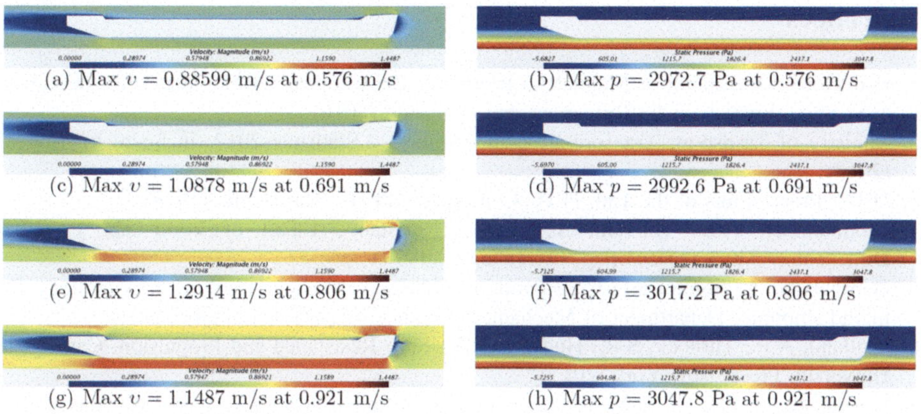

(a) Max $v = 0.88599$ m/s at 0.576 m/s
(b) Max $p = 2972.7$ Pa at 0.576 m/s
(c) Max $v = 1.0878$ m/s at 0.691 m/s
(d) Max $p = 2992.6$ Pa at 0.691 m/s
(e) Max $v = 1.2914$ m/s at 0.806 m/s
(f) Max $p = 3017.2$ Pa at 0.806 m/s
(g) Max $v = 1.1487$ m/s at 0.921 m/s
(h) Max $p = 3047.8$ Pa at 0.921 m/s

Fig. 14. Velocity magnitude (left) and static pressure (right) of 2700 TDW self-propelled inland vessel at different velocities (symmetry view).

7 Conclusions and Future Work

Since the partial results for this study have been presented, the conclusions could be listed as follows:

- The empirical formulations were satisfactory only in case of vessel navigating in middle river.
- The results obtained through numerical simulation are satisfactory only in case of low velocity. As the speed increases, the relative error also increases up to 18%.

The numerical simulation results can be corrected with more precision in order to obtain results similar to those in cases of navigation in shallow waters. For this purpose, it is important to present the future work, specifying aspects that must be analyzed.

The future analysis will aim the improvement of the numerical simulation results. For this purpose, the factors or aspects that could improve the simulation such as time step, numerical damping, dimensions changes on the volume domain, changes the turbulence model or the configurations of the mesh must be identified. The relative error must be reduced by, approximately, 7% or 11% for the higher velocities and 3% or 4% for the lower velocities. Once this is done, the final step is to perform the verification study for the discretization error analysis applying Grid Convergence Index (GCI) that is developed by Roache [46] and described by Celik et al. [35].

Acknowledgments. To Technological Research Institute (ITP, acronym in Portuguese) for providing the report of 2700 TDW inland vessel self-propelled test.

References

1. Encyclopædia Britannica: Magdalena River. In: Encyclopædia Britannica Ultimate Reference Suite. Encyclopædia Britannica, Inc., Chicago (2016)
2. Ministry of Transport: Transportation in Numbers - Statistics 2017 (in Spanish), Bogotá (2018)
3. ITTC: Proceedings of the 15th ITTC Conference, ITTC, The Hague (1978)
4. Guldhammer, H.E., Harvald, S.A.: Ship Resistance Effect of Form and Principal Dimensions. Akademisk Forlag, Copenhagen (1974)
5. Georgakaki, A, Sorenson, S.C.: Report on collected data and resulting methodology for inland shipping. Department of Mechanical Engineering, DTU, Lyngby (2004)
6. Molland, A.F., Turnock, S.R., Hudson, D.A.: Ship Resistance and Propulsion. Cambridge University Press, New York (2011)
7. Latorre, R., Luthra, G., Tang, K.: Improvement of Inland Waterway Vessel and Barge Tow Performance: Translations of selected Chinese, German and Russian Technical Articles (Report No. 249). Ann Arbor, United States (1982)
8. ITTC: Proceedings of the 18th ITTC Conference, ITTC, Kobe (1987)
9. Pompée, P.-J.: About modelling inland vessels resistance and propulsion and interaction vessel - waterway. Key parameters driving restricted/shallow water effects (Paper 180). In: 7th International PIANC-SMART Rivers Conference, PIANC, Buenos Aires (2015)
10. Schlichting, O.: Ship resistance in water of limited depth – resistance of seagoing vessels in shallow water (in German). Janhrb der Schiffbautechnischen Gesellshaft, vol. 35, pp. 127–148 (1934)
11. Landweber, L.: Test of a Model in Restricted Channels. DTMB, Washington (1939)
12. Lackenby, H.: The Effect of Shallow Water on Ship Resistance. Shipbuild Mar Engine-Builder 70 (1963)
13. Karpov, A.B.: Calculation of Ship Resistance in Restricted Waters (in Russian). Tr GII T IV 2 (1946)
14. Artjushkov, L.S.: Wall effect correction for shallow water model tests. Trans North East Coast Inst Eng Shipbuild, vol. 85, pp. 35–42 (1968)
15. Euler, L.: General Principles Concerning the Motion of Fluids (in French). Mémoires lLacadémie des Sci des B-lett, pp. 217–273, Berlin (1755)
16. Navier, C.-L.-M.-H.: Memory on the laws of the movement of fluids (in French). Mémoires l'Académie R des Sci, pp. 389–416, Paris (1822)
17. Stokes, G.G.: On the theories of the internal friction of fluids in motion and of the equilibrium and motion of elastic solids. Trans. Camb. Philos. Soc. **8**, 287–305 (1845)
18. Reynolds, O.: On the dynamical theory of incompressible viscous fluids and the determination of the criterion. Philos. Trans. R. Soc. Lond. **186**, 123–164 (1895)
19. Launder, B.E., Spalding, D.B.: The numerical computation of turbulent flows. Comput. Methods Appl. Mech. Eng. **3**, 269–289 (1974). https://doi.org/10.1016/0045-7825(74)90029-2
20. Yakhot, V., Orszag, S.A., Thangam, S., et al.: Development of turbulence models for shear flows by a double expansion technique. Phys. Fluids **4**, 1510–1520 (1992). https://doi.org/10.1063/1.858424
21. Shih, T.-H., Liou, W.W., Shabbir, A., et al.: A new k-∈ eddy viscosity model for high Reynolds number turbulent flows. Comput. Fluids **24**, 227–238 (1995). https://doi.org/10.1016/0045-7930(94)00032-T
22. Wilcox, D.C.: Reassessment of the scale-determining equation for advanced turbulence models. AIAA J. **26**, 1299–1310 (1988)

23. Menter, F.R.: Zonal two equation k-ω turbulence models for aerodynamic flows. In: 23rd AIAA Fluid Dynamics, Plasmadynamics, and Lasers Conference, p. 21, Orlando (1993)
24. Noh, W.F., Woodward, P.: SLIC (Simple Line Interface Calculation). In: van de Vooren, A. I., Zandbergen, P.J. (eds.) Proceedings of the 5th International Conference on Numerical Methods in Fluid Dynamics, pp. 330–340. Springer, Berlin (1976)
25. Hirt, C.W., Nichols, B.D.: Volume of fluid (VOF) method for the dynamics of free boundaries. J. Comput. Phys. **39**, 201–225 (1981). https://doi.org/10.1016/0021-9991(81)90145-5
26. Courant, R., Isaacson, E., Rees, M.: On the solution of nonlinear hyperbolic differential equations by finite differences. Commun. Pure Appl. Math. **5**, 243–255 (1952). https://doi.org/10.1002/cpa.3160050303
27. Gentry, R.A., Martin, R.E., Daly, B.J.: An Eulerian differencing method for unsteady compressible flow problems. J. Comput. Phys. **1**, 87–118 (1966). https://doi.org/10.1016/0021-9991(66)90014-3
28. Barakat, H.Z., Clark, J.A.: Analytical and experimental study of the transient laminar natural convection flows in partially filled liquid containers. In: 3rd IHTC, IHTC, pp. 152–162, Chicago (1966)
29. Runchal, A.K., Wolfshtein, M.: Numerical integration procedure for the steady state Navier-Stokes equations. J. Mech. Eng. Sci. **11**, 445–453 (1969). https://doi.org/10.1243/JMES_JOUR_1969_011_056_02
30. Warming, R.F., Beam, R.M.: Upwind second-order difference schemes and applications in aerodynamic flows. AIAA J. **14**, 1241–1249 (1976)
31. Hodge, J.K., Stone, A.L., Miller, T.E.: Numerical solution for airfoils near stall in optimized boundary–fitted curvilinear coordinates. AIAA J. **17**, 458–464 (1979). https://doi.org/10.2514/3.61155
32. Tamamidis, P., Assanis, D.N.: Evaluation of various high-order-accuracy schemes with and without flux limiters. Int. J. Numer. Methods Fluids **16**, 931–948 (1993). https://doi.org/10.1002/fld.1650161006
33. Thompson, C.P., Wilkers, N.S.: Experiments with Higher-Order Finite Difference Formulae. Harwell, Oxfordshire (1982)
34. Patankar, S.V., Spalding, D.B.: A calculation procedure for heat, mass and momentum transfer in three-dimensional parabolic flows. Int. J. Heat Mass Transf. **15**, 1787–1806 (1972). https://doi.org/10.1016/0017-9310(72)90054-3
35. Celik, I.B., Ghia, U., Roache, P.J., et al.: Procedure for estimation and reporting of uncertainty due to discretization in CFD applications. J. Fluids Eng. **130**(7), 078001 (2008). https://doi.org/10.1115/1.2960953
36. Ji, S.C., Ouahsine, A., Smaoui, H., Sergent, P.: 3-D numerical simulation of convoy-generated waves in a restricted waterway. J. Hydrodyn. **24**, 420–429 (2012). https://doi.org/10.1016/S1001-6058(11)60263-1
37. Liu, Y., Zou, L., Zou, Z.J., et al.: Numerical predictions of hydrodynamic forces and squat of ships in confined waters. In: Liu, G., Han, X., Li, Z. (eds.) Proceedings of the 8th ICCM, pp. 1095–1110. Scientech Publisher LLC, Guilin (2017)
38. IPT: Resistance Test of Inland Vessel 2700 Self-propelled in Sallow Water (in Portuguese), São Paulo (1974)
39. Siemens PLM Software: Simcenter STAR-CCM+ Documentation Version 13.04, 11628 (2018)
40. ITTC: ITTC - Recommended Procedures and Guidelines. ITTC, Lyngby (2014)
41. ITTC: Proceedings of the 11th ITTC Conference, ITTC, Madrid (1957)
42. Holtrop, J.: A statistical re-analysis of resistance and propulsion data. Int. Shipbuild. Prog. **31**, 272–276 (1984)

43. Holtrop, J., Mennen, G.G.J.: An approximate power prediction method. Int. Shipbuild. Prog. **29**, 166–170 (1982)
44. Raven, H.C.: A computational study of shallow-water effects on ship viscous resistance. In: Proceedings of 29th Symposium on Naval Hydrodynamics. Chalmers University of Technology, Gothenburg (2012)
45. Linde, F., Ouahsine, A., Huybrechts, N., Sergent, P.: Three-dimensional numerical simulation of ship resistance in restricted waterways: effect of ship sinkage and channel restriction. J. Waterw. Port Coast. Ocean Eng. **143**, 1–11 (2017)
46. Roache, P.J.: Verification and Validation in Computational Science and Engineering. Hermosa Publications, Albuquerque (1998)

Essential Aspects in the Supervision of the Ship Construction Projects

Pedro P. Martínez Villa[1(✉)] and Romilio G. Carralero Garcés[2]

[1] IPIN Cuba, Dolores No. 9 Lawton, Havana, Cuba
pedromartinezvilla2014@gmail.com
[2] IPIN Cuba, Guadalupe No. 6S Miguel, Havana, Cuba
carraleroromilio@gmail.com

Abstract. In the present article, shows up the integral process of the ship's newbuilding supervision, beginning by the analysis of the initial projects documentation, the legal and normative aspects, together with the technical elements; that it necessarily incorporate, in the construction process inside the Shipyards. Equally, are exposed some practical examples of the ship's newbuilding process, from the point of view of the class's and owner's supervisor, based on the authors's experience, to have participated during their professional work in this specific type of supervision.

Keywords: Ship's newbuilding · Supervision · Shipyards

1 Introduction

The topics that are showed in this work, were selected all time that, its summarize; in great measure, the essential aspects that conform the supervision of the ship's newbuilding and its have as objective, to provide to the interested of a quick and better understanding of this so important process, because it belongs to all acquaintance that it guarantee in the first place, the quality of the ship's newbuilding.

The topics; for chapters, are the following ones:

Section 2 - Ship's newbuilding projects. Legal and normative aspects.
Section 3 - Type of supervisions and the supervisor's functions.
Section 4 - Supervision's Process.
Section 5 - Project's Certification. Final documentation of the project.
Section 6 - Conclusions.

As the chapters are developed; together with some basic elements of the ships supervisions's process, some references about the moments of this work's type, are given; that in the practice, were carried out by the authors of this work, in the construction of 10 bulkcarrier ships, in Shanghái Shipyards in the P.R. of China and in others ships's constructions executed in some Shipyards in Spain. See this labor experience; as fundamental support to achieve the objective outlined in this work.

V. A. J. E. Carreño Moreno et al. (Eds.): CIDIN 2019/COPINAVAL 2019, pp. 153–162, 2020.
https://doi.org/10.1007/978-3-030-35963-8_13

2 Ship's Newbuilding Projects. Legal and Normative Aspects

The ship construction projects generally are carried out by the Shipyards, that plannnig a group of activities to develop, to sum up the demands of different Owners, related with a specific design of ship to build, as can be, a bulkcarrier ship, a general cargo ship, a tankers ship, etc. The design can be executed by the Shipyards, or by another designer's entity and it could be based on specific's elements requested by Owners, or general's design according to the ship type, that is built in that Shipyards.

Starting from the design, are derived the general and constructive plans, the outlines, charts of data and contents, and other related documents. We could say that the design is the base to unchain all the activities of the Project and it is one of the elementary points of the same one (Fig. 1).

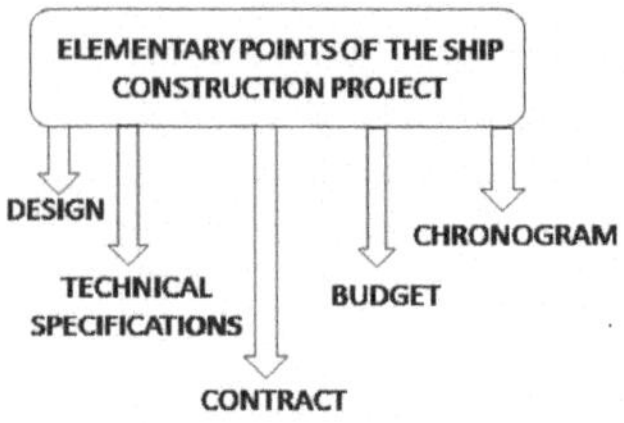

The elaboration of the ship construction projects is obligatory before beginning the constructive process of the ship and each Shipyards will be able to have an own definition of how to make this project, according to that requested by owner.

Fig. 1. Elementary points of the project

In this case, the present work is based; exclusively, in the Ship's Newbuilding Projects of merchant ships, requested by a Shipping Agency Company (Owner), executed inside the facilities of a Shipyard or Naval Shop (Manufacturer) and certified by a certain Classification's Society (Class). These three mentioned entities, will be the essential group to reach the objectives and the reason of the project that it is to achieve the construction of the ship in the planned time and with the quality and the budget, established.

These ship's construction projects, are considered for its content and complexity as technician - productive Projects, and they look for; much for the Shipyards as for Owner to obtain economic benefits (earnings) and in their execution are used plans, outlines and other documents that define the conditions of the construction of the ship, the distribution of spaces, the use of materials and technologies, the assembly of the different elements and the technical justification of the execution of the specifications required by the applicable normative technique.

The normative aspects in the ship's newbuilding projects should be very clear, so much in the Specifications of the same one, as in the Manufacturer's procedures. The norms to use, should be known by all the responsible for its execution and its shall be able to be changed or modified, but this process shall be notified previously and agreed among the involved parts.

The materials of the hull and compartments (steel laminate, profils, electrodes, paint, pipes, etc.), materials of general outfitting, the dissimilar equipment that will settle, and many other inputs that are used in the construction, could be bought by the Shipyards or given by Owner according to what is made a pact in the contract. Nevertheless, both should have, very good defined the budget to make sure of the financial control of the Project, and they also have to designate people to work in the different construction's stages.

In all ship construction project, should exist certain conditions that its must be previously discussed and then completed as much for the Shipyard (Manufacturer) as for Owner (Buyer) that it will be the part that receives or buys the finished ship. These conditions are reflected in a document, of exhaustive and obligatory character, manifested in common between both part as legal aspect that is denominated Contract and is elementary function of this, to originate the rights, obligations and responsibilities that are demanded for the Manufacturer and for the Buyer.

In the Contract, should appear; as main points, those defined in the Fig. 2:

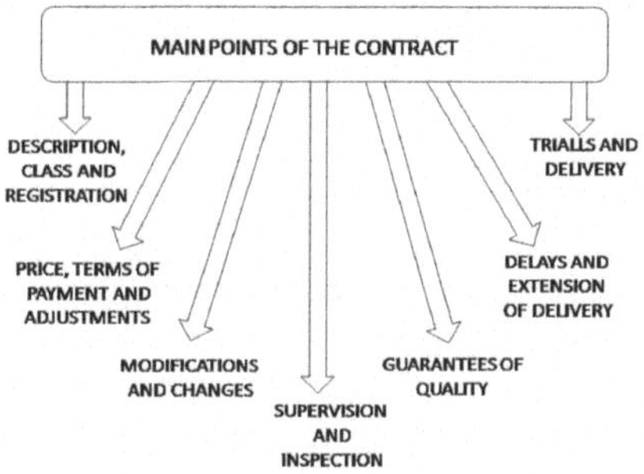

Other considerations in the Contract can be related with the Interpretation, Disputes and Arbitration, Insurances, Taxes, etc. The points of the Contract that are more closely related with the topics that are approached work presently they are: Description, Class and Registration – Trials and Delivery and Supervision and Inspection,that are described next.

Fig. 2. Main points of the contract

Description, Class and Registration. The ship's characteristics are indicated, its form of construction in agreement with the plans, the technical specifications, the general conditions of the construction, the materials, equipment, machineries, instruments, etc. to use, the normative, legal and administrative aspects, the forms of measuring the executed parts and the payment stages, among other indications. The Classification's Society (CS) will be specified for whose Norms and Regulations the ship will be built and which of the parts; it will be the signatory of the Contract of Supervision and Classification, with this CS.

In any way, the Manufacturer should guarantee that the Specifications agree with the requirements of CS and it will collaborate and to complete its petitions, that as part of the supervision of SC, they are indispensable, to obtain required Class's Notación for the ship.

Trials and Delivery. All the concerning aspects to the warnings or notification of realization of the final tests, as well as the possible variations of dates, the participation of the Buyer's supervisors, the process of these tests, the methods of acceptance or rejection and other that could be explanatory, they will be very defined in the Contract. Equally they will be defined the types of documents of acceptance that should be emitted by the Buyer and given to the Manufacturer.

The delivery date and the place where the same one will be made they should be fixed, as well as the documents that must be emitted and other questions related with the possession of the ship for the Buyer.

Supervision and Inspection- It should be left clear in this point, which are the rights of the supervisors that will be nominated by the Buyer, so that they carry out the necessary inspections during the construction of the ship and the form of permanency of these in the Manufacturer's facilities. Equally the facilities are notified that will be provided to the supervisors and other personal additional, either as Representative of the Buyer or Client of this last.

The necessary inspection of the ship, their machineries, equipment and the outfitting, should be taken respectively by the Supervision's teams of the Manufacturer and the Buyer's and for CS and the Supplier if the aspects to inspect have requirements of Class or they are related with the guarantees of equipment and given products.

3 Type of Supervisions and the Supervisor's Functions

In the of ship's construction process, it's important from the first moments to define all the concerning one to the inspections that should be carried out as guarantee to obtain the wanted result, as for the quality and the operability of the ship. This is translated in inspecting the quality of the materials, of the equipment, of the instruments, the competition of the manpower, the certification of the productive processes and many other questions that they compose the universe of the construction.

To inspect, it's part of the supervisors's work, but it's not limited to this process. The inspection; regarding the work's quality in the ship's construction consists on to examine and to measure the characteristics of quality of a process, that it is part of the combined works and mensuration instruments, comparison patterns or test's equipment could be used, to see if it completes or not the specified requirements.

The supervision, begins from the verification and validation of the ship's construction project until the execution of the final tests of operation of all the equipments machineries, instruments, etc. and the delivery of the corresponding documentation. Also the contribution of ideas could be inside the supervision, for the improvement or modification of a process, parameter, or until even the design, in strict agreement with the involved parts and the professionals that executed this design.

As the supervisor's elementary characteristic it should be, possessing theoretical and practical wide knowledge of the own activity of naval construction that allow to give solutions to the problems that arise during the construction's process, because in general, they are accidental situations that are not contemplated in any manual or procedure. Equally; during their work, it should be based on the logic and the order, with good behavior and appropriate manners.

The supervision tasks for the entities that participate in the ship's construction process are multiple, and to carry out them, in function of the complexity of the work, it can be required that the supervision of the work is carried out by an entire multidisciplinary team, with a Boss of the Supervision's Team.

This way, we could classify the supervision's types as: Manufacturer's Supervision, Buyer's Supervision, Class's Supervision and Supplier's Supervision. Exist the supervisors' functions that; for any supervision of those mentioned, they should complete and other specific functions; according to the entities that these they represent, they are committed to execute. In the Figs. 3 and 4 are show the most important general and specific functions of the supervisors.

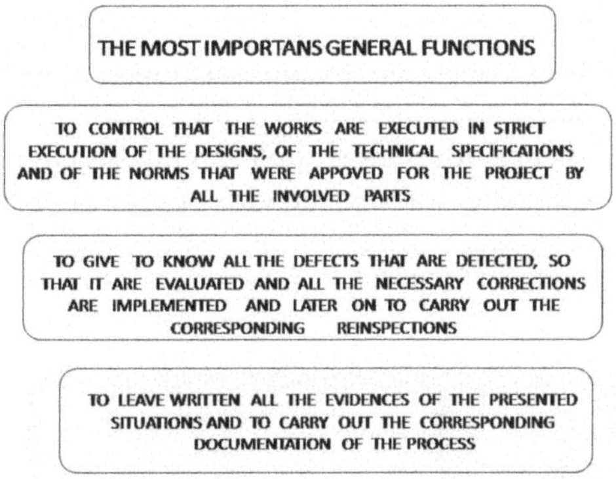

Fig. 3. The most important general functions of the supervisors

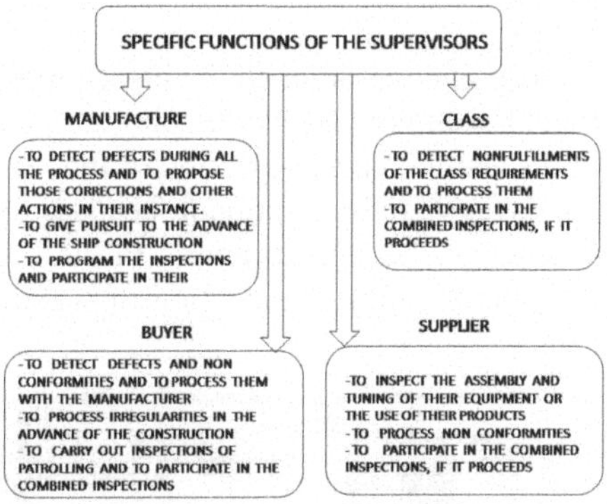

Fig. 4. The most important specific functions of the supervisors

4 Supervision's Process

The knowledge and the supervisor's experience will allow to carry out all the tasks that are executed during the ship's construction process. Their work, not alone its in the moment to carry out the inspection, should also give permanent pursuit to the work, to make constant walked for the working's shops and other areas where works are executed and make observation in patrolling's inspections that will allow to foresee and to alert of defects or alteration of what is executing.

The process of any supervision includes; the verification of the certificate of quality of the materials that are used and of the competition of the workers and technical personnel, the confirmation of the technical state of the machineries, equipments and work tools and if it is necessary of its certification. Also the control of the works in the strict execution of the design and of the technical specifications for the ship and also the approval of the beginning of the works, according to the chronogram, controlling in all moment that the sequence is completed for its execution.

In the Fig. 5 is appreciated, the verification of the designs of the steel structure, of the test's parameters of equipment in the control bridge, a product certification, and of the confirmation of the center line, before the blocks assembling process in the dry-dock. All the aspects; above mentioned, related with the construction supervision of the ten 35 thousand tons bulkcarrier in Shanghái Shipyards, P.R of China.

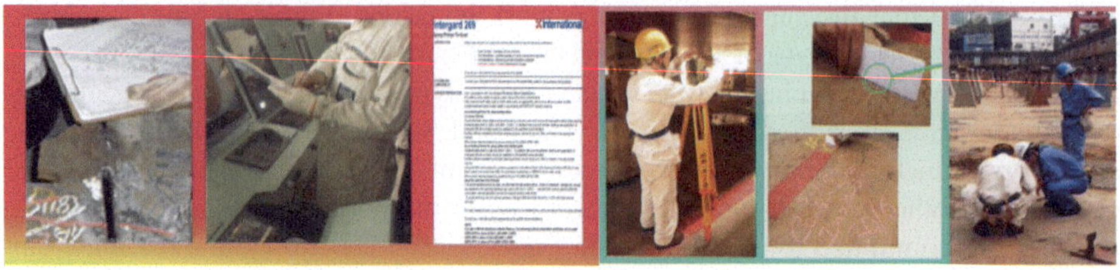

Fig. 5. *Verification of the design, parameters and product certification Confirmation of the Center Line

The participation in all the inspections to check the correct fixation and union of the different sections and blocks, the correct preparation of these unions for the welding process and the later inspection of the welding seams together with the non destructive tests, is extremely important because of the quality of this process, depends the resistance of the ship structure. In the Fig. 6, can see, the fixation of sections and the non destructive tests realization of welding seams.

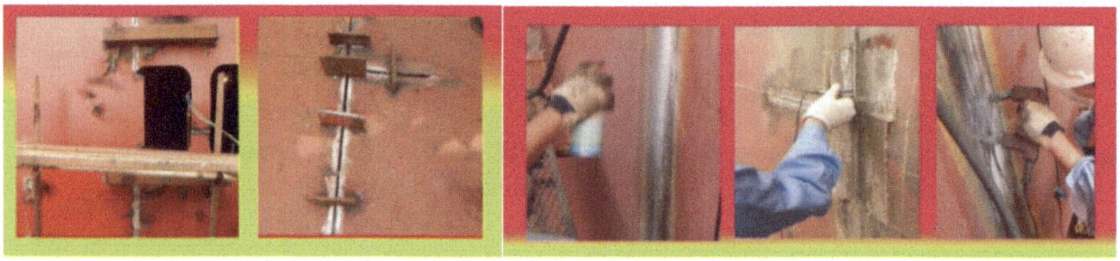

Fig. 6. *Fixation of sections and realization of non destructive tests to welding seams

Also, the effective execution of the normative, in the topic of work security and in the environment care, are verified. In the Figure 7, can observe the correct distribution of the scaffolding in the blocks assembly works, that it guarantees the security of the workers, of the technical personnel and of all the supervisors, among other people that frequent the ship.

Fig. 7. *Scaffolding distribution

In the supervision process, in general; firstly, the Manufacturer's supervisors carry out their daily inspections and they indicate the defects that exist to solve them before the combined inspections and on the other hand, the Buyer's supervisors carry out inspections of patrollings to have a preliminary evaluation of the work situation. Later on, the Manufacturer summons to the realization of the combined inspections including to the Class and the Supplier if it proceeds, where approaches are joined and they take the most consequent decisions (Fig. 7).

In the Fig. 8, can appreciate the realization of combined inspections, during the bulkcarrier construction, mentioned previously.

Fig. 8. *Combined inspections

In this same ship construction project, in these or other inspections each supervisor marked his signalizations with a certain chalk color to identify the same ones well. In the Fig. 9 can appreciate the identification for colors of each supervisor's signalizations.

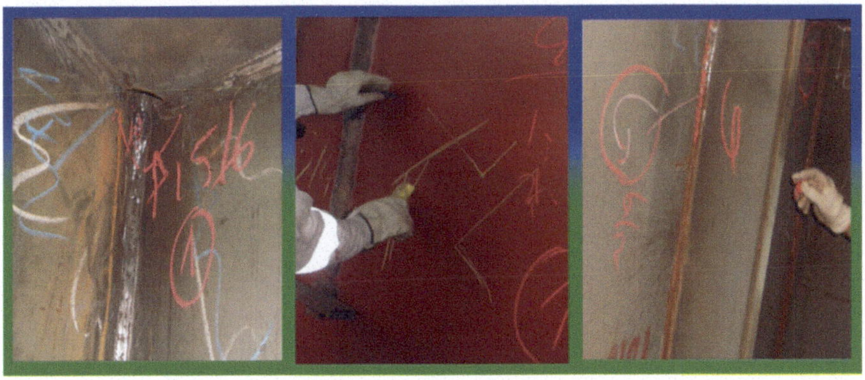

Fig. 9. *Identification for colors of the signalizations

If during the patrolling inspections; that the Buyer carries out, some deficiencies are detected, these can be correspondents to the Manufacturer, in writing and of being possible with a graphic reference. In the Fig. 10, deficiencies can be appreciated, found in these inspections and correspondents by the Buyer to the Manufacturer.

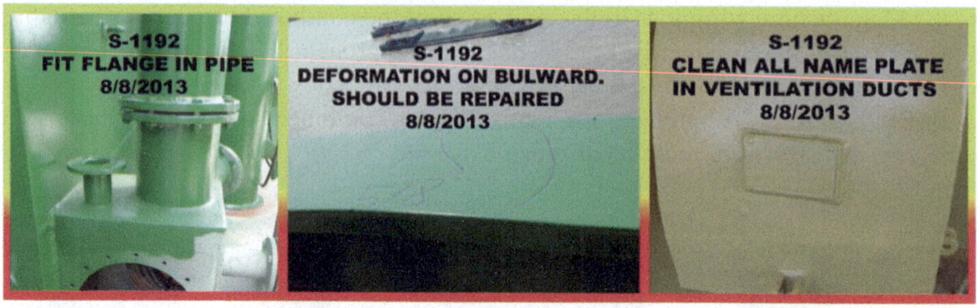

Fig. 10. *Deficiencies found in patrolling inspections

If; in an inspection object, several signalizations have taken place, on the part of the Buyer's supervisors and the solutions given by the Manufacturer they don't satisfy the established demands, it will proceed to carry out a Non Conformity Report (NCR). This to proceed will be officialized in a specific model, to be given to the Manufacturer.

LOGO	SHIPBUILDING PROJECT :			
SHIPCONSTRUCTION AND TEST EQUIPMENT NO CONFORMITY REPORT	NCR Ref:	DATA :	Page of	Folio:
OBJECT/PLACE : REPORT FINDING:				

ELABORATED:	CHECKED :	APPROVED :

Fig. 11. NCR model

In the Fig. 11, model's example is shown to be used in these cases. It's the supervisors' responsibility to take the control of NC and of the terms that are given for the corrections.

When there is several construction project for oneself series of ships, it can happen that the first given ships begin to report flaws in their exploitation and then it becomes necessary a design change, equipment or another technical question for the rest of the ships in process. As example, can show the change of fairleaders design during the construction supervision of mentioned bulkcarrier in the Fig. 12.

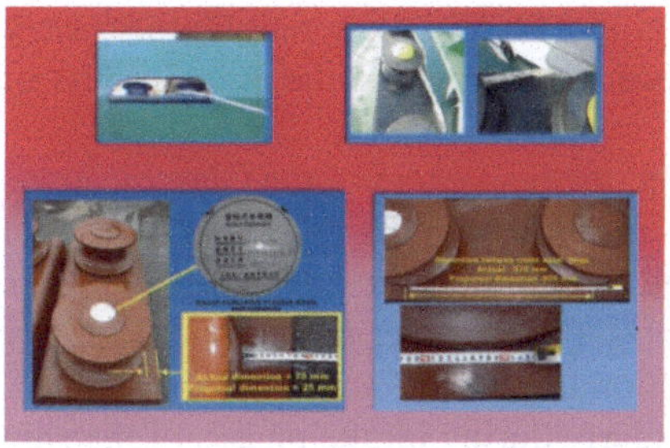

The first ship of the series, presented problems with the ropes that broke up due to the touch with the openings border in the bulwark. The Buyer proposed to the Constructor, to modify the design of the fairladers in orden to avoid this situation and this proposal; carried out by the supervisors, was sended and accepted and implemented by the Constructor.

Fig. 12. *Proposal of design change

Another of the essential aspects, during the supervision process, is the use of mensuration instruments (See Fig. 13), to value different parameters in the inspections. These should be exact and precise, and their readings should be registered for to make the valuations that are needed.

Fig. 13. *Mensuration instruments

The supervisors of the Class and of the Buyer, they will be able to request the certifications of these instruments if is presented some doubt of their efficiency or they check shortcomings in the readings of the parameters that are measured.

Others of the essential aspects in the supervision process, its the elaboration of periodic reports; that could be weekly or biweekly, with the objective of to make the corresponding analyses and to be able to take measures if in certain moments, all that made a pact is not executed.

5 Project's Certification. Final Documentation of the Project

As final stage of the ship construction project, their integral revision is included, where all the results are valued, so much from the technical point of view as financial and the pertinent valuations are made, according to that planned and executed, as well as their results, in consideration to the achievement of the outlined objectives. In these valuations they should be present the supervisors that should elaborate the necessary informations, to take the decisions with the purpose of improving the effectiveness of other ships construction projects.

In all the stages of the execution of the project, the inspectors of the Classifications Society (CS) should check that the ship is built according to the approved plans and according to the requirements of the Rules as well as that the materials and equipments be approved for its use in the project and certificates. On the other hand it should verify that the personnel that participates in the construction is qualified and properly certified. The inspector of CS executes the inspections regularly during all the phases of the ship conformation, of installation of his equipment and of the termination of the construction. Concluded this stage and carried out all the partial test; included its related with the certification of the welding processes, the inspector certifies the jetty tests and of sea and before satisfactory results, he gives for good the works of construction of the ship so that it is assigned the Class, with what will be certifying that the conditions of the same one are the appropriate ones.

The certification granted to a ship by CS; after having finished their construction, it is of vital importance in the marine sector. In the prestige, reputation of honesty, ability, professionalism and capacity that have the inspectors of CS, are based the positive point of view from Manufacturer's and the Buyer, to know that the built ship will be excellent from the point of view of reliability, structural efficiency and for the sailing security.

6 Conclusions

The new constructions supervision process, is of great importance non alone to guarantee the quality of the works; that rebounds in the quality of the finished ship at the end, but also that the daily chore, the reports of the supervisors and its recommendations, the constructive discussions of the construction topics, because these they are a base for the improvement of the acting of the shipyard in general sense and in particular for all the specialists that face new challenges that they make grow their experticia in the constructive following tasks of the new projects.

Note: *The Pictures were taken by the author of this work, during the construction of ten bulkcarrier ships in Shanghái Shipyards in P R. of China from September of the 2010 to September of the 2013, and were edited by the same author, in December of the 2018.

References

1. Martínez Villa, P.: Personal point, carried out during the supervision work, Shanghai P. R of China, September 2013–September 2015
2. Lloyds Register: Organization of the new constructions supervision by owner, Document (2010)
3. Nordstrand Maritime and Trading Co Ltd.: Control and Pursuit of the non conformities. Procedure P01 (2010)

Noise and Vibration Control Program for Warship: The New Spanish Frigate F110

J. M. Riola[1(✉)], J. C. Díaz-Cuadra[2], and Publio Beltrán[3]

[1] REAL ARMADA ESPAÑOLA, Madrid, Spain
jriorod@fn.mde.es
[2] NAVANTIA, Madrid, Spain
jcdiaz@navantia.es
[3] TSI, S.L., Avda. PIO XII, 44, Torre 2, bajo izda, 28016 Madrid, Spain
publiobp@tsisl.es

Abstract. Among the great challenges that every naval designer of modern warships must face, it stands out, due to its strategic importance with regard to the Control, from the early phases of the project, of the "acoustic signature" of the new design of the ship. Specifically, its "minimization" to achieve the objective of making the ship as "stealthy" (stealth) as possible.

In parallel, the experience of the most advanced National Navies has shown the strategic importance of controlling the levels of vibrations and noise that are generated on board the ships, not only from the point of view of structural integrity of the platform, but from the point of view of guaranteeing the Protection of the Health and the Comfort of the Endowments, and in the European context, required by Directives and Local Regulations of Obligatory Compliance. Indeed, the experiences of one of the most advanced navies in the world have revealed the high costs derived from the consequences of "hearing losses" on part of the specialized personnel of their staff, as well as the "high risk" that this implies for the missions of surveillance and combat the fact that the personnel does not have a conditions of comfort guaranteeing the maximum performance and attention required.

Fully aware of the aforementioned aspects, the Spanish Navy, in close collaboration with NAVANTIA, and advised by experts in the matter, has required for the new design of its Frigate F-110, the establishment of a "Noise Control Plan, Vibrations and Acoustic Signature", that, applied from the early phases of the design, allows us to give a satisfactory answer to these challenges.

In the present work, and with support of previous experiences, the general lines of the aforementioned "Noise and Vibration Control Program and Acoustic Signature" are described, which is based on, among others, the Comfort Class Notations of the main Classification Societies as well as the most recent experiences of experts and specialized consultants.

Keywords: "Underwater signature" · Noise and vibration control techniques · Reduction of underwater signature of ships · "Silent ships"

V. A. J. E. Carreño Moreno et al. (Eds.): CIDIN 2019/COPINAVAL 2019, pp. 163–174, 2020.
https://doi.org/10.1007/978-3-030-35963-8_14

1 Introduction

The Spanish Navy's F-110 Frigate, designed by Navantia in conjunction with The Spanish Navy aims to replace the country's six Santa Maria class vessels as they approach the end of their operational life from 2020. With the construction contract signed on December 14, 2018 AND a full service date 2024, these vessels will seek to provide a multipurpose capability. The frigates will work as an additional solution the BAM (Buque de Acción Marítima) ships and Álvaro de Bazán class frigates, providing maritime security, mine countermeasures, escort duties and many other possibilities.

This new F-110 frigates will operate in high-threat scenarios forming battle groups at sea or may act alone in areas near the coast.

Anti-aircraft, asymmetric warfare, maritime safety, naval power projection, military-civilian cooperation, deployment flexibility, hybrid propulsion and space for unmanned vehicles are features and missions that will be carried out by the Spanish Navy with these new frigates.

These modern frigates have a displacement of about 6200 t, an overall length of 145,0 m, a beam of 18,6 m and constructed mainly in steel grade DH 36. They will have accommodation for 187 people. The following Fig. 1 shows the General layout of the Frigate.

Fig. 1. General layout

The main propulsion system is a combined diesel-electric (CODELAG). This system employs two (2) electric motors which are connected to the propeller shaft. The electric motors are powered by four (4) diesel generators installed in separate locations. These two electric motors directly drive a controllable pitch propeller (CPP). For further speeds, a gas turbine powers the shaft via a cross-connecting gearbox.

This arrangement combines the diesel engines (diesel generating set) used for propulsion and for electric power generation. The electric motors are connected directly to the propeller shaft and the gearbox can be used to combine the mechanical output of a gas turbine and diesel-electric system.

They will account with a multi-purpose space in order to provide several capabilities in accordance with the mission profile, some of these include: naval special warfare, unmanned vehicles operations (UAV, USV, etc.), medical support and others.

Due to the sophisticated equipment installed on board these Frigates along with the activities entrusted to it and the fact that some of the operational tasks may be compromised by the low performance of its equipment, the ship shall be designed and built minimizing the levels of Noise and Vibration on board. The design seeks a multi-purpose vessel with low operating costs and reduced radar signature.

In Spain, for every programme, the capability development process has four phases: Conceptual; Definition and Decision; Execution; and Service. The 'Conceptual' phase has already finished. This phase includes the development of two documents – the 'Mission Need Document' and the 'Staff Target Document'. Conceptual phase has finished yet and Definition phase has starting right now.

The noise and vibration levels on board of the ships should be controlled to avoid the detection of the warships, improve the comfort of the vessel crew and mitigate the negative effects of radiated noise to the marine fauna. For this purpose, the following mitigation actions are recommended to limit the levels of noise and vibration levels in the ships."

(1) The main noise and vibration sources of the ship need to be assembled on elastic mounts to avoid the transmission of vibrations in the ships.
(2) The cavitation of the propeller is the main source of the noise. Therefore, an appropriated design of the propeller is of vital importance to avoid the presence of undesirable cavitation phenomenon's which generates high levels of radiated noise.
(3) The insulation of the machinery and auxiliary equipment is necessary to limit the transmission of vibrations and minimize the structural noise of the ships. To achieve this goal, the selection of materials plays a critical role in the attenuation of vibrations and noise. Furthermore, It should be noted that the increment of the weight due to the insulation need to be considered in the design stage of the ships.
(4) The exhaust ducts and the insulating elements of the pipes need to be properly supported using an array of elastic supports.
(5) It is important to guarantee that the noise levels generated by the HVAC system are 5 dB lower than the noise limits promoted by the environmental regulations. This behaviour should be fulfilled in the multiple locations of the ship (e.g. workshop, galley, captain salon, outdoor main deck, …).
(6) With the aim to control the levels of noise and vibrations, it is necessary to install silencers in the exhaust ducts and inlets/outlets of the fans which are commonly located on the decks or open areas.
(7) It is recommended to use a floating type design instead of rigid connections to suppress the vibrations of critical vibration sources.
(8) The suppliers are selected on the basis of these considerations: the contractual commitment with the shipyards and their capability to carry out advanced dynamic studies.

(9) The characterization of the vibration and noise behaviour of the ships is of vital importance to avoid resonances and calculate the amount of insulation which need to be applied.
(10) The shipyard should require the assistance to witness and control the noise insulation installation to verify that the mounting process is properly done and in compliance with the noise criteria required to both items: accommodation and noise insulations.

The achievement of all these points allows selecting the proper sources: Machinery and propulsion, their proper assembly as well as the contract requirements for the suppliers, even for the accommodation.

In parallel, a complete program of simulations activities is developed: Noise, Vibration and Underwater Radiated Noise Prediction, which identifies the possible deviations and the necessary corrective actions. The "private tests" allow introducing the last adjustments just before the official sea trials. This way, the shipyard avoids penalties for incompliance with the requirements.

2 Vibration Control Program

In this kind of ships, vibration levels on board the ship shall be controlled due to the following reasons:

- In order to assure the proper comfort of the crew on board. In this regard, the vibration levels on the ship shall be in-compliance with limits defined in BV Rules for Naval Ships Pt. E, Ch. 6, Sec. 1 November 2016.
- To guarantying the correct performance of electronic equipment sensitive to external induced vibration as well as assuring a long operational life with maintenance cost as low as possible. Unless otherwise specified, vibration levels defined in the MIL-STD-167-1 are taken as a reference. In this regard, it is important to note that these vibration levels are defined for accelerated tests and corresponding transformations shall be made according to the expected life of the ship.
- To contribute to a low acoustic signature of the Frigate. Apart from the noise induced by the propeller, machinery noise contributes to the acoustic signature through the vibration propagated over the ship structure.

In order to assure low vibration levels on board the ship, the structure shall be as stiff as possible and free of any resonance phenomenon; local and global, and, on the other hand, the dynamic excitations shall be minimized.

With regard to the latter, main pieces of machinery such the diesel generators or turbines are resiliently mounted with a properly designed elastic bench to minimize the vibration transmitted to the structure. On the other hand, the propeller due to its dynamic forces acting on it and the hull pressure fluctuations is also capable of exciting the ship and cause significant vibration levels if they are not under controlled.

Regarding the structure, this type of ships has flexible structures that are prone to vibrate and have low natural frequencies that can be excited by the dynamic forces mentioned above. In this regard, it is very important to verify that the structural design is free of any risk of resonance leading to high vibration levels.

As important as the design, during the shipbuilding phase, a series of good practices as well as a proper quality control will enable the shipyard to fulfil with what is initially specified with the aim to guarantee low vibration levels.

The first step for any vibration control programme is to analyse the basic design of the structure at very early stages of the project. During this analysis, "weak points" are identified and possible modifications are proposed. This step aims to identify lack of continuity of the main ship structure that, not being a problem from a structural integrity point of view, it is likely to cause undesirable vibration levels. It is particularly important the identification of large "spans", areas with low natural frequencies due to the initial specified scantling as well as assuring a proper stiffness of the stern area of the ship and the foundation of the main engines and propulsion motors.

Additionally, specific dynamic and acoustic requirements are set for the most important pieces of machinery, such as the electric motors or the diesel gen-sets. According to those requirements, specific solutions are to be implemented in order to guarantee the compliance with them.

In a more advance step, numerical FEM based models are employed to:

- Identify the main natural frequencies and assess their coincidence or not with the frequency of the main excitations. In this stage, not only the global bending modes are quantified, but also, local natural frequencies of superstructures and decks as well as natural frequencies associated to the masts.
- Predict the expected vibration levels in each space along the whole operational profile under different loading conditions.
- Coupled with an acoustic model, it enables the prediction of the acoustic signature at low and mid frequencies.

In order to achieve that, a global model of the whole ship structure is performed. This model is then loaded with all masses associated to machinery, electronic equipment and fittings, which will have an impact, at least locally, on the dynamic behaviour of the ship. Finally, the added water mass, which has a dramatic effect on the dynamic behaviour of the ship, is also accounted numerically by the application of boundary elements.

Once the model is completed, the first step consists on performing a modal analysis in which the wet natural frequencies are computed. At this stage, areas with risk of high vibration levels can be already detected (Figs. 2 and 3).

Fig. 2. First vertical bending mode

Fig. 3. Some results of the modal analysis

Then, the final step consists on applying the dynamic forces caused mainly by the propeller expected for this ship and obtain the vibration levels for each space over the whole propeller speed range. If vibration levels obtained in this prediction are above or close to the limits, the results are analysed to identify the nature of the vibration and the structure participating on it so structural modifications can be proposed and tested in the model (Fig. 4).

Fig. 4. Vibration levels on the decks at the nominal condition

In particular, for this ship, a natural frequency of the stern area of the ship close the blade pass frequency at nominal condition was identified. Even though, there is no accommodation or electronic equipment room in this area, so there were no applicable vibration limits, this area is directly excited by the hull pressure of the propeller. Having a natural frequency in this area close to the frequency of the excitation applied on it can lead to vibration amplification in all spaces of the ship causing undesirable vibration levels and probably undesirable high noise radiation. Therefore, the structure of this area was modified to stiffen it and shift out its natural frequency of the blade pass frequency (Fig. 5).

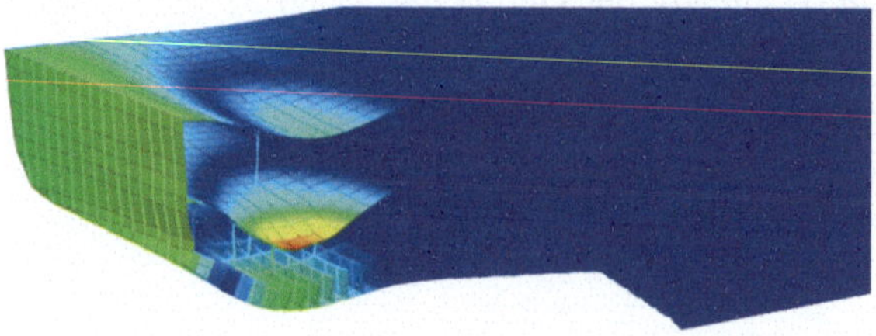

Fig. 5. Natural frequency detected in the stern area of the ship

Apart from a global good design of the ship structure, in order to control the acoustic signature of the ship, a good design of the machinery foundation is required in order to minimize the vibrational power that the diesel engines, propulsion motors and turbine can input into the structure. For that, detailed dynamic analysis of this important ship structures are to be carried out. This local analysis is then coupled in an FEM/BEM acoustic model to quantify the underwater radiated noise signature of the machinery on board the ship.

3 Noise Control Program

Navies are sensitized with the importance of the vibro-acoustic designing of their vessels, not only from the point of view of structural integrity of the platform and stealth, also from the point of view of guaranteeing the Protection of the Health and the Comfort of their Endowments. It has been demonstrated through the years and the most advanced navies in the world have revealed the high costs derived from the consequences of "hearing losses" on part of the specialized personnel of their staff, as well as the "high risk" that this implies for the missions of surveillance and combat the fact that the personnel does not have a conditions of comfort guaranteeing the maximum performance and attention required.

The noise pollution generated by the shipping activity is attracting considerable attention due to their negative environmental impacts on marine crew. This can be explained because of the fact that the noise radiated by the ship has a direct impact on the quality of the workplace and the living conditions for crew and passengers on board. For example, the exposure to high level of noise can result in loss of the hearing ability or deaf of the marine crew. The loss of hearing ability is directly affected by the degree of exposition to the noise pollution which is related to the time of exposure and the hearing protection of the vessel crew.

Noise control program is aimed to guarantee comfort noise levels at all operational conditions and to fulfil frigate specification noise requirements. In this particular case Frigate F-100 has to be in compliance with the noise limits specified in Bureau Veritas Rules for COMFORT ON BOARD, especially its section **BV Rules for Naval Ships Pt. E, Ch. 6, Sec.** 1 **November 2016** [1] with regard to noise. Additionally, the future fulfilment of the BV Rules for Naval Ships was considered. Draft April 2016. And also with **Standard of Habitability for New Construction Vessel (June 2010) of Ministry of Defence – Navy** [2].

Finally, for those spaces that do not appear in the abovementioned regulations, **the Resolution MSC.337(91)** [3] was applied.

The fulfilment of these limits in the different accommodation spaces imply that the 24 h equivalent continuous sound level shall not exceed 80 dB(A). This indicator is expressed as Leq (24) < 80 dB(A). Take care of this aspect is very important form a health point of view, there are medical studies that shown that if a person is exposed to noise levels higher than 80 dB(A) for 8 h each day he or she may develop hearing loss in one or both ears over time. This damage tends to get worse as the exposure is extended in time. The damage may even affect the high frequencies around 4000 Hz. This aspect is mandatory and is regulated by the **Directive 2003/10/EC of the**

European Parliament and of the Council of 25 June 2002 [4] on the minimum health and safety requirements regarding the exposure of workers to the risks arising from physical agents (noise) 2003.

Another acoustic aspect that has been taken into account in the Specification is the one related to speech privacy level required between the different partitions. For these vessels the apparent weighted sound reduction index (R'w) measured in situ between two adjacent accommodation spaces has to be higher than the values defined in the BV Rules for Naval Ships.

The first step that has to be done at any vessel from an acoustic point of view is to analyse the General Arrangement from a noise point of view. The purpose is to identify those spaces that due to its noise level limit and arrangement with respect to the main noise sources or the activities developed in the adjacent space are susceptible to present noise problems. The methodology applied is based on the Connectivity Matrix. This methodology classifies the spaces of the vessel in "source spaces" and "receiver spaces". A source space is an area that generates high noise levels due to its machinery and/or activity. A receiver space is the area that receives the noise of the source spaces.

This analysis is very important to be carried out in the early stages of the design phases, because a modification of the general arrangement of the vessel may be required. It should be Keep in mind that the possibilities of achieving predetermined noise levels in certain locations only through insulation materials have technical limitations. Sometimes vessels' general arrangements have been defined without a previous acoustic analysis. So, in some cases it is known that a vessel is not going to fulfil noise limit defined in the Specification without the need to carry out a noise level prediction calculation.

In the particular case of the Frigates some modifications were required in the in the hospital and infirmary area were required, due to its proximity to the casing, see Fig. 6.

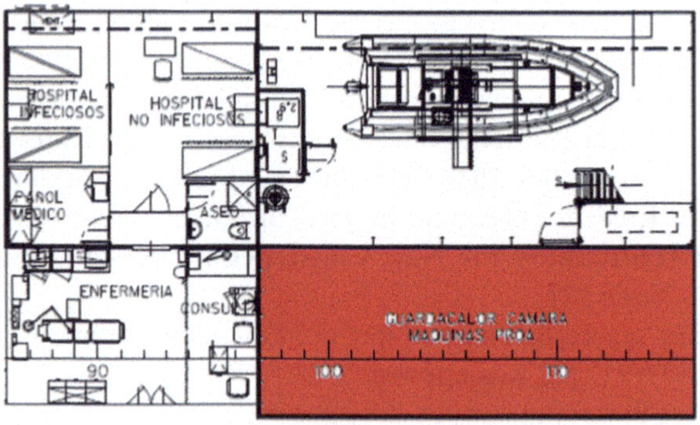

Fig. 6. Hospital and infirmary area

The hospital area has 60 dB(A) noise limit and it is next to a noisy local, FWD Engine Room Casing which noise limit is 110 dB(A). To achieve 60 dB(A) in the hospital only adding insulation materials in the bulkhead between both locals was a risk that should not been assumed in a critical space as the hospital is. So it was agreed a modification in this are as it can be seen in Fig. 7. Another local with higher nosie limit was located, as acoustic barrier, between the hospital and the casing.

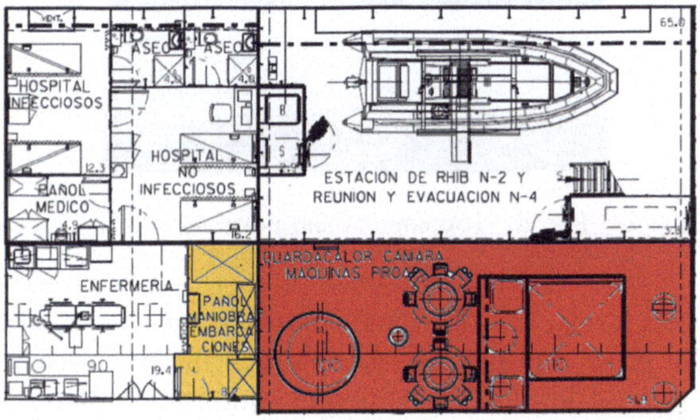

Fig. 7. Hospital and infirmary area modified

Another advantage of this analysis is the identification, without doing any calculation, of all areas that could present any noise problem. In this sense, sometimes is possible to make any small modifications or even modify some noise source characteristic, and therefore reduce, minimize or even eliminate the risk of noise non-fulfilment in advance. In the worst case, all problem areas are identified and can be studied in detail during the execution of the noise prediction calculation.

The second step is to carry out a noise prediction calculation. It is recommended to use a SEA Method (Statistical Energy Analysis) that is use for predicting the transmission of sound and vibration through complex structural acoustic systems. The method is particularly well suited for predicting responses at higher frequencies.

SEA software takes into account the exact location of the rotary and alternative machinery/equipment of the Frigate, the transmission path to the receiver spaces and the vibro-acoustic characteristics of the structural elements. The main noise transmission paths analysed by the software are:

- Airborne noise.
- Structureborne noise.

The following Fig. 8 shows the Topological and Noise models of the **Frigate F-110** used for assessing the noise levels on board. The Topological model has the information about the location of the different spaces. The Noise model has the structural information, the definition of the noise sources, the receiver and source spaces and the insulation treatments. Once created the Topological model and assigned the physical properties such as material types and thicknesses, the acoustical properties and treatments are incorporated into the Noise model.

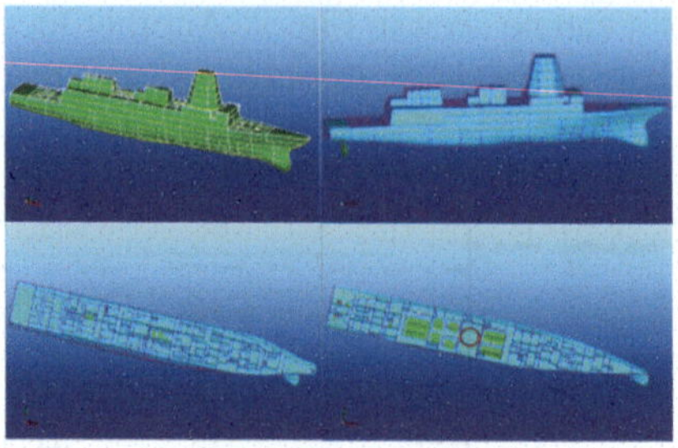

Fig. 8. Topological and noise models

To calculate noise levels, the following parameters have been taken in account: **(1)** the thickness of the plates and the distance between the reinforcement. **(2)** Air noise level and structural noise level of the different equipment. **(3)** Isolation drawings in bulkheads and roofs, pavement and sub-pavement. **(4)** Plan of the General Disposition of the vessel and situation of focus and locals.

To carry out the noise prediction calculation, the starting point is the original vessel model. In the original model is analysed the initial acoustic insulation defined by the shipyard; thermal and fire insulation are also included in this initial study.

Results obtained from the Noise Prediction in the original model for different operational conditions defined in the specification are compared to those stated in the Specification (Fig. 9).

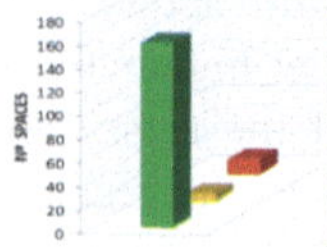

ORIGINAL MODEL. SUMMARY OF NOISE EVALUATION. NORMAL NAVIGATION CONDITION (17,2 kn)			
DECK	EVALUATION		
	GOOD	ACCEPTABLE	OVER LIMIT
Tank Top	15	0	0
First Platform	32	0	0
Second Deck	39	2	5
Main Deck	32	0	4
Level 01	22	4	2
Level 02	16	0	1
Total:	156	6	12
Nº Total of spaces:	174		

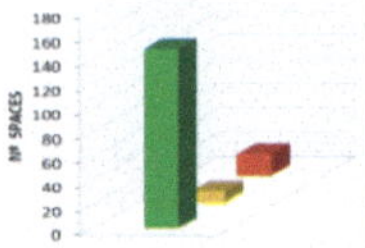

ORIGINAL MODEL. SUMMARY OF NOISE EVALUATION. MAXIMUM SPEED CONDITION (27 kn)			
DECK	EVALUATION		
	GOOD	ACCEPTABLE	OVER LIMIT
Tank Top	14	1	0
First Platform	30	2	0
Second Deck	34	2	10
Main Deck	32	0	4
Level 01	22	4	2
Level 02	16	0	1
Total:	148	9	17
Nº Total of spaces:	174		

Fig. 9. Noise evaluation original model

In those areas with noise levels exceeding the limits, corrective actions or complementary noise insulations are proposed and agreed with the Shipyard in order to reduce these levels. Then a new calculation process will be made for verifying that new expected levels meet the limits required by the Specification (Fig. 10).

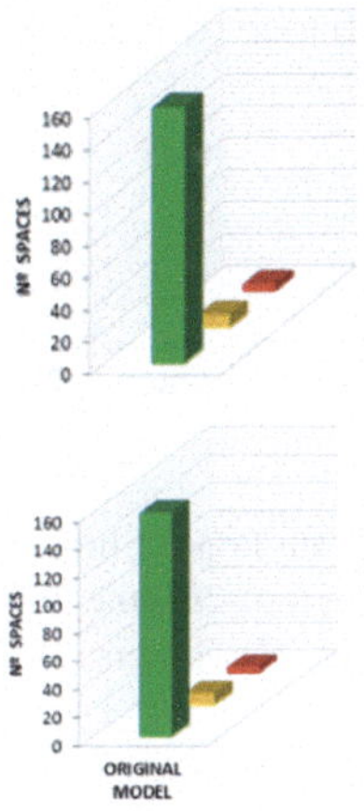

MODIFIED MODEL. SUMMARY OF NOISE EVALUATION. NORMAL NAVIGATION CONDITION (17,2 kn)			
DECK	EVALUATION		
	GOOD	ACCEPTABLE	OVER LIMIT
Tank Top	15	0	0
First Platform	32	0	0
Second Deck	41	3	2
Main Deck	32	3	1
Level 01	26	2	0
Level 02	16	0	1
Total:	162	8	4
N° Total of spaces:	174		

MODIFIED MODEL. SUMMARY OF NOISE EVALUATION. MAXIMUM SPEED CONDITION (27 kn)			
DECK	EVALUATION		
	GOOD	ACCEPTABLE	OVER LIMIT
Tank Top	15	0	0
First Platform	32	0	0
Second Deck	41	3	2
Main Deck	32	3	1
Level 01	26	2	0
Level 02	16	0	1
Total:	162	8	4
N° Total of spaces:	174		

Fig. 10. Noise evaluation modified model

These were the results obtained in different locations of the Frigate during in the Conceptual phase. In the Definition phase it will try to improve the locals with noise levels higher than the ones defined in the Specification.

The reasons for using in the Noise Prediction Calculation a software based on SEA-Statistical Energy Analysis- among other, are the following:

- To know the Expected Noise Level at each space at the early stage of the Project.
- To Apply the "proper noise insulation solution" in the "right partition", and only that due that "type of noise: Airborne or Structural one" that cause the deviation is well-known at each space of the ships.
- To optimise the "noise insulation weight": "It is only applied what is needed in the most effective partitions".
- Flexibility to look for the more effective solution as well as the most feasible one for execution for the shipyard.
- Definition of commercial solutions using materials existing in the market.

4 Conclusions

The vibration and noise control programme in ships with specific vibration and acoustic requirements shall be implemented from the early stages of the project and taken into account in the definition of the specification. During the specification phase, those requirements for the overall ship have to be translated into specific requirements for the main suppliers and the shipyard. Additionally, basic design has to consider this for a proper basic design of the structure, selection of the propulsion system and an adequate general arrangement. Finally, the current state-of-the-art in simulation tools enables experts on the field to predict and identify problems or risks in very early stages of the project and test solutions before it is too late or too expensive to implement. The benefits of applying these methodologies from the very beginning of the project has been clearly proven in terms of minimizing risks of incompliance at the lowest cost possible.

References

1. Standard of Habitability for New Construction Vessel of Ministry of Defence – Navy, June 2010
2. BV Rules for Naval Ships Pt. E, Ch. 6, Sec. 1 November 2016
3. Resolution MSC. 337(91) Adoption of the Code on Noise Levels on board Ships
4. Directive 2003/10/EC of the European Parliament and of the Council of 25 June 2002 on the minimum health and safety requirements regarding the exposure of workers to the risks arising from physical agents (noise) (2003)

A Study on the Effect on Fatigue Damage Due to Different Shipping Routes

Luis C. De Gracia[1(✉)], Hitoi Tamaru[2], Naoki Osawa[1], Adan Vega[3], and Hector O. Ruiz[1]

[1] Osaka University, Osaka 565-0871, Japan
luis_de_gracia@naoe.eng.osaka-u.ac.jp
[2] Tokyo Maritime Science and Technology, Tokyo 135-8533, Japan
[3] Zoartech Consulting, Panama City 0819-03568, Panama

Abstract. An important tool in optimizing the operation of ships on long voyages is weather routing, that is a method used to avoid rough weather and to find the minimal fuel, time or cost route between ports. Recently, weather routing of ships is increasingly recognized as an important contribution to safe, economical and reliable ship routes. Consequently, most ships follow weather routing and those ships are not challenged by extreme wave loads. Although that in the maritime industry fatigue failure is one of the most significant failure modes for ships structures, in the classification society rules, the ship structure fatigue assessment is performed without the consideration of weather routing. In addition, the ships that operate worldwide shipping routes would face different oceanographic conditions, i.e. North Atlantic or North Pacific routes, and those ships will face differences on the cumulative fatigue damage depending on the trade route.

This paper presents a comparative study based on the S-N fatigue assessments for a welded joint in a container carrier that follows two different shipping routes. It is assumed that the target ship sails following a weather route in the North Atlantic and North Pacific Oceans. Fatigue damages accumulated by the ship following a ship weather route are calculated. Short-sea sequences are generated by a wave load sequence model. Based on these results, the effect of different shipping routes on the cumulative fatigue damage is discussed.

Keywords: Storm model · Cumulative fatigue damage · Weather routing

1 Introduction

The ship is a complex system consisting of multiple subsystems that need a synchronized coexistence so that the ship can operate largely independently once it is at sea. The random motion of ocean waves is the main cause of the fatigue damage in ships due to the varying wave loads acting on the ship structure. The variable encounter sea state is characterized by the significant wave height and wave period. In order to evaluate the fatigue lives of ships accurately, it is necessary to include the load histories induced by a wave, which may comprise a random sequence with different stress range. The ship design process needs to ensure the integrity of the hull, where wave-induced loads contribute significantly. These loads have to be representative of the

V. A. J. E. Carreño Moreno et al. (Eds.): CIDIN 2019/COPINAVAL 2019, pp. 175–184, 2020.
https://doi.org/10.1007/978-3-030-35963-8_15

environmental and operational conditions encounter through the entire life of the ship [8]. Nowadays there are more concerns for the ship owners, classification societies, and shipping companies that the ships follow a weather routing to maximize the safety. Those vessels do not experience extreme wave load, and this will have an important effect on the ship structural member's response that needs to be examined. Tomita [9, 10] proposed 'storm model' which can simulate the wave load sequence experienced by ocean-going ships. Kawabe [4] and Prasetyo [5] modified Tomita's model to improve the emulation capability of real sea state sequence. The latest version was modified so that it can consider the stochastic nature of the wave direction [1].

Fatigue assessments of a container carrier's welded joint are performed in this study. The target ship is assumed to face two different shipping routes: A North Atlantic and a North Pacific Ocean route, following a ship weather routing, called Minimum Time Route. Short sea sequences are generated by using the Japan Weather Association (JWA) hindcast data, and those for the minimum time route are simulated by adopting Tamaru's weather routing algorithm [7]. SN-based fatigue assessments are performed for the minimum time routes, and the effect of the different shipping routes on the fatigue damage is examined. Section 2 discusses the oceanographic data along the shipping routes. A brief review of the weather routing algorithm adopted, a comparison of the wave statistics encounters by the target ship and the wave load model applied to generate the stress sequence is presented. Section 3 describes a numerical example of S-N analyses based on the storm model stress sequences. As a result is found that the difference on the cumulative fatigue damage of the target ship, in the two different trade routes is around 5%. It is considered that this difference is the effect of different wave environment, uncertainties on the operational aspects (ship heading, ship speed reduction), selection of the S-N curve, among others. Section 4 are founded the conclusions and future task are proposed.

2 Ocean Data Within the Routes

2.1 Weather Routing

One of the major contributions of the ship routing is that it can help to decrease the fatigue-caused problems by scheduling a ship's route which causes the lowest possible fatigue damage to a vessel. Among other benefits are time and cost reductions and increased crew and structural safety. For ship owners and operators, the economic aspect is of equal importance as safety. The reduction in transit time, fuel consumption, extreme weather encounters, and hull damage is directly related to saving in operating cost reductions, maintenance and repair costs. Tamaru [7] proposed a ship weather routing algorithm adopted in this study. The algorithm can determine the minimum time route (MTR) between two ports from a spatiotemporal distribution of sea states (significant wave height and wave direction). The correlation between operational aspects, i.e. ship speed reduction and relative heading angle, and the significant wave height are considered in the analysis. The Japan Weather Association (JWA) hindcast data is adapted to generate the spatiotemporal sea state data and the shipping route is optimized by analyzing the isochrones.

2.2 Shipping Routes

Hereafter, the ship weather routing explained in Sect. 2.1 is adopted. Two shipping routes are examined: A North Atlantic (NA) and a North Pacific (NP) route between US/UK and Japan/US, respectively. Figure 1 show examples of the assumed minimum time routes. A 6000TEU container carrier is chosen as the target ship. It is considered that she sails for a period of 10 years. The sea state (significant wave height *Hs*, mean period *Ts* and wave direction θ) sequence experienced by her is determined by those at the nearest JWA hindcast data grid point. The arrangement of the data grid points interval is 2.5° in latitude and longitude direction.

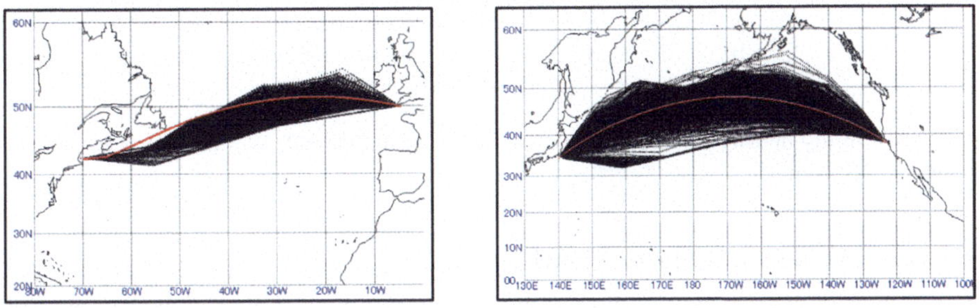

Fig. 1. An example of minimum time routes for US/UK (left figure) and Japan/US routes (right figure) [7].

First, the wave environment along the routes are examined. Figure 2 presents the probability density function (p.d.f.) of *Hs*, P_{Hs}, from the DNVGL and the JWA hindcasting spatiotemporal along the NA and NP routes. In (Fig. 2) it is shown that *Hs*'s p.d.f. of the two different routes nearly equivalent for the high wave region ($Hs > 5$ m). In the NA waves smaller than 2.5 m has a larger probability of occurrence than those encountered in the NP. Further, the p.d.f. in the NP is higher than that of the NA at their maximum ($Hs = 2.5$ m). Further, it is noted that the JWA hindcasting along the different routes differ from that recommended wave environment for design, e.g. DNV rules [3]. The main reason is that the wave environment recommended by class rules do not contemplate operational conditions, e.g. weather routing systems effects or operational decisions by captains such as ship speed reductions or adjust the course along different shipping routes. Aspects related to captain decisions with respect to maneuvering in heavy weather and heavy weather avoidance can be found in [6]. Alternatively, to the recommended wave environment, the wave load models could provide a reliable description of the stochastic characteristics of the wave environment expected to encounter by the marine structure, e.g. ships, offshore structures, etc., during her lifetime based on various oceanographic wave data sources, such as hindcast data, visual observation, satellite or buoy data. In the following section, a wave load model to emulate the random nature of the waves is presented.

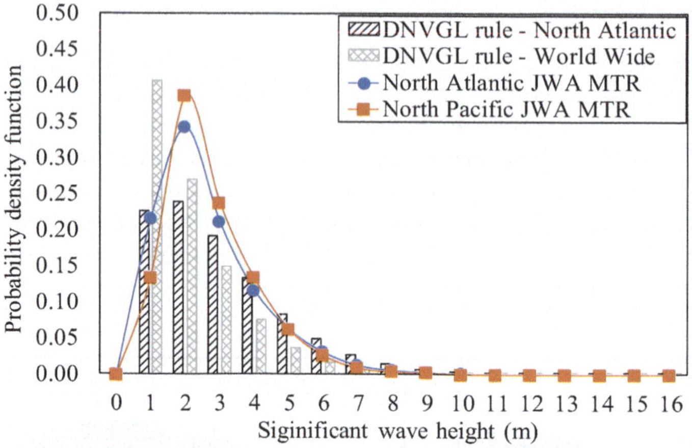

Fig. 2. Probability density function (pdf) of *Hs*, P*Hs*, from DNVGL Weibull, JWA in MTR for the NA and NP routes.

2.3 Wave Model - 4G Storm Model

The basic of the first generation (1G) storm model was proposed by Tomita [9, 10]. This model was applied to examine fatigue crack growth behavior in the generated random loading, based on collected wave data of ships that sailed on the Pacific Ocean. The model assumes that the oceanography phenomena can be divided into calm sea and storm conditions. Furthermore, the random sea state history can be modeled with significant wave height *Hs* as random and time-independent for calm sea and crescendo/de-crescendo of *Hs* for a storm. These two conditions appear randomly, see Fig. 3. A more recent and advanced storm model has been developed by De Gracia [1], which is called 4G-storm model. This model takes into account the occurrence probability distribution of the wave direction to evaluate the calculating stress response.

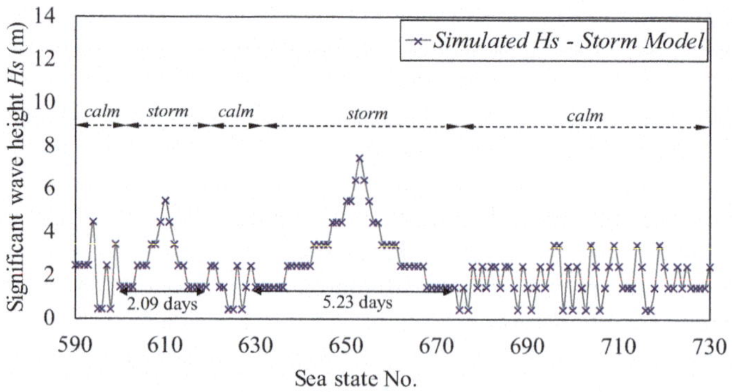

Fig. 3. Example of generated significant wave height sequences based on storm profiles with different storm duration.

Storm model consists of 'storm profile' and *Hs*'s probability distribution in calm seas. 'Storm profile' consists of a series of storm waveforms and the occurrence probability of storms. In order to generate the *Hs* probability distributions, first, the joint frequency distributions of (*Hs*, *Ts*), called as wave scatter diagram, are generated for the encountered wave conditions on both NA and NP routes. The long-term distribution of *Hs* is assumed to follow the Weibull distribution [10] and the joint probability distribution of the significant wave height *Hs* and mean wave period *Ts* is refitted by obeying the conditional of the lognormal distribution of *Ts* given *Hs* [12]. The *Hs* profile in a storm is determined based on the assumption that this profile is distributed by the following the tail of the Weibull distribution [9]. Once *Hs* is determined, the mean period *Ts* is determined using the conditional probability of *Ts* given *Hs* proposed in [12]. Then, the joint probability distribution *p(Hs,Ts)* is calculated by (Eq. 1):

$$p(H_s, T_s) = p(H_s)p(T_s|H_s) \tag{1}$$

Individual waves for each *Hs* is assumed to follow the Rayleigh distribution, and the energy spectrum can be defined by the ISSC spectrum.

It is considered that the ship fatigue assessment might be deteriorated under the all-heading model [11]. In this study, the heading angle of the ship is determined by considering the occurrence probability distribution of the wave direction [1]. Consider θ, α, and χ as the wave direction, the ship's heading angle and relative heading angle. The 4G-storm model evaluates the stress response taking into account the χ's occurrence probability. Considering θ's occurrence probability f_θ, θ is chosen by random selection number and χ is calculated by (Eq. 2) each time. The averaged f_θ is determined by JWA hind-cast data. Figure 4 shows an example of the "real-headings model" compared to the conventional "all-headings model" in the North Atlantic. It is observed that the average wave direction in the North Atlantic is predominant between 210 and 330°.

$$\chi = \alpha - \theta \tag{2}$$

Sea state sequences are generated from the storm model once the storm profiles are configured. Figures 5 and 6 show the simulated significant wave heights patterns by the storm model and JWA hindcast *Hs* along the NP and NA routes. Ten years of simulated waves along the measured routes are compared with that of the JWA hindcast for the ship routes. The horizontal axis denotes the sea states duration (1 h) along the routes and the vertical axis the significant wave height. From the hindcast waves time history, it is clear that the significant wave height increases over time and reach a maximum value at one point, then decrease gradually. A typical storm condition reproduced by the storm model simulation is shown in Fig. 3. This tendency is successfully expressed by the storm wave model simulated waves. This means that the storm model is capable of reproducing the statistic nature of ocean waves.

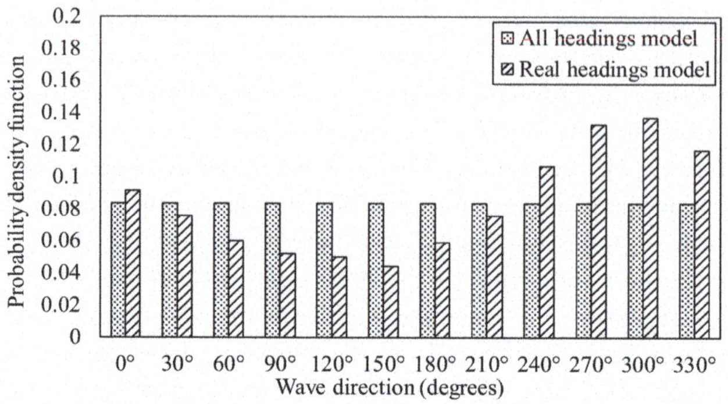

Fig. 4. Comparison between the wave direction's occurrence probability distribution determined from JWA hindcast data in the North Atlantic and the "all-headings assumption".

Furthermore, let ΔS denotes the nominal stress range and P_{EX}, ΔS be the ΔS's exceedance probability. Let $P_{EX,\Delta S\_NA}$ and $P_{EX,\Delta S\_NP}$ be $P_{EX,\Delta S}$ of the storm model stress sequences for NA and NP routes. Figure 7 shows the comparison of $P_{EX,\Delta S\_NA}$ and $P_{EX,\Delta S\_NP}$. It is shown that a substantial difference in $P_{EX,\Delta S}$ is observed in the NP route compare to that found in the NA, especially in the high ΔS. It is considered that the difference in the stress exceedance probability is the effect of the correlation between the relative headings angle of each route, the stress amplitude operator (RAO) of the target ship, in addition to the significant wave height's exceedance probability encounter by the target ship along the shipping routes in her total ship's life.

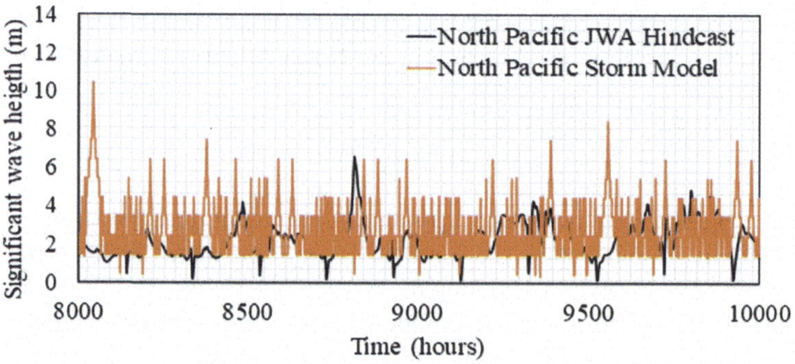

Fig. 5. Simulated significant wave heights patterns by the storm model and JWA hindcast *Hs* along the NP route.

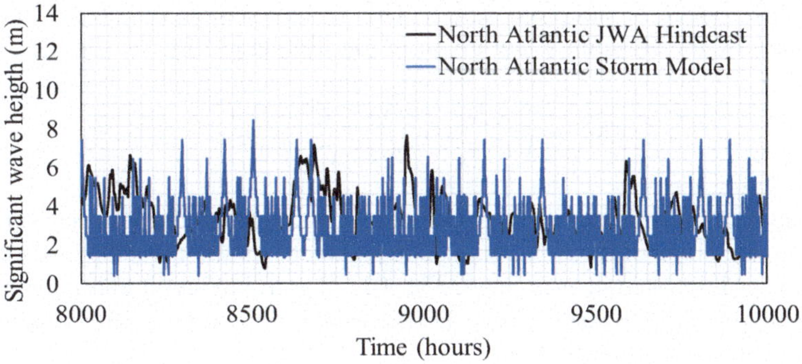

Fig. 6. Simulated significant wave heights patterns by the storm model and JWA hindcast *Hs* along the NA route.

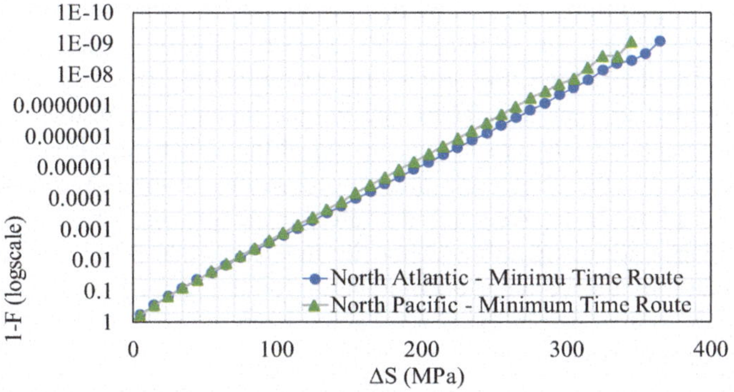

Fig. 7. The comparison of the stress range's exceedance probability, $P_{EX,\Delta S}$, for the storm model stress sequence for NA and NP routes.

3 Fatigue Assessment Approach

3.1 Cumulative Fatigue Damage

Under variable amplitude loading, the life is estimated by calculation of the total damage done by each cycle in the stress spectrum. The cumulative fatigue damage is evaluated by the Palmgreen-Miner rule as represented by (Eq. 3).

$$D_{10yr} = \sum\nolimits_{i=1}^{S} \left(\frac{n_i}{N_i} \right) \tag{3}$$

where S: number of blocks, ni: number of cycles of stress range in the i-th block, and Ni: fatigue life corresponding to the stress range in the i-th block taken from the S-N curve. In this study the linear cumulative damage is examined during 10 years, $D_{10years}$.

The thickness effect is not considered. The relationship between $\Delta\sigma$ and N_b is approximated by the Basquin's law given in the following form as (Eq. 4):

$$\Delta\sigma = \begin{cases} CN_b^m \; ; & \Delta\sigma \leq C \cdot (10^7)^m \\ C'N_b^{m'} \; ; & \Delta\sigma \geq C' \cdot (10^7)^{m'} \end{cases} \tag{4}$$

where N: predicted number of cycles to failure for stress range $\Delta\sigma$, $\Delta\sigma$: stress range, m: negative inverse slope of the S-N curve. The parameter of the two-slope S-N curve are chosen as C = 1.1494e+04, m = −3.3333e−01, C' = 1.3405e+03 and m' = −0.2000e −01 according to the UK-DeN Curve D (for welded joints) [13]. The structural concentration factor SCF is set to 1.0.

3.2 Fatigue Assessment Results

This section presents the damage accumulated during sailing in different routes. The structural stress concentration factor, SCF, is set to 1.0. The objective is to present how the analysis methodology can be adapted to the evaluation of the fatigue reliability of ship structural members. First, the stress time histories are generated by storm model procedure presented in Sect. 2.3. Then, the fatigue damage accumulated during each sea state is examined for 10 years sailing period of the target ship. Figure 8 shows the fatigue damage accumulation per sea state during sailing in the NA and NP routes for 10 years (right plot). It is observed that in the early stage of sailing period the cumulated damages are comparable. After the first year (first 4200 sea states), a considerable increase of the damage accumulation for the ship that sails in the North Atlantic compare to than that in the North Pacific is presented in Fig. 8 (left plot). This is because harshest weather environment during her sailing in open sea are found in the North Atlantic. Furthermore, the mean and standard deviation of accumulated fatigue damages during 10-years are presented in Table 1. It is shown that the differences on the statistical properties of the cumulative damage due to different shipping routes are nearly ten percent larger when sailing on the NA compared to that of the NP. In addition, these results indicate that the storm model can describe well the wave variation experienced by the ship.

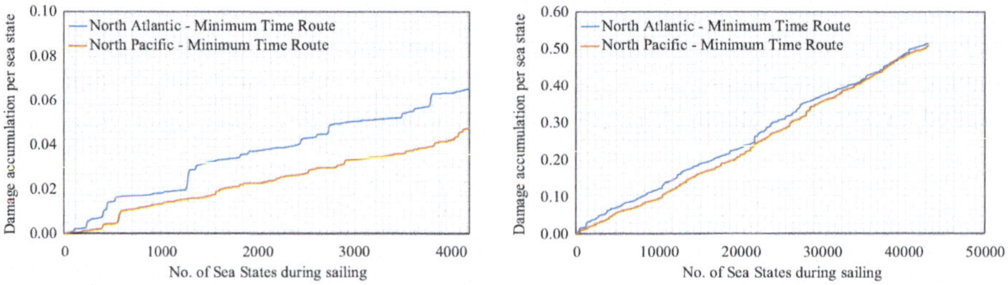

Fig. 8. Fatigue damage accumulation per sea state during 1-year sailing (left plot) and 10-years (right plot) sailing in the NA and NP routes.

Table 1. Cumulative fatigue damage results based on the S-N fatigue assessment.

Route	S-N fatigue damage	
	Mean	STD
North Atlantic	0.2715	0.1518
North Pacific	0.2471	0.1521

4 Conclusions

In this study the statistical storm wave model is briefly introduced, and the application to generate stress sequence for fatigue assessment is presented. A 6000TEU container carrier is chosen as a target ship. It is considered that the sails in two different ocean routes. It is considered that the ship was assisted by a weather routing system to avoid harsh weather environment during her sailing. Sea states are generated by a storm model simulation. Stress sequences are generated considering the occurrence probability of sea state and the relative heading angle to examine the effect of the wave environment encountered by ships. The statistics properties of the cumulative damage based on S-N based fatigue assessment shows that a difference of ten percent could be expected when the ship sails in the North Atlantic compared to that sailing in the North Pacific, under the condition chosen. This is considered because harsh weather environment during sailing are found in the North Atlantic. Furthermore, the rate of damage accumulated by the target ship during sailing substantially differ depending on the ocean route. Finally, research on the development of a more advanced wave load sequence model which can consider the effects of the high-frequency response, e.g. whipping and springing, is necessary.

Acknowledgment. The authors would like to acknowledge Dr. Kuniaki Matsuura of Japan Weather Association (JWA) for providing JWA hindcast data.

References

1. De Gracia, L., Tamaru, H., Osawa, N., Fukasawa, T.: A study of the influence of weather routing on the preciseness of ship structure's fatigue assessment. In: Proceeding of 27th International Offshore and Polar Engineering Conference (ISOPE), San Francisco, California, pp. 957–963 (2017)
2. De Gracia, L., Tamaru, H., Osawa, N., Fukasawa, T.: Effect of encountered wave condition on fatigue life prediction of ship structures. In: Vega Sáenz, A., Pereira, N., Carral Couce, L., Fraguela Formoso, J. (eds.) Proceedings of the 25th Pan-American Conference of Naval Engineering—COPINAVAL. COPINAVAL 2017, part 1, Panama City, pp. 15–25. Springer, Cham (2019). ISBN 978-3-319-89811-7
3. DNV (Det Norske Veritas): Recommended Practice DNV-RP-C205 Environmental conditions and environmental loads, April 2014

4. Kawabe, H., Syuuji, O., Masayoshi, O.: The study of storm loading simulation model for fatigue strength assessment of ship structural member: 1st report new storm loading simulation model which consistent with a wave frequency table. J. Soc. Nav. Arch. Jpn. **193**, 39–47 (2003). ISSN 05148499
5. Prasetyo, F., Osawa, N., Kobayashi, T.: Study on preciseness of load history generation based on storm model for fatigue assessment of ship structures members. In: Proceeding of 22nd ISOPE Conference, pp. 709–712 (2012). ISSN 1098-6189
6. Prpić-Oršić, J., Parunov, J., Šikić, I.: Operation of ULCS - real life. Int. J. Nav. Arch. Ocean. Eng. **6**(4), 1014–1023 (2014). ISSN 2092-6782
7. Tamaru, H.: About the optimum route by the weather routing. In: Proceeding of Japanese Society of Naval and Ocean Engineers – JASNAOE, Japan, pp. 109–114 (2016)
8. Temarel, P., Bai, W., Bruns, A., Derbanne, Q., Dessi, D., Dhavalikar, S., Fonseca, N., Fukasawa, T., Gu, X., Nestegård, A., Papanikolaou, A., Parunov, J., Song, J.H., Wang, S.: Prediction of wave-induced loads on ships: progress and challenges. Ocean. Eng. J. **119**, 274–308 (2016). ISSN 0029-8018
9. Tomita, Y., Kawabe, H., Fukuoka, T.: Statistical characteristics of long-term wave-induced load for fatigue strength analysis for ships. In: Proceeding of 6th PRADS, vol. 2, pp. 2792–2805 (1992)
10. Tomita, Y., Matoba, M., Kawabe, H.: Fatigue crack growth behavior under random loading model simulating real encountered wave condition. Mar. Struct. **8**, 407–422 (1995)
11. Tomita, Y., Hashimoto, K., Osawa, N., Terai, K., Wang, Y.: Study on fatigue design loads for ships based on crack growth analysis. J. ASTM Int. **1**(9), 1–15 (2004). ISSN 1546-962X
12. Wan, S., Shinkai, A.: The statistical characteristics of global wave data and appraisal for long-term prediction of ship response. J. Soc. Nav. Arch. Jpn. **178**, 289–296 (1995). ISSN 05148499
13. UK Department of Energy: Offshore Installation: Guidance on Design, Construction, and Certification, 4th edn. HMSO, London (1990)

Offshore Patrol Vessel (OPV) Interceptors Evaluation Using Computational Fluid Dynamics (CFD)

E. Marcelo Cali Y.[1(✉)] and Marcos A. Salas[2]

[1] COTECMAR, Cartagena, Colombia
mcali@cotecmar.com
[2] University Austral of Chile, Valdivia, Chile
msalas@uach.cl

Abstract. In this paper, a comparative study of the performance of stern interceptors of an OPV type vessel is presented. The study was developed using numerical CFD simulations to estimate the resistance, which allowed calculation of the fuel consumption for the operating profile. Six interceptors, with chords between 65 and 340 mm, were analyzed for 7 speeds between 8 and 22 knots. The results indicate there is resistance reduction for speeds above 18 knots. For 20 knots, corresponding to the maximum speed of the operational profile, the 200 mm chord interceptor was the one that produced the greatest reduction in resistance, however this speed corresponds only to 10% of the operating time. The sum of the results for the operational profile, showed that the 65 mm interceptor was more efficient for economic benefits over the speed range, however fuel saving is modest, considering the high budgets required to operate these vessels. Other aspects are to be considered for decision making of the inclusion of the interceptor.

Keywords: Fuel consumption · Advance resistance · Interceptors · Numerical analysis · Computational fluid dynamics

1 Introduction

This paper investigates the effect on the hydrodynamic performance of a ship with a vertical plate that protrudes in the stern of the hull of an offshore patrol vessel (OPV). The action of the interceptor is mainly due to high pressure forces that originate in the plate due to the stagnation region and that can reduce the drag and produce consequent energy savings in terms of fuel consumption of the main engines, obtaining a reduction in the operating costs of the vessel.

The degree of improvement or deterioration in the performance of a vessel due to the inclusion of an interceptor, is given by the operating conditions of the vessel and the geometric characteristics of the interceptor, mainly its chord.

The project of the new generation of offshore patrol vessels for the Colombian Navy, called OPV93C, represents a challenge for the Colombian naval industry and the implementation of a stern interceptor would be a simple solution that could help to obtain a good performance and develop a successful product, not only for Colombia,

V. A. J. E. Carreño Moreno et al. (Eds.): CIDIN 2019/COPINAVAL 2019, pp. 185–200, 2020.
https://doi.org/10.1007/978-3-030-35963-8_16

but also to be exported to the region. For this, an analysis is required to determine the feasibility of its implementation and the appropriate configuration of it.

The methodology used in this study, for the estimation of the advance resistance of the naked hull was computational fluid dynamics (CFD), which allowed developing numerical simulations of tests with the naked hull model and with interceptors, in a virtual tank, in order to evaluate the performance.

In the first part of this document, a general chronological review of studies and implementations that have been developed for this type of appendices is made, as well as the hydrodynamic theory of the physical phenomenon and the numerical analysis proposed to solve the problem.

In the second part of the study results are analyzed. First, the CFD model used for the simulations is described and the validation of the mesh is carried out. Then the results of the simulations for the naked hull without and with interceptor are presented, and the extrapolation of the results to full scale is performed. Finally, a cost estimation of fuel consumption projected for a year is made, considering the operational profile of the vessel, which allowed evaluating, from the economic point of view, the best configuration of interceptor for the hull of OPV93C.

Finally, in the third part, conclusions and recommendations regarding the evaluation are presented, which allows to use this numerical tool in the decision making process, regarding the inclusion, or not, of this appendix in the hull of OPV93C.

2 State of Art and Theoretical Framework

In the seventies, the effect of stern wedges and interceptors in the performance of planning vessels began to be studied, from experimental investigations, establishing empirical formulations to estimate the addition of lift (Bannikov and Luckashevsky 1976) and reduction of resistance (Bannikova et al. 1978).

Towing tank tests of models with interceptors have also been carried out, obtaining reductions in resistance of up to 17% with interceptors (Tsai and Huang 2003), (Tsai et al. 2003).

One of the most detailed studies on interceptors hydrodynamics was developed by Brizzolara in 2005, he used a simplified CFD model in two dimensions to study the local flow around an interceptor, applying a flat boundary to represent the hydrodynamic flow around the bottom of a ship, this being the most idealized model for interceptors that had been applied until then. This model allowed for viscous effects to be tested for a limited series of simulations, determining the relationship between the hydrodynamic forces and the main interceptor parameters (Brizzolara 2003).

Later a new study with considerable changes, in context and approach, in comparison with previous ones was presented, the hydrodynamic models based on CFD considered the complete description of the hull form with aft appendages, using a CFD tool with a numerical solver of RANS equations (Reynolds Averaged Navier Stokes), (Villa and Brizzolara 2009).

At University Austral of Chile (Salas and Tampier 2013), the performance of an interceptor on a fishing vessel was evaluated using CFD methodology, resulting in 10%

of reduction on the resistance with the inclusion of this appendix for Froude numbers greater than 0.40, while for lower values there was an increment in resistance.

The action of the interceptor is mainly due to pressure forces: a high pressure peak is expected just in front of the plate due to the stagnation region, while forward of it, the pressure decays to undisturbed values, with a certain related tendency with the divergence caused in the input current lines. Figure 1 presents the theoretical schematization of the hydrodynamic problem that must be solved.

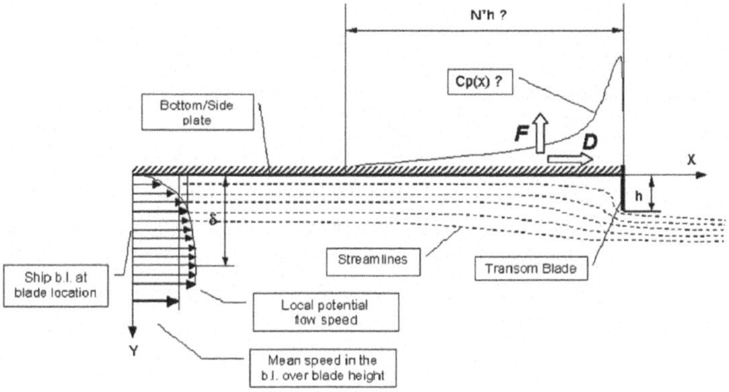

Fig. 1. Diagram of the 2D hydrodynamics of the interceptor (Brizzolara, 2003).

Generally the chord length (height of the plate protruding from the hull) of the interceptors is well below the thickness of the boundary layer at that position of the hull; in this way the relation between the global forces (F and D) in terms of the main parameters of influence can be considered as the following function.

$$\frac{F,D}{h\rho V^2} = f\left(Rn_L, \frac{L}{h}, H, Fn_{ht}\right) \tag{1}$$

Where $H = \delta^*/\theta$, is the form factor of the boundary layer, Rn_L is the Reynolds number, and Fn_{ht} is the Froude immersion number at the stern.

From the dimensional analysis, leaving aside the less significant dimensionless terms, relationship (1) can be reduced to:

$$F, D = f(h, V, \delta^*, \theta, \gamma, L, \rho, g, ht) \tag{2}$$

Where h is the interceptor chord length. δ^* and θ are the thickness and the momentum of the boundary layer (these parameters define a measure of the effects of fluid movement around the boundary layer), V is the ship speed, γ is the kinematic viscosity, L is the length of the vessel and ht is the stern draft.

The widely accepted idea is that the interceptors must be completely contained in the boundary layer, i.e. the chord of the interceptor would be a fraction of the boundary layer, thus an initial approach for the boundary layer near the stern, would be to select the chord of the interceptor equal to the thickness of the boundary layer. The thickness of the boundary layer, in turbulent flow, can be calculated with the following equation (Schlichting 1979).

$$\delta(x) = 0.37\, x \,\mathrm{Re}_x^{-1/5} \tag{3}$$

3 Analysis and Results

Using the CFD methodology, numerical simulations were carried out to solve the hydrodynamic problem around the hull and estimate the advance resistance of the bare hull, and then with the addition of the interceptor. For this, the hull surface of the OPV93C was modeled with a scale of 1: 11,607, equal to the hull model used in towing tank tests (Hamburg Ship Model Basin HSVA 2017). The characteristics of the prototype hull are presented in Table 1.

Table 1. Main particulars of the vessel (OPV-93C)

Characteristic	Symbol	value	Units
Length over all	Loa	92,30	m
Waterplane length	Lwl	85,10	m
Beam	B	14,03	m
Draught	T	4,10	m
Displaced volume	V	2.598,20	m^3
Maximum speed	v	20	knots

3.1 CFD Model

To simulate the virtual tank required for this study, a block was created to define the domain, which together with the hull, provided the limits for the volume mesh, and the symmetry condition allowed to define the block and perform the calculations only for half of the hull.

As a base size for the generation of the mesh, 1.72% of the waterplane length was chosen. In addition, four control surfaces were created: hull, deck, tank contours and the interceptor. The target size for the faces of these elements with respect to the base size was 25% for hull elements, 200% for the deck, 1600% for the tank contours and 12.50% for the interceptors. Figures 2 and 3 show the meshing generated in the domain.

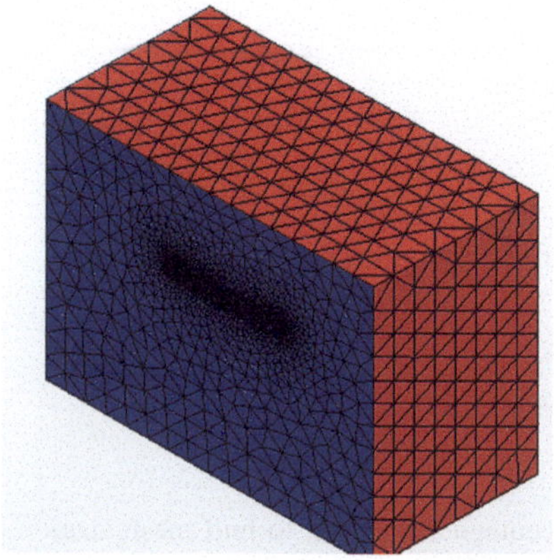

Fig. 2. Virtual tank meshing (naked hull)

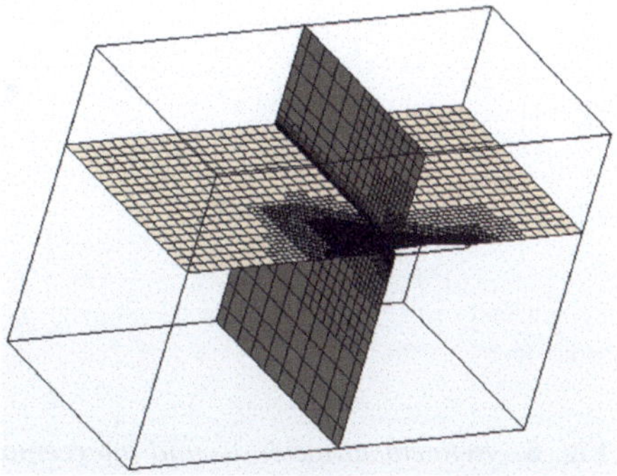

Fig. 3. Isometric cuts of the volume mesh

The results of the simulation will depend largely on the domain meshing, for this reason it is very important to validate the influence of this on the results and to determine an adequate mesh size that allows obtaining reliable results in a reasonable computational time. In the case of the hull meshing, experimental results available from towing tank tests were used to compare results. The difference between the numerical (CFD) and experimental results was less than 4%, which allowed validating the mesh for the naked hull. In the case of the hull with interceptor, it was tested to validate the meshing, refinements of 10, 5, 2.5 and 1.25% with respect to the mesh of the hull in the area around an interceptor. The refinement of 2.5% was chosen because with a greater refinement, considerable changes in the results are not expected. In Fig. 4 the refinements are presented and in Fig. 5 the results obtained for an interceptor of 160 mm (full size) at 18 knots.

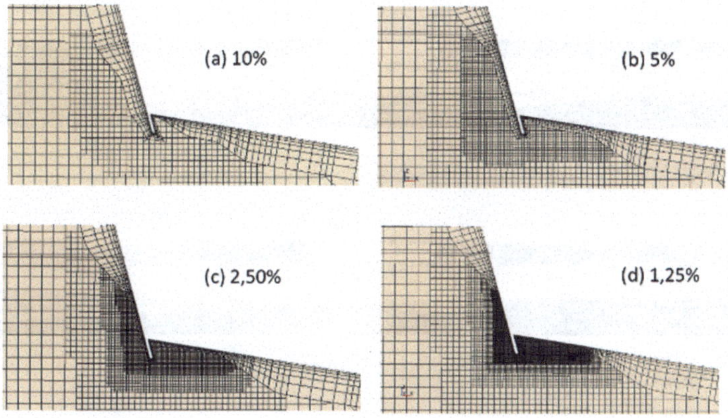

Fig. 4. Mesh sizing around the interceptor (relative percentage compared to base size).

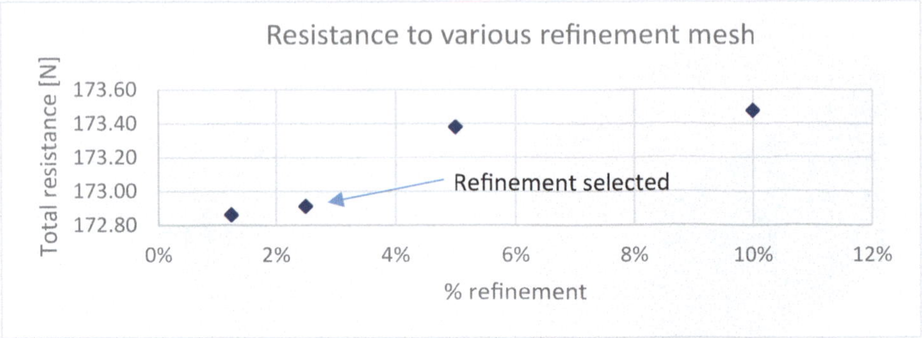

Fig. 5. *Refinement mesh around interceptor* (Percentage in relation to hull mesh size).

3.2 Naked Hull Results at Model Scale

The first simulations developed in this study considered free heave and pitching movements. The simulations correspond to Froude numbers between 0.21 and 0.39. During the tests; drag forces, dynamic trim and sinkage were recorded. These were compared to experimental results obtained at HSVA. Figure 6 shows photographs comparing dynamic sinkage and trim for speeds corresponding to 12, 16 and 22 knots at full scale. The similarity between the towing tank tests and CFD simulations is observed. The shapes of the bow wave and stern wake also are in good agreement.

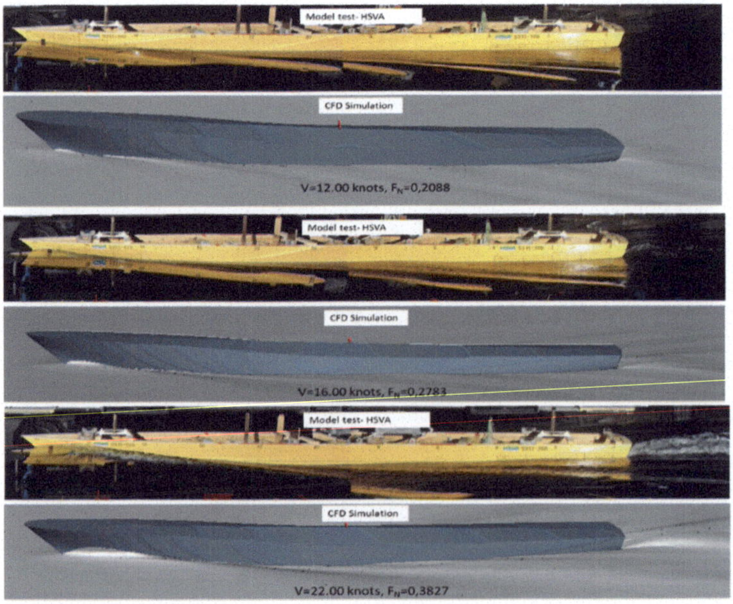

Fig. 6. Towing tank and CFD wave patterns. Equivalent full scale speeds 12, 16 and 22 knots.

Figure 7 graphically depicts both, CFD and experimental, results of forward resistance for the bare hull of the scale model. The difference between these results was less than 2% for Froude numbers between 0.2137 and 0.3917 corresponding to model speeds of 1.81 and 3.32 m/s. For a speed of 1.21 m/s reached 3.77% difference, this is possibly due to the fact that, for low speeds, the friction forces are more important than those of the waves and in this case, to obtain more approximate results, it would be advisable to fine-tune the meshing of the skeg zone. However, this was not done, since differences of 5% are acceptable in this type of studies and the interest was focus in measuring the reduction of resistance when the interceptor was included.

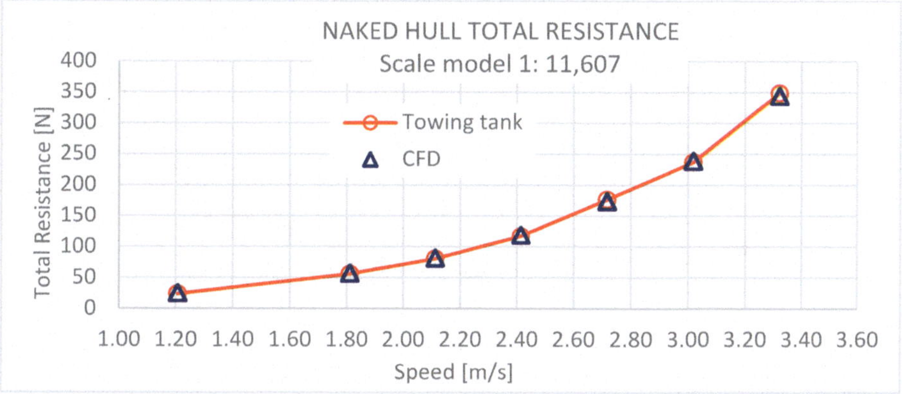

Fig. 7. Naked hull total resistance for CFD and towing tank tests (scale model).

Differences in results for sinkage of the center of gravity of the hull are also below 2% for all speeds except for 1.21 m/s which reaches 5.74%, as shown in Fig. 8.

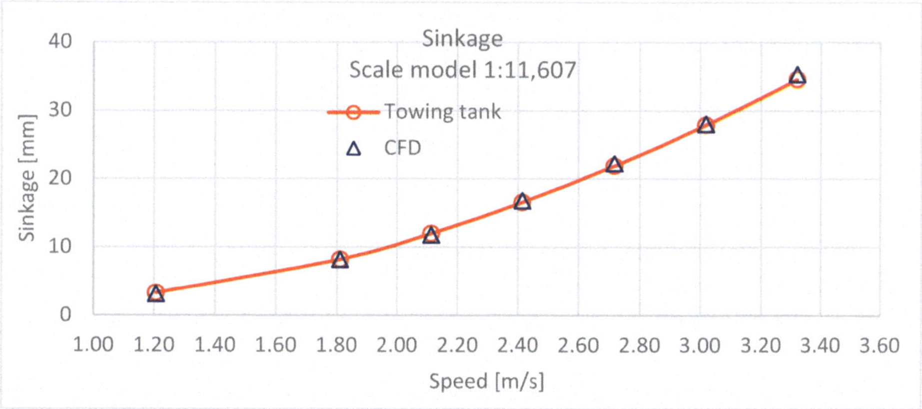

Fig. 8. Sinkage of the center of gravity of the hull from CFD simulations and towing tank tests (scale model).

Regarding the dynamic trim, greater differences were obtained in percentage terms, for example for the speed of 2.11 m/s, difference was 12.96% and for 2.72 m/s was much higher, reaching 92.66%, however these numerically correspond to differences of 0.009 and 0.015 degrees respectively. It must be noticed that, a displacement hull does not substantially modify its dynamic trim with speed. In relative terms, this change of trim is negligible and would not produce significant differences in the results of hydrodynamic forces acting on the hull. Figure 9 shows the results obtained through CFD simulations and experimental results.

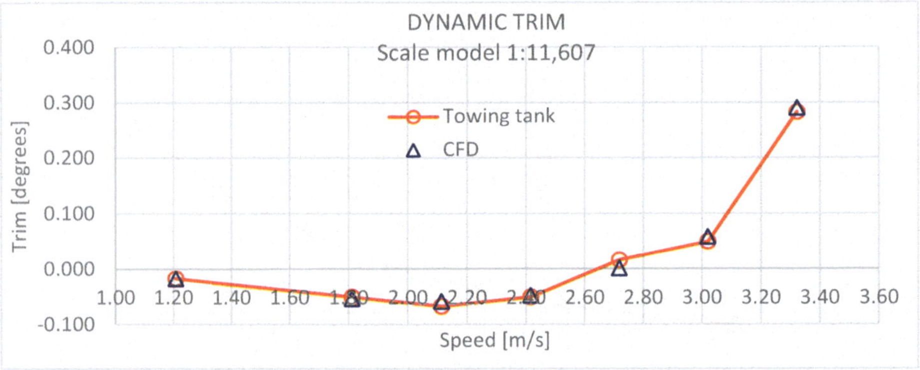

Fig. 9. Dynamic trim from CFD simulations and towing tank results (scale model).

3.3 Full Scale Resistance Results Extrapolation

Total resistance results of the model (RTM) moving at constant forward speed were obtained at full scale, based on the assumption that this resistance is represented by two components; friction and the so-called residual resistance, produced mainly by wave generation. The calculation was made using dimensionless coefficients of resistance, so, the dimensionless total advance resistance C_T coefficient was found using the following equation:

$$C_T = \frac{R_T}{\frac{1}{2}\rho S V^2} \quad (4)$$

Where:

R_T = Total ship or model resistance [kN] or [N]
ρ = Fluid density [kg/m^3]
S = Wetted surface [m^2]
V = Speed in m/s

For the viscous component, a three-dimensional extrapolation was applied, using ITTC'57 ship-model relationship to obtain the friction resistance coefficient, C_F, where the Reynolds number (Rn) is related to turbulent flow.

$$C_F = \frac{0,075}{(\log R_n - 2)^2} \quad (5)$$

Additionally, it was assumed that the residual resistance component in dimensionless form C_R, is identical for the model and the prototype, at the same Froude number (F_n), as follows.

$$C_{RM} = C_{TM} - C_{FM} = C_{RS} \quad (6)$$

Where:

C_{RM} = Model residual resistance coefficient
C_{TM} = Model total resistance coefficient
C_{FM} = Model friction resistance coefficient
C_{RS} = Ship residual resistance coefficient

Taking advantage of the unchanged residual coefficient, the full scale residual resistance can be obtained from the following equation.

$$C_{TS} = C_{RS} - C_{FS} + C_A \quad (7)$$

Where:

C_{TS} = Ship total resistance coefficient
C_{RS} = Ship residual resistance coefficient
C_{FS} = Ship friction resistance coefficient
C_A = Correlation allowance, 0.000207 was used, same value as used by HSVA

In Fig. 10, results for the total resistance obtained from CFD simulations are presented and compared with towing tank results, the differences were less than 3% except for 8 knots, which was 5.26%.

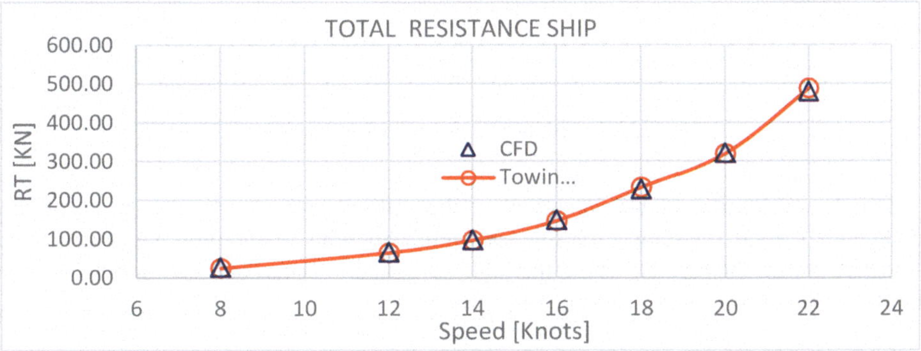

Fig. 10. Total resistance Ship from CFD simulations ant towing tank tests.

3.4 Effect of the Interceptor in Decreasing Resistance

Simulations were performed for the bare-hull model at different speeds, including interceptors with h/L ratios between 0.076 and 0.40% as shown in Table 2.

Table 2. Interceptor chord lengths evaluated.

Chord length		h/LBP
Full size [mm]	Model [mm]	
65	5,60	0,076%
160	13,78	0,188%
200	17,23	0,235%
220	18,95	0,259%
240	20,68	0,259%
340	29,29	0,400%

The resistance results, with and without interceptor, are shown in Fig. 11, where it is observed that the interceptor is effective in reducing resistance at speeds over 18 knots.

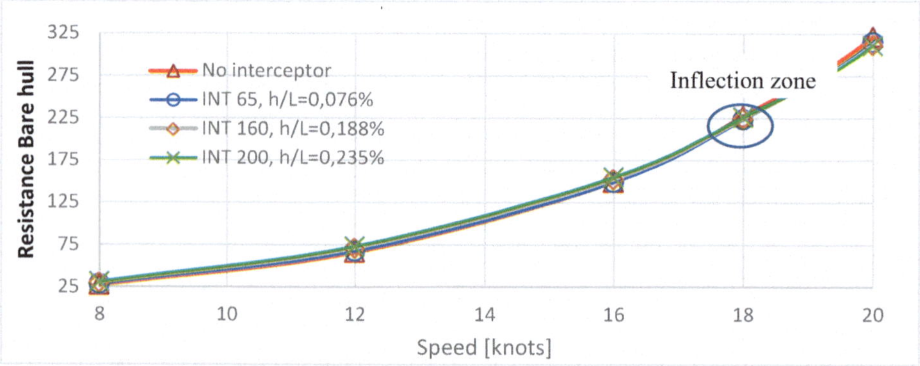

Fig. 11. Resistance evaluation of interceptors.

For the evaluation of the interceptors, the difference in percentage terms of the coefficient of total resistance for the bare hull and with interceptor was calculated. Figure 12 shows the results, where the highest reduction found are 3.23 and 3.29% for chord lengths of 200 and 220 mm, respectively.

The effectiveness of the interceptor in the case of the hull of the OPV93C, is above 18 knots, for these speeds, as the chord is increased, the reduction also increases, however it reaches a limit around 220 mm, where the effectiveness begins to decline, at 18 knots however, there is a penalty when the chords lengths are higher.

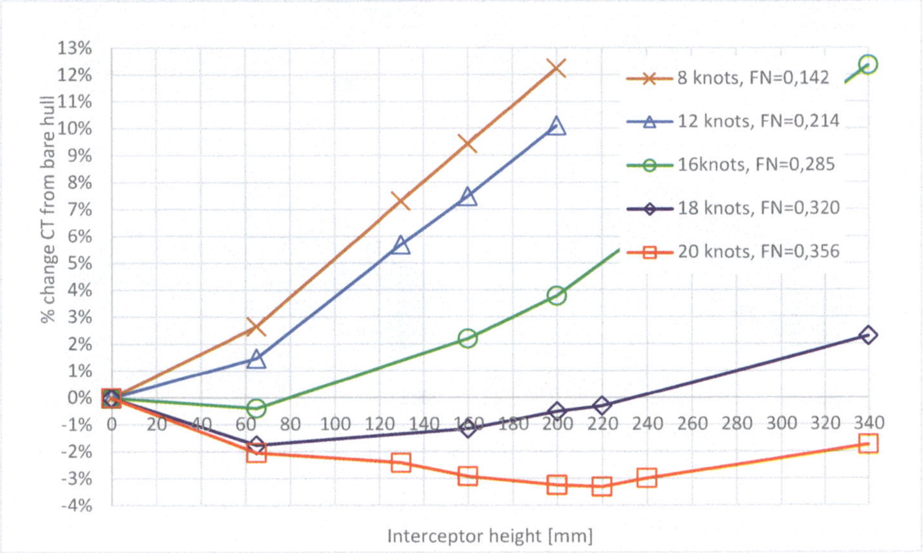

Fig. 12. Resistance variation for several speeds

3.5 Effect of Interceptor in Pressure Distribution

When interceptors are implemented at the stern on a hull, the change in pressure distribution is the main responsible for the reduction of advance resistance. Figure 13 shows a pressure map distribution at 20 knots. It is observed that in the stern pressure of 550 pa is reduced when the interceptor is introduced and therefore the drag force (D) is also reduced. On the opposite side, in the aft bottom area, pressure increases from 800 pa to 1200 pa when the interceptor was introduced.

For 200 mm chord length, the area with 550 pa pressure was completely reduced, being around this chord size where the greatest effectiveness was obtained in terms of the resistance reduction. For larger interceptor chords, although the aft low pressure zone is reduced, the pressure in the bottom is also increased due to the increase in the area of pressure at 1200 pa, resulting from the stagnation of the fluid. This causes a reduction in the effectiveness in reducing the resistance of the hull.

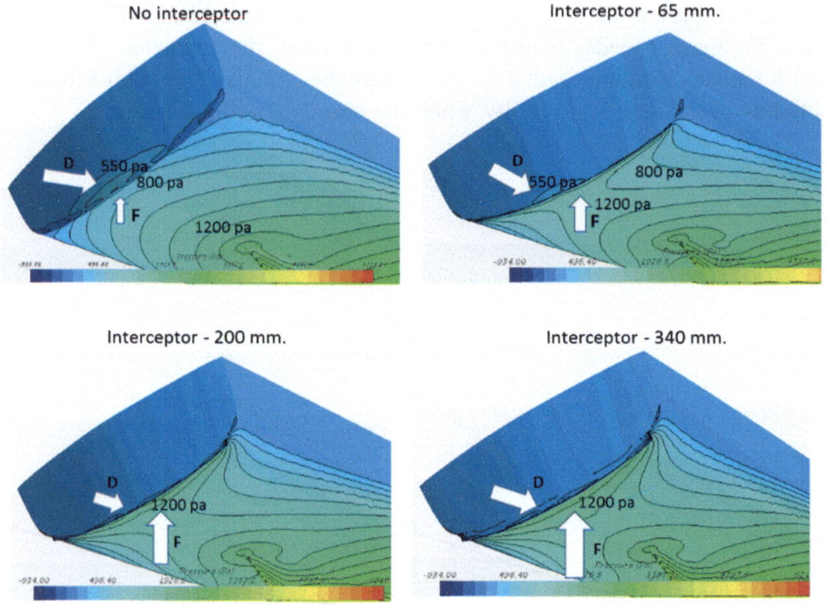

Fig. 13. Pressure distribution at the stern at 20 knots speed.

3.6 Economic Evaluation

The variation in fuel consumption was evaluated in a simplified way, in order to have parameters to allow a better criterion for the selection of the appropriate interceptor. For this, it was assumed the installation of a propulsion system composed of two main engines, reduction gear and fixed pitch propeller, to fulfil the following operational profile.

Table 3. Operational profile

Mission	Speed	Time
Maneuvering, patrol	8	45%
Economic	12	45%
Maximum	20	10%

The percentage of variation of the coefficient of total resistance (CT) will be the same for the variation of power at the same condition. For this reason to estimate the variation in fuel consumption, the same percentages obtained for the coefficient of resistance (CT) for interceptors with chord lengths of 60, 160 and 200 mm were applied to calculate the delivered power curve PD (Fig. 14).

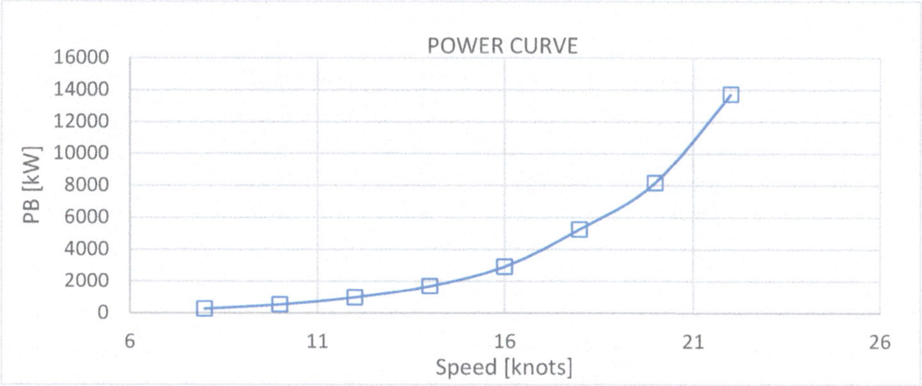

Fig. 14. Brake power (BP)

The installation of 2 MTU 16V-1163-M74 engines of 4800 kW each was assumed and the specific consumptions were taken as follows:

- 283 kW (8 nudos): 300 kg/kW.h
- 1000 kW (12 nudos): 240 kg/kW.h
- 8177 kW (20 nudos): 205 kg/kW.h

In addition, it was considered 2500 h of operation per year and the percentage of operating time for the speeds, according to the operational profile. The average cost of fuel considered was $2 per gallon (Ecopetrol 2018).

The results are presented in Fig. 15, where, only for the interceptor of 65 mm chord, a decrease of 569 gallons is observed, while for 160 and 200 mm consumption was higher than for the hull without interceptor in 5,010 and 7,463 gallons. The saving for the smallest interceptor was 1,139 dollars and for the 160 and 200 mm interceptors, the increase in fuel costs was 10,109 and 14,925 dollars respectively, in a period of one year.

If we consider the operational profile of Table 3, the option of the interceptor with chord of 65 mm would be the one of greater benefit from the economic point of view, however considering the high percentage of operating time (90%) at low speeds, the saving of 1,139 annual dollars is not very encouraging, if we consider the high budgets involved in operating these vessels.

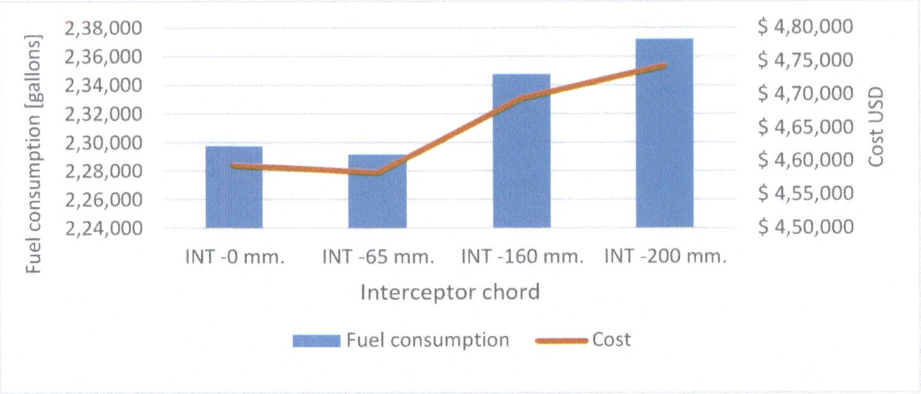

Fig. 15. Fuel consumption OPV 93C. 10% operational time at full speed.

From the hydrodynamic point of view, the 200 mm chord for the maximum speed of 20 knots, could be a good option to be implemented in the hull of OPV93C, however, economically it would produce increases in fuel consumption costs. It would be appropriate not to rule out this option and instead consider other factors that allow a better judgment in decision-making. These factors to consider could be the following.

- Consider the possibility of implementing a retractable interceptor that can operate with the highest chord at maximum speed and be retracted at low speeds.
- Reduction in initial costs, due to the installation of a smaller engine when less power is demanded for the same maximum speed.
- Greater reachable speed, due to reduction in drag forces.

In other types of vessels, mainly commercial, the percentage of operating time at maximum speed is generally greater and substantial savings in fuel consumption would be expected, even for small percentages of decrease in resistance, as a result of the implementation of interceptors. As an example and as a reference to note the advantages of an interceptor, the costs were estimated by changing the operating time for the maximum speed of 20 knots to 50%. Results are presented in Fig. 16.

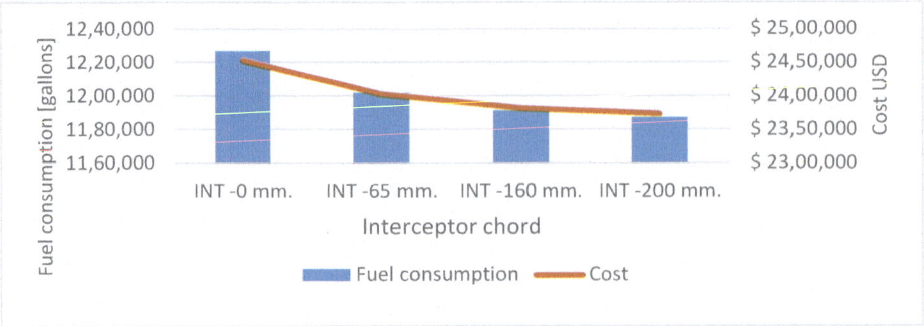

Fig. 16. Fuel consumption. 50% operational time at full speed.

The results with the change of this operational profile are more encouraging, the fuel consumption is reduced as the chord length increases, accounting for savings of $79,320 in one year operation, for the 200 mm interceptor. That is to say, the hydrodynamic performance is combined with economic considerations that support the implementation of the interceptor of better performance in the maximum speed.

4 Conclusions

Through numerical simulations and using the CFD methodology, the bare hull resistance of the OPV93C was estimated for different speeds and compared with results obtained from towing tank tests. The maximum difference found between the two methodologies was 3.77% for the scale model and 5.26% when extrapolating to full scale at 8 knots. For other speeds, the difference did not exceed 3%.

An interceptor was incorporated to the scale model of the OPV93C hull and the resistance was estimated with the same methodology used for the hull without interceptor, varying the speed and chord of this appendix. For the maximum speed (20 knots) the chord of greatest effectiveness was 220 mm full scale, with a percentage of reduction in the total resistance coefficient (TC) of 3.29% in relation to the hull without interceptor. For the chord of 200 mm the percentage of reduction is 3.23%, while for 130 mm it falls to 2.40% and for 65 mm to 2.03%. From the practical point of view and considering only the maximum speed, a chord of 220 or 200 mm would be a good implementation option, because the difference between the two is only 0.06%.

Despite the fact that at speed of 20 knots corresponding to a Froude number of 0.356, the interceptor produces benefit, for other speeds that complete the typical operational profile for this type of vessels, it was found that the application of the interceptor generated an increase in total resistance and tends to be greater as the chord of the interceptor increases. For example, for the 200 mm interceptor, the increase was 12.25 and 10.11% for speeds of 8 and 12 knots respectively.

An economic analysis was made from a preliminary estimate of fuel consumption for the percentage of operation of the speeds for 2500 h in a year: 45% of the time at 8 knots, 45% at 12 knots and 10% at 20 knots. The obtained results indicated that only the interceptor with chord of 65 mm, produced savings and that for other chords, the costs would increase in relation to the hull without interceptor. The saving is of 1,139 dollars, value not very significant, considering the high budgets involved in operating these vessels.

The low percentage of operating time for 20 knots is the reason for the increase in fuel consumption and consequently costs, despite the fact that power demands at speeds of 8 and 12 knots are around 8 and 28 times less than 20 knots, respectively.

As a reference to get an idea of the benefits of the implementation of the interceptor, the calculation of costs was made varying the percentage of operating time at 20 knots to 50%. With this change, a reduction in costs was obtained for the interceptors with chords greater than 65 mm, for example, 200 mm, savings would be $79,230, a value that can be considered significant for one year and which would need to be considered in the decision towards the implementation of this appendix.

References

Bannikov, Y.M., Luckashevsky, V.A.: Experimental investigation of lift and drag of planing plates. TSAGI Report, vol. 7, no. 1 (1976)

Bannikova, T.A., Bannikov, Y.M., Luckashevsky, V.A., Zeitlin, M.Y.: The investigation of the hydrodynamic characteristics of the planing surface with a spoiler on the trailing edge. TSAGI Report No. 1908, pp. 1–22 (1978)

Brizzolara, S.: 7th International Conference on Fast Sea Transportation. Hydrodynamic Analysis of Interceptors With CFD Methods, Ischia, pp. 49–56 (2003)

Ecopetrol, 01 de marzo de 2018. https://www.ecopetrol.com.co/wps/portal/es/ecopetrol-web. Obtenido de https://www.ecopetrol.com.co/wps/portal/es/ecopetrol-web/productos-y-servicios/precios/precios-vigentes/combustibles-y-zonas-de-frontera

Hamburg Ship Model Basin HSVA: Model Tests for the OPV 93 COTECMAR. Hamburg: HSVA (2017)

Salas, M., Tampier, G.: Assessment of appendage effect on forward resistance reduction. Ship Sci. Technol. **7**(13), 4–7 (2013)

Schlichting, H.: Boundary Layer Theory. Mc Graw Hill Inc, New York (1979)

SIEMENS: STAR-CCM+(13.02.011) [Software] (2018)

SPAR Associates Inc.: Estimating Commercial Ship Life Cycle Cost & Required Freight Rate (3-Port Model). Annapolis, Maryland, Estados Unidos, 21 de 12 de 2012

SPAR Associates Inc.: Naval Ship Life Cycle Cost (LCC) Model. Annapolis, Maryland, Estados Unidos (2014)

Tsai, J., Huang, J.K.: Study on the effect of interceptor on high speed craft. J. Soc. Naval Architects Mar. Eng. **22**, 95–101 (2003)

Tsai, J., Hwang, J., Chou, S.: 7th International Conference on Fast Sea Transportation. Study on the Compound Effects of Interceptor with Stern Flap for Two Fast Monohulls with Transom Stern, Ischia, pp. 23–28 (2003)

Villa, D., Brizzolara, S.: 10th International Conference on Fast Sea Transportation, Fast 2009. A Systematic CFD Analysis Of Flaps/Interceptors Hydrodynamic Performance, Athens, pp. 1032–1038 (2009)

Evolution and Present of Modularity in Warships

Luis Andrés Piñeros Bello[1] and Camilo Ernesto Segovia Forero[2(✉)]

[1] ARC-DIPROE-PES Program, Cartagena D T y C, Cartagena, Colombia
luis.pineros@armada.mil.co
[2] ARC-Fuerza Naval del Caribe, Cartagena D T y C, Cartagena, Colombia
camilo.segovia@armada.mil.co

Abstract. The results of this research seeks to understand the concept of modularity in engineering applications, strengthening previous works of the authors and focusing on its necessary implementation in naval architecture, as a practical and efficient method to optimize time and resources in shipbuilding, likewise seeks to demonstrates the relevance of other modularity concepts as it would be specifically that of mission modules, intended to fulfill different operational capabilities through the adaptability of warships for the accomplishment of the missions assigned to the Navies, optimizing the use of defense systems and economic resources.

Keywords: Modularity · Standardization · Adaptation · Flexibility · Payload

1 Introduction

In the field of engineering, the concept of modularity is seen as a strategy for the improvement of construction processes in different types of industry and has captured the attention of designers, builders, ship owners and researchers associated with the maritime industry . Likewise, the adaptation to this trend in the naval architecture field, especially in the shipbuilding processes of warships, allows focusing the value of this analysis in the exploration of different practical applications implemented by Navies in the world (Piñeros Bello et al. 2018) .

The world progress and it evolution to globalization need to agile process and focus the industry projection towards the implementation of modularity as a basis to achieve improvements in the production chain and construction time, as well as being a growing trend for Navies worldwide adopting flexibility in the ship configuration to accomplish different missions.

Modularity can be seen as a technological advance, and as a measure to deal with the dilemma of budget constraints in military expenditure faced worldwide and the raise in design and production of military systems. Navies are looking modularity as a way to reduce costs associated with the design, production, acquisition, operation and modernization of their ships.

V. A. J. E. Carreño Moreno et al. (Eds.): CIDIN 2019/COPINAVAL 2019, pp. 201–210, 2020.
https://doi.org/10.1007/978-3-030-35963-8_17

2 Adaptability Concept

The concept of adaptability and its active relationship with the main topic of this study must be contextualize. Before exploring the different environments that lead the logical efforts to the understanding of the subject in engineering, therefore, the ability to adapt the projected design to a future technological development and systematic change of needs must be described, which in turn give rise to two concepts as it would be the Modularity and Flexibility (Piñeros Bello et al. 2018). Figure 1 shows the relationship established by RAND Corporation for the aforementioned concepts.

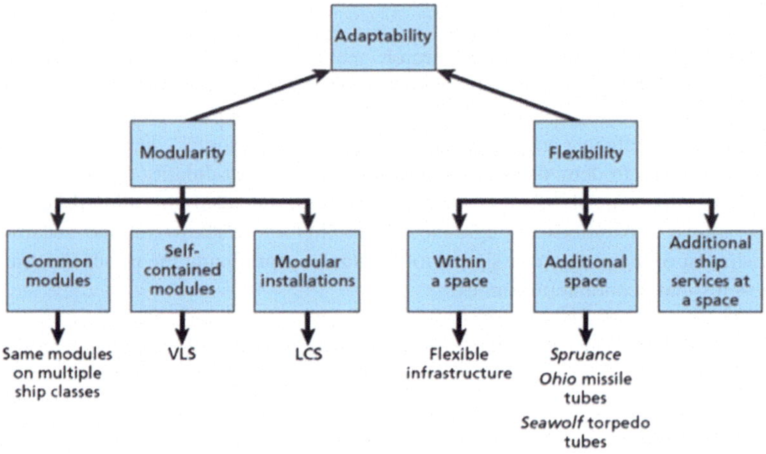

Fig. 1. Adaptability and its relationship with modularity and flexibility Source: Designing Adaptable Ships 2017. Shrank et al. p. 17.

The term modularity is generally used with three different scopes. In design of complex systems with naval application is meant to interchangeable systems to be integrated into specific locations with restrictions in area, volume, and auxiliary services. In regards to construction and architecture, modularity refers to the construction of systems by standardized components. In manufacturing, modularity refers to the use of interchangeable units to create product variants (Largiader 2001).

Adaptability Criteria

Deepening the adaptability necessary to be able to consider modular and flexible vessels with extended capacities should establish minimum criteria for their application; in this way we want to establish the modular requirements and their transversal criteria, as can be seen in the Fig. 2.

At this point we can begin exploring the applicability of modularity in warships, which will be the main topic of this research, explaining different types of modularity applied to the increase in capabilities of a Navy in two ways, modularity of configuration and modularity of mission, without being exclusive since as will be seen below, some relevant developments in this topic can be achieved by the use of both concepts at the same time while other ones will involve only one of them.

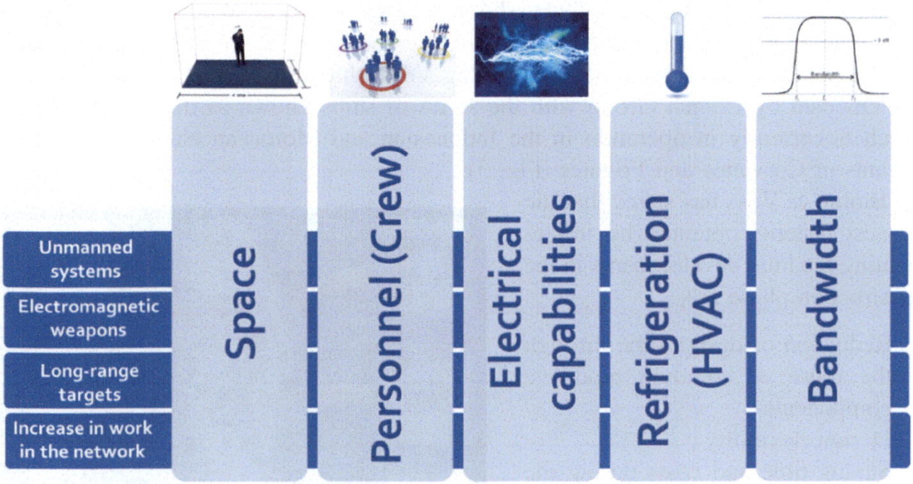

Fig. 2. Adaptability criteria

3 Modularity Cases

It should be noted that these concepts should be established since the early stages of needs definition, for the design and construction. The modularity during the process of the vessel can be applied in different ways that can be understood as follows:

- Common modules used in multiple classes of ships: They are structural parts of the ship built and tested in similar environments. Potential applications include functional modules such as kitchens, medical facilities and laundries.
- Self-contained modules that provide a plug-and-play capability for the equipment within the module. These modules have defined interfaces and boundaries and are design for a task, such as firing a missile. Common modules can be used in different classes of vessels, for example vertical launch system (VLS).
- Modular facilities that provide a basic structure of the ship and services that allow the installation of several mission packages and exchanged as necessary. Modular installations, such as self-contained modules, have defined interfaces but with widely defined limits. The US LCS (Littoral Combatant Ship) and the ships of the Absalon class of the Danish Navy are examples of this type of modularity (Schank 2016).

3.1 MODULARITY IN MEKO FAMILY SHIPS (Multi-purpose Combination)

It refers to a family of ships that through the design and construction of weapon modules, with standardized interfaces and connections, allowing exchanging or installing new capabilities. The MEKO has become a reference for this concept in the shipbuilding industry, being applied to different types of vessels, as well as being used in more than 43 frigates and corvettes, partially or completely (Piñeros Bello et al. 2018).

The concept developed by Blohrm & Voss shipyards has been adopted and implemented by Damen Group with the series of ships known as the SIGMA Class, which is currently in operation in the Indonesian and Moroccan Navies in different variants of Corvettes and Frigates (Fig. 3).

Blohm & Voss has stated that the greatest benefits obtained by implementing modular developments in the construction phase are:

- Reduction of design time through the reuse of common modules/ components.
- Design flexibility
- Saving time and costs during the production process.
- Clear division of responsibilities between the shipyard as the main contractor and the manufacturer of weapons, electronic systems and machinery (Largiader 2001).

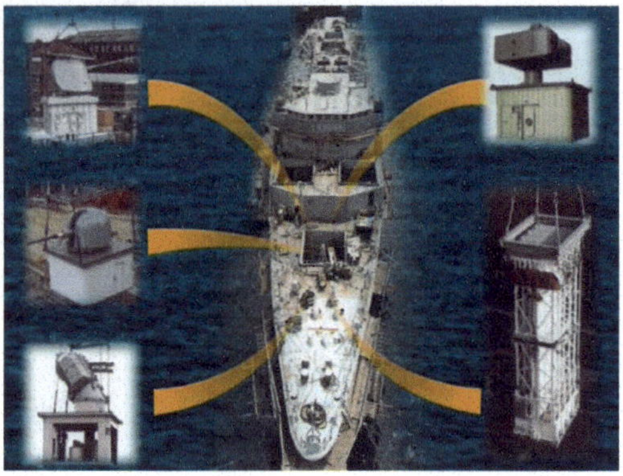

Fig. 3. Concept Blohm + Voss MEKO Source: STSSD, 2004. NATO Naval Group 6. p. 77.

3.2 Standard Flex (NATO NAVAL GROUP 6, 2004)

There is talk of a self-contained concept that implements mission modules standardizing the design and applying flexibility in order to supply specifically the operational needs.

The standardized containers and the associated interfaces would then allow the role of the vessel to be adapted in a very short time and thus be able to fulfill the planned operations, optimizing the specialization of tasks to be performed and reducing maintenance costs largely (Fig. 4).

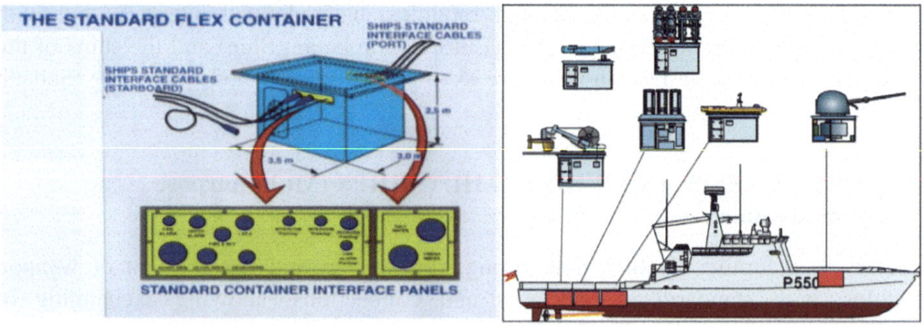

Fig. 4. Standard interface panel for the module "Standardflex" and concept for Small Ships Source: STSSD, 2004. NATO Naval Group 6. pp. 78–79.

3.3 Littoral Combat Ship (LCS)

It covers a concept of spaces generated by which modularity is expanded, but with more emphasis on manned and unmanned external systems. Sensors and weapons make up what they called mission packages with an exceptional exploitation of flexibility, achieving that the hull can be separated from the capacity to the point where the ships could be configured in a time of 48 h from one mission to another (Fig. 5).

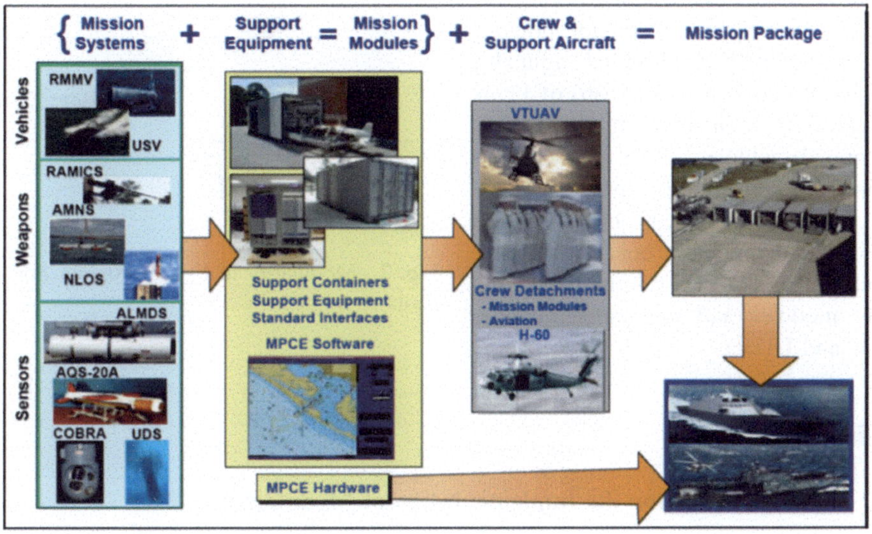

Fig. 5. LCS Units Mission Package. Source: Rodríguez, C. E. (2011). LOS LCS Y SU IMPACTOEN LA GUERRA DEL LITORAL. REVISMAR (2), 130–145.

3.4 Global Combat Ship (GCS)

This same concept has been evaluated by other navies such as the United Kingdom Royal Navy (RN), exploring the concept of modularity in its new project Global Combat Ships, class 26 (GCS) in which a mission bay located below the flight deck can accommodate and deploy additional boats, unmanned vehicles (aerial, surface or underwater), or a maximum of ten self contained modules (Piñeros Bello et al. 2018) (Fig. 6).

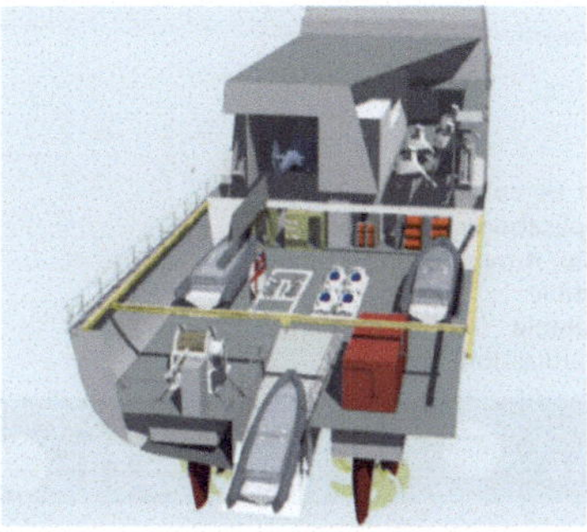

Fig. 6. Mission Bay Frigate type 26 (Global Combat Ship) Source: www.defenseindustrydaily.com

3.5 F110 Program

In the 70s and 80s, Navantia (then Bazán) built two classes of frigates for the Navy, the Balearic Class (F-70) and the Santa María Class (F-80) both with American design and technology derived from the Knox and Oliver Hazard Perry respectively, adapted by Navantia to the requirements of the Spanish Navy.

It incorporates concepts of mission modularity, with the idea of working in a multi-mission space on its side housing the boat (RHIB rigid-hulled inflatable boat by its acronym in English) of 11 m, standardized container modules and unmanned vehicles (USV and UUV). having a multi-mission space to allow the accommodation of boats (RHIB "rigid-hulled inflatable boat"), (Variable Depth Sonar (VDS), standardized container modules and unmanned vehicles (USV and UUV).

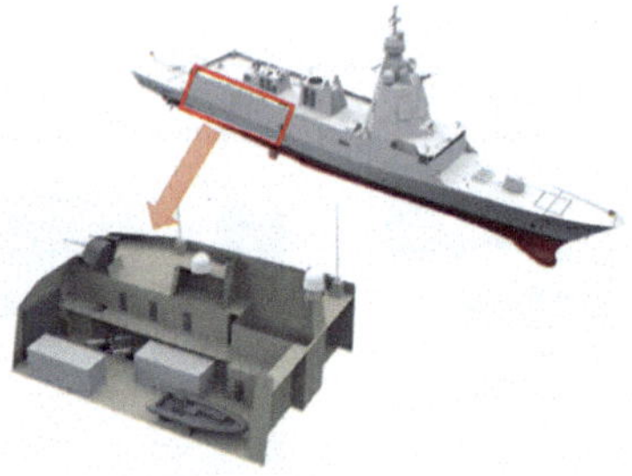

Fig. 7. Modularity in F 110 program Source: Riola Rodríguez José M.

4 Results

4.1 Considerations for the Implementation of Modularity

Initial it must be considered for the use of mission modules into a ship, its compatibility in such a way that allows an easy and fast integration into the ship as a system of systems, taking into account that the complexity of the module will require different types of integration, some modules will only require mechanical and electrical connections to the ship services, but some others will require connectivity with the management systems as is the case of sensors or weapons modules like the VDS (Fig. 8).

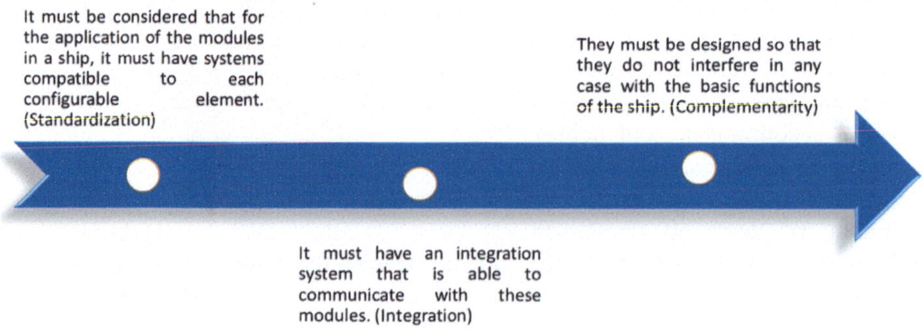

Fig. 8. Important elements to implement modularity.

It is important, to ensure the ease of integration of this modular capacities since the early stages of the warships acquisition programs, therefore there must exist a close relation between the requirements definition authorities and the design authorities in order to adequately establish the requirements associated to different services needed to guarantee the integration of these modules as an integral part of the sips operational capabilities. The most common of these services are:

- Electric power supply
- Data transfer (necessary bandwidth)
- Drinking water
- HVAC
- Cooling by transfer (cold water).
- Support for personnel.

Completing these requirements for the implementation of this concept, the modules must be designed in such a way that they do not interfere in any case with the basic functions of the vessel, according to the type of mission for which it was designed. Figure 7 shows how a vessel can be configured by integrating the required capacities for the mission accomplishment through mission modules.

By means of mission modularity, operational requirements can be met with ships that can share mission modules of defense systems related to sovereignty and national maritime interest's protection with the possibility to reduce acquisition an operational costs, and without reducing ship robustness. The possibility of designing highly complex vessels with mission modules that make them more versatile and that can, if necessary, contain the appropriate capabilities to meet strategic needs to maintain power projection capabilities as a deterrence force.

One of the most important references in recent naval programs of strategic defense warships is the US Navy DDG 51, an initiative to apply the concept of modularity and flexibility to Arleigh Burke class destroyers. This program is in turn subdivided into the SSES program, which explores the concept of multi-mission ships with the Aegis Combat System (Schank et al.2016) (Fig. 9).

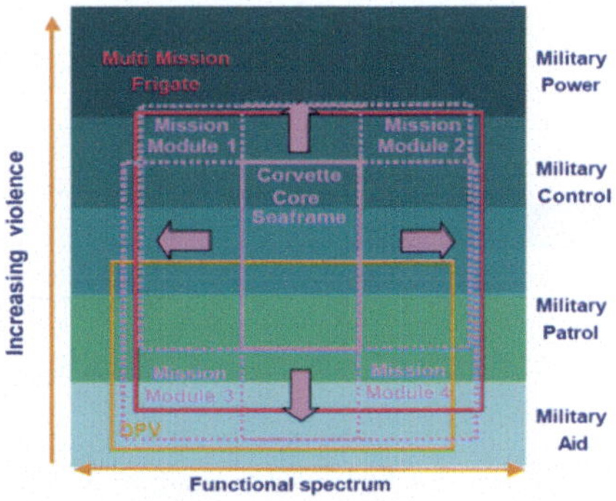

Fig. 9. Tasks of both a multi-mission frigate and an OPV executed by a corvette using mission modularity. Source: STSSD, 2004. NATO Naval Group 6. p. 81.

Designing modular and adaptive features in response to changing and unpredictable national security needs allows navies to avoid the two unattractive alternatives of short life or extremely expensive modernizations; however there must be a compromise between these design features, in which the needs are met with the most efficient use of the allocated budget (Doerry and Koeing 2017):

1. Greater effort of initial design.
2. Reduction of design freedom (possibly delaying technological progress).
3. Usually more displacement.
4. Usually more space required.

According to these premises, it can be established that the real drawback is to establish what type of modularity characteristics must be incorporated into the design of a warship, to allow it to remain operationally relevant during the service life and taking into account the premise of the operations and tasks that will perform according to the operational concept for which it was designed (Figs. 10 and 11).

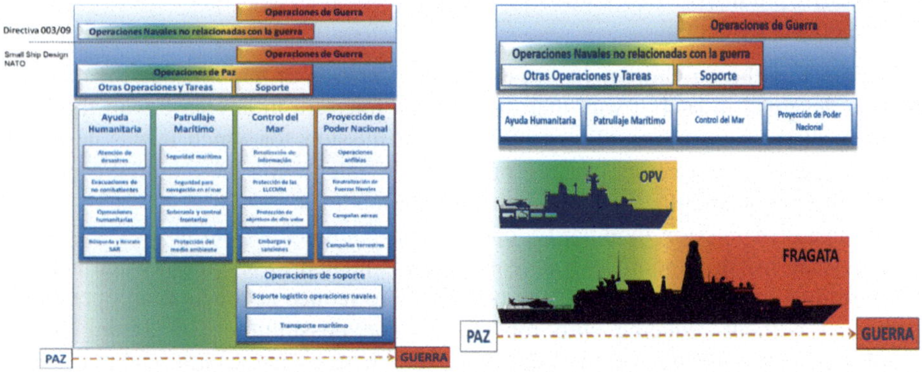

Fig. 10. Colombian Naval Operations and Ship projection requiments. Source: PES Program, from Directive 003/09 and STSSD, 2004. Nato Naval Group 6, (Piñeros Bello et al. 2018).

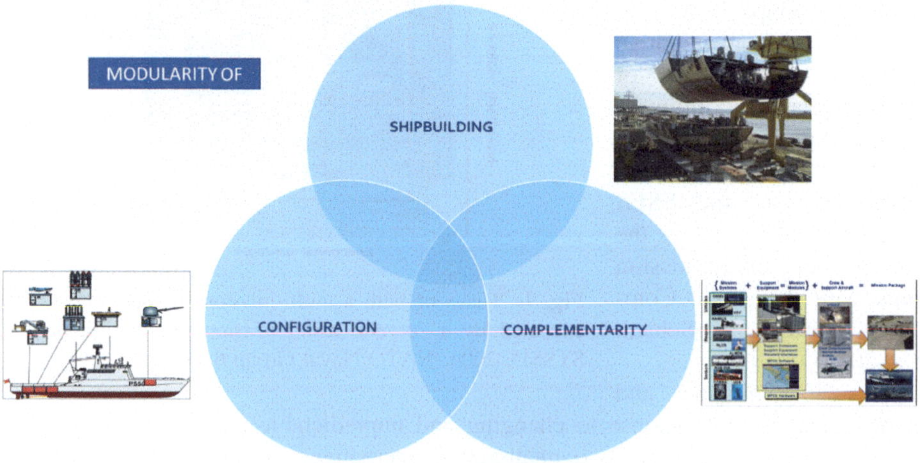

Fig. 11. Types of Modularity

5 Conclusions

It is crucial to establish the performance specifications, the design constraints and the appropriate growth margins considering a modular construction strategy.

By understanding, each of the proposed types of modularity you can understand the fundamental contribution they have in the design and naval architecture, streamlining processes and potentiating capabilities required by ships with limitations.

Therefore, it is possible to declare three great practical types of modularity with the purpose of centering the knowledge and the conceptualization developed by experts and applied in the different shipyards and navies of the world.

Modularity associated to the shipbuilding process optimizes time and resources in the construction, due to the possibility to build modules in parallel until a final integration of the structure of the ship; modularity associated to mission modules seeks to improve operational performance giving the Navies the possibility to adapt a ship according to the stated mission, making warship more versatile, optimizing the use of limited resources and reducing time needed to make the ship mission reconfiguration.

The modularity by configuration is that the hull, the machinery and a certain number of elements installed on board can have longer service time than others, such as electronic equipment and weapons, so if they are applied in a modular way, it is possible to reduce obsolescence at a relatively lower cost since they are easy to exchange or upgrade.

It is important to conclude the common parameters and attributes that make modularity possible to implement:

- The design should consider the flow of the construction modules or construction strategy and their tasks in parallel to optimize the construction time and reduce costs in this phase.
- Standardization of the components that will be used in both constructive modules and mission modules, since they must comply with the installation requirements for their integration, either at the time of modules changes of configuration and later in the maintenance during the life cycle of the ships.
- Again, the issue of growth margins that must be defined since the conceptual phase and future phases is addressed, in which the space available for the configuration must be considered, as well as the one that should be contemplated for future adaptations in three important areas: which would be integration, modernization or the replacement of new or existing systems, elements or components without affecting the characteristics of the vessels.

Mission modularity or complementarity can be considered a technological advance in naval warships that allows the improvement of operational capabilities especially for national defense; this trend must be considered with special care in the new warships acquisition programs as a way to enable navies to maintain or improve its deterrence capabilities for defense purposes.

The modularity of mission is in many cases an optimal solution for the complement of capabilities of a medium size Navy, but should never be consider as the complete solution to a poor planning or in a certain case insubstantial errors through the phases of acquisition of defense systems or likewise of errors of design and construction of ships.

References

Doerry, N., Koeing, P.: *Doerry Home Page*. Obtenido de (2017). http://www.doerry.org/norbert/papers/20171010%20Doerry-Koenig%20Modularity%20distro%20A.pdf

Schank, J.F., Savitz, S.: Designing Adaptable Ships, Modularity and Flexibility in Future Ships Designs. RAND Corporation, Santa Monica (2016)

Largiader, C.A.: Modular Platform Based Surface Ship Design. MIT Libraries Document Services, Cambridge (2001)

Majchrzak, L.B.: Real options valuation in the design of future surface combatants. NAVAL POSTGRADUATE SCHOOL. Dudley Knox Library, Monterrey (2017)

Piñeros Bello, L.A., Segovia Forero, C.E.: LA MODULARIDAD Y SU APLICACIÓN EN BUQUES DE GUERRA. DERROTERO, pp. 9–26 (2018)

Rodríguez, C.E.: LOS LCS Y SU IMPACTOEN LA GUERRA DEL LITORAL. REVISMAR **2**, 130–145 (2011)

Schank, J.F., Savitz, S., Munson, K., Perkinson, B., McGee, J., Sollinger, J.M.: Design Adaptable Ships. RAND Corporation, Santa Monica, California (2016)

Evaluation Methodology for the Selection of the Combined Propulsion System for an Offshore Patrol Vessel (OPV93C)

Eduardo Insignares(✉), Bharat Verma, and David Fuentes

Science and Technology Corporation for Naval, Maritime and Riverine Industry Development, Cotecmar, Cartagena, Colombia
{einsignares, bharat, dfuentes}@cotecmar.com

Abstract. In accordance with Colombian Navy Force Planning Document for 2030 and supported by the Pentagonal Strategy of the Colombian Navy, the need for a new OPV-type vessel design for the Colombian Navy has been identified. For this reason, it is very important to know the mission requirements of these ship types to guide its design. This has led the Colombian shipbuilding industry, especially COTECMAR to acquire capabilities for the design and construction of these vessels in the country. During the development of conceptual design, an aspect extensively analyzed is the propulsion system. In the present study CODAD and CODELOD combined propulsion systems are analyzed, identifying their technical advantages. The performance criteria to be evaluated are defined and a multi-criteria evaluation is carried out to recommend the most convenient configuration for the vessel OPV93C. Apart from the technical parameters, the life cycle cost analysis for each of the alternatives is also addressed, taking in account: acquisition cost, fuel consumption and maintenance costs of the propulsion system and electric generators. Combined propulsion systems represent a favorable technological solution that seeks to reduce these costs, since they are usually associated with fuel consumption and maintenance of the platform.

Keywords: CODAD · CODELOD · Multicriteria evaluation · Combined ship propulsion system · Life cycle cost

1 Introduction

Cotecmar[1], aiming to comply with new mission requirements, capacities and growth margins, initiated a new design of an offshore Patrol Vessel (OPV93C). One of the most important challenges in the design of the new OPV93C is to select the most suitable combined propulsion configuration as per the project requirements. The present study corresponds to the preliminary design phase for the vessel OPV93C, which was completed in 2017.

Several studies have been carried out for the selection of propulsion system for ships. For example, the studies conducted by MS Shamasundara, BS Arora and AS Par-wekar in Sep. 2014, where they express that during the decision-making process that essentially involves the qualitative and quantitative selection of multiple criteria, a

V. A. J. E. Carreño Moreno et al. (Eds.): CIDIN 2019/COPINAVAL 2019, pp. 211–222, 2020.
https://doi.org/10.1007/978-3-030-35963-8_18

compromise between tangible and intangible factors is essential to select the best among themselves [12].

Similarly, in the work presented by Pinnekamp, Hoppe and Heger, RENK AG, which describes the evaluation process for an optimized combined marine propulsion system and its validation by dynamic simulation [13].

The purpose of this document is to show the methodology used to generate a negotiation space (Cost-Effectiveness) that allows the decision maker to select the most appropriate propulsion system configuration for the vessel OPV93C.

Having said that, the development of the OPV93C project will help the Colombian Navy to reduce the knowledge gap in terms of design and engineering capabilities for highly complex vessels.

1.1 Problem Description

It is necessary for COTECMAR, as a Science and Technology Corporation and as the main supplier of design, construction, maintenance and repair services for the Colombian Navy, to acquire the capabilities for the analysis and selection of the combined propulsion systems to comply with the specific demands of the Colombian Navy. For this purpose, it is necessary to propose a methodology that allows the decision maker to define the most appropriate propulsion configuration for the OPV93C vessel.

1.2 Methodology Implemented

In the conceptual phase of the project, a study was developed with 7 propulsion configurations. After analyzing the results of these configurations and in common agreement with the Colombian Navy, it was decided to restrict the study only for CODAD (Combined Diesel And Diesel) and CODELOD (Combined Diesel Electric Or Diesel) configurations.

The propulsive analysis is carried out considering the principal characteristics of the ship - OPV93C, as shown in Table 1.

Table 1. Principal characteristics - OPV93

Characteristic	Value	Unit
Length overall	93.00	Meters
Molded breadth	14.00	Meters
Depth	7.00	Meters
Design draft	4.10	Meters
Displacement @ design draft	2660	Tons
Maximum velocity	20	Knots

The methodology used for the evaluation of propulsion system configurations is summarized in Fig. 1.

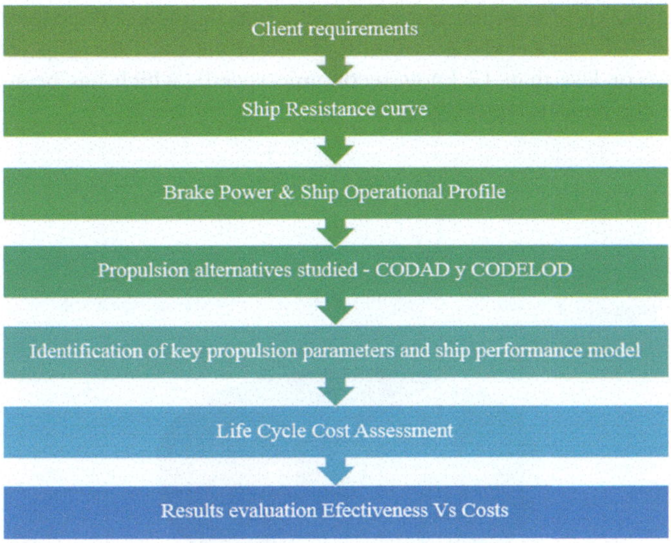

Fig. 1. Methodology implemented

The starting point for the analysis is the ship resistance curve, which was obtained from the canal testing results of the OPV93C ship model made in HSVA (Hamburgische Schiffbau-Versuchsanstalt GmbH) in december 2017. From this curve, the required brake power to be installed on the vessel is calculated. Figure 2 shows the Brake Power - Speed curve of the ship.

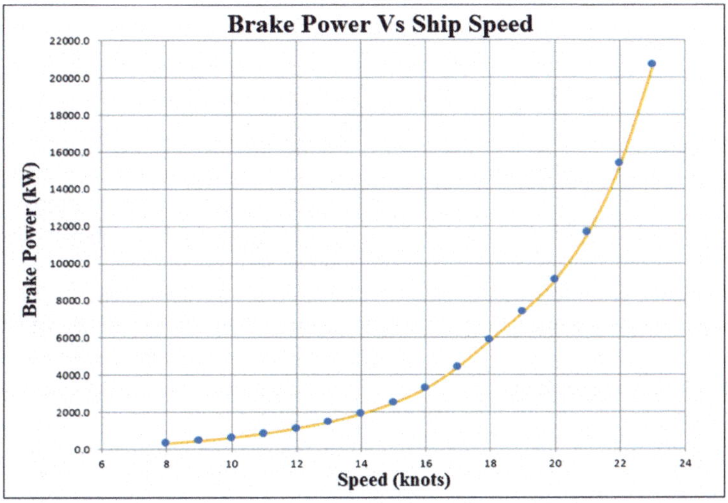

Fig. 2. Power – speed curve - OPV93C

The operational profile of the ship (OPV93C) was defined in the conceptual design phase, see Fig. 3. It can be observed that for 90% of the time, the ship will operate at speeds equal to or less than 12 knots (economic speed), which has been used as the optimal speed for the present analysis.

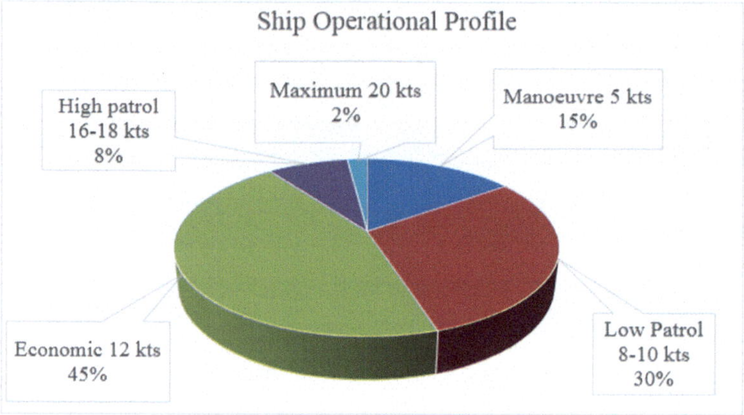

Fig. 3. Ship operational profile - OPV93C

As already being discussed that the two alternatives selected for the present study are: CODAD (Combined Diesel and Diesel) and CODELOD (Combined Diesel Electric or Diesel). A brief comparison between the two is given in Table 2 and in Fig. 4.

Table 2. CODAD vs CODELOD - OPV93C

CODAD	CODELOD
Diesel engines running – 100% time	90% of operating time with Electric Motor (as per the ship operational profile)
Low-load operation of diesel engines	Avoid the low-load operation of the Diesel Engine
Higher maintenance cost (100% main diesel engines and generators for hotel load in operation)	Reduced maintenance costs due to sharing of operating hours between Diesel Engines, Diesel Engine/PTI and generators
No need for additional generators or frequency drives	Additional generators and frequency drives are required
Typical vibration levels	Lower vibration levels in the drive train PTI (Power Take In) mode

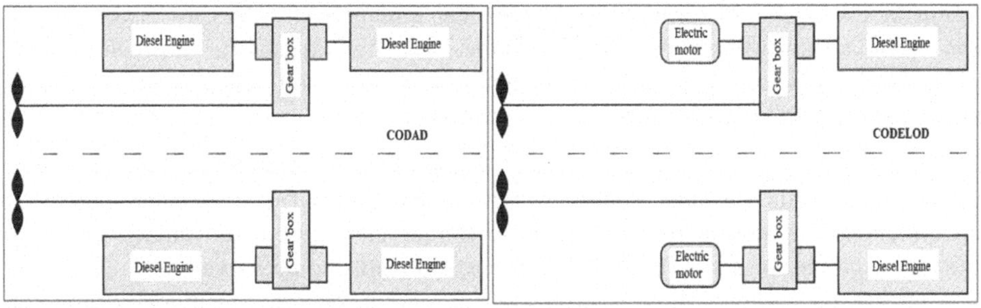

Fig. 4. General arrangement for CODAD and CODELOD system configuration

2 Analysis

For the analysis, sixteen propulsion systems configurations were evaluated considering six (06) suppliers of diesel/electric motors and two (02) suppliers of propeller shafts, gear-boxes and propellers, with configurations CODELOD and CODAD. To facilitate the identification of each propulsion system analyzed, each configuration is associated with a capital letter as shown in Table 3 below.

Table 3. Propulsion system configurations

Configuration	A	B	C	D	E	F	G	H	I	J	K	L	M	N	O	P
Type	CODELOD	CODAD	CODELOD	CODAD	CODELOD	CODELOD	CODAD	CODELOD	CODAD	CODELOD	CODAD	CODELOD	CODELOD	CODELOD	CODELOD	CODELOD

2.1 Identification of Propulsion Parameters and Ship Performance Model

To evaluate the technical characteristics of the propulsion configurations and quantitatively compare the different proposals, the propulsion parameters are identified, which are grouped into four criteria: Performance, Power, Footprint and Availability.

Performance Criteria: This criterion considers parameters like: Specific fuel oil consumption, Mean Piston Speed (MPS), Brake Mean Effective Pressure (BMEP) and Stroke/Diameter (S/D) ratio of the piston for every propulsion configuration.

Power Criteria: With this criterion, the available power in each configuration is measured to reach the cruising speed (economic speed) and the maximum speed, with the purpose to define the configuration with better utilization of installed power. It allows to understand the adaptability of each configuration to the ship's operational profile based on the capacity to use efficiently the available power to carry out its operations.

Footprint Criteria: This criterion measures the dimensional properties of the propulsion configurations offered and how they impact the ship's platform. This measurement is obtained from two parameters: weight/power ratio and area occupied by the propulsion system configuration.

Availability Criteria: This criterion seeks to identify the availability that the engine will have during its life time. Time between overhauls (TBO) is a statistical parameter recommended by engine manufacturers to define the number of running hours or calendar time before a ship engine or other component requires overhaul.

After having reviewed how each propulsion parameter could technically influence the vessel performance, the weights for each propulsion parameter are assigned by expert judgment according to their importance based on Analytic Hierarchy Process (AHP). Table 4 lists the measures of merit assigned to each parameter according to their importance.

Table 4. Propulsion parameters and weights

Criteria	Performance parameters of the propulsion system	Weights
Performance	Specific fuel oil consumption	20%
	Mean Piston Speed (MPS)	10%
	Brake Mean Effective Pressure (BMEP)	9%
	Piston Stroke/Diameter (S/D) ratio	10%
Power	Power utilization	10%
Footprint	Weight/power ratio	17%
	Engine space area	6%
	Engine volume	7%
Availability	Number of maintenance for engine	7%
	Number of maintenance for generator	4%
		100%

For the vessel, the Overall Measure of Effectiveness (OMOE) is constructed from Measures of Performance (MOP), which are specific to each of the missions that the OPV93C type vessel is expected to develop. Table 5 shows the 3 affinity groups, i.e. Mobility, Performance and Availability affected by the propulsion parameters, that will be analyzed and within which MOPs have been classified.

Table 5. Measure of performance – OPV93C

Affinity group	MOP
Mobility	Maneuverability
	Velocity
Permanence	Endurance
	Accommodations
	Range
Availability	Operational time

Table 6 shows the Measure of Performance (MOP) of the affinity group and how it's been affected by each of the propulsion system parameter within the ship's effectiveness model.

To quantify the influence of each performance parameter of the propulsion configuration on each of the Measure of Performance (MOP) of the vessel, it must be considered that the weight assigned to each of the affinity groups is dependent on the mission to be evaluated.

The data of each parameter obtained in each configuration are normalized in such a way that the most desirable value is unity.

Table 6. Ship performance measures affected by the propulsion parameters

	AFFINITY GROUP				
	MOPS				
	MOBILITY		PERMANENCE		AVAILABILITY
Key Parameters for Propulsion Performance	**Maneuverability**	**Velocity**	**Endurance**	**Range**	**Operational time**
Specific fuel oil consumption					
Mean Piston Speed (MPS)					
Brake Mean Effective Pressure (BMEP)					
Piston Stroke/Diameter (S/D) ratio					
Power utilization					
Weight / Power ratio					
Engine space area					
Engine volume					
Number of maintenance for engine					
Number of maintenance for generator					

Table 7 shows the results obtained from the measure of merit of the MOPs affected by the MOEs in the OPV93C project.

Table 7. Measures of merit of the MOP and the MOE

	AFFINITY GROUP					Measure of merit - MOE	Ship missions	
	MOBILITY		PERMANENCE		AVAILABILITY			
	Maneuverability	Velocity	Endurance	Range	Operational time			
MOP	0%	83%	56%	11%		7%	Peace Operations and Humanitarian Aid	MILITARY AID
	9%	10%	0%	6%		16%	Search and rescue	
	50%	10%	43%	43%		4%	Environmental control	MILITARY PATROL
	5%	4%	8%	7%		22%	Security and Control of Maritime Traffic	
	0%	9%	0%	5%		51%	Maritime Interdiction; Intelligence, Surveillance and Recognition	
Weighted sum	5%	13%	7%	8%	10%	100%		
Normalization	11%	31%	17%	18%	23%	100%		

The final measure of effectiveness (MOE) that represents the effect of each mission of the ship over each measure of performance (MOP) parameter of the ship is obtained by the method of WSM (weighted sum model). If there are M alternatives and N criteria, then the best alternative is the one that satisfies (in the case of maximization) the following expression [6];

$$A_{WSM} = \max \sum_{j=1}^{N} (a_{ij} w_j)$$

Where: A_{WSM} is the WSM score of the best alternative, N is the number of decision criteria, a_{ij} is the actual value of the ith alternative in terms of the jth criterion, and w_j is the weight of the jth criterion.

Table 8. Measures of merit of propulsion parameters affected by the MOP and the MOE

		AFFINITY GROUP MOP						
Key Parameters for Propulsion Performance	Measure of merit	Maneuverability	Velocity	Endurance	Range	Operational time	Final measure of merit	Normalized Weighted sum
Specific fuel oil consumption	20%			17%	18%		6,99%	25,11%
Mean Piston Speed (MPS)	10%					23%	2,34%	8,40%
Brake Mean Effective Pressure (BMEP)	9%	11%	31%				3,75%	13,48%
Piston Stroke/Diameter (S/D) ratio	10%			17%	18%		3,49%	12,55%
Power utilization	10%			17%	18%		3,49%	12,55%
Weight / Power ratio	17%			17%			2,94%	10,57%
Engine space area	6%			17%			1,04%	3,73%
Engine volume	7%			17%			1,21%	4,35%
Number of maintenance for engine	7%					23%	1,64%	5,88%
Number of maintenance for generator	4%					23%	0,94%	3,36%
	100%						28%	100%

The final measure of merit for each configuration is the result obtained from the contribution of the ship's performance values to the estimated merit measures for the propulsion performance parameters referenced in Table 4. With the final measure of merit shown in Table 8, we proceed to evaluate the propulsion parameters of the engines offered by each supplier. Figure 5 summarizes the results of effectiveness obtained for each configuration of propulsion systems evaluated.

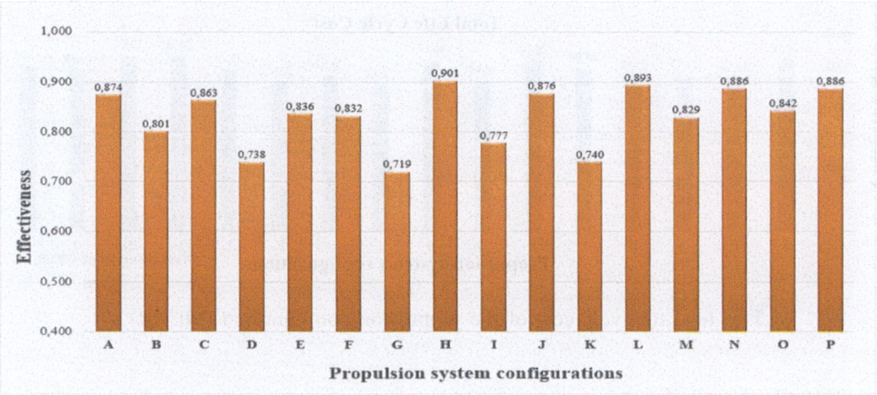

Fig. 5. Results - effectiveness of each configuration (OPV93C)

2.2 Life Cycle Cost Assessment

The life cycle cost analysis is intended to estimate the relative costs to acquire and operate each alternative. The calculation of the total cost is based on the following equation;

$$OMOC = MOC1 + MOC2$$

OMOC: It is the measure of the total cost.

MOC 1: Acquisition Cost, which includes technical services, technical assistance and the materials and equipment package.

MOC 2: Operation and Maintenance Cost (Fuel consumption cost + Maintenance cost).

The fuel consumption is calculated from the specific fuel consumption data according to the Engine load percentages corresponding to the speeds defined in the ship's operational profile presented in Fig. 2. In the same way, the maintenance cost of each arrangement is provided by the suppliers. This is calculated for 30 years of operation and includes the preventive maintenance of diesel and electric diesel engines. For the consumption of lubricant, 0.3% of the fuel consumption was established for the main diesel engines and generator sets.

By adding the acquisition cost, fuel consumption cost, lubricants and maintenance cost, we can determine which is the most economical option throughout the 30 years of ship operation. Figure 6 summarizes the total life cycle cost for each configuration.

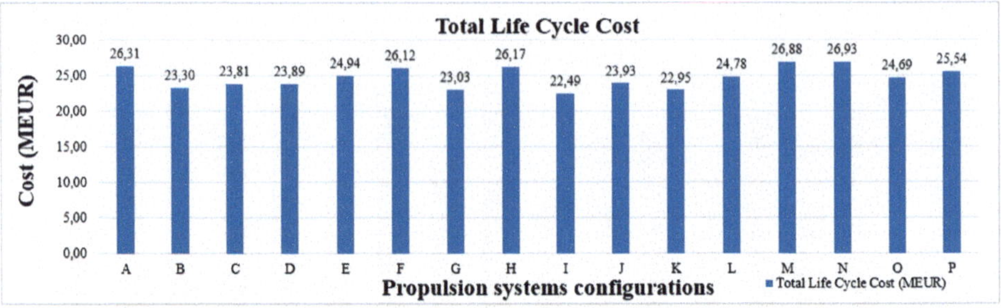

Fig. 6. Life cycle cost of the propulsion configuration (MEUR).

2.3 Results Analysis

To complete the comparative analysis between the proposed propulsion configurations, by means of dispersion charts, the performance measures obtained will be compared against the life cycle costs for each propulsion configuration, as shown in Fig. 7. The figure shows all the feasible propulsion configurations considered in the study and helps to differentiate the non-dominated alternatives from the dominated ones in the trade-off space.

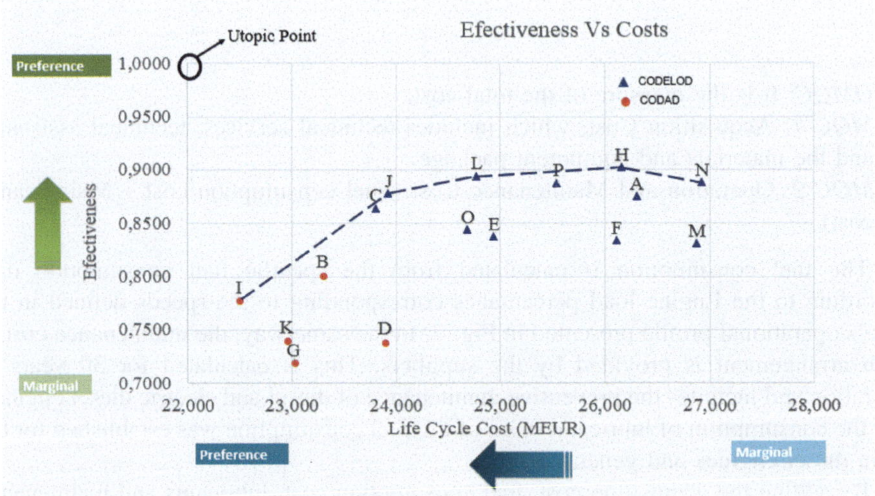

Fig. 7. Evaluation effectiveness vs costs of propulsión systems configurations

It is in this trade-off space in which the analysis of the different alternatives is made to provide necessary information to the decision maker. The use of the trade-off-space and its analysis using different methodologies, corresponds to multiple-criteria decision making problems (MCDM) and are diverse (Topsis, Vikor, utopian point distance, etc.). The methodology proposed with the application of multiple-criteria decision analysis (MCDA), for the evaluation process, is what generates added value to the process of selection of the propulsion system for the OPV93C.

3 Conclusions and Future Work

- The presented evaluation methodology for the analysis and selection of the propulsion system for the OPV93C can also be implemented to other types of vessels. However, the behavior of the model will depend on the type of vessel, its operational profile and the expectations of the shipowner related to the ship under investigation. All these factors will influence the overall measure of effectiveness, the cost measure, the trade-off space and hence the decision.
- In this study, only CODAD and CODELOD configurations were evaluated. If a different propulsion configuration evaluation is required, it is recommended to review the "performance parameters" for the respective configuration to be evaluated.
- The presented methodology facilitates the decision maker to have an overview of the propulsion configurations studied, as it can identify which alternatives have more technical advantages along with their life cycle cost and aid to compare each of the configurations in the trade-off space through multiple-criteria decision analysis.
- The multiple-criteria decision analysis is part of the future work that will be developed as part of the OPV93C project design. It includes the selection, use and application of various MCDA methodologies (Topsis, Vikor, utopian point distance, etc.), not only restricted to a specific vessel type as discussed in this paper but in a more generalized way, to any other vessel type that is being evaluated.

References

1. North Atlantic Treaty Organization: Cost Structure and Life Cycle Cost (LCC) for Military Systems. de RTO Meeting Proceedings 96, Paris, 24–25 October 2001 (2001)
2. Barringer, H.P.: Life Cycle Cost and Good Practices. de NPRA Maintenance Conference, San Antonio, Texas (1998)
3. Parámetros de selección de un motor diésel parte 1, octubre 2011. www.atmosferis.com
4. Z. P. F. - S. C. C.: Diseño Preliminar de una lancha patrullera rápida para ríos colombianos, Cartagena (2004)
5. Antoni, C.H.: Predimensionamiento de la cadena propulsiva y de un yate de 30 metros de eslora. Facultad Náutica de Barcelona/Universidad Politécnica de Catalunya, Catalunya (2011)
6. Dedes, E., Hudson, D., Turnock, S.: Design of hybrid diesel-electric energy storage systems to maximize overall ship propulsive efficiency. In: 11th International Symposium on Practical Design of Ships and Other Floating Structures, pp. 703–713. Universidad de Federal de Rio de Janeiro, Rio de Janeiro (2010)
7. Jeng, H.-G., Huang, J.-J.: Multiple Attribute Decision Making - Methods and Applications. CRC Press Taylor and Francis Group, Boca Raton (2011)

8. Precios de combustible en Colombia Año 2017 (2017). https://www.minminas.gov.co/precios-de-combustible
9. ciadelubricantes.com (2017). http://ciadelubricantes.com/shop/flotas/mobil-delvac-1-esp-5w-40/
10. MAN: Basic Principles of Ship Propulsion, MAN Diesel & Turbo. https://marine.man.eu/docs/librariesprovider6/propeller-aftship/basic-principles-of-propulsion.pdf?sfvrsn=0
11. Rolls-Royce: In-service maintenance for fixed speed hybrid electric propulsion for ships (2016)
12. Shamasundara, M.S., Arora, B.S., Parwekar, A.S.: Analytic hierarchy process approach for selection of ship propulsion system – case study, September 2014
13. Pinnekamp, B., Hoppe, F., Heger, M.: Combined marine propulsion systems: optimization and validation by simulation, Alexandria, Virginia, October 2012

A Practical Approach for Decision-Making on Preliminary Naval Ship Cost Estimating Using Multiple Cost Estimation Methods

Kevin E. Patrón(✉), Luis D. Leal, and Omar D. Vasquez

Design and Engineering Office, COTECMAR, 130013 Cartagena, Colombia
kpatron@cotecmar.com
http://www.cotecmar.com

Abstract. Reliable cost estimating is a key element to effectively manage complex naval ship projects. To conduct the estimates, selecting the appropriate method is of great importance. There may be more than one applicable method, depending on the level of the project definition and the availability of cost, technical and economic data. For instance, analogy-based and parametric estimating methods are commonly used during concept development phases. Thus, multiple cost estimating results could be obtained. It becomes necessary to apply an additional rational method to get a conclusive cost estimation that best represents the results of the involved methods. Since the results of each estimation can be presented as a set of probability distribution functions, we suggest for this a practical approach based on the derivation of an aggregated cost probability distribution function. This approach exploits an existing methodology which aims the maximization of the consensus among the assessed cost probability distribution functions. Herein that consensus is defined as the sum of the overlapping areas between each cost probability distribution function and the aggregated cost distribution function. To solve the corresponding optimization problem, a modern heuristic optimization algorithm is used to ensure global optimal solutions. We believe the proposed approach may enhance the decision-making process on naval ships project management at early stages.

Keywords: Preliminary cost estimation · Naval ship · Decision-making

1 Introduction

1.1 Ship Cost Estimating

In design and shipbuilding projects, the cost estimating process is intended to predict an estimate of the future cost of a naval ship. In this process, historical data, relevant to the type of vessel under analysis (and available at the time of estimation) are generally collected and analyzed [1]. For the analysis of such historical data, quantitative estimation methods are used and implemented either in spreadsheets or specialized software.

Cost estimating is a systematic process of paramount importance in the effective management of naval projects, and it is required to carry out a realistic planning of such projects. In addition, it allows making decisions about these projects in a convenient

V. A. J. E. Carreño Moreno et al. (Eds.): CIDIN 2019/COPINAVAL 2019, pp. 223–232, 2020.
https://doi.org/10.1007/978-3-030-35963-8_19

way from early phases of project acquisition, until the delivery of the final product. For instance, it allows optimizing the use of limited financial resources in government entities, favoring the appropriate allocation of budgets for ongoing projects in a given time period, and controlling the costs of these throughout their life cycle [2].

Before the development of the cost estimation, it is convenient to make the selection of the method or methods to be considered according to the stage of execution of the project and the available support information. Likewise, it is important to select the suitable computational tools that implement these methods. In general, during early phases of naval ship projects, analogy-based and parametric methods are used, which allow the obtention of results quickly. On the other hand, in more advanced stages of the project, engineering build-up methods (also known as "bottom-up") can be successfully used. The usage of these methods results into more precise estimations as compared to parametric methods, although the former are more time-consuming and demand much more effort [3]. Usually, over time, preliminary estimation methods are increasingly refined as long as more naval vessels are constructed and tested, resulting in new information available for estimation purposes [2]. The selection of cost estimating methods is influenced by various aspects, such as: the level of project definition, the purpose of the estimate, the availability of technical and economic support information, the quality of the support information available, and the project time restrictions [2].

1.2 Motivation

The application of more than one cost estimating method leads to the obtention of several estimates. If the applied methods are equally valid and produce reliable results, and each provides with relevant value judgments to the decision makers, then it becomes necessary to have an additional rational method that facilitates the obtention of a single conclusive result. This result should represent in a global sense the results of the methods involved.

As different estimation methods inherently present different judgments about the estimation, the aggregation of these different "opinions" will effectively be made up of a consensus [4]. Thus, it is of interest to attempt to compute a single estimate that is derived out of the consensus of various cost estimates. According to the authors' knowledge, to date no specific methodology has been reported in the literature to assist in making decisions regarding the cost of naval vessels in the early design phases, considering the results of various cost estimation methods. It is presumed that it is a common practice for cost estimators to lean towards a single estimation method at the beginning of naval projects. However, it is worth mentioning that some authors claim that the combination of results from various prediction techniques might result in an improvement of the estimate [5–7]. In fact, several methods have been proposed to combine expert probability distributions for risk analysis. This has facilitated the obtention of a distribution that represents a summary of the current state of experts' opinion about the uncertainty of a topic of interest [8]. The results of the cost estimation methods can be represented as a set of probability distribution functions. Therefore, in this article we suggest a practical approach to assist the decision makers on the cost estimate of naval vessels during early stages of the project, when using more than one cost estimating method.

This approach consists essentially of the derivation of a new cost probability distribution function that best represents the cost estimating results. This new function, which will be called the aggregated cost probability distribution function, arises as a result of the aggregation of the cost probability distribution functions (PDF) of the methods involved. The aggregation of these cost PDFs is carried out following the methodology established in [4]. In the following sections of this document we present: the methodology based on the maximization of consensus as proposed in [4], and an example of application of this methodology for estimating the cost of a naval ship.

2 Methodology

There are several methods that can be used to perform naval ship cost estimating. The level of confidence of these methods is determined not only by the project design phase, but also by the nature of the model itself, and by the quantity and quality of support information for the estimate (for instance, historical records of ships similar to the ship whose cost will be estimated).

In general, the results of the application of a cost estimating method can be reduced to a probability distribution function (PDF). This PDF reflects the level of uncertainty of the cost associated with the estimation method used. As an example, a triangular probability distribution can model the results of a cost estimating method that outputs the most likely, the maximum (pessimistic) and minimum (optimistic) values. The problem of obtaining a single conclusive estimate is reduced to the aggregation of a set of cost probability distribution functions to a consensus distribution, or aggregated distribution [4].

Let $f(x)$ be the cost aggregated probability distribution function. The simplest way to model this function is by means of a linear combination of the cost probability distributions of the different cost estimation methods used. Then,

$$f(x) = \sum_{i=1}^{n} f_i(x) w_i \tag{1}$$

In Eq. (1), $f_i(x)$ is the function of the cost probability distribution function corresponding to the i-th estimation method; w_i is the priority level of the i-th method (also known as the weight or relative importance), and n is the number of methods used. The following conditions must be met for the weights:

$$\sum_{i=1}^{n} w_i = 1 \tag{2}$$

$$0 \leq w_i \leq 1 \tag{3}$$

There are several techniques to determine the weights corresponding to each of the functions f_i, mainly in the field of decision making under the judgment of multiple decision makers. In the context of cost estimation problems, these techniques attempt to address two main issues: the criteria to be applied to evaluate the results of the estimation methods considered, and the assignment of priority levels to each of these methods. For instance, there is a technique that assigns weights to the estimation methods, based on the rating (ranking) that the method has in relation to the event in which the estimation is intended. Thus, if there are m estimation methods involved, and the i-th method has a ranking r_i, its corresponding weight can be computed as $w_i = \frac{r_i}{\sum_{i=1}^{m} r_i}$. The Delphi method has also been used for this purpose, as well as the AHP method through pairwise comparisons [4].

In this document, we propose the application of an existing methodology presented and described in [4] for the aggregation of probability distribution functions, in the context of group decision making. In the field of preliminary naval ship cost estimation, the methodology aims to maximize the consensus among the used cost estimation methods, or minimize the difference between their judgments, based on the following premises.

2.1 First Premise

The suitability of any cost estimation method is represented by the area of intersection between the its probability distribution function and the cost aggregated probability distribution function.

In the area of decision making, this area represents the expertise of a decision maker in relation to the decision problem [4]. In such a case, the greater this area of intersection is, the more expertise the decision maker has. Before obtaining the aggregated cost distribution function, it is assumed that the cost probability distribution functions corresponding to each of the estimation method (used during the early phases of the project) are already available.

2.2 Second Premise

The numerical weight (or relative importance) assigned to any method is given in proportion to the corresponding area of intersection or overlap between its cost probability distribution function (PDF) and the aggregated cost probability distribution function (see Fig. 1).

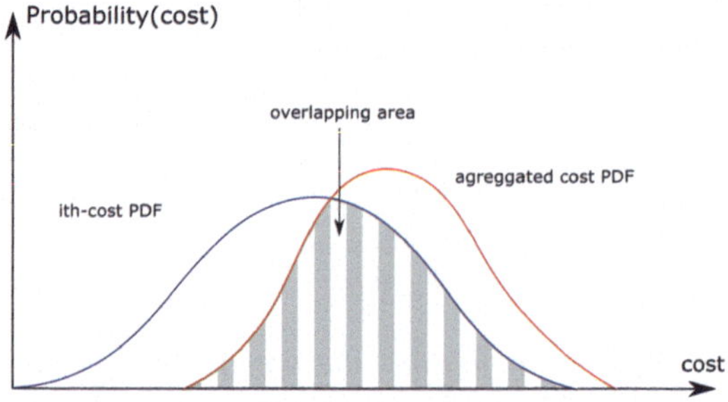

Fig. 1. Area of intersection between the i-th cost probability distribution function (PDF) and the aggregate cost probability distribution function [4].

Therefore, the weight of the i-th estimation model w_i is

$$w_i = \frac{A_i}{\sum_{i=1}^{m} A_i} \tag{4}$$

In Eq. (4), A_i is the area of intersection between the distribution of the i-th model and the aggregated distribution. This area A_i can be determined by numerical integration.

2.3 Third Premise

The aggregated cost PDF is a function to be determined, since its weights are unknown. *The problem of determining this PDF then consists in finding the weights that maximize the consensus between the cost estimates results. Therefore, the problem is an optimization problem, where the goal is the maximization of the sum of the overlapping areas between the PDFs of the cost estimation methods and the cost aggregated PDF, as follows*

$$Maximize\, A_{total} = \sum_{i=1}^{m} A_i \tag{5}$$

$$Subject\ to:$$

$$\sum_{i=1}^{m} w_i = 1 \tag{6}$$

$$0 \le w_i \le 1 \tag{7}$$

In this problem, the design variables are the coefficients w_i. Various values of w_i lead to different values of A_{total}. To solve this problem, classical (gradient-based) optimization algorithms can be used, as well as heuristic algorithms (such as, for example, simulated annealing, or the particle swarm method). In this work, the particle swarm optimization (PSO) is used to solve the optimization problem previously described, with a linearly decreasing inertial term, and a number of particles as recommended in the literature [10].

3 Illustration of the Proposed Approach

In order to illustrate the application of the proposed approach, this section presents an example of cost estimating of a naval vessel during the early stages of the project. In this stage, four different parametric methods (or models) were used: Carreyette's cost estimation model [10], Lamb's model [11], the CGT model [12, 13], and a multiple regression model based on information collected from databases. Their results are represented by three-point estimates, i.e., triangular probability distribution functions.

Table 1 shows the minimum, maximum and or most likely cost values by using each of these methods. For Carreyette's, Lamb's and CGT models, the dispersion of the cost distribution function is defined according to the expected level of uncertainty stated by the authors of these methods.

Table 1. Estimated costs for a naval vessel by using several cost estimating models

Cost estimating model	Minimum cost (USD)	Most likely cost (USD)	Maximum cost (USD)
Carreyette	62,048,269	68,942,521	75,836,773
Lamb	62,304,721	73,299,672	84,294,622
CGT	68,546,855	78,957,125	89,293,750
Multiple regression	72,844,500	90,443,000	108,041,000

For the multiple regression model, the dispersion of the cost distribution corresponds to the 95% confidence interval for the average cost. The classical interpretation of this interval is that, in repeated samplings, this interval will contain the true mean of the population from which the samples were taken, 95.0% of the time. In this distribution, the wide range of possible cost values is a consequence of the high standard deviation of the costs of the database used for the regression.

The objective is to determine the cost aggregated probability distribution function that maximizes the consensus among the cost probability distributions of the cost estimation models considered.

4 Results

By using Eq. (1) with the above-mentioned cost probability distributions, and considering a set arbitrary weights according to Table 2, an aggregated cost distribution function is obtained, as shown in Fig. 2. Since such weights have not been obtained with the proposed approach, it is expected that the resultant cost aggregated probability distribution function will not maximize the consensus of the cost distributions of the considered cost estimation models.

Table 2. Arbitrary weights for the cost estimation models

Cost model	Weight (w)
Carreyette	0.1
Lamb	0.1
CGT	0.4
Regresión múltiple	0.4

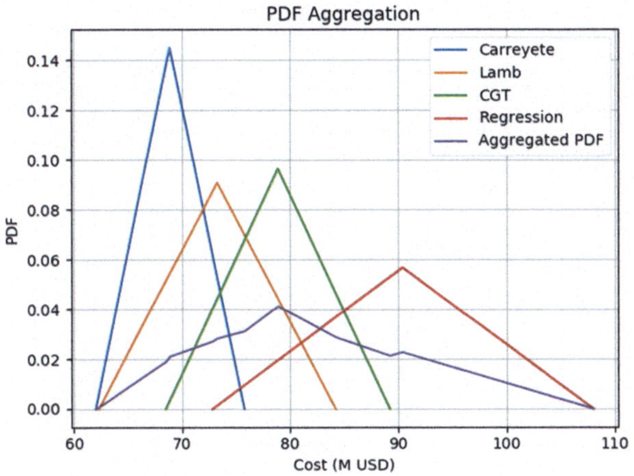

Fig. 2. Cost aggregated PDF (probability distribution function) with non-optimal arbitrary weights, and the cost estimation models PDFs.

Using the Eq. (5), the total area is $A_{total} = 0.89$, which is not the optimal (maximum) total area. The shape of the cost aggregated distribution function corresponds to the numerical values of the weights for each method, and varies according to Eq. (5). Using the method of consensus maximization, after 57 iterations, the weighting values are obtained for the cost estimating models PDFs, according to Table 3.

Table 3. Weight values for the cost estimation models, obtained as a result of the maximization of the consensus among the cost estimating models PDFs

Cost estimating model	Weight (w)
Carreyette	0.215
Lamb	0.425
CGT	0.358
Regresión múltiple	0.000

It is observed that these weights maximize the total area of intersection, making it equal to $A_{total} = 1.0$. Figure 3 shows the tendency of the cost aggregated distributed function to have its peak in the area where most distributions converge, covering overlapping areas with all the cost probability distribution functions.

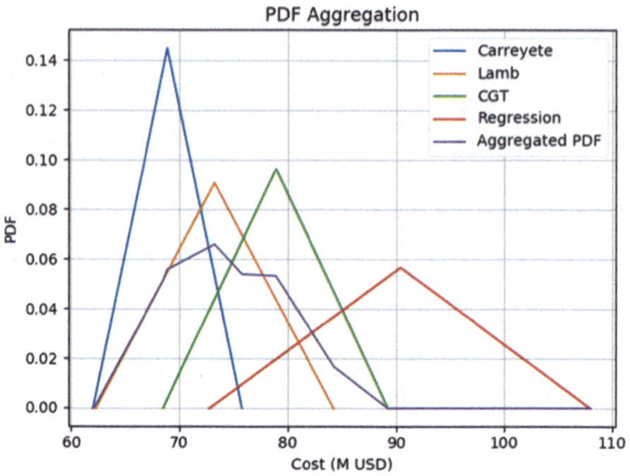

Fig. 3. Cost aggregated probability distribution function (PDF), as a result of the maximization of the consensus among the results of the cost estimating models

As seen in Table 3, the optimization model obtains a practically zero weight for the regression model, since its cost probability distribution function is the most distant from the rest of the distributions. For any cost probability distribution function $p(x)$, the probability that the cost is in the range $[a, b]$ can be computed by the following equation:

$$p(a \leq x \leq b) = \int_a^b p(x)dx \tag{8}$$

In addition, the probability that the cost is less than a given value x can be computed by using the cost cumulative distribution function $P(x)$:

$$P(x) = \int_{-\infty}^{x} p(t)dt \tag{9}$$

The above definitions from Eqs. (8) and (9) can be used for a cost analysis on the obtained cost aggregated distribution function. The results of this cost analysis can be used to assist the project manager and stakeholders for decision-making around the project definition.

For instance, for the illustrated example presented in the previous section, the corresponding cost cumulative distribution function is as shown in Fig. 4.

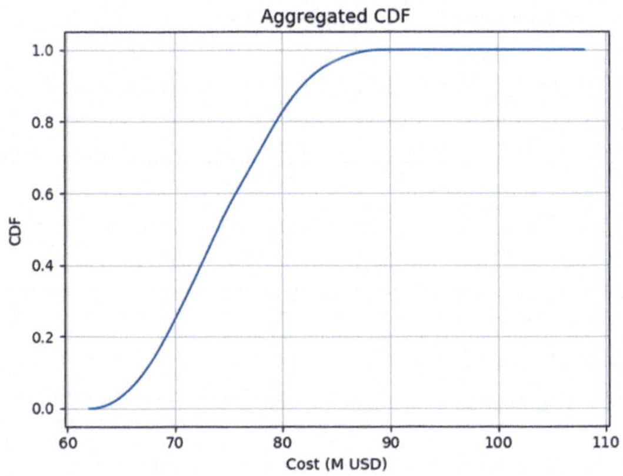

Fig. 4. Cost cumulative distribution function for the aggregated cost distribution function corresponding to the illustrated example

From Fig. 4, it is possible to assess the likelihood of the cost to be less than a given value for our example. For instance, it is observed that the probability for the cost to be less than or equal to 78.56 MUSD is equal to 0.75, and the probability for the cost to be less than or equal to 79.52 MUSD is equal to 0.8.

5 Concluding Remarks

In the present work, we have suggested the application of an existing rational methodology for the aggregation of probability distribution functions [4], in order to estimate the cost of a naval ship during early stages of the project. This approach becomes useful when multiple cost estimation methods are used simultaneously.

Although this methodology was originally formulated for decision making problems under multiple decision makers [4], in this document it was possible to demonstrate that it can also be applicable for the cost estimating process of naval vessels. The usage of this approach demands the various cost estimating methods to have results presented in a probabilistic-fashion. The foregoing has been illustrated in this document through an example of cost estimation of a naval vessel, making use of several parametric cost estimating methods.

It is evident that the presented approach can be used to assist decision makers and project stakeholders on decision-making around the definition of naval projects.

References

1. Mislick, G., Nussbaum, D.: Cost Estimation: Methods and Tools. Wiley Series in Operations Research and Management Science. Wiley, Hoboken (2015)
2. USA DoD, NAVSEA Cost Estimating Handbook (2005)
3. NASA, NASA Cost Estimating Handbook. Version 4.0 (2015)

4. Liu, X., et al.: A novel approach to probability distribution aggregation. J. Inf. Sci. **188**, 269–275 (2012)
5. Makridakis, S., Winkler, R.: Averages of forecasts: some empirical results. Manag. Sci. **29** (9), 987–996 (1983)
6. Clemen, R.: Combining forecasts: a review and annotated bibliography. Int. J. Forecast. **5**, 559–583 (1990)
7. Winkler, R., Makridakis, S.: The combination of forecasts. J. R. Stat. Soc. **146**, 150–157 (1983)
8. Clemen, R., Winkler, R.: Combining probability distributions from experts in risk analysis. Risk Anal. **19**(2), 187–203 (1999)
9. Rao, S.: Engineering Optimization: Theory and Practice. Wiley, Hoboken (2009)
10. Carreyette, J.: Preliminary ship cost estimation. The Royal Institution of Naval Architects (1977)
11. Lamb, T.: The development of gross tonnage compensation coefficients for naval shipbuilding based on direct productivity calculations (1994)
12. Craggs, J., Bloor, D., Tanner, B., Bullen, H.: Methodology used to calculate naval CGT factors. Britannic Majesty's Stationery Office (2003)
13. Firs Marine International: Findings for the global shipbuilding industrial base benchmarking study, part 1: major shipyards (2005)

The Viability of Implementing a Container Terminal at Port of Açu – A Preliminary Analysis

Luiz Felipe Bretas Rozo and Luiz Felipe Assis(✉)

Federal University of Rio de Janeiro, UFRJ, Rio de Janeiro, Brazil
luizfelipe@poli.ufrj.br

Abstract. The Port of Açu is a new private port located in the Southeast region of Brazil, which accounts for approximately 75% of the country's GDP and exports. The port has many bulk facilities ant an available area for setting up a new container terminal. The study made a preliminary analysis of the viability of implementing a Container Terminal at Port of Açu. Based on available demand studies, cargo statistics and data from the Port of Açu itself, scenarios were developed for the evolution of cargo flow in the East Coast of South America in order to establish capacity of a container terminal in the Port of Açu. Through a study of demand, the investments and operational costs required to implement the terminal were estimated. Considering demand data and cost estimates of equipment and infrastructure, a discounted cash flow analysis was performed in order to establish the cost/box for different scenarios.

Keywords: Container terminal · Port governance · East Coast of South America Container Trades

1 Introduction

Over the last decades, the intensification of international trade has been accompanied by profound changes in the organizational and technological structure of maritime transport, especially involving general cargo. These changes, coming from the containerization process, affected ports, retroportuary terminals and logistics systems in general, and allowed large cost reductions through economies of scale (Cullinane and Khanna 1999).

In order to withstand these changes, the Brazilian port sector underwent a process of privatization of services, which began in the last decade of the last century (Law 8630/1993 and Law 12815/2013). This process involved changes in the search for efficiency gains, with the privatization of port operations and changes in labor legislation. These changes contributed to a significant reduction in port costs, increasing the productivity of terminals and inter-port and intra-port competition (Assis and Canen 2001).

Under these perspectives, the Port of Açu, positioned with a Private Landlord Port management model (Brooks 2004), and therefore motivated by the identification and exploration of opportunities in the logistics chain, shows itself as a potential player to operate as a port for containerized cargo. The hinterland of the Port of Açu covers southeastern Brazil, responsible for 75% of the GDP and about 75% of exports.

V. A. J. E. Carreño Moreno et al. (Eds.): CIDIN 2019/COPINAVAL 2019, pp. 233–245, 2020.
https://doi.org/10.1007/978-3-030-35963-8_20

2 Methodology

The objective of this paper is to evaluate the implementation of a container terminal in Port of Açu, focusing on capturing cargo of potential competitors in the region where the port is inserted. It is understood that the Private Landlord Port management structure has some cost advantages, mainly in manpower management, and in investment decisions when compared with potential competitors. It is important to note that the ports located in the region of influence of Port of Açu operate basically with a Landlord governance system, with the port infrastructure under the responsibility of the government and a labour management still divided between port operators and public authorities.

In order to carry out the analysis, the Container port throughput, and its potential growth in the Brazilian market was surveyed, with differentiation of the participation of the main ports located in the Eastern cluster, where Port of Açu is located.

The evaluation of the potential of a container terminal could be considered under two aspects: cargo capture from the hinterland of the port or insertion as a typical transshipment port.

Although the operation of Port of Açu as a container terminal for feeder services is not ruled out, the present analysis focused on the potential for capturing regional cargoes.

Based on the potential demand for cargo handling at Port of Açu, a sizing of the facilities required from international benchmarks and from the local market was carried out. The costs of setting up and operating the terminal were surveyed. The costs of port services for tugboats and pilotage were also considered, since they could affect the competitiveness of the port.

In order to verify the terminal's viability, a discounted cash flow model was developed based on the revenues obtained by the tariffs and handling curve, its CAPEX, OPEX and SG&A (Sales, General & Administrative) costs, as well as taxes (PIS, COFINS and ISS), depreciation and income tax. A sensitivity analysis of the main variables involved was performed as well as a comparison of the results with the values practiced by potential competitors.

3 Container Throughput in the Brazilian Market and the Eastern Cluster

Container shipping in Brazil is carried out mainly by container liner services in deep-sea operations and cabotage.

In the case of deep-sea services there are many domestic and foreign ports serviced by feeder services.

In Brazil, cabotage are held for Brazilian flag vessels (Law 9.432/1997) and involve: cargo transportation between Brazilian ports; cargo transportation between Brazilian, Argentine and Uruguayan ports (Mercosur cargoes); as well as cargo transshipment for connection with the long haul.

Considering also this national scenario, the containerized cargo transport market in Brazil can be divided into 5 main clusters: North, Northeast, East, Santos and South, as indicated in Fig. 1. The *container throughput evolution* in each of the clusters is presented in Table 1.

As it can be seen in Fig. 1, Port of Açu may have the potential to attract cargoes from regions covered by the Brazilian states of Espírito Santo, Rio de Janeiro and Minas Gerais, with emphasis on the latter due to its strategic proximity to Port of Açu and its economic relevance.

On the other hand, the Eastern Cluster accounts for only 8% of container throughput in Brazil. That is, a relatively small volume in national terms. Santos, the main Brazilian port, is responsible for about 40% of the market.

In the Eastern Cluster, the terminals of ports of Vitória, Itaguaí and Rio de Janeiro were identified as potential competitors of Port of Açu. In addition, Port of Santos, adjacent to the Eastern Cluster, was also considered in the analysis since, as mentioned, it represents nearly half of the container throughput in the country and accounts for many transshipment cargoes that may eventually migrate to a more efficient port.

Figure 2 presents a comparison between Port of Açu and its main competitors. A first advantage to be highlighted concerning Port of Açu is the fact that it is a private port. The container terminals located in competing ports are owned by the government, responsible for the port infrastructure and have areas leased for port operators, that are in charge of the terminal superstructure and operation. That is, they follow a Landlord Port model. The exception to this model is the DP World container terminal which, in spite of being located in the geographical area of the Port of Santos, functions as a Private Port.

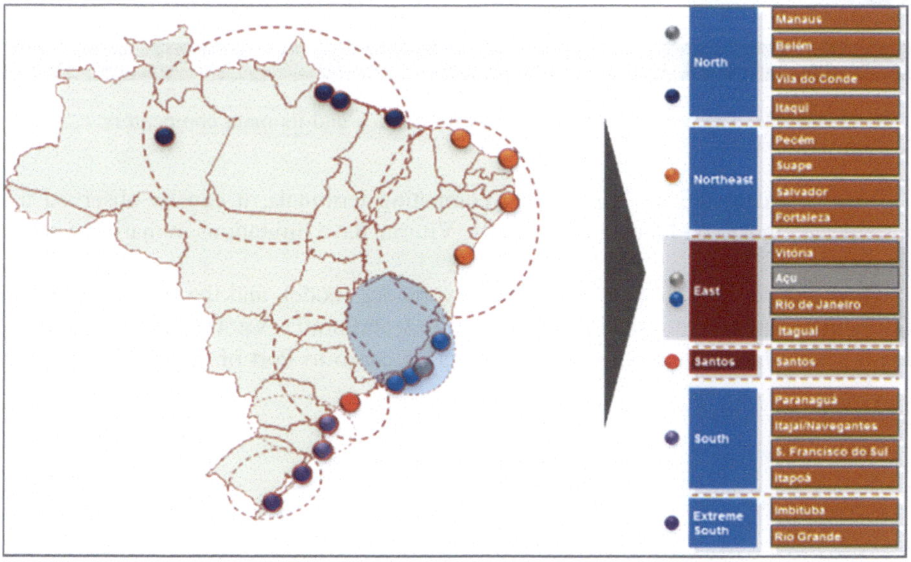

Fig. 1. Brazilian container ports clusters and its hinterlands.

Table 1. Brazilian container ports clusters throughput in teu

Main port clusters	2015	2016	2017
North Cluster	677,645	655,449	658,103
Northeast Cluster	998,148	989,372	1,111,564
East Cluster	742,611	676,301	777,259
Santos Cluster	3,762,565	3,502,895	3,761,633
South Cluster	3,171,107	3,130,949	3,155,209
Brazil/Total	9,352,076	8,954,966	9,463,768

Source: Antaq (2018)

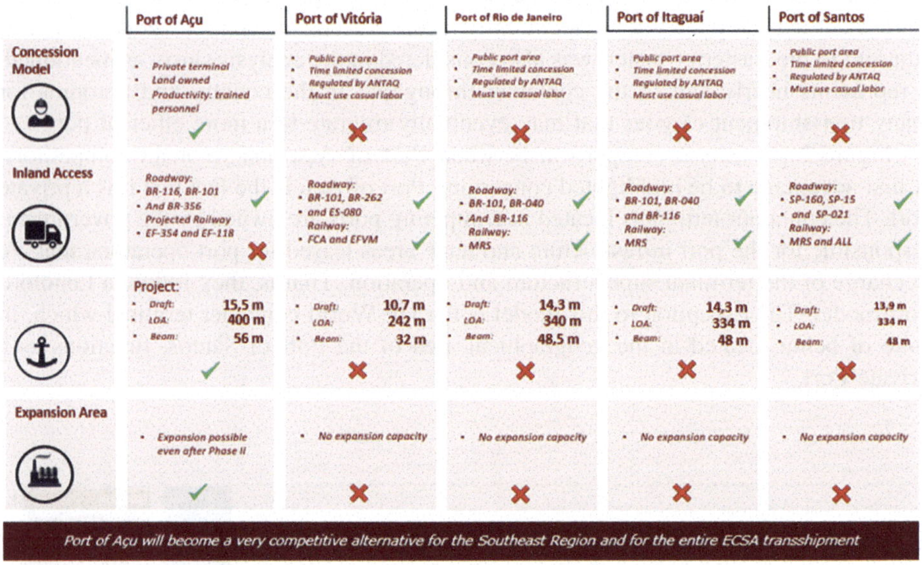

Fig. 2. The comparison between Porto do Açu and its main competitors.

With regard to the facilities of the competing terminals, it can be observed that those located in Rio de Janeiro and, mainly, Vitória, have limitations in maritime access and area.

Thus, considering the proximity, the governance model, and the limitations of the port facilities, mainly physical restrictions (LOA and draft), Vitória can be considered the port with the greatest potential for attracting cargo to Port of Açu.

4 Port of Açu

Port of Açu is a private port venture located in the northern state of Rio de Janeiro, in the city of São João da Barra, approximately 150 km from Campos Basin, where 85% of the Brazilian oil is produced.

The Industrial complex of Port of Açu under construction and development since October 2007 relies, as seen in Fig. 3, on two port terminals: T1, offshore terminal for the handling of iron ore and oil; and T2, an onshore terminal, with potential for 14 km of docks, and which is currently 6.5 km long, 300 m wide, with up to 14.5 m depth, and 90 km^2 backspace. T2 is able to receive industrial renters and handle various types of cargo, including containers.

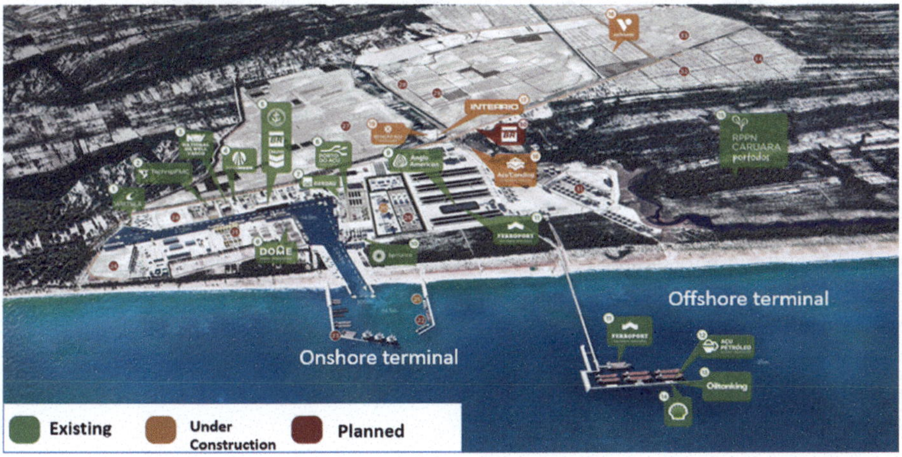

Fig. 3. Port of Açu layout. Source: www.portodoacu.com.br

Port of Açu has already an approved area in the onshore terminal directed to container handling, with quay of 350 m in length and depth of 14.5 m. However, the terminal can be expanded to up to 1,300 m of dock length with a total area of 390,000 m^2 (Fig. 4). It is noteworthy that the terminal has contracts of Adhesion and Term of Liberation of Operations valid in the regulatory agency of the ports and waterways transport - ANTAQ, in addition to an Environmental Operation License with the environmental regulator INEA. Therefore, it is possible to deploy a container terminal to compete with other regional operators.

Fig. 4. Port of Açu - installation site of the container terminal. Fonte: Google Earth

5 Feasibility Analysis

For the feasibility analysis, the cargo demand for the region adjacent to Port of Açu and its hinterland was estimated. Next, a sizing of the facilities of Port of Açu was carried out considering the estimated potential demand and international and local market benchmarks. Finally, the terminal's competitiveness was analyzed in detail, considering its implementation costs and estimated cargo handling fees, as well as the cost of maritime access to the terminal.

5.1 Estimation of the Cargo Handling Curve

For the calculation of terminal feasibility a container handling curve was estimated in teu over time. The estimation considered the container throughput of the ports of the Eastern Cluster and also of Santos (except transshipment cargo) and compared its variation with the Brazilian GDP. Based on projections of the Brazilian GDP from Brazilian Central Bank up to 2022 (for the following years an annual growth rate of 2.5% was used), an aggregate cargo handling curve was established. The market share of Port of Açu was estimated based on market studies, focusing on the cargoes generated and attracted by the region, including those that are transshipped in Port of Santos. As can be seen in Fig. 5, they are cargoes generated by Port of Açu complex itself, coffee, marbles & granites, chemical cargoes, among others. Since the analysis only consider the potential for regional cargoes, the transhipment cargoes have not been included in the forecasts.

The growth projections of the containerized cargo for the Region and for Port of Santos and the potential demand for Port of Açu are presented in Fig. 6.

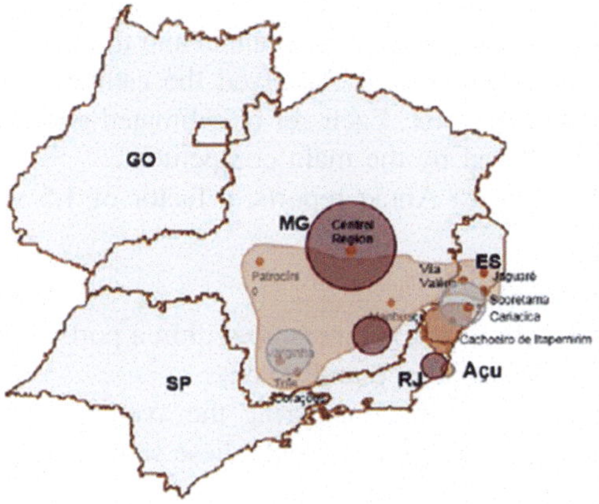

COFFEE Main Clients	Natonal Volume (TEU)	Current Port
Cooxupé, Terra Forte, Outspan, Nestlo, Stockler, Louis Dreyfus, Atlantica	100.000	Santos & Vitória (feeder to Santos)
GRANITE Main Clients	**National Volume (TEU)**	**Current Port**
Pemagran, Marbrasa, Thor, Gramazini	65.000	Vitória (feeder to Santos)
INDUSTRIES Main Clients	**Volume (TEU)**	**Current Port**
Campos Indutries: Dorel, Purac	3.600	Vitória, Rio & Itaguaí
MG's Industries: Iveco, Dow, Unif, Plaster MG, Mercedes Benz, Fiat Chrysler, Votorantim, GE, Philips	44.000	Vitória, Rio & Itaguaí
OTHERS Main Clients	**Volume (TEU)**	**Current Port**
EADIs	26.600	Vitória
Oil and Gas Industry	20.000	Rio

Fig. 5. Port of Açu – potential cargoes and cargo generation centres.

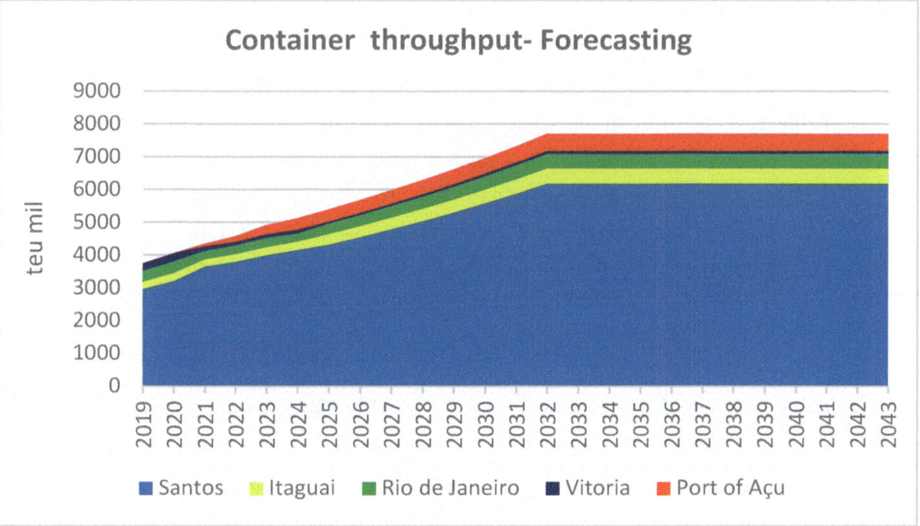

Fig. 6. Container throughput forecasting.

5.2 Port Cost Estimators Per Box

Estimates of the costs per box of the Port of Açu were divided into two parts: (i) the first one covered the cost of one access to the maritime access channel and the berthing of one container vessel per box, and the second one (ii) involved the estimation of handling cost in the terminal also calculated per box. Each set of estimated costs for Port of Açu was compared with the costs offered by the main competitors.

For estimating the costs by box, according to Antaq reports, a factor of 1.5 was applied to the values in teu.

Cost of Maritime Access Channel and Berthing

The costs of access to the channel and berthing of a containership within a port can be separated into 4 large groups: access, tugboats, pilotage and quay.

Table 2 presents indicators of these cost groups, considering the average of the operations, for the main competitors of the port. Table 3 presents these same costs for different consignment patterns. It is observed that these are tariff values provided by the terminal ports themselves, including Port of Açu.

Based on the analysis performed by the study, considering only this cost center, Port of Açu would be competitive in relation to the ports of Vitória, Rio de Janeiro and Itaguaí, in case it had a handling of more than 750 boxes per ship.

On the other hand, in order to be competitive with the Port of Santos, the average consignment would have to approach 1,200 boxes per ship. In order to achieve these values, it would certainly be essential to have an expressive volume of transshipment cargoes in Port of Açu.

Table 2. Cost of maritime access channel and berthing – values in R$/box

Port	Access	Pilotage	Tugboats	Quays	Total cost
Itaguai	88.52	43.25	96.30	6.61	**234.68**
Rio de Janeiro	88.52	32.51	96.30	4.07	**221.40**
Santos	71.56	17.23	67.64	0.02	**156.45**
Vitória	67.89	28.24	91.42	0.00	**187.55**

Table 3. Cost of maritime access channel and berthing – values in R$/box

Boxes/ship	Access	Pilotage	Tugboats	Quays	Total cost
500	51.19	47.37	187.25	0.02	**285.83**
600	51.19	39.48	156.04	0.02	**246.73**
750	51.19	31.58	124.83	0.02	**207.62**
800	51.19	29.61	117.03	0.02	**197.85**
1000	51.19	23.69	93.63	0.02	**168.52**
1500	51.19	15.79	62.42	0.02	**129.42**

Implementation and Operational Costs and Return for the Investor

In this section, all the infrastructure costs (CAPEX) that were foreseen for the construction of the terminal were contemplated, as well as a market analysis to understand the terminal's operation cost per box and the cost of personnel (OPEX and SG&A), in a way to obtain an indicator for an average rate per box to be charged by the terminal. In the case in question, an aggregate calculation was made of the values coming from dock handling, yard handling and storage.

In order to survey the costs, it was necessary to perform a sizing of the terminal from international and local benchmarks.

Based on this information, a discounted cash flow analysis was performed in order to obtain the net present value of the terminal deployment and operation project for a period of 25 years, including a two-year period for physical deployment of the terminal.

The analysis also involved obtaining the values collected by the competing ports in order to verify the competitiveness of Port of Açu in the container market.

– *Infrastructure (CAPEX) and operational (OPEX) costs*

Aiming to estimate the costs of deploying the container terminal, it was necessary to dimension the terminal involving civil works and equipment acquisition.

Considering the load handling benchmark indicators from Drewry (2010), which presents standards for estimating container terminal capacity based on quay, Port of Açu, under current dock conditions (350 m), in the worst case, could handle 280 thousand teu/year. This value satisfies solely the handling anticipated in the first years of operation. Thus, the evaluation considered an expansion of the dock to 650 m in order to contemplate 2 berths, a standard used, for example, by the DP World terminal in Santos. In this way, it is possible to serve, in the worst case scenario, 650 thousand teu/year.

The container yard area was defined based on dock length and a depth of 300 meters, which resulted in 195,000 m^2 of container handling yard, a smaller value than that of the DP World terminal, although approximately 2 times the container terminal of Port of Vitória, the main potential competitor of Port of Açu. According to indicators of Drewry (2010), this area is sufficient to meet the expected demand for the terminal.

Based on the studies performed for competing terminals and international bechmarks, necessary equipment sizing and capital costing survey were carried out. The values are shown in Table 4.

For the operating and personnel costs, the available 2017 annual statements from 3 different terminals were used. These terminals were Santos Brasil (Port of Santos): Container Terminal of Paranaguá (Port of Paranaguá) and Terminal of Vila Velha (Port of Vitória). It is noteworthy that the Paranaguá Terminal is not within the area of the present study, but it has similar characteristics to the terminal intended to be deployed in Port of Açu. The results are summarized in Table 5.

Similarly to what has been done with operating costs, the revenue per box of each of these terminals and a weighted average were calculated (Table 6).

Table 4. Port of Açu – equipment and civil works investments

Year	2019	2020	Year	2019	2020
1. Equipment - Total (R$ MM)	**61.1**	**138.7**	**2. Civil Work - Total (R$ MM)**	**276.2**	**160.2**
1.1 Portainer (Qtity)	6	–	2.1 Quay (m)	650	–
Value (R$ MM)	47.8	111.6	Value (R$ MM)	222.4	148.3
1.2 Reach Stacker (Qtity)	–	2	2.2 Container Courtyard (m^2)	195,000,00	
Value (R$ MM)	–	4.2	Value (R$ MM)	47.9	12
1.3 Transtainer (Qtity)	4	–	2.3 Gate (unit)	2	–
Value (R$ MM)	10.5	7	Value (R$ MM)	2.7	–
1.4 Trucks (Qtity)	–	24	2.4 Warehouse (m^2)	500	–
Value (R$ MM)	–	5.4	Value (R$ MM)	1,1	–
1.5 Scanners (Qtity)	–	1	2.5 Utilities Infra (m^2)	195,000,00	–
Value (R$ MM)	–	7.4	Value (R$ MM)	2.1	–
1.6 TOS (Terminal Operating System)	1				
Value (R$ MM)	2.8	2.8			
1.7 Power plug Reefer (Qtity)	–	150			
Value (R$ MM)	–	0.2			
1.8 Road Balance	–	1			
Value (R$ MM)	–	0.1			

Table 5. Operating and personnel costs

	Total cost/box
Container Terminal of Paranaguá (Paranaguá)	R$ 478.27
Santos Brasil (Santos)	R$ 363.17
Terminal of Vila Velha (Vitória)	R$ 394.16
Port of Açu	R$ 411.87

Table 6. Container terminals revenue

	Revenue per box
Container Terminal of Paranaguá (Paranaguá)	R$ 1.344,39
Santos Brasil (Santos)	R$ 695,06
Terminal of Vila Velha (Vitória)	R$ 985,71
Port of Açu	R$ 1.008,39

– *Financial model*

In order to verify the viability of the terminal and its net present value, a financial model was performed based on the revenues obtained by the tariffs and handling curve, its CAPEX, OPEX and SG&A costs, as well as taxes (PIS, COFINS and ISS), depreciation and income tax/CSLL.

For such a model, the variation of working capital was not considered, since it is understood that the competence regime is in line with the flow regime. Finally, the terminal net present value (NPV) was obtained and its feasibility discussed.

In the Income Statement for the Year, the income was calculated as follows:

$$Revenue = Volume(mil\,boxes) * Average\,Tariff \tag{1}$$

The amount used for average tariff was previously calculated and it is R$ 1,008.39.

After calculating the revenue, it is necessary to discount the service taxes, namely PIS, COFINS and ISS, which add up to 11.25% of the revenue.

The next line will be the net revenue, which is equal to:

$$Net\,Revenue = Revenue - (PIS + COFINS + ISS) = 0.8875 * Revenue \tag{2}$$

After that, the total annual cost of the net revenue should be withdrawn and this will result in the EBITDA (Earnings Before Interest, Taxes, Depreciation and Amortization) of the terminal.

$$EBITDA = Net\,Revenue - Cost \tag{3}$$

The next step was to include depreciation, calculated as approximately 4.35% of CAPEX annually, and the calculation of EBIT - Earnings Before Interest and Taxes.

$$EBIT = EBITDA - R\$\ 27{,}661{,}000.00 \tag{4}$$

Finally, we find the net profit that is given by:

$$Net\,Profit = EBIT - IT/CSLL \tag{5}$$

IT/CSLL is the income tax or social contribution over the net income and is equal to 34% of EBIT.

The NPV was calculated over the cash flow. In this case, the CAPEX, which had not been taken into account in the DRE since it is not part of the result of the fiscal year, but an initial investment that generated flow, does not directly impact the annual profit.

The free cash flow from the terminal was calculated as follows:

$$Cash\,Flow = EBITDA - (IT) \pm Variation\,of\,Working\,Capital - CAPEX \tag{6}$$

Finally, the NPV was calculated and for such it was considered a discount rate that is the capital cost used in a return analysis. It can be calculated in several different ways, because it is not an exact science. One of the best-known forms is WACC - Weighted

Average Capital Cost. This rate indicates the level of minimum attractiveness of the investment, i.e. it is the return one would expect to have on other investments safer than the current one.

In the context of this analysis a 10% rate was considered since it is a value that, as stated above, can vary widely and depends on a series of attenuating factors. Recently, when it comes to a lease bid, the regulatory agency ANTAQ has been setting the discount rate. The last disclosed amount of this rate for leases was 8.03%.

For the discount rate of 10%, the obtained NPV was R$ 120,303.00.

Table 7 presents a panel with NPV values obtained from 3 different tariffs and 2 discount rates.

It can be observed that even reducing the tariff to the price of Vitória, it is possible to remain with a positive NPV, which guarantees to the terminal more competitiveness without loss of profitability. With regard to the Santos Brazil terminal, the negative NPV is due to the fact that it is transshipped, which guarantees a considerable reduction in the cost of operating the terminal and a possibly higher profit margin, even with the reduction of the tariff.

Table 7. Port of Açu container terminal NPV analysis

Rate	WACC	NPV
Average rate (R$ 1,008.39)	10.00%	R$ 120,302,587.18
Average rate (R$ 1,008.39)	8.03	R$ 278,711,867.73
Terminal of Vila Velha/Vitória (R$ 985.71)	10.00%	R$ 93,085,556.48
Terminal of Vila Velha/Vitória (R$ 985.71)	8.03	R$ 245,277,867.62
Santos Brasil/Santos (R$ 695.06)	10.00%	(-) R$ 255,809,426.86
Santos Brasil/Santos (R$ 695.06)	8.03	(-) R$ 183,312,416.13

6 Conclusions

The result of the feasibility analysis of local container terminal in Port of Açu presented satisfactory results, since by charging the average tariff of 3 different terminals a positive NPV was obtained. In addition, a positive result was also obtained when charging the value of the main competitor.

As far as the maritime access to the port is concerned, it can also be considered competitive. However, the conditions of tugboats and pilotage can be improved, since the volume of access of cargo ships is still low.

Although Port of Açu does not operate container terminal yet, it presents favorable aspects for this segment. The fact that it is a new, developing and totally private port located in southeastern Brazil, responsible for most of the national GDP, brings many advantages for the installation of a port for containerized cargo in this place. However, there are still negative factors, such as: poor and limited road access and disconnection with the Brazilian rail network, which is already inefficient for transport of containers. Logistics on land is essential for the flow of containers.

Finally, it is believed that the Açu container terminal would boost the expansion and consolidation of the Açu industrial district and the development of the EPZ, and a first step in what could become a hub port in the future due to its political and management conditions, its wide area and its favorable draft.

References

Antaq: Anuário estatítico aquaviário. http://web.antaq.gov.br/Anuario. Accessed 15 Sept 2018

Assis, L.F., Canen, A.G.: Brazilian port logistics: going global. In: The 6th International Symposium on Logistics (6th ISL), Salzburgo (2001)

Brooks, M.R.: The Governance structure of ports. Rev. Netw. Econ. **3**(2), 168–183 (2004)

Cullinane, K., Khanna, M.: Economies of scale in large container ships. J. Transp. Econ. Policy **33**(2), 185–207 (1999)

Drewry Shipping Consultants. Container Terminal Capacity and Performance Benchmarks. Drewry Publishing, London (2010)

Analysis of the Performance of FLNG Vessels According to Basic Design Parameters

Daniel P. Vieira(✉), Kam Yu Kang, and Claudio M. P. Sampaio

Numerical Offshore Tank, Department of Naval Architecture and Ocean Engineering, University of São Paulo, São Paulo, Brazil
dpratavieira@gmail.com

Abstract. Natural gas is an alternative energy to oil, which has gained attention and investments in the offshore market recently. For exploration, storage and transport, due to its unique characteristics, it requires to be processed at Floating Liquefied Natural Gas platforms (FLNG). The main differences from the traditional oil FPSOs platforms are the presence of a larger processing plant and of thermal insulated specialized tanks, both intrinsically related to the need to liquefy natural gas at extremely low temperature. Besides the challenges of installing an LNG processing plant onto a floating structure, the FLNG design process presents several challenges due to maritime requirements (stability and structural) as well as to operational characteristics. The present work, based on a parametric synthesis model development, aims to compare single viable FLNG platforms obtained by numerical simulations of different dimensional and arrangement parameters or cargo and ballast conditions and evaluating its performance regarding production and storage capacity and operational downtime. Given a set of input parameters, varying for a range of typical values, the numerical program performs the FLNG design in agreement with established international rules considering topside layout, volume and weight distribution, cargo and ballast filled conditions, intact stability conditions, longitudinal and transversal structural specifications, and seakeeping analysis. The results are compared and analyzed for a typical production level and storage capacity, highlighting that the parametric model approach is usually applied to the conceptual design with the objective to provide basic performance indicators on viable solutions for the client's decision process.

Keywords: FLNG · Model test · LNG tanks

1 Introduction

Natural gas is an alternative fossil energy source to petroleum, relatively clean, abundant and widely used in the domestic, commercial, industrial and transportation sectors. Due to such characteristics, the natural gas sector has attracted more and more investments with many new natural gas reserves located in deep water and, therefore, encouraging the development of floating design units, called FLNG (Floating Liquefied Natural Gas) capable of producing, processing and storing liquefied gas (LNG) offshore.

V. A. J. E. Carreño Moreno et al. (Eds.): CIDIN 2019/COPINAVAL 2019, pp. 246–257, 2020.
https://doi.org/10.1007/978-3-030-35963-8_21

In Brazilian pre-salt condition, oil reserves present a high gas-oil ratio (GOR), which represents a possibility of natural gas monetization. Nowadays, due to restrictions imposed by the inexistence of export lines, most companies reinject the associated gas into the reservoirs. Therefore, if the natural gas is to be monetized, the use of FLNG platforms with adequate processing and storage capacity seems to be the only viable solution to put natural gas on the market.

Due to the complexity of all processes involved in LNG production and storage, these novel designs are pointing towards a solution considering huge barges, capable of supporting a large process plant on the deck and large storage capacity with insulated tanks into the hull, as presented in Vieira et al. (2016). Since the process and storage units are floating structures at sea, they are under the action of environmental conditions (waves, winds and current) which adds extra complexity and increased risks to its operation, both technical and economically.

In 2016, extensive research undertaken at the Numerical Offshore Tank (TPN) laboratory of the University of São Paulo evaluated several factors and parameters of a FLNG floating unit design. The research investigated experimental and numerically various inherent physical processes as LNG sloshing, side-by-side offloading, wind, and current forces, etc. Tests were carried out on ocean tanks and tanks and presented in Rocha et al. (2015) and Vieira et al. (2018).

As a first step in the previous research study, it has been necessary to define the main characteristics of the floating structure, that is a parametric synthesis model was developed (Vieira et al. 2016) in which different dimensional parameters/variables were coupled to provide global performance requirements (cargo DWT, annual production volume, etc.) and to attend maritime and safety standards.

The present work corresponds to an improvement of that previous synthesis model, where an annual production range varying from 1 MTPA to 4 MPTA has been considered and, unlike the previous work that assumed only one tank load condition (90% full), the current version evaluates three different tank filling conditions. Besides, this synthesis model also includes a verification of the natural frequency of tank sloshing and some improvements of the computational model for the stability, structural and sea behavior evaluation.

As a fact for carrying out a design process through a parametric synthesis model, the objectives were to evaluate, as many as possible, feasible solutions. Therefore, to take full advantage of the computational cluster available in the TPN laboratory, special attention was focused on the computer program structure organization in order to distribute the numerical simulation by different processors and reduce the required computational time.

The present article proposes to present a brief review of the new capabilities of the present synthesis model and applied them, taking as a reference basis the solution obtained in the first model, for comparing and evaluating the effects of new and extended parameters on the performance on viable solutions. In this sense, the characterization of deck spaces, cargo/ballast volumes and weight and centers directly affect final displacement, freeboard, intact stability, and structural resistance and, consequently, the behavior of the floating FLNG under the proposed metocean conditions, the downtime percentage being the most important index associated with the FLNG's performance (Molin et al. 2002).

2 Synthesis Model (SM) Overview

In a resumed form, the first process of SM is the definition of the design premises where it should be defined the oil/gas capacity, the installation area with the water depth, environmental conditions, and equipment operational limits. Figure 1 presents a flow chart of the synthesis model that was used during the hull sizing process.

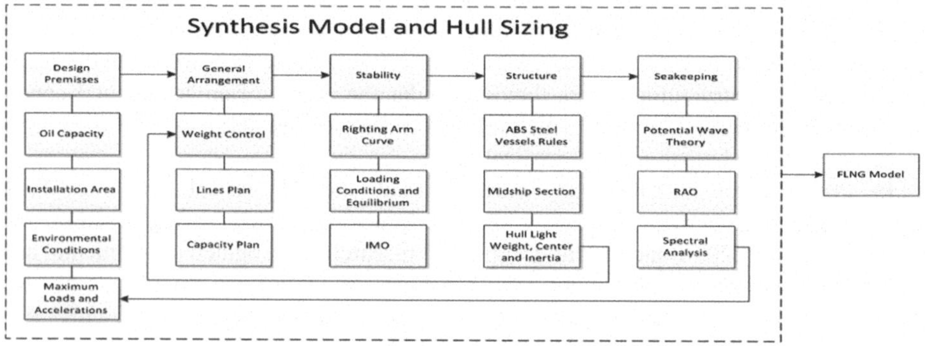

Fig. 1. Synthesis model

It is possible to define a general arrangement with the outputs of the hull lines plan and the tanks capacity plan. These properties are the input of the stability model to define the loading conditions and the ship equilibrium. Also, in a second evaluation, this information will be used to verify the rules for ship stability.

Together with the stability routines are the main structure code with the output of the midship section, lightweight hull estimation, hull center of gravity and inertia. These are the input of the seakeeping analysis through a potential wave theory is evaluated the response amplitude operators that after a spectral analysis the accelerations and maximum loads are defined.

In regards of natural gas storage and management, the synthesis model adopted the norms and rules established by Veritas (2012).

2.1 Synthesis Model (SM) Optimization Procedure

The main role of this procedure is evaluating each solution set of performance parameters. After that, other techniques must be applied to rank the solution among the others, and this topic will be discussed further.

A set of parameters, or a potential solution, goes through the SM and each mathematical model will evaluate it and produce performance data. Eventually, a solution may be unfeasible, and the whole process is interrupted to give place for a new solution. The sequence of mathematical models is presented in Fig. 2.

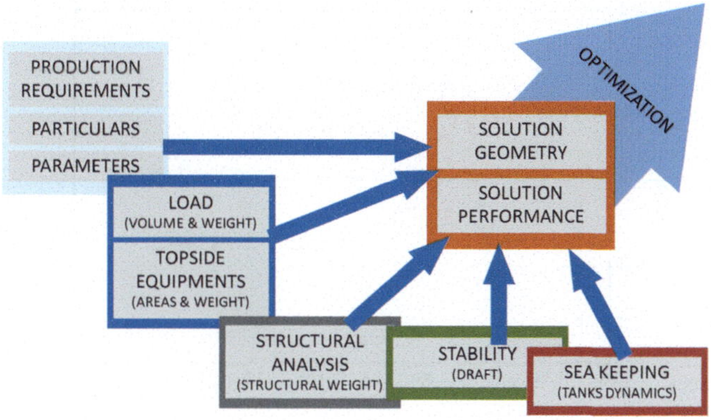

Fig. 2. Synthesis model structure

The starting storage volume is calculated and divided into tanks. Besides, load mass is calculated, considering LNG and LPG volumes and densities. Later, an estimate for the topside modules (areas, volume, and position) is calculated, based on the available deck area using the load mass and deck mass to estimate the total mass and draft.

This information is fed into the structural routine that designs longitudinal, transversal and the bulkhead arrangement, aiming at meeting classification criteria. The same routine estimates the structural mass that is added to the load mass and deck mass to produce total mass with good precision.

All the loads and position are input for the stability routine that will computes the final draft, the transversal metacentric height (GM_t) and the righting arm curve (GZ). The last routine is the one to evaluate seakeeping. It uses the hull and tanks dimensions to create a panel mesh that is solved in the frequency domain by WAMIT, producing response amplitude operators (RAOs) of the evaluated solution. Some values such as maximum acceleration on the liquefaction area as well as roll and heave natural period are calculated. All the performance parameters are stored, and a new solution is evaluated. In the end, a family of feasible solutions is available and can be displayed as a cloud, Fig. 3. Table 1 shows the set of basic parameters that describes the solution:

Table 1. FLNG basic parameters description

Parameter	Description	Parameter	Description
LOA	Length overall	Lpr	Length of bow chamfer
B	Beam	Ntanks	N° of storage tanks
D	Depth	NtanksT	N° of storage tanks rows
Hpo	Height of stern chamfer	hDD	Height of double deck
Hpr	Height of bow chamfer	hDB	Height of double bottom
Lpo	Length of stern chamfer	wDS	Width of double shell

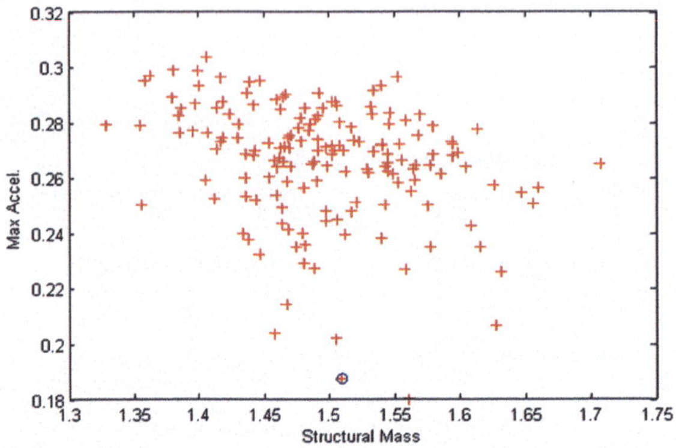

Fig. 3. Cloud of the possible solutions

3 Numerical Model

A numerical model in the frequency domain was used to evaluate the response amplitude of these hulls. The simulation was conducted using WAMIT, a boundary element method code based on the potential linear theory (Newman 1977). The main properties evaluated by WAMIT are the wave forces and the hydrodynamic coefficients (additional mass, potential damping and hydrostatic restoring) as well as the Response Amplitude Operator (RAO). It is possible to evaluate these properties for an N floating body system, considering the interference between them.

The input to the numerical model is basically the mesh of the wet surface of the vessel, the information about mass and inertia, a vector of periods (or frequencies) of the waves and a vector of wave incidences.

In addition, some configurations parameters dependents on the type of simulation are also provided. To evaluate the first order forces, the code standard method was used, as presented in Newman and Sclavounos (1988). First, the wet surface of each vessel was designed in a computational routine environment that created a geometrical data file (GDF) for each loading condition. Figure 4 shows a wet hull surface to 90% percent loaded ship that was used in the analysis.

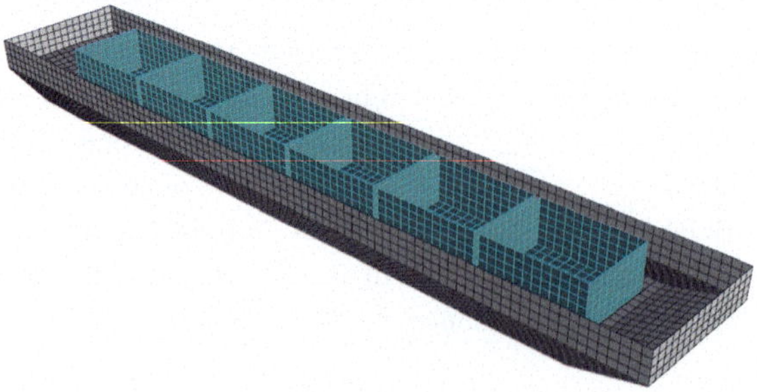

Fig. 4. 90% loaded ship wet surface: Hull (grey) and Tanks (blue).

The tanks wet surfaces were also provided to evaluate the influence of liquid cargo motion on ship motions (Newman 2005). The setup was made accordingly WAMIT (2016).

4 Study Case Parameters

4.1 Configuration A

The object of the study presented here is the FLNG designed in CH- TPN in Brazil in association with Frade Japão Petróleo Ltda. Utilizing the SM procedure described previously, the main characteristics proposed for the FLNG with 4 MPTA productions are described in Table 2. While the proposed solution shows reasonable performance to the environmental conditions of Santos Basin, it is possible to perform an optimization of the internal space between the LNGC tanks and the double side.

Table 2. FLNG main dimensions (Config. A)

Property	Value	Units	Property	Value	Units
Length	463	m	N° tanks	6	Units
Breadth	80	m	Double deck	2	m
Depth	38	m	Double bottom	3	m
Load. Cap.	240.000	ton	Double side	2.5	m

Regarding the double side characteristics, it leaves an empty space between the LNG tanks and the ballast tank which could be utilized to increase the vessel's total displacement to reduce its freeboard and improve performance. Figure 5 shows the tank configuration.

This means that there is a space of nearly 36 m wide of empty space between the LNG tanks and the double side. As mentioned earlier, this space will be filled with ballast, and the system behavior in waves evaluated. Table 3 shows the transversal section dimensions of the FLNG in configuration A.

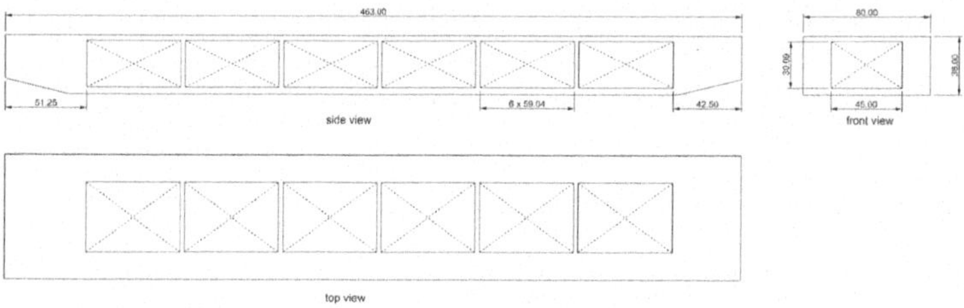

Fig. 5. FLNG tank configuration (Config. A)

Table 3. FLNG secondary dimensions (Config. A)

	Case 1	Case 2	Case 3	Case 4	Case 5	Units
Tank length	59	59	59	59	59	m
Tank width	45,0	52	59	66	73	m
Tank height	30,1	26.1	23	20.5	18.6	m
Double deck	2.9	5	6.5	7.7	8.7	m
Double bottom	3.9	6	7.5	8.7	9.7	m
Double side	17	13.5	10	6.5	3	m

Aiming to reduce the empty space between the tanks and the hull a study is done by means of a sensibility analysis, that is, by increasing the double side dimensions and keeping the loading capacity constant, new tank configurations are created and the wave behavior tendency can be observed.

Molin et al. (2002) describe that during its long-term production cycle, the FLNG is commonly located in remote locations and usually subjected to complex meta-ocean environment conditions. The loading condition of FLNG is constantly changing during the production and offloading processes. For this reason, the FLNG's loading conditions were evaluated at 30%, 60% and 90% of the maximum loading capacity respectively.

4.2 Configuration B

In the same way, a second case is verified with two tanks rows arrangement while keeping the same loading capacity of 240,000 t; it is important to note that this configuration also has a very small space destined for the ballast tanks, which also results in a considerable freeboard for the FLNG. Table 4 shows the main characteristics and Fig. 6 describes the tank arrangement for the configuration B. Finally, Table 5 describes the model's secondary dimensions for configuration B.

Table 4. FLNG main dimensions (Config. B)

Property	Value	Units	Property	Value	Units
Length	450	m	N° tanks	20	Units
Breadth	81	m	Double deck	8	m
Depth	38	m	Double bottom	5	m
Load. Cap.	240.000	ton	Double side	3	m

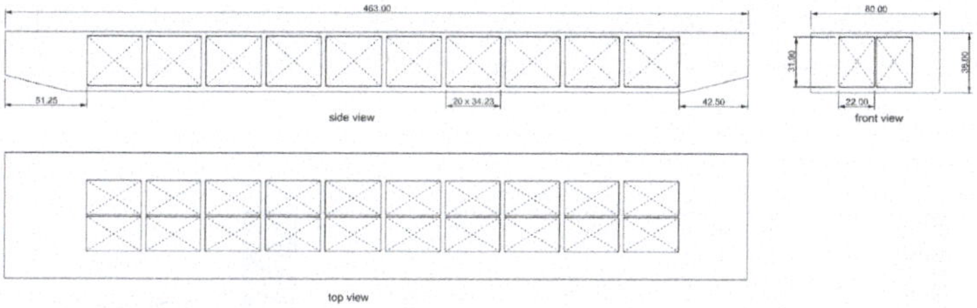

Fig. 6. FLNG tank configurations (Config. B)

Table 5. FLNG secondary dimensions (Config. B)

	Case 6	Case 7	Case 8	Case 9	Case 10	Units
Tank length	34.2	34.2	34.2	34.2	34.2	m
Tank width	22	25.5	29	32.5	36	m
Tank height	31.9	27.5	24.2	21.6	19.5	m
Double deck	2.1	4.2	5.9	7.2	8.3	m
Double bottom	3.1	5.2	6.9	8.2	9.3	m
Double side	17	13.5	10	6.5	3	m

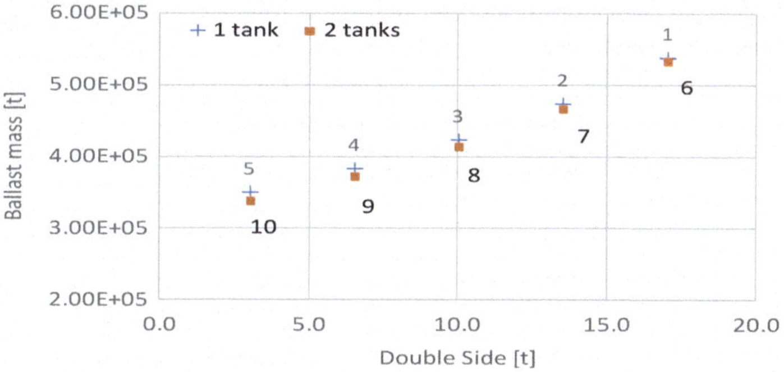

Fig. 7. Double side *versus* Ballast mass

5 Results

The results evaluation was done comparing each case proposed using the implemented interface analysis. The solution was analyzed according to structural mass, freeboard and seakeeping/downtime performance. The structural mass is directly related to the hull costs. The freeboard is another key parameter because an elevated freeboard implies in difficulties to carry operations as side-by-side offloading. Downtime is a percentage of the time it is expected that the system is not operational. This factor is related to the ability to operate in adverse environmental conditions especially in rough seas.

Figure 7 shows the relation between the double side parameter (*wDS*) and the total ballast mass. As expected, the increase of the double side dimensions and consequently filling the vertical empty space (see Fig. 8) between the LNG tanks and the hull with ballast contributes significantly to reduce the vessel's freeboard due to its displacement increase.

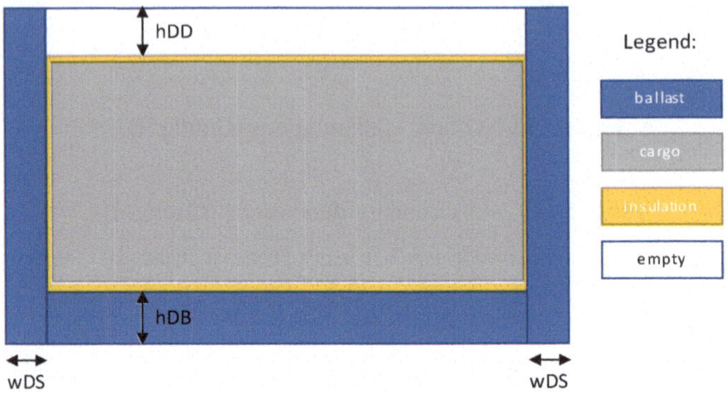

Fig. 8. Schematic of cargo and ballast tanks arrangement

This means that from the freeboard perspective, it is favorable to build narrow and high tanks. Figure 9 shows the freeboard reduction mentioned and Table 6 shows the ballast mass used for each case.

Table 6. FLNG Total ballast

Case id	Ballast mass [t]	Case id	Ballast mass [t]
1	539,115	6	534,970
2	474,317	7	466,946
3	424,138	8	414,679
4	384,103	9	373,075
5	351,243	10	339,030

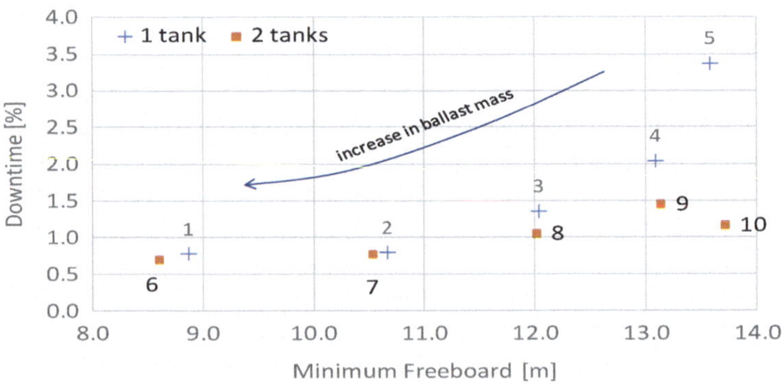

Fig. 9. Minimum Freeboard *versus* Downtime evaluation

Figure 9 also shows the reduction of downtime associated with the reduction of the displacement. From wave behavior point of view, the additional displacement due to the ballast increase contributes to reducing roll amplitudes as shown in a comparison between Case 1 and Case 5 respectively in Fig. 10. The reduction is mainly in the region between 10 and 15 s (operational seas).

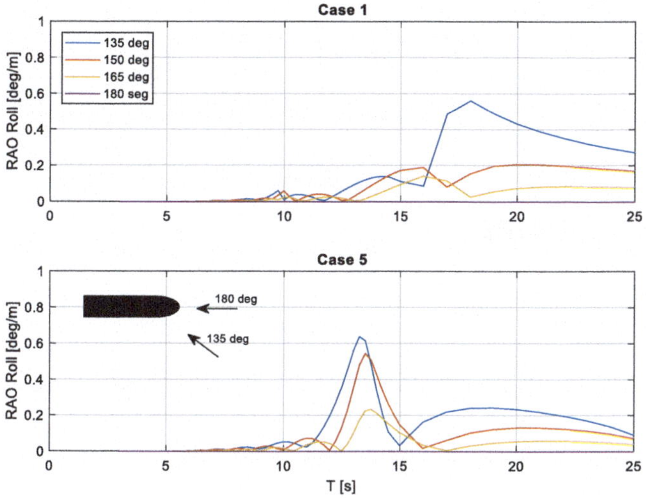

Fig. 10. Case 1 and 5 roll motions

Another point is the comparison between configuration A and B tank arrangements where, due to smaller tanks, it is observed a reduction of the sloshing effect which explains the downtime differences between Case 1 and Case 6 in Fig. 9. This effect is also observed in Fig. 11 comparing de downtime with the tank width. It is observed that for smaller tanks (cases 6 up to 10) the downtime varies less than for large tanks (cases 1 up to 5). Tanks with high height/width ratio have less influence of the tank width than tanks with low height/width ratios.

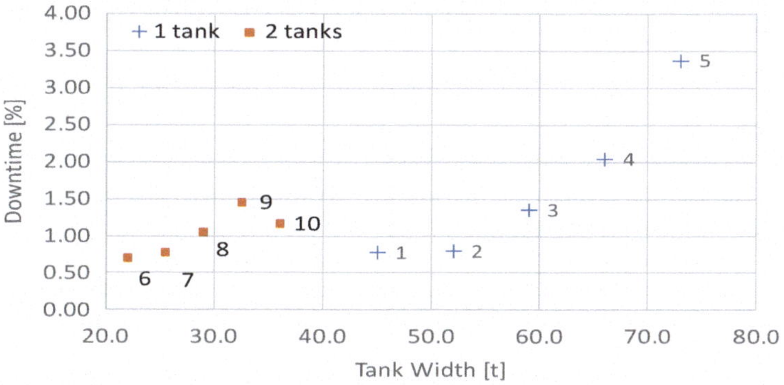

Fig. 11. Tank width *versus* Downtime

This result can be visualized in the roll motion comparison between cases 5 and 10, as showed in Fig. 12.

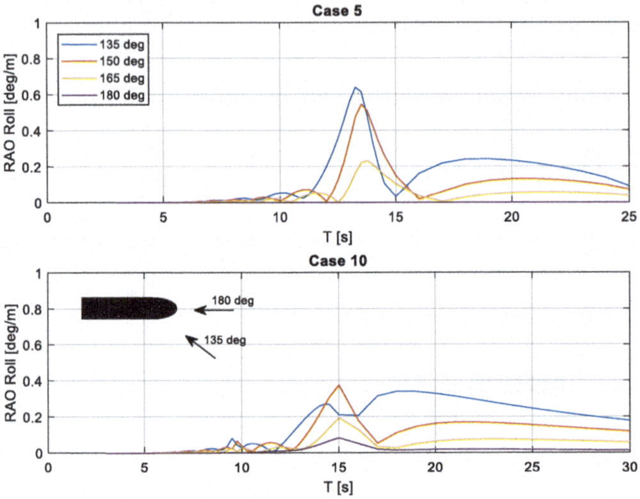

Fig. 12. Case 5 and 10 roll motions

6 Conclusions

Both proposed tank arrangement cases showed a possibility to reduce freeboard and optimize its behavior in waves and reduce downtime. The synthesis model allowed not only to create the desired case condition but also provided a powerful tool to evaluate each case and verify the best solution.

Finally, the proposed model allows the designer to evaluate and, according to the given environmental conditions, design and study the pertinent features for a vessel in each metocean condition.

Acknowledgments. The authors acknowledge Frade Japão Petróleo Ltda for the financial support and for the motivation of this work.

References

Veritas, D.N.: Rules for Classification of Floating LNG/LPG Production, Storage and Loading Units (DNV-OSS-103). DNV (2012)

Newman, J.N.: Marine Hydrodynamics. MIT Press, Cambridge (1977)

Newman, J.N., Sclavounos, P.D.: The computation of wave loads on large offshore structures. In: 5th International Conference on the Behavior of Offshore Structures (BOSS), Trondheim, Norway (1988)

Newman, J.N.: Wave effects on vessels with internal tanks. In: 20th International Workshop on Water Waves and Floating Bodies (IWWWFB), Spitsbergen, Norway (2005)

Molin, B., Remy, F., Rigaud, S., de Jouette, Ch.: LNG-FPSO's: frequency domain, coupled analysis of support and liquid cargo motions. In: 10th International Congress of International Maritime Association of the Mediterranean (IMAM), Greece (2002)

Rocha, T.P., Dotta, R., Vieira, D.P., Mello, P.C.D., Malta, E.B., Nishimoto, K.: Experimental investigation on the influence of liquid cargo in floating vessels motions. In: OTC Brasil. Offshore Technology Conference (2015)

Vieira, D.P., Lavieri, R.S., Rocha, T.P., Dotta, R., Rampazzo F., Nishimoto K.: A synthesis model for FLNG design. In: 35th International Conference on Ocean, Offshore & Arctic Engineering (OMAE), Busan, Korea (2016)

Vieira, D.P., de Mello, P.C., Dotta, R., Nishimoto, K.: Experimental investigation on the influence of the liquid inside the tanks in the wave behavior of FLNG vessels in side-by-side offloading operations. Appl. Ocean Res. **74**, 28–39 (2018)

WAMIT Inc.: WAMIT User manual, Version 7.2 and 7.2PC (2016)

Optimization of the Emissions Profile of a Marine Propulsion System Using a Shaft Generator with a MMPT Based Control System and the Use of EEDI and EEOI

Joel R. Pérez Osses[1(✉)] and Carlos A. Reusser Franck[2(✉)]

[1] Universidad Austral de Chile, Valdivia, XIV Región de los Ríos, Chile
joelperez@uach.cl
[2] Universidad Técnica Federico Santa María, Valparaíso, V Región de Valparaíso, Chile
carlos.reusser@usm.cl

Abstract. Nowadays marine propulsion systems based on thermal machines that operate under the diesel cycle have positioned themselves as one of the main options for this type of applications. The main comparative advantages of diesel engines, compared to other propulsion systems based on thermal machines, are the low specific fuel consumption, residuals and their higher thermal efficiency. However, its main disadvantage lies in the emissions produced by combustion, such as carbon dioxide (CO_2), oxide sulphur (SOx) and oxide nitrogen (NOx). These emissions are directly related to the operating conditions of the propulsion system [1].

Over the last decade, the International Maritime Organization (IMO), has adopted a series of regulations to reduce these emissions [2, 3], based on the introduction of an energy efficiency design index (EEDI) and an energy efficiency operational indicator (EEOI). EEDI is mandatory for any new ship and the EEOI is optional to be applied. In this context, adding a shaft generator [4] allows to reduce the nominal design power of the auxiliary generation system and, under nominal operating conditions, the propulsion plant, favouring lower EEDI and EEOI values, which means lower CO_2, SOx and NOx emissions [5, 6].

However, the use of shaft generators can only be justified if the propulsion system operates, most of the time, under 75%–80% of the maximum continuous rating (MCR) design. In addition, the incorrect operation of the shaft generator can result in overloading the main engine, which means an increase of CO_2, NOx and SOx emissions.

The present work proposes a selective control system with a Maximum Power Point Tracking (MPPT) that allows operating the shaft generator in Power Take Off (PTO) or Power Take In (PTI) mode ensuring that the main engine operates, always, at the optimum point to generate minimum CO_2, SOx and NOx emissions.

Keywords: Marine propulsion system · Shaft generator · EEDI · EEOI · Emissions and control system

V. A. J. E. Carreño Moreno et al. (Eds.): CIDIN 2019/COPINAVAL 2019, pp. 258–274, 2020.
https://doi.org/10.1007/978-3-030-35963-8_22

1 Introduction

Even though marine propulsion systems have been in constant development since the 18th century nowadays the most common system used on board large carriers i.e. container ships and tankers, is a system considering a diesel engine as the prime mover. Most of these engines are of the crosshead type, operating on the two-stroke cycle at low speed having long strokes, turbocharged and directly coupled to a single fixed-pitch propeller. The power installed vary from 10 MW and up to 80 MW [7].

Diesel engines for marine applications have many advantages when compared to other prime movers such as turbines. They have a higher thermal efficiency and a low fuel oil consumption of low-cost residual fuels. The disadvantage of consuming residual fuels is the high amount of CO_2, SOx and NOx emissions, which are related to the ship's operational condition [8]. The ship's design power requirement represents the main constrain when the operational condition of the ship is assessed and compared to the power demands at normal operating conditions. The prime mover is forced, most of the time, to operate under underrated conditions increasing its fuel oil consumption therefore the amount of emissions. Diesel engines, as the one described, found their optimum fuel oil consumption point around 75%–80% of the MCR [9].

High fuel oil consumptions lead to an increase of emissions, which have been monitored and since 2011 have been measured using an index called EEDI [2]. This index is part of a mandatory regulation coming from the IMO and applicable to every new ship. An operational indicator called EEOI has also been considered but this is not mandatory yet, although is integral part of the ship energy efficiency monitoring plant (SEEMP), which is mandatory [10]. The EEOI is used as a measuring tool to voluntary assess the efficiency of new and existing ships.

The EEDI encourage the use of technologies such as shaft generator to reduce the use of the power installed on board through auxiliary generator sets. This reduction of the use reduces the fuel oil consumption, therefore, as was mentioned before, reduces the amount of emissions. The shaft generator force to use the diesel engine in a loading range, quite close to the optimum fuel oil consumption point. The use of shaft generators above this point has been considering unjustified because of the possibility to overload the diesel engine leading to increase the fuel oil consumption.

The present work presents a marine diesel engine propulsion system with a direct driving shaft generator and a back to back converter based on the use of a selective control scheme. This scheme enables for the diesel engine to operate at its optimum fuel oil consumption point, which has been renamed as its minimum emissions operating point (MEOP). The scheme considers the use of the shaft generator as a power take-off (PTO) drive when the diesel engine operates below the MEOP and as a power take-in (PTI) when the diesel engine operates above the MEOP. The shaft generator, at PTO, generates enough power to turn-off the generator set of the ship.

2 Background

Emissions are generated during the process of converting the chemical energy of the fuel into mechanical work, Eq. (1) represent the stoichiometric reaction of the fuel and the consequence emissions generation, CO_2 emissions are the higher amount of all of them.

$$C_mH_n + \left(m + \frac{n}{4}\right)O_2 + pN_2 \rightarrow mCO_2 + \frac{n}{2}H_2O + pN_2 \tag{1}$$

The stoichiometric air to fuel ratio (AFRst) is the minimum amount of air required to burn a kilogram of fuel and when compared to the actual air to fuel ratio (AFR), the stoichiometric ratio λ presented in Eq. (2) can be found [11]. The AFR can be considered at any engine load for the purposes of analysis.

$$\lambda = \frac{AFR}{AFR_{st}} \tag{2}$$

The engine's output power suffers when operating at lower loads condition because at this condition less fuel is available and the engine while trying to achieve a higher load demands more fuel to be provided to overcome the demand. The engine trying to maintain the power output at the desired operational condition increases the specific fuel oil consumption until it reaches the desired engine load, which is related to its speed as can be seen in Fig. 1. At low engine load the λ is a low value but when the engine goes at high loads, above 75%–80% load, the λ values goes above 1, which means the consumption of higher amounts of air and fuel to achieved higher engine power outputs.

Figure 1 shows these λ conditions and the optimal fuel oil consumption point or MEOP. The MEOP has been considered in this form because is a representation of the specific fuel oil consumption (SFC), which is one of the factors to evaluate the amount of emissions generated by the engine when using the EEDI and EEOI. Figure 1 also differentiate the operating regions of the shaft generator when considering the MEOP over the entire engine load range. Before MEOP the shaft generator operates as PTO and after MEOP operates as PTI.

The engine efficiency at the MEOP is the higher and has been used as the evaluation point of the EEDI mandated by the IMO for every new ship constructed. The EEDI started with a minimum value established by 2013 and reduced by a percentage over the next 12 years [12], a low EEDI value means a more efficient ship.

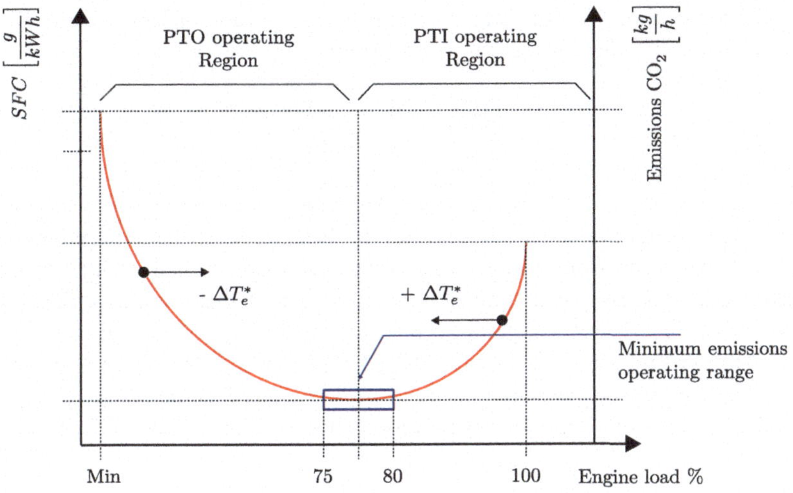

Fig. 1. PTO/PTI operating regions

The EEDI can be defined as a technical measure of CO_2 emissions per ship's capacity per nautical mile applied to new ship designs [2]. The equation to calculate EEDI is presented in Annex A and here a modified version to be applied for the purposes of this paper is presented in Eq. (3). This modified version accounts only for the main engine influence of emissions generation, therefore it is an approximation that is going to be modified to establish and represent the influence of the shaft generator over the entire main and auxiliary systems of the ship.

$$\mathrm{EEDI} = \frac{\mathrm{P_B} * \mathrm{SFC} * \mathrm{C_F}}{\mathrm{DWT} * \mathrm{V_S}} \tag{3}$$

The P_B corresponds to the 75% rated installed brake power in kW, C_F[1] is the carbon factor in g CO_2/g fuel, *DWT* is the capacity of the ship in tonnes and V_s is the ship's design speed in knots.

The MEOP has been also used to calculate the EEOI. The EEOI is the monitoring tool supporting the Ship Energy Efficiency Management Plan (SEEMP) applied to new and existing ships to measure the amount, in grams, of CO_2 per tonne cargo transported per nautical mile for a single voyage [10]. The equation to calculate EEOI is presented in Annex B and here a modified version to be applied for the purposes of this paper is presented in Eq. (4). This modified version uses the SFC instead of the total amount of fuel consumed for a single voyage to relate the indicator with the operational performance described to evaluate EEDI and have a comparison point of the ship's design performance and the actual operation at the required MEOP at low and high loads.

[1] The Carbon factor is a conversion factor between fuel consumption and CO_2 emissions accounting as a ratio between CO_2 and the carbon content of the fuel used by the ship.

The m_c factor accounts for the mass of cargo transported in tonnes and D to the distance, in nautical miles, of the cargo transported.

$$\mathrm{EEOI} = \frac{\mathrm{SFC} * \mathrm{C_F}}{\mathrm{m_C} * \mathrm{D}} \tag{4}$$

Having a comparison point between the design and the operational behavior of the ship allows for a better understanding of these tools to evaluate the current efficiency of the ship. Also allowing to consider technological and operational options to improve ship's efficiency. The use of a shaft generator is part of these improvements and its influence into of ship's efficiency is presented when analyzing the Shaft Generator/Motors Emissions Factor into the EEDI equation, this factor is presented in Eq. (5), where the power generated accounts and is related to the power generated by the auxiliary engines or generator set to support the service of the ship.

$$(\mathrm{f_i} * \mathrm{P_{PTI}} - \mathrm{f_{eff}} * \mathrm{P_{AE}})\mathrm{C_{FAE}} * \mathrm{SFC_{AE}} \tag{5}$$

Equation (5) can also be contrasted with the Efficiency Technology Factor presented in Eq. (6), which relates the reduction in power requirements that any technology generates and is used to improve the ship's efficiency.

$$\mathrm{f_{eff}} * \mathrm{P_{eff}} * \mathrm{C_F} * \mathrm{SFC} \tag{6}$$

The efficiency technology factor (f_{eff}) in Eqs. 5 and 6 is the percentage of influence of the power output of the technology and relates its efficiency. The background presented allows for the introduction of the control scheme selected as the most appropriate to represent the influence of a shaft generator into the propulsion system of a ship because relates its fuel consumption and its emissions in accordance with known and validated indexes for ship's efficiency evaluation.

3 PTO/PTI Control Scheme

The ship propulsion system under study consists of a single low speed diesel engine, driving a permanent magnet synchronous machine, PMSM, coupled to the main diesel engine using a single-stage amplifying gearbox. The synchronous machine is connected to the main ship's grid, by a back-to-back two-level voltage source power converter, 2L-VSC, as shown in Fig. 2.

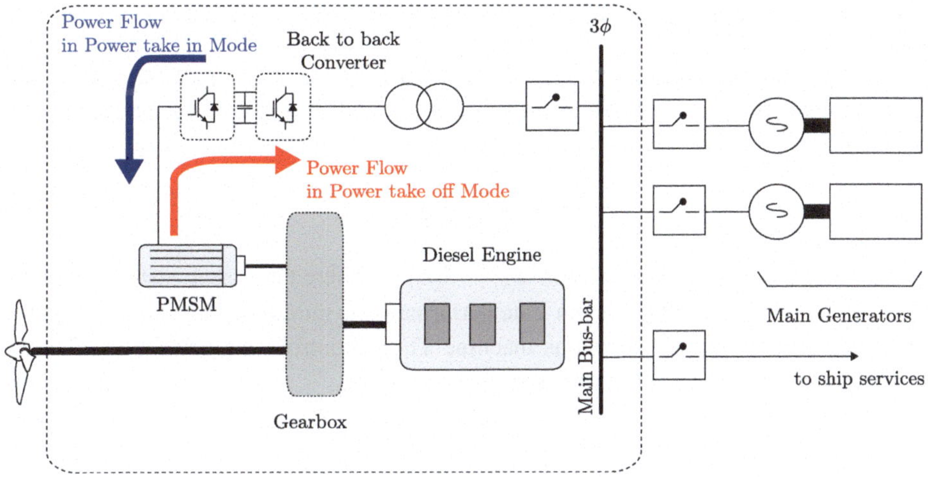

Fig. 2. Propulsion system configuration and PMSM connection to main ship's grid

This arrangement enables the synchronous machine to operate within their four quadrants, motor or generator. This is achieved implementing a torque field-oriented control (TFOC) scheme, which is based on the decoupling of the current space vector into a flux producing current and a torque producing current. Despite the classical field-oriented control (FOC) scheme, which is used to control the drive shaft speed, in this case to control objective is the torque developed by the electric machine, which became the aim of the optimization problem.

3.1 Optimization Strategy

The optimization strategy is based on a perturb and observe (P&O) local search approach. Let us define an optimization problem ϕ, defined as in Eqs. (7) to (10).

$$\phi = \langle C, S, v, f(x) \rangle \tag{7}$$

$$C = x; x = \{x_0, x_1 \ldots x_n\} \tag{8}$$

$$S = x \pm \delta; 0 < \delta < \epsilon \tag{9}$$

$$v = \min\{f(x)|_{x \pm \delta}\} : x \in C \tag{10}$$

Where; $\mathbb{C}$ corresponds to the set of candidate solutions, $\mathbb{S}$ is the set of solutions with $\mathbb{S} \subseteq \mathbb{C}$; v is the optimization sense, thus $v = \{min, max\}$; $f(x)$ the objective function, $f(x) \ : \mathbb{S} \rightarrow \mathbb{R}$, and δ is the system state perturbation.

The optimization objective is to minimize the CO_2 emission given a certain required output power, thrust, by perturbations on the torque reference to the TFOC scheme, which finally sets the operation condition for the synchronous machine; thus operating as a PTO increasing the load to the main propulsion engine, or as PTI, supporting the output power of the main diesel engine, as presented previously in Fig. 1.

The specific fuel consumption, SFC, denoted as g can be characterised as a non-linear discrete function depending on the engine load at a given instant $Q_{B(k)}$ as in Eq. (11).

$$g_{(k)} = h_d\left(Q_{B(k)}\right) \tag{11}$$

Becoming $g_{(k)}$ the objective, thus $g_{(k)} = f\left(x_{(k)}\right)$, where the system state is given by $x_{(k)} = \left[x_{0(k)} \ldots x_{n(k)}\right]$ and the system state perturbation is introduced by variations on the torque reference to the synchronous machine TFOC control scheme by $\delta = \Delta T_e^*$; the optimization solution can be expressed as in Eq. (12).

$$S = \min\left\{ f\left(x_{(k)}\right)\Big|_{x_{(k)} \pm \delta} \right\} \tag{12}$$

The direction of the optimization trajectory, will depend whether the synchronous machine is operating in PTO or PTI mode, thus giving the sign of the system perturbation δ. The optimization algorithm can be expressed as in Fig. 3.

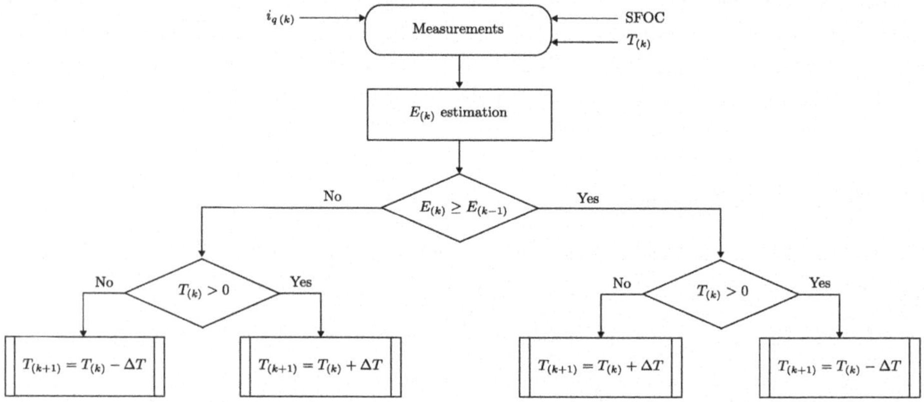

Fig. 3. Minimum emissions tracking point, P&O, algorithm

3.2 Oriented Control Strategy

The characteristics of Field Oriented Control (FOC) have made this control strategy the most widely used for high demanding industrial applications [13]. FOC is based in the decoupling of the current space vector into a flux producing current and a torque producing current [14]. Despite the classical FOC control scheme, which is used to control the drive shaft speed, in this case to control objective is the torque developed by the electric drive [15].

4 Simulation Results

Results are presented differentiating the performance of the control scheme of the shaft generator and the results of the fuel consumed by the engine when simulating the PTO and PTI conditions. The methodology considers the use of the IMOs EEDI and EEOI tools to show the benefits of the scheme providing a reference to evaluate the design efficiency and the operational efficiency.

Results are plotted over the two simulation conditions considered but presented over a power range to simulate specific operational conditions such as slow steaming. This operational condition has been considered because represents the ability of the simulation to show the performance of the ship at low ship's speed, which can be used to evaluate a ship design over a higher range of options to get an efficient design. A more accurate evaluation of the EEDI could be necessary but still results are providing a great assertiveness of the methodology selected.

The data used to simulate the performance of the control scheme considers the use of a ship, which specific information includes; its capacity, speed, cargo transported, distance navigated, power installed, and the type of fuel consumed. The type of ship considered was selected from a worldwide database of ships [16]. When analyzing the database and the specific information needed to evaluate the design and operational efficiencies of a ship, very large crude oil carriers were the type of ships more reluctant to be used because of the simplicity of their propulsion system and the significance of the amount and type of fuel consumed. The propulsion system consists of a diesel engine directly coupled to a fixed pitch propeller. Table 1 presents the open source data used for simulation that were fixed as the initial conditions. The range of data is only a reference of the type of ship found in the database and none consideration to the operational profile of them has been considered for simulation purposes [16].

Table 1. VLCC data used for simulation

Type of ship	Fuel consumed	Capacity (dwt)	Speed (kn)	Cargo transported (tonnes)	Installed power (MW)
Very Large Crude Oil Carrier (VLCC)	Heavy Fuel Oil (HFO)	300.000 320.000	15 21	315.000 330.000	25 36

The engine selected to be modelled and evaluated is an engine from MAN [17]. The 7G80ME-C9.2-TII diesel engine was selected having a specified maximum continuous rating (SMCR) power of 33 MW at 72 rpms. The specific fuel oil consumptions (SFC) vary from 187.1 g/kWh at low engine load to 166 g/kWh at high engine load. The normal continuous rating of the engine was considered around 75% of the SMCR. When consuming HFO, a carbon factor of 3.114 g CO_2/g Fuel was used to estimate the emissions. With this information, the simulation of the control scheme was carried out and the results are presented in Sect. 4.1.

4.1 Control Scheme Performance at PTO Operating Region

The control scheme was evaluated considering a navigation time period long enough to simulate a consistent increase of the brake power of the engine over a step time to reflect its performance and SFC variation to get a steady evaluation of the efficiencies. Blue curves in Fig. 4 shows the results of a segment of the PTO operating region while simulating a period of navigation time of 5 h where in the left-hand side of the figure is possible to appreciate how the brake power lineally increases from 5 MW to 12 MW. The right-hand side of the figure shows the decrease of the SFC from a maximum value at low engine load, the power delivered as PTO is stable enough to allow for the ship to turn-off the generator set. The red curves are the results of not having a shaft generator installed. The increase in the brake power when using a shaft generator is around 5% of the SMCR.

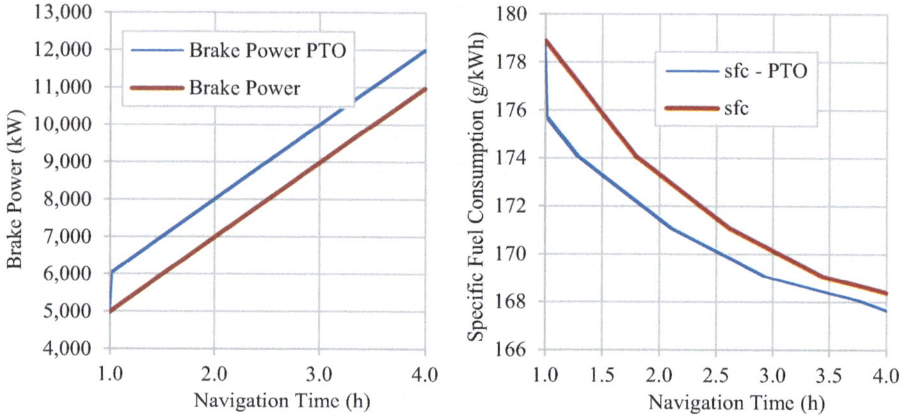

Fig. 4. PTO evaluation performance

When evaluating the EEDI, a specific value of 2,7412 g CO_2/g Fuel was calculated. This value represents the amount of CO_2 emissions that by design this ship can generate and because is a design value can be modified at design stages when the main and auxiliary machinery is selected and allocated to it. A better approach could be i.e. to install a less powerful engine but that means a completely different approach of the design spiral of the new ship. Following this evaluation, EEOI has been considered because represents the ability to calculate the emissions of the ship when in service navigating different routes. Operational emissions are of great value when trying to compare or establish an analysis of the design emissions. EEOI allows to check the variations of the same parameters that EEDI uses to be evaluated, therefore provides with a more comprehensive form to understand the operational behavior of a well design ship.

EEDI provides a fixed value at the design conditions yet EEOI can be used to evaluate the performance at every variation of engine load because considers the amount of fuel consumed no matter the velocity, although the consideration of using mass of cargo transported and distance navigated is just another representation of cargo capacity and ship's speed.

One of the main objectives of this work was to evaluate, through the use of EEDI and EEOI, the influence of a shaft generator when applying a control scheme of its operation. The purpose is looking for reduction into the SFC at different engine loads. Also, to prove that the reduction of the SFC compared to the increase into the necessary power to be developed, to overcome the extra necessary brake power to be produced, to propel the ship and to use the shaft generator as PTO and PTI respectively. Figure 5 shows the results of the SFC variation at PTO operating region when applying the proposed control scheme, red curve, and the blue curve shows the SFC variation when not having a shaft generator. The SFC differences are between 0,1% to 2% over the whole engine load range plotted, the difference even though can be considered small is quite significant when evaluating the EEOI. The difference is barely noticed because of the scale of the plotted results.

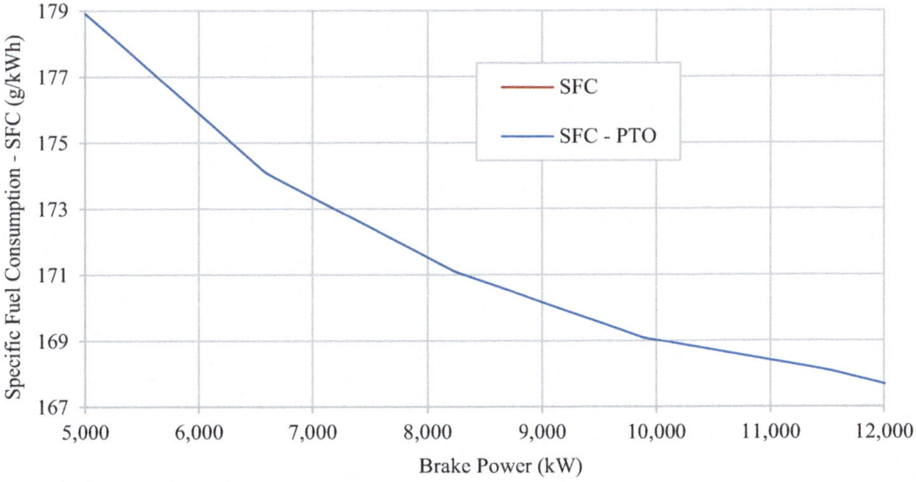

Fig. 5. SFC difference at PTO and without shaft generator

The EEOI values are presented in Fig. 6 against the SFC obtained when applying the shaft generator control scheme. The SFC decreases while the engine load increases, which gives a set of EEOI values in accordance with the amount of fuel consumed over the navigation period and the navigated distance, as was described before as the initial conditions for simulation. The increment of EEOI is considered low over the engine load and reflects the decrease of the SFC. Although the decrease of the SFC is not quite significant allows for a decrease of the EEOI value, which leads to a reduction of the amount of operational CO_2 emissions.

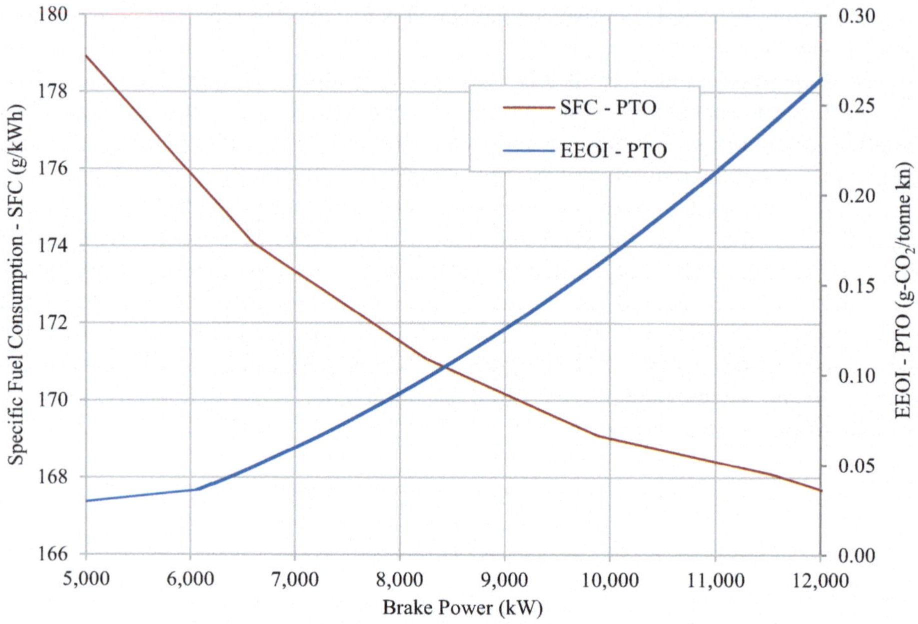

Fig. 6. EEOI against SFC - PTO

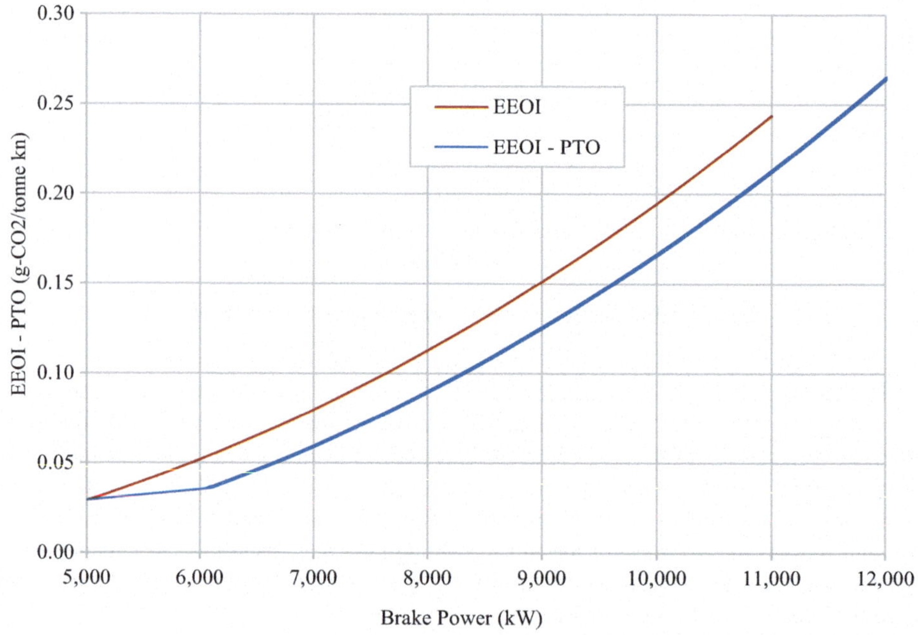

Fig. 7. EEOI comparison

Figure 7 shows the results of comparing the EEOI values of the applied control scheme, red curve, against not having a shaft generator installed, blue curve. Results are showing that the control scheme applied makes the ship to reduce the amount of operational CO_2 emissions even though an extra amount of brake power is necessary to be generated providing great assertiveness of the methodology and the SFC reduction over the period when operating at the PTO operating region.

4.2 Control Scheme Performance at PTI Operating Region

Following the same procedure to describe the results of the PTO operating region, the PTI operating region results are shown. Blue curves in Fig. 8 shows the results of a segment of the PTI operating region while simulating a period of navigation time of 2,5 h where in the left hand side of the figure is possible to appreciate how the brake power lineally increases from 24 MW to 27 MW when not having a shaft generator, red curve, which also means an increase in the SFC as can be seen in the right-hand side of the figure. When applying the control scheme to operate the shaft generator, the PTI reduces the brake power, blue curve on the left-hand side of the figure therefore, a reduction of the SFC as can be seen in the right-hand side of the figure. The PTI reduces the brake power around 5% of the SMCR.

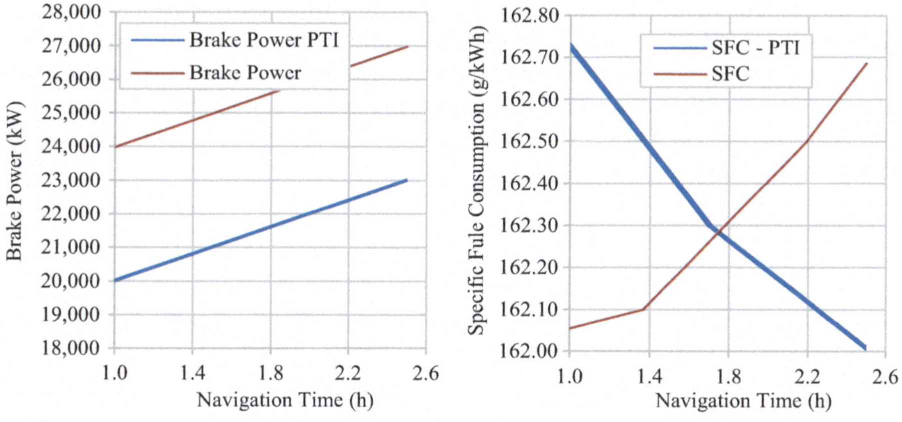

Fig. 8. PTI evaluation performance

Figure 9 shows the results of the SFC variation at PTI operating region when applying the proposed control scheme, blue curve, and the red curve shows the SFC variation when not having a shaft generator. The SFC differences are between −0,4% to 1,5% over the whole engine load range plotted. The difference reflects that even though the engine load increases, for PTI operation, the SFC decreases because the engine goes back to the high efficiency region operation or MEOP described before. On the other hand, the engine without a shaft generator increases the SFC, as expected, because of the operation away of the MEOP.

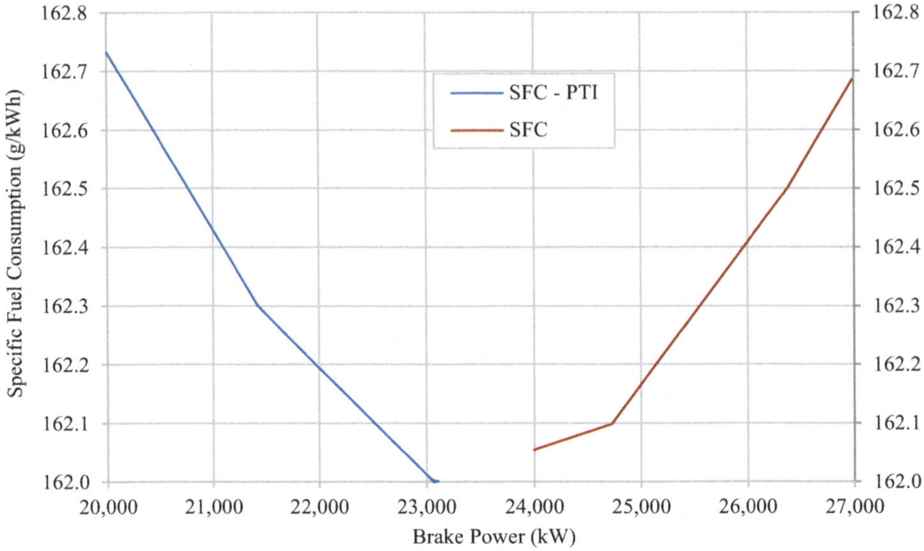

Fig. 9. SFC difference at PTI and without shaft generator

The EEOI values are presented in Fig. 10 against the SFC obtained when applying the shaft generator control scheme. The SFC decreases while the engine load increases, which gives a set of EEOI values in accordance with the amount of fuel consumed over the navigation period and the navigated distance. The lineal increment of the EEOI is considered low over the engine load and reflects the decrease of the SFC because of the control scheme applied. The operational emissions are reduced in accordance with the reduction of the SFC.

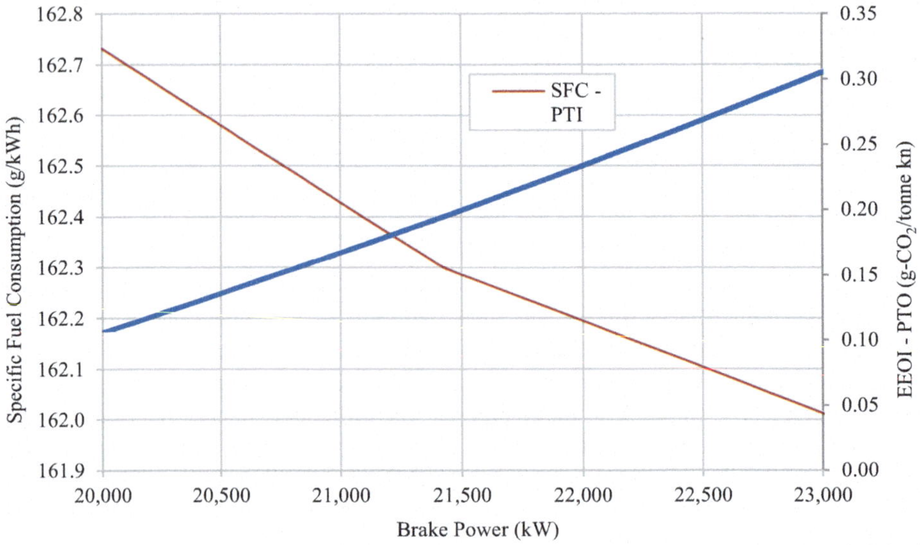

Fig. 10. EEOI against SFC – PTI

Figure 11 shows the results of comparing the EEOI values of the applied control scheme, blue curve, against not having a shaft generator installed, red curve. Results are showing that the control scheme applied makes the ship to reduce the amount of operational CO_2 emissions because is making the engine to work closer to MEOP providing great assertiveness of the methodology.

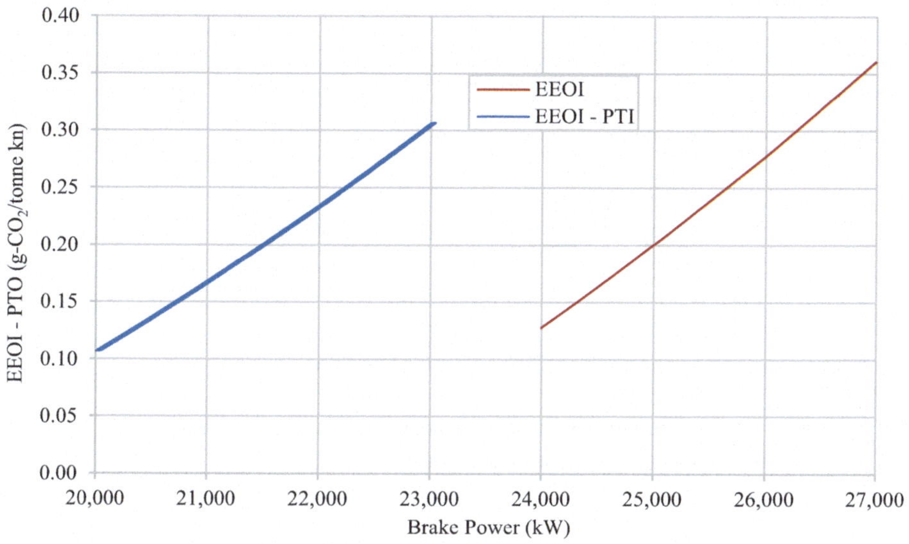

Fig. 11. EEOI comparison

5 Conclusions

Results presented, differentiating the performance of the control scheme of the shaft generator, are quite significant when compared to not having a shaft generator installed.

Results of the fuel consumed by the engine when simulating the PTO and PTI conditions are not quite significant yet are helping to reduce EEOI values at both operating regions, which means less operational CO_2 emissions.

The control scheme when simulating the engine performance at PTI operating region makes the engine to work closer to the MEOP, which leads to low SFC therefore, low operational emissions.

The control scheme shows assertiveness and accuracy at both operating regions reducing the SFC.

The perturb and observe optimization function presents an accurate performance to obtain a local search for the minimum emissions point, starting at a random state. Future results may include the use of an adaptive perturbation function, in order to ensure full convergence when reaching the minimum emissions point.

Slow steaming was mentioned because of the application of the scheme yet needs to be worked out separately from a design stage to provide more accurate conclusions to this work.

The ship and engine data considered for the simulations is open source and provides great value to continue to be used in this research.

Annex A – EEDI Ship's Equation

$$\frac{\left(\prod_{j=1}^{M} f_j\right)\left(\sum_{i=1}^{nME} P_{ME(i)} * C_{FME(i)} * SFC_{ME(i)}\right) + \left(P_{AE} * C_{FAE} * SFC_{AE*}\right) + \left(\left(\prod_{j=1}^{M} f_i * \sum_{i=1}^{nPTI} P_{PTI(i)} - \sum_{i=1}^{neff} f_{eff(i)} * P_{AEeff(i)}\right) C_{FAE} * SFC_{AE}\right) - \left(\sum_{i=1}^{neff} f_{eff(i)} * P_{eff(i)} * C_{FME} * SFC_{ME}\right)}{f_i * f_c * Capacity * V_{ref} * f_w}$$

- Main engines emissions:

$$\left(\prod_{j=1}^{M} f_j\right)\left(\sum_{i=1}^{nME} P_{ME(i)} * C_{FME(i)} * SFC_{ME(i)}\right)$$

Where:

f_j Correction factor for ship specific design elements.
P_{ME} Power of main engines.
C_{FME} Main engine conversion factor between fuel consumption and CO_2 emission.
SFC_{ME} Main engine specific fuel consumption.

- Auxiliary engines emissions:

$$\left(P_{AE} * C_{FAE} * SFC_{AE}\right)$$

Where:

P_{AE} Power of auxiliary engines.
C_{FAE} Auxiliary engine conversion factor between fuel consumption and CO_2 emission.
SFC_{AE} Auxiliary engine specific fuel consumption.

- Shaft generators/motors emissions:

$$\left(\left(\prod_{j=1}^{M} f_j * \sum_{i=1}^{nPTI} P_{PTI(i)} - \sum_{i=1}^{neff} f_{eff(i)} * P_{AEeff(i)}\right) C_{FAE} * SFC_{AE}\right)$$

Where:

f_j Correction factor for ship specific design elements.
P_{PTI} Power of shaft motor divided by the efficiency of shaft generator.
f_{eff} Availability factor of innovative energy efficiency technology.

P_{AEeff} Auxiliary power reduction due to individual technologies for electrical energy efficiency.
C_{FAE} Auxiliary engine conversion factor between fuel consumption and CO_2 emission.
SFC_{AE} Auxiliary engine specific fuel consumption.

- Efficiency technologies:

$$\left(\sum_{i=1}^{neff} f_{eff(i)} * P_{eff(i)} * C_{FME} * SFC_{ME}\right)$$

Where:

f_{eff} Availability factor of innovative energy efficiency technology.
P_{eff} Output of innovative mechanical energy efficient technology.
C_{FME} Main engine conversion factor between fuel consumption and CO_2 emission.
SFC_{ME} Main engine specific fuel consumption.

- Transport work:

$$f_i * f_C * Capacity * V_{ref} * f_w$$

Where:

f_i Capacity factor.
f_c Cubic capacity correction factor.
V_{ref} Ship speed.
f_w Weather factor
Capacity or dwt rating for VLCCs.

Annex B – EEOI Ship's Equation

$$EEOI = \frac{\sum_j F_j * C_{Fj}}{m_{Cargo} * D}$$

Where:

j Fuel type.
FC_j Mass of consumed fuel j.
C_{Fj} Fuel mass to CO_2 mass conversion factor for fuel j.
m_{cargo} Cargo carried (tonnes)
D Distance in nautical miles corresponding to the cargo carried.

References

1. Bilgin, B., Magne, P., Malysz, P., Yang, Y., Pantelic, V., Preindl, M., et al.: Making the case for electrified transportation. IEEE Trans. Transp. Electrif. **1**(1), 4–17 (2015)
2. MEPC.212(63) Annex 8 - 2012 Guidelines on the Method of Calculation of the Attained EEDI for New Ships (2012)
3. Circ.681 - Interim Guidelines on the Method of Calculation of the Energy Efficiency Design Index for New Ships (2009)
4. Prousalidis, J., Patsios, C., Kanellos, F., Sarigiannidis, A., Tsekouras, N., Antonopoulos, G. (eds.): Exploiting shaft generators to improve ship efficiency. In: 2012 Electrical Systems for Aircraft, Railway and Ship Propulsion, 16–18 October 2012 (2012)
5. Sarigiannidis, A.G., Chatzinikolaou, E., Patsios, C., Kladas, A.G.: Shaft generator system design and ship operation improvement involving SFOC minimization, electric grid conditioning, and auxiliary propulsion. IEEE Trans. Transp. Electrif. **2**(4), 558–569 (2016)
6. Sarigiannidis, A., Kladas, A., Chatzinikolaou, E., Patsios, C. (eds.): High efficiency shaft generator drive system design for Ro-Ro trailer-passenger ship application. In: 2015 International Conference on Electrical Systems for Aircraft, Railway, Ship Propulsion and Road Vehicles (ESARS), 3–5 March 2015 (2015)
7. Molland, A.F., Turnock, S.R., Hudson, D.A. (eds.): Ship Resistance and Propulsion: Practical Estimation of Ship Propulsive Power. Cambridge University Press, Cambridge (2011)
8. Shi, W., Grimmelius, H.T., Stapersma, D.: Analysis of ship propulsion system behaviour and the impact on fuel consumption. Int. Shipbuild. Prog. **57**, 30 (2010)
9. Woud, H.K., Stapersma, D. (eds.): Design of Propulsion and Electric Power Generation Systems. IMarEST, Institute of Marine Engineering, Science and Technology, London (2002)
10. Circ.684 - Guidelines for Voluntary Use of the Ship Energy Efficiency Operational Indicator (EEOI) (2009)
11. Griffiths, D.: Marine Low Speed Diesel Engines. Institute of Marine Engineers (2006)
12. MEPC.215(63) Annex 11 - Guidelines for Calculation of Reference Lines for Use With the EEDI (2012)
13. Wu, B., Narimani, M.: High-Power Converters and AC Drives. Wiley, Hoboken (2016)
14. Holtz, J.: The representation of AC machine dynamics by complex signal flow graphs. IEEE Trans. Ind. Electron. **42**(3), 263–271 (1995)
15. Novotny, D.W., Lipo, T.A.: Vector Control and Dynamics of AC Drives. Clarendon Press, Oxford (1996)
16. Clarksons. Clarksons World Fleet Register (2019). www.clarksons.net/wfr/
17. MAN Diesel & Turbo. MAN Diesel & Turbo Engine Builder (2019). www.mandieselturbo.com

Knowledge Management Applied to Productive Process

Pedro Gómez[1(✉)] and Marlis Angulo[2(✉)]

[1] ASMAR, Talcahuano, Chile
pgomezs@asmar.cl
[2] MUTUAL SER EPS, Cartagena, Colombia
maangulo@mutualser.org

Abstract. Considering that the naval industry is increasing in complexity and competitiveness and according to the new technological challenges that are projected in the Shipyards, it is necessary to innovate in the production processes, in order to bolster the company and make it more competitive with respect to their peers.

It is believed that the different existing methods of Knowledge Management, from now on K.M., transforming this from tacit to explicit, will enable the Shipyard to innovate and therefore, save on costs of the different processes of the shipbuilding industry.

Regarding the previous point, the creation of a new K.M. model is proposed, from known and previously chosen models. There are historical precedents about different procedures previously carried out. It is recognized in other shipyards where they agree that the greatest difficulty in construction stages is the lack of explicit knowledge and implementation within an established organizational culture, as well as the satisfaction of the staff consulted.

Thus a diagnosis of 18 models of K.M. is made, evaluating one by one, analyzing the possibility of implementing and intervening one of these in the processes of the production department of a Shipyard.

For this investigation, a brief survey was conducted at all levels, with the objective of determining which are the most critical and most interesting points of each productive process and analyzing the procedures that existed previously as well as finding the weaknesses of it and implementing a new work methodology based on a model of K.M. and take it to the reality of the Shipyard. This data collection was made to a universe of 110 people, including operators, supervisors and engineers of the workshops dependent on the production department of the Shipyard.

The results show evidence of dissatisfaction in one of the existing procedures, which is why, along with the development of this survey, the study and investigation of the methods of K.M. continue, and three methods were selected, that were close to the requirements established to solve this problem. Finally, the model proposed by Nonaka and Takeuchi was chosen; they propose that tacit knowledge be transformed into explicit knowledge.

From this new knowledge, documentation must necessarily be developed, so that it remains as evidence for the organization, complying with the integrated management system based on one of the standards in force at the Shipyard, ISO 9001.

V. A. J. E. Carreño Moreno et al. (Eds.): CIDIN 2019/COPINAVAL 2019, pp. 275–286, 2020.
https://doi.org/10.1007/978-3-030-35963-8_23

The correct use of information and the transformation of the K.M., is key in the evolution of the projects; This knowledge and evaluation criteria were applied to the construction strategy of the shipyard under study, developing a document capable of contributing to the next projects to improve planning levels, obtaining as a result through its use, More detailed, efficient and effective production levels.

The methodology applied for the development of this document was to categorize the different stages of the construction of any naval artifact, knowing all the limitations of the Shipyard. Once identified, they are separated into three main parts, which allows the development of a report oriented towards productive processes, taking into account the sequence of assembly, installation and integration of each block or zone in the different stages of the Project.

Keywords: Knowledge Management · Shipbuilding

1 Introduction

The history of Naval Construction in Chile goes back to the time of the Spanish Conquest, although according to historical background, it began with the manufacture of handmade artefacts made by native peoples such as. So in Chile, it is during the conquest when this industry find, where this industry in Chile, finds its most important impulse due to the policies imposed by Pedro de Valdivia, who commands the construction of the first ships, establishing Valparaíso as the main port of our country.

With the advance of the conquest and the passing of the years, small shipyards were established throughout the national territory, giving rise to a tradition that until today prevails in our country.

Globally, the shipbuilding industry has been characterized by its intensity in the Use of labor and high entry barriers that require large investments in infrastructure and Staff training to maintain high quality standards and their level of local competitiveness, regionally and internationally.

A final study conducted in 1995 by the Directorate-General of Maritime Territory and Merchant Marine (DIRECTEMAR), from that year to 2016, showed that the shipyards defined as "constructor shipyard" increased from 6 to 17 throughout our territory, being the "Los Lagos" Region which presented a greater increase in its capacity to manufacture naval artifacts.

One of the companies with the greatest tradition and international presence in the area of Shipbuilding and Ship Repair is the ASMAR Shipyard, which since 1895, maintains the tradition of both Naval Construction and Ship Repair. ASMAR' goal is to have an administration, infrastructure, processes and trained personnel Capable of maintaining at a high level the units of the Navy, making available to the Chilean Navy and the shipping companies of the world Their repair and naval construction capabilities, in order to achieve full employment, reaching costs that transform it into a more efficient and competitive company.

The internationalization of the Naval Construction industry has forced the development of increasingly competitive and complex technologies and methods. Usually, this evolution is somewhat more laborious in its initial stages, due to the large number of details and preliminary technical information that a project has. A high and finished definition in this phase will make it possible to lower the overall costs of any unit under construction, due to A proper planning of the works and a controlled management of materials and project equipment supply, reflecting this management in its design stages and production, decreasing productivity costs and increasing the volume of elements assembly in the early stages of construction.

2 Basic Design Processes

Many lessons have been obtained from the experiences derived from important European shipyards, especially when it comes to the efficiency with the efficiency of methods of the productive processes and above all the philosophy of continuous improvement. In order to carry out this new way of thinking, an important qualitative leap must be made in terms of Knowledge Management, oriented necessarily towards production centers, transferring that knowledge from the shipyard's design offices and making that leap that is sought with the "production-oriented design".

A brilliant answer is not enough to solve the client's needs; in addition, the technical directions must provide the most efficient and simple solutions from the economic point of view which will benefit the Shipyard.

It is not just about considering certain factors of a Naval Construction project, such as the acquisition of quality components which adjust to the cost of the project, the elimination of traditionally used elements, replacing them with other equivalents of lower cost and same quality, but to focus fully on the previous planning of the project, with the aim of anticipating all the current and future problems which the shipyard will face in any work it plans to execute.

The information that will derive from the design to the different production areas must be reduced and simplified, but like any change, the transition towards this new philosophy, both for the technical offices and the production centers, will always be difficult to adopt; therefore, we must always have trained personnel willing to implement this new work condition that enables the transference of tacit to explicit knowledge.

The traditional attitude in the projects had been to limit oneself to achieving a good technical solution, focusing most of the time on executing and finishing a construction project at all costs. Now the idea is to introduce the planning and economic components and together with them, new roles for the engineers of the production department of the Shipyard. The profile of this professionals requires more work, more technical preparation, more knowledge of new technologies and processes present in the market that benefit their management, in oder words, a polyvalent professional.

Knowledge is identified as a key factor to achieve competitive advantages. Within the organization, it is an intangible asset of high value and this is where the concept of "Knowledge Management" comes to light, a tool that represents in a simplified, summarized and symbolic way, the description of processes and structures; it seeks to create strategies and provide relevant data for productive processes.

The following is the revision of some models developed for the K.M. and their principles and contributions.

3 Selection of Methods for the Knowledge Management

According to the RAE, method is the orderly and systematic way to proceed to obtain a result or a determined goal. There are numerous studies on K.M. in the literature, as well as many authors present in these investigations.

On its 20th anniversary, the organization CEGESTI (Knowledge Management Center), and in its edition No. 135, 2001, publishes the article called "Implementation of Knowledge Management in the Enterprise", a study by Mr. Humberto Pereira Alfaro, where he defines knowledge as: "the whole set of cognitions and abilities with which individuals usually solve problems, includes both theory and practice, daily rules as well as instructions for action, knowledge it is based on data and information, but unlike these it is always linked to people; it is an integral part of individuals and represents their beliefs about causal relationships" (Probst 2001).

There are three important classifications of knowledge:

(a) Explicit - Tacit Knowledge:
The difference between these two was presented by Ikujiro Nonaka, known for his study on knowledge management; He argues that explicit knowledge or also called coded knowledge, is the best known, since this knowledge is finally transformed into transmittable information, something more tangible; unlike the tacit knowledge, which has a much more individual character and has to do directly with the subject, which makes it much more difficult to formalize and therefore transmit.

(b) Generative, Productive and Representative Knowledge:
In their book "Knowledge and Value: A New Perspective on Corporate Transformation", (Solveig Wikstrom 1994), they focus their research towards an organizational culture based on learning; they take this knowledge, create value and finally put it into practice.
This generative knowledge speaks of the process and the creation of new knowledge, which has happened from an eventual solution of problems or the identification of new proposals or alternatives for new opportunities to generate ideas; This is used later in the production or service processes, which, in turn generate a type of applied knowledge, that is specified in the products or results, with an explicit nature and with use value (Núñez Paula 2004).

(c) Conceptual, operational and instrumental knowledge:
Conceptual knowledge encompasses everything that is through the study of knowledge, obtaining from the experts the knowledge base (Núñez Paula 2004).
Operational knowledge carries knowledge from the different methods and procedures that can be combined as implementation alternatives. This knowledge also consists of knowing how operations should be carried out; it is an opportunity to develop other skills (Núñez Paula 2004).
The instrumental knowledge: to apply this knowledge, you must have all the available tools and different techniques, since basically, this type of knowledge are based on knowing how to develop and apply the instruments, consequently (Núñez Paula 2004).

3.1 Selection of Models for the KM

Sánchez Díaz 2005, in his summary called "Brief inventory of models for knowledge management in organizations", describes 18 different numbers of K.M.:

1	Balanced Scorecard model (Kaplan y Norton, 1996)	10	Knowledge Practices Management model, (Tejedor y Aguirre, 1998).
2	Technology Broker model (Annie Brooking, 1996)	11	Nova model, (Nova Care, 1999)
3	Canadian Imperial Bank model (Hubert Saint-Onge, 1996)	12	Andersen model, (1999)
4	West Ontario University model (Bontis, 1996)	13	Knowledge Management Assessment Tool, (Arthur Andersen y APQC, 1999).
5	Skandia Navigator model (Leif and Malone, 1997)	14	EFQM Model of Excellence
6	Intangible Assets model, (Sveiby, 1997)	15	Operations Intellectual Capital Benchmarking System, (OICBS Viedma, 2001)
7	Nonaka y Takeuchi model, (1995)	16	Innovation Intellectual Capabilities Benchmarking System, (IICBS Viedma, 2001)
8	Intelect model, (Euroforum, 1998)	17	Social Capital Benchmarking System Model, (SCBS, 2001)
9	Strategic Management model by competition: Intangible capital (Bueno, 1998)	18	Cities Intellectual Capital Benchmarking System. (CICBS, 2001)

For this study, three models are pre-selected to deepen the study so that only one model of K.M. is selected to be applied to the production processes of ASMAR Shipyard.

3.1.1 Model of Nonaka and Takeuchi

It makes the distinction between two types of knowledge, the tacit and the explicit; in simple words is the knowledge that is transferred from an intellectual base and is carried to something tangible (See Fig. 1).

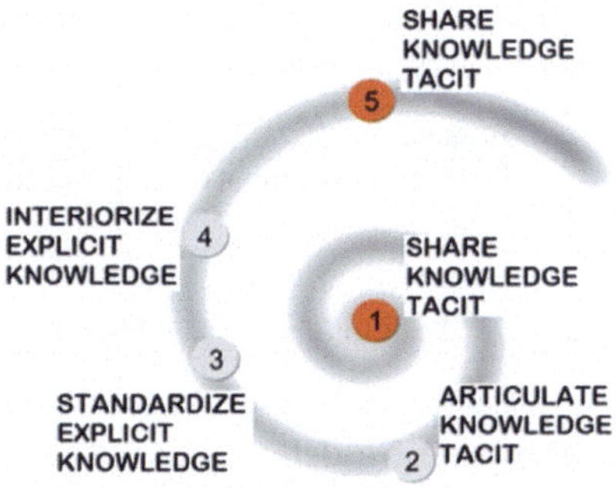

Fig. 1. Nonaka and Takeuchi Model

The relationship that exists between these two knowledge forms the basis of this proposed model and the combination of these leads us to: From Tacit to Tacit (1–5): People acquire knowledge directly from others, from experiences, training, observation, imitation, practice, etc.

From Tacit to Explicit (2): This knowledge model is acquired through dialogue, reading, etc., with the ultimate goal of expressing it in some kind of tangible document.

From explicit to explicit (3): Different types of forms of this knowledge converge, whether documents, bibliographic sources, databases, etc. Finally, this explicit knowledge is exchanged from this documents and something tangible is formed.

From explicit to tacit (4): People internalize the knowledge of the explicit, taking it based on their own experience in tacit knowledge through mental forms or models or usual work practices.

3.1.2 Technology Broker Model

It is based on the concept of market value of companies, the sum of intangibles and intellectual capital. This model gives greater importance to qualitative than quantitative issues. Brooking deduces the need to develop a methodology to audit information related to intellectual capital.

Intangible assets are classified into four categories, which constitute the intellectual capital (See Fig. 2).

Market assets

- Intellectual property assets
- Human assets
- Infrastructure assets.

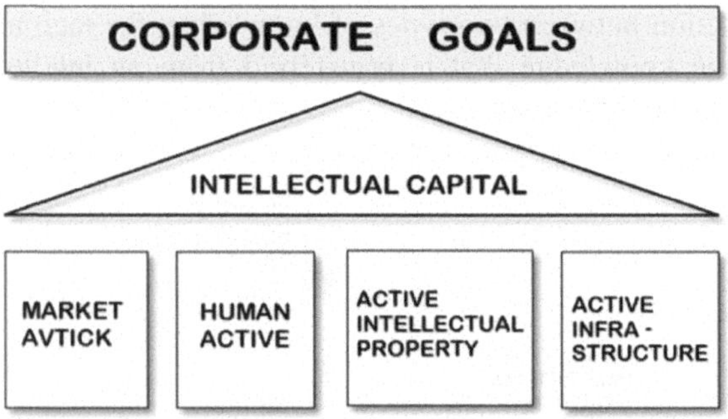

Fig. 2. Technology Broker Model

3.1.3 University of West Ontario Model

The capital of knowledge is composed of a holistic system of three elements: human capital, structural capital and client capital. This model studies the relationship between intellectual capital and its measurement, as well as organizational learning (See Fig. 3).

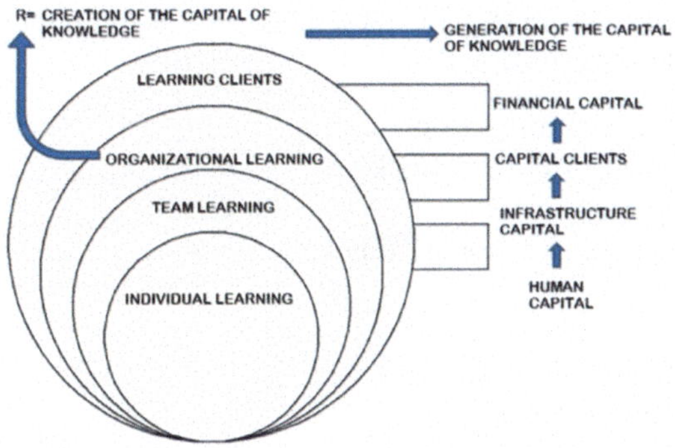

Fig. 3. University of West Ontario Model

4 New Model Proposal: L.O.R.E (Logistics, Organization, Resources, Strategy)

Finally and after investigating the three models of the K.M. previously selected the creation of a new own model and method of K.M. is proposed, but focused on the processes and realities of the Shipyard. This new model is the combination of the central focuses of Nonaka and Takeuchi, Technology Broker and the University of West Ontario. This model focuses on the Vision and Strategy, fundamental axes that the Production Department of the Shipyard seeks to exploit and develop through the specialty of Constructive Strategy, and the proposed model L.O.R.E (See Fig. 4).

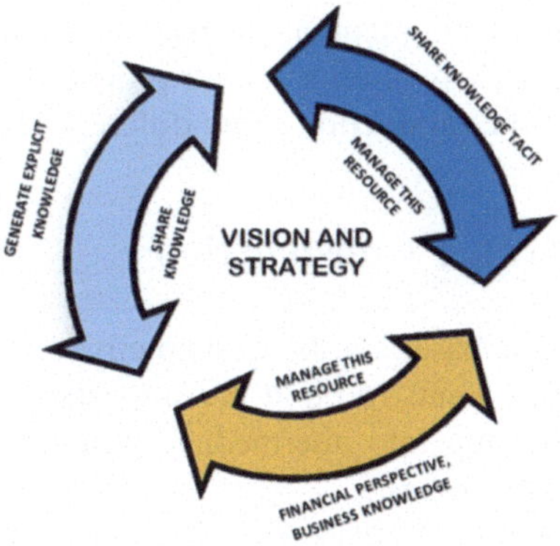

Fig. 4. Proposed model L.O.R.E. (Source: own development)

The expected benefits with this model applied in the shipbuilding processes are: MATERIALS MANAGEMENT

- Supply of materials according to plan.
- Efficient management of critical stock.
- Reduction of inventory costs.
- Optimization of spaces and storage times.

PLANNING

- Maximize the satisfaction of demand.
- More detailed planning levels.
- Efficiency in critical equipment guarantees times.

PRODUCTION

- Efficiently supply the requirements flow.
- Maximize productive efficiency.
- Reduction of time in dispatch of information to the work areas.
- Order In the activities dispatch.
- Reduction of accident rate.
- Improvement in ergonomics in the execution of works.

From the point of view of material management, a production plan must balance the fulfillment of three commonly conflicting objectives:

- COMMERCIAL, maximize demand satisfaction.
- FINANCE, reduce inventory levels.
- PRODUCTION, maximize productive efficiency.

From the point of view of the planning management, it consists basically of the previous analysis of the client's requirements; all proposed in the high level requirements and programming of the production in its different levels.

5 Definition of Integrated Construction

The Integrated Construction is based on the Constructive Strategy or a pre-established plan with which the shipyard wants to build the ship; therefore, it defines as soon as possible, the division of the ship into intermediate products and the classification of production work in zones and stages.

Given that the definition of the Constructive Strategy is made prior to the beginning of the project, it requires a work with preliminary documentation and, therefore, the process is iterative.

These Definition activities of the Constructive Strategy are based on the application of Concurrent Engineering techniques, where all the departments involved in the construction of a ship participate: Planning, Production, Engineering, Purchasing Department, Supply (Warehouse), Quality Control and Project Managers.

6 The Constructive Strategy

It is an activity that is carried out by a multidisciplinary team, which is established by resolution of the company and directed by a specialist engineer of the production department. The following areas participate in this team:

- Design Department.
- Logistic unit of N.C. of the design department.
- Planning. (Dept. of design, Dept. of Planning and Control and Dept. of Production)
- Logistic monitor of the dept. of planning and control.
- Shipyard production centers.
- Other shipyard workshops in support of production.
- Purchases and supply center.
- Quality Control.

7 Organizational Culture and Job Satisfaction

Organizational culture is an idea in the field of organizational and management studies that describes both personal and cultural psychology, attitudes, experiences, beliefs and values of an organization.

Job satisfaction has been defined as the result of various attitudes that a worker has towards his job, the concrete factors (such as the company, supervisor, co-workers, salaries, promotions, working conditions, etc.) and life and workplace in general (Blum 1990).

Implementing a K.M. system and taking it from the tacit to the explicit is easy; what is difficult is to bring this new concept to the personnel who will carry out the work in the different work areas. The predominant culture in the organization is to keep this information and not share it, so the barrier to overcome was the fear of sharing knowledge and experiences.

It is important to discover how knowledge develops in the company, taking into account individuals and groups, in order to consider the spread of ideas in processes of innovation and improvement.

Through induction training to the personnel of the different workshops dependent on the production department of the Shipyard, this barrier was broken down and a different, more friendly, more cooperative culture was developed and willing to take on this new knowledge and transmit it to those who did not have the possibility to attend the talks. That is why after these sessions were finished representatives who met certain characteristics of leaders were appointed in each workshop, and were designated as "knowledge facilitators", they will be responsible for assuming this commitment to instruct the staff with the new K.M. tool for the constructor shipyard, the new model of knowledge management, L.O.R.E.

8 Work Contribution

There are associated risks, related to the establishment of new management systems; the most frequent risk happens when trusting excessively on the theory and imposing a "data base", relying on people to make their contributions spontaneously, without having first generated the appropriate internal policies.

Another risk factor is given to start with a program too ambitious, since sharing knowledge is something unnatural in some people or organizations. It is therefore preferable to start with a pilot project that allows the results to be clearly measured in a short period of time.

Finally, the greatest danger is to isolate the project, turn it into the domain of a few enlightened people and not involve the entire organization in it. For this, the active participation of the Management and the exercise of the leadership constitute fundamental elements to achieve success.

Although the K.M. it is an intangible, it is generating a new type of capital, the intellectual capital, and in the sense that it is feasible to build indexes for the qualitative and quantitative measurement of it.

With this new method of K.M., seeks to contribute to the optimization of the manufacture times of various mechanical systems, miscellaneous and/or structures of any product.

The chart of Fig. 5 reveals the benefits of K.M. in the organization. Two very similar projects are compared in terms of structure and their mechanical systems, but one is more complex than the other in their communication and armament systems.

Both projects were scheduled in the same period; it is possible to determine at the end of the programmed time that the intervention with the K.M. in the "A" project with the new L.O.R.E. reduced the delivery times by 4 months, generating significant savings in manpower, equipment, supplies, etc., averaging a total cost reduction close to 0.93%. In addition, the chart indicates that the construction period of the vessel was kept according to the initial planning of the project, not having greater deviation during its entire execution period.

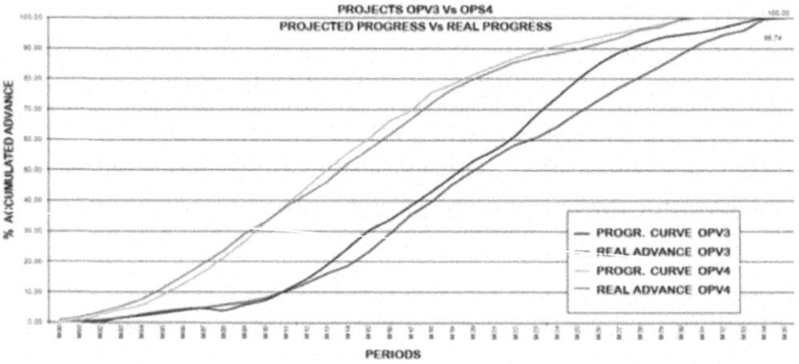

Fig. 5. Comparative graph per period Vs cumulative%, OPV3 v/s OPV4

9 General Objective

TO PROPOSE that through the K.M., the productive processes of the production department of a shipyard can be increased in a positive way, reducing time and costs applied to a project in execution.

10 Specific Objectives

TO INVESTIGATE the different models of the K.M. existing in the literature.

TO SELECT a model from K.M. that is applicable to the Naval Construction industry and applies it to a real project.

TO IMPLEMENT this model of the K.M. to the production processes of the ASMAR Production Department.

TO GENERATE through tacit knowledge, explicit knowledge demonstrating the proposed information.

TO CREATE a new model of K.M. focused on the productive processes and idiosyncrasy of Asmar Shipyard.

11 Discussion of Results and General Conclusions

The evolution of the manufacture of the ship under this methodology, which is based on the elaboration of intermediate products with a high level of integration of the structural elements with an advanced pre-outfitting, makes necessary to consider in the organization a department that has under its responsibility the outfitting process and a large part of the steel processes.

The most relevant aspects obtained were:

- Connecting all the departments involved in the construction of a unit with a single objective in common.
- Follow up of the construction at a individualized programming level per block, allowing greater control of delays and progress.
- It obliges the respective departments and/or workshops to carry out their own planning, assuming responsibility for their compliance.
- The project is kept updated, allowing a more efficient and effective feedback of the costs and degree of progress of the works.
- Deviations are detected and informed in time so that action can be taken and delays in the final term can be avoided.
- Erroneously budgeted items are reviewed when the uncertainty is large.
- The estimation of the project focuses on preparing tasks packages according to the planning and sending them to the workshops so that they comply with the requirements.
- This new model of knowledge management, L.O.R.E. they brought multiple benefits such as:

- The construction times are reduced, the remade works are almost completely eliminated, a working methodology is created and the whole shipyard is educated for its implementation; furthermore, the costs associated to the previously described items are reduced.
- The C.E., comprehendseverything. Through his staff and the different tools that this spatiality possesses, it is able to carry out the sequence and logical integration of the works in different stages of assembly of the unit under construction.
- It allows developing a pre-Outfitting at panel levels, establishing criteria for the assembly of these, with the aim of reducing interferences between the different executing workshops of the production department.
- There is evidence of a decrease in programming time and a greater efficiency ordering the dispatch of equipment and materials to the different productive fronts.
- It achieves a decrease of welding work at the level of the slipway, thus avoiding rework, burns in spaces already finished and reducing accident risks to our employees.
- Ergonomics is improved in the production works, avoiding risks associated with pre-occupational health.

Acknowledgments. The main author of this paper thanks the workers of "Astilleros y Maestranzas de la Armada ASMAR" who have contributed with the research on the application of Knowledge Management in the production processes of a shipyard, and who have inspired the Knowledge Management Model "L.O.R.E". Likewise, the author thanks the co-author Marlis Angulo, whose contribution and passion made this work possible.

References

Armasur, A.G.: Astilleros en Chile: Historia, problemas y proyecciones de un actor relevante en la economía nacional, Julio de 2015. Obtenido de www.armasur.cl, http://www.armasur.cl/noticias/astilleros-en-chile-historia-problemas-y-proyecciones-de-un-actor-relevante-en-la-economia-nacional/

Blum, M.L.: Psicología industrial: Sus fundamentos teóricos y sociales, México, D.F. (1990)

Núñez Paula, I.: 10 de mayo de 2004. http://www.scielo.cl, http://scielo.sld.cu/scielo.php?script=sci_arttext&pid=S1024-94352004000300004&lng=es&tlng=es. Recuperado el 01 de octubre de 2017

Núñez Paula, I.: La gestión de la información, el conocimiento, la inteligencia y el aprendizaje organizacional desde una perspectiva socio-psicológica (2004). http://www.scielo.cl/, http://scielo.sld.cu/scielo.php?script=sci_arttext&pid=S1024-94352004000300004&lng=es&tlng=es. Recuperado el 01 de octubre de 2017

Palancar, P.L.: 90 AÑOS DE CONSTRUCCIÓN NAVAL INTEGRADA Y MODULAR. Ingeniería naval (863), 74–91 (2008). ISSN 0020-1073

Probst, G., Raub, S.: Administrando el conocimiento, pág. 24. En Pearson Educación, México DF (2001)

Sánchez Díaz, M.: Breve inventario de los modelos para la gestión del conocimiento en las organizaciones (2005). http://www.scielo.cl, http://scielo.sld.cu/scielo.php?script=sci_arttext&pid=S1024-94352005000600006&lng=es&tlng=es. Recuperado el 01 de octubre de 2017

Solveig Wikstrom, R.N.: Knowledge and Value: A New Perspective on Corporate Transformation. Routledge, London (1994)

Comparison of Serrated Helical Strakes in Suppressing the Vortex-Induced Vibrations of a Circular Cylinder

Gustavo R. S. Assi[1(✉)] and Tommaso Crespi[1,2]

[1] Department of Naval Architecture and Ocean Engineering, University of São Paulo, EPUSP, São Paulo, Brazil
g.assi@usp.br

[2] Department of Civil and Environmental Engineering, Politecnico di Milano, Milan, Italy

Abstract. The present work presents an experimental investigation of the suppression of vortex-induced vibrations of a circular cylinder by means of helical strakes. Experiments have been per- formed in a water channel at moderate Reynolds numbers. Three different types of helical strakes are compared: (i) a continuous helical strake, (ii) a helical strake with serrated blades and (iii) a helical strake with serrated blades but inverted (negative) in relation to the helical pitch of the strake. While the serrated- blade strake appeared as an improvement over the conventional model, the novel negative-bladed strake produced worse results, allowing for the return of resonant VIV. The most effective VIV suppressor was the bladed strake, achieving a reduction of 88% in the peak amplitude of vibration when compared to that of a bare cylinder during resonance. We conclude that simply inverting the local angle of attack of the individual blades (or fins) in relation to the helical pitch of the strake does not improve the performance of the strake in terms of VIV suppression.

Keywords: Vortex-induced vibration · Suppression · Circular cylinders · Helical strakes

1 Introduction

The vibration provoked by the external flow around slender structures of a bluff cross-section poses a problem to submarine and offshore cables, flexible lines and pipes, drilling and production risers and other elastic structures exposed to sea currents. The flow-induced excitation originates in the shedding mechanism of alternating vortices occurring in the wake of bluff bodies, so the hydroelastic phenomenon is called vortex-induced vibration (VIV). Flexible lines exposed to vibrations for a long time or too many cycles may be damaged by structural fatigue [1]. The amplification of drag due to the vibration of the body is also of considerable concern, since it increases static and dynamic loads at the joints, platform and other fixtures.

One way to mitigate the effects of VIV is the installation of suppressors along the riser, or at least on the length of the line where currents are most intense. Helical strakes and fairings, for example, have been widely employed by the offshore industry as VIV

V. A. J. E. Carreño Moreno et al. (Eds.): CIDIN 2019/COPINAVAL 2019, pp. 287–297, 2020.
https://doi.org/10.1007/978-3-030-35963-8_24

suppressors [2]. On one hand, significant VIV suppression of low-mass-damping structures requires wider strakes, which increases drag. Fairings, on the other hand, tend to be more efficient in VIV suppression as far as drag is concerned, but may suffer from hydroelastic instabilities that may increase the response [3]. With the improvement of the plastic industry applied to offshore systems, helical strakes and fairings have indeed become sturdy devices, but they still take considerable time to install and occupy large areas on the deck. Other devices based on the disruption of the wake by interfering control surfaces (as explored by [4], for example) may suffer from similar problems.

During the last decades many devices have been investigated and offered as commercial products. Following the industry's demand for more efficient, robust and easy-to-install devices, the technological development for suppressing VIV has been under pursuit by both the scientific and industrial communities. Inspired by the conventional helical strakes, many have investigated other variations of this family of three-dimensional VIV suppressors [5]. While some have invested in new geometries to reduce the drag penalty, others invested in the robustness and ease of installation of the devices. New shapes, sizes and installation solutions appeared on the market. However, as far as offshore helical strakes are concerned, it appears that the height of the blades (with the consequent drag increase) is required to promote the appropriate three-dimensional flow interaction in the near wake to disrupt the formation of coherent vortices and suppress VIV.

Among so many different geometries of strakes (many achieved by empirical variations) there remain the scientific question on the fundamental fluid-structure interaction behind the suppression [6]. This paper is part of a wider investigation on the physical mechanisms behind the suppression of flow-induced vibrations. By proposing small variations on the geometry of helical strakes, we hope to shed new light on the intricate hydrodynamics behind the device.

1.1 Objective

Starting with the geometry of a conventional helical strake, we have divided the continuous blades into smaller, serrated blades. On a second step, the local angle of attack of each individual fin was inverted, without changing the helical pitch of the device. All geometrical parameters, except blade angle, were kept constant. Three models of strakes have been tested and their responses to VIV compared to that of a bare cylinder.

2 Method

Three slightly different models of helical strakes have been produced for this experimental investigation, as shown in Fig. 1. The first model represented a conventional helical strake – with three continuous blades helically wrapped around a bare cylinder – also known as a "three-start strake". This first model is referred to as the CS (from *continuous strake*) and served as a reference for the other two. Its dimensions have been inspired by commercial strakes available for the offshore industry. The helical pitch was set to $P/D = 5$ and the blade height to $h/D = 0.2$.

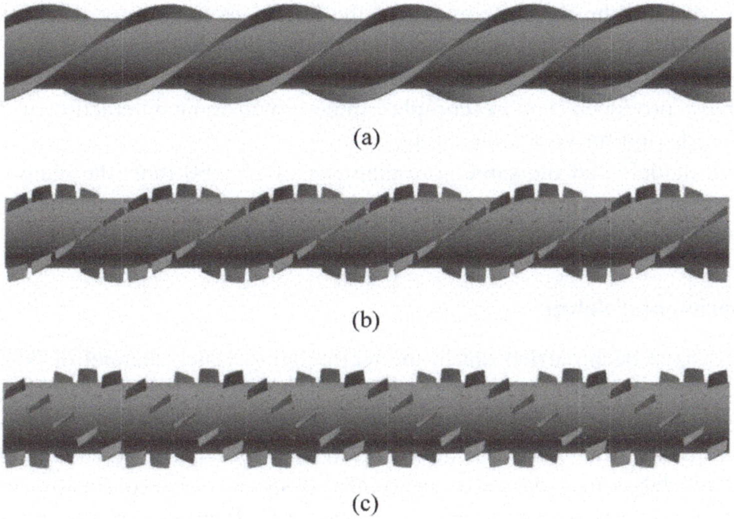

Fig. 1. Models: (a) CS, continuous strake; (b) BS, bladed strake; (c) NBS, negative-bladed strake.

The second model was a variation of the first: the continuous blade of the CS was broken into smaller segments, equally distributed along the cylinder span, producing identical serrated blades (or "fins"), now called the BS (from *bladed strake*). The span of a single segment containing one fin was $b_1 = D/3$, thus one full pitch of $5D$ contained 15 fins. The small fin had a trapezoidal geometry, with a longer root and a tapered tip defined by $b_2 < b_1$. Lengths b_1 and b_2 were measured parallel to the axis of the cylinder, as seen in Fig. 2. Helical pitch, blade height and overall length for the BS were kept the same as those for the CB. Serrated strakes are also available as commercial solutions for the industry.

The third model was produced by a modification of the BS. The segments of each individual fin were inverted so that the blades are orthogonal to the pitch path, as shown in the detail of Fig. 2. This model was called NBS, from *negative-bladed strake*, meaning that the blades are in the opposite orientation as the pitch. Again, *P/D, h/D* and *L/D* for the NBS were kept the same.

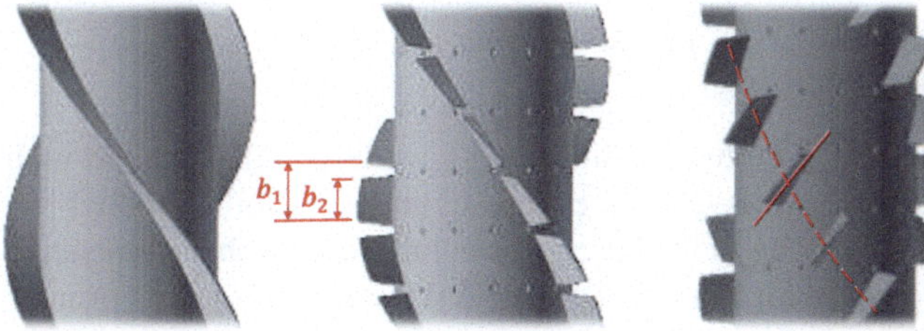

Fig. 2. Details of the CS, BS and NBS models shown in Fig. 1.

The idea behind the novel geometry of the NBS was to transfer momentum from the incoming flow to the spanwise direction in two different ways: (i) following the conventional pitch line and (ii) following the new angle of the individual blades. We hoped to have produced a more complex three-dimensional interaction in the near-wake region, disrupting vortex shedding.

All three models had the same core diameter of D = 60 mm (the diameter of the bare cylinder) and the same underwater length of L/D = 10, corresponding to at least 2 fully submerged pitch lengths.

2.1 Experimental Setup

Experiments have been carried out in the recirculating water channel of NDF (Fluids and Dynamics Research Group) at the University of São Paulo, Brazil. The water channel has a free- surface test section which is 0.7 m wide, 0.9 m deep and 7.5 m long. Good quality flow can be achieved up to 1.0 m/s with turbulence intensity less than 3%. This laboratory has been especially designed for experiments with flow-induced vibrations. For further details the apparatus, validation and information on the facilities please refer to [7–9].

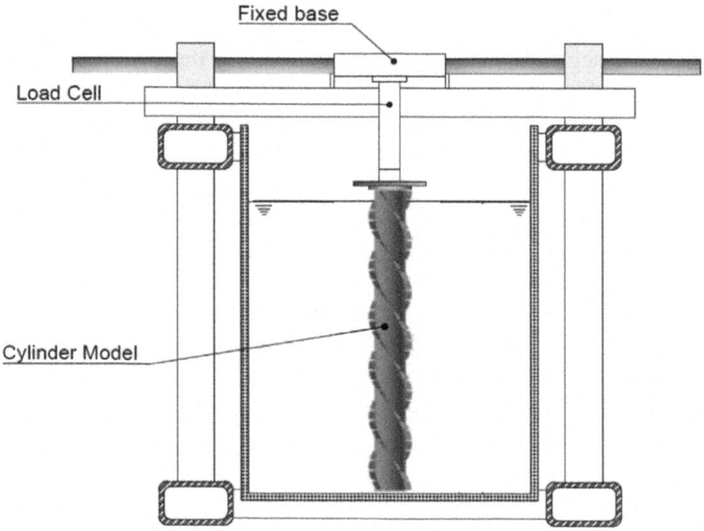

Fig. 3. Experimental setup.

A rigid section of a circular cylinder was made of ABS plastic with an external diameter D = 60 mm to serve as the bare cylinder. Another model with a slightly smaller diameter was made to be the core structure to hold the segments for the strakes in place. The segments for the CS, BS and NBS were rings 3D - printed out of ABS plastic.

Models were mounted on a one-degree-of-freedom elastic rig especially built by [10] for experiments on flow-induced vibration. A cross-view of the experimental setup installed in the water channel is presented in Fig. 3. The rig consisted of two long carbon-fibre pipes sliding through four air-bearings. The light structure allowed for displacements only in the cross-flow direction with very low structural damping. A pair of coil springs provided the necessary stiffness to the system and optical sensors measured the cross-flow displacement without adding extra damping.

Table 1. Parameters for the present investigation

Bare-cylinder diameter	D	60 mm
Underwater span Helical	L/D	11.7
Pitch	P/D	5
Blade height	h/D	0.2
Segment span	b_1	$D/3$
Mass ratio	m^*	0.6
Structural damping factor	ζ	0.75%
Mass-damping parameter	$m^*\zeta$	0.0045
Reynolds number	Re	5×10^3 to 5×10^4
Froude number	Fr	5×10^{-2} to 5×10^{-1}

The reference bare cylinder and the other three strake models were mounted under a load cell designed to deduce the instantaneous and time-averaged hydrodynamic forces on the sub- merged bodies. (Hydrodynamic loads will not be discussed in this paper).

The only flow variable changed during the course of the experiments was the flow velocity U, which alters the Reynolds number ($Re = UD/v$, based on the diameter D of the bare cylinder and the viscosity of water v) between 5,000 and 50,000.

Froude number ($Fr = U/\sqrt{gD}$, where g is the acceleration of gravity) was verified to be very low and wave effects have been neglected following [11] (see Table 1).

The reduced velocity $U/(f_N D)$ was defined by the natural frequency of the system (f_N) determined by decay tests performed in still water. As expected, the natural frequencies of the models with strakes were slightly lower than that of the bare cylinder. The structural damping factor of $\zeta = 0.75\%$ (expressed as a fraction of the critical damping) was determined during similar decay tests performed in air. The ratio of the total mass of the system to the mass of the displaced water was $m^* = 0.6$ for all models, thus producing a mass-damping factor of only $m^*\zeta = 0.0045$. Mass and damping were intentionally kept to a minimum to promote severe vibrations.

A summary of all the geometrical, structural and flow parameters investigated in the present experiment is presented in Table 1.

3 Results and Discussion

Preliminary VIV tests have been performed with a bare cylinder to serve as a reference for comparison. The typical VIV response – with three branches along the reduced-velocity axis – is clearly identified in the top plot of Fig. 4. In this paper, $\hat{y}/D$ is the RMS of displacement multiplied by $\sqrt{2}$, also known as the "harmonic amplitude of vibration". The synchronisation range occurred between $U/(f\_ND) = 3$ and 12, with a peak of amplitude of $\hat{y}/D = 0.8$ around $U/(f_N D) = 5$. As shown by [8, 12] and others in the literature, this response is typical of low-mass-damping systems.

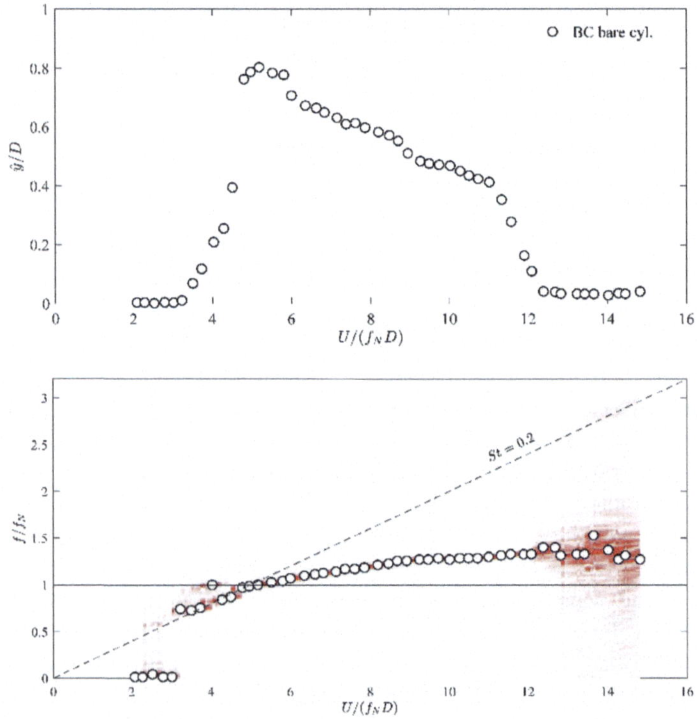

Fig. 4. VIV response for bare cylinder (BC): amplitude of displacement (top plot) and dominant frequency of oscillation (bottom plot).

The bottom plot of Fig. 4 shows f/f_N, the dominant frequency of oscillation normalised by the natural frequency of the system. The dashed line represents the frequency of vortex shedding associated with a Strouhal number of 0.2. The lighter and darker shades of red behind the symbols represent the peaks of the normalised spectra of vibration for each reduced velocity. The narrow peaks are hidden behind the symbols during the lock-in range, but the broader spectra are noticeable after the end of the synchronisation for $U/(f_N D) > 12$ (please refer to [7] for further details on this technique).

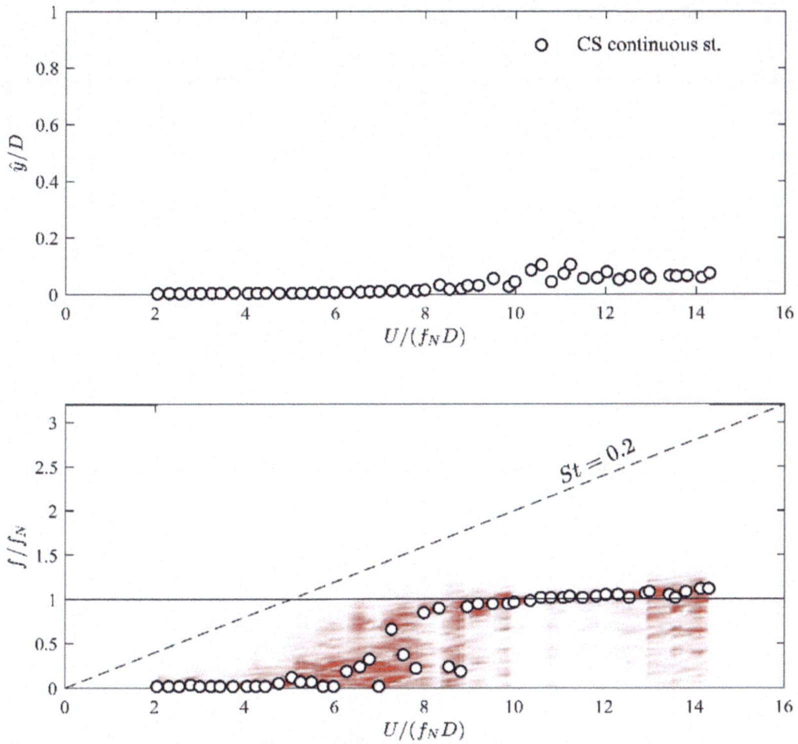

Fig. 5. VIV response for continuous strake (CS): amplitude of displacement (top plot) and dominant frequency of oscillation (bottom plot).

Figure 5 presents the VIV response of the CS (continuous strake) model. The conventional CS kept the amplitude of displacement below $\hat{y}/D = 0.15$ for the entire series, with minor vibrations appearing for higher reduced velocities past the resonance for $U/(f_N D) > 9$. The frequency signature (bottom plot) shows the vibrations for $U/(f_N D) > 9$ occur predominantly at a frequency near the natural frequency of the system ($f/f_N \approx 1$), which is typical of turbulence buffeting [13].

The response of the BS (bladed strake), presented in Fig. 6, revealed the most effective device in suppressing VIV, only producing a maximum of $\hat{y}/D \approx 0.10$ near the end of the synchronisation range. The frequency signature produced a much broader spectrum without a clear trend for the dominant frequency. There is no clear dominant frequency near $f/f_N = 1$, suggesting that the serrated blades also delay or even prevent buffeting.

The response o VIV of the novel NBS (negative-bladed strake) is shown in Fig. 7, highlighting that it was the less effective in terms of suppression. The NBS reached a considerable value of $\hat{y}/D \approx 0.47$ at resonance for reduced velocities around 5 and 6. The device was not as effective in suppressing VIV as the other two straked models and a resonant response was indeed observed. The frequency signature produced a much more predictable behaviour, with f/f_N following parallel to the $St = 0.2$ line. Turbulence buffeting was not observed as the response was predominantly a type of resonant VIV.

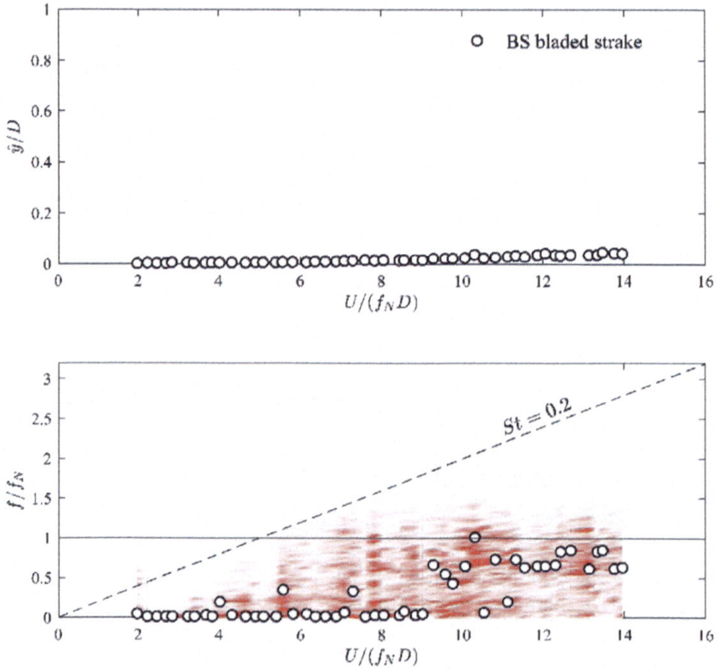

Fig. 6. VIV response for bladed strake (BS): amplitude of displacement (top plot) and dominant frequency of oscillation (bottom plot).

The CS and BS models performed very similarly as far as the amplitude of vibration was concerned. They all managed to keep $\hat{y}/D$ below 0.05 at the resonance (around $U/(f_N D) = 5$), suggesting that the greater displacements observed for higher reduced velocities could indeed be attributed to turbulence buffeting.

The frequency signatures of the straked models are rather interesting (Fig. 4). As reduced velocity was increased past the resonance at $U/(f\_ND) = 5$, CS and BS models tended to vibrate near the natural frequency in still water ($f/f_N \approx 1$). Once the incoming flow became more energetic (as U was increased) the CS model quickly jumped from a slow-drift regime ($f/f_N \approx 0$) to $f/f_N \approx 1$, a behaviour typical of buffeting. The BS model took a more erratic path with dominant frequencies between $f/f_N = 0$ and 1.

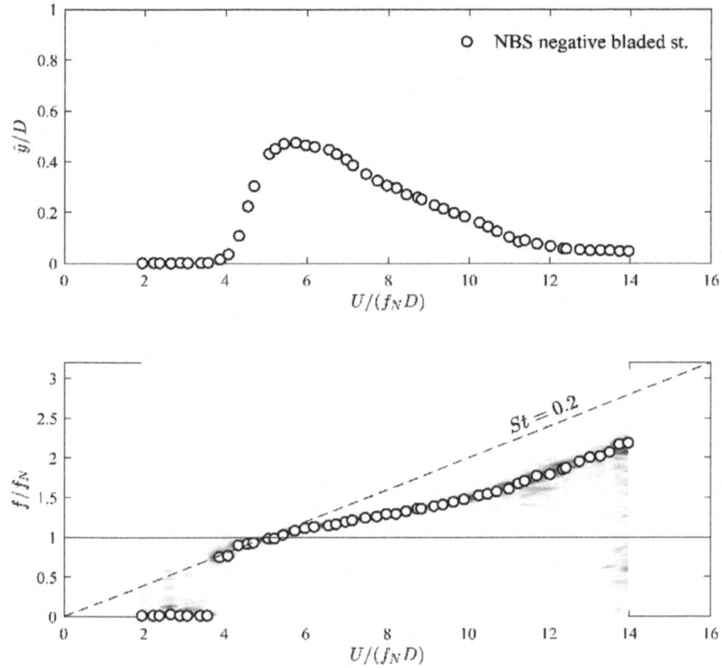

Fig. 7. VIV response for the negative-bladed strake (NBS): amplitude of displacement (top plot) and dominant frequency of oscillation (bottom plot).

4 Conclusion

In the present investigation we have evaluated the effective- ness to suppress cross-flow VIV of two types of helical strakes with serrated fins when compared to a conventional, continuous strake and a bare cylinder with a low mass-damping parameter. While the serrated BS (bladed strake) appeared as an improvement over the conventional CS model, the novel NBS (negative-bladed strake) model produced a worse result, allowing for the return of resonant VIV.

Table 2 summarises the main results comparing the maxi- mum amplitude of vibration produced by each model and the corresponding suppression ratio. The most effective VIV sup- pressor was the bladed strake (BS), achieving a reduction of 88% in the peak amplitude of vibration when compared to that of a bare cylinder during resonance. The novel negative-bladed strake (NBS) was the least effective VIV sup-pressor, achieving only a 42% reduction in the peak amplitude of vibration.

We conclude that simply inverting the local angle of attack of the individual blades (or fins) in relation to the helical pitch of the strake does not produce a favourable result in terms of VIV suppression. Perhaps the twisted blades annihilate the effect that the continuous blades produce in the near wake, allowing for the formation of more coherent vortices in the near wake to drive. Of course this hypothesis could not be tested in the present investigation and is left as an open question.

This paper is part of an ongoing research work to investigate the physical mechanisms behind the helical strakes. Future experimental investigation will concentrate on the complex vortex interaction in the near wake region.

Table 2. Summary of the results.

Model	$(\hat{y}/D)$max	Suppression
Bare-cyl.	0.81	Ref.
CS	0.15	81%
BS	0.10	88%
NBS	0.47	42%

Acknowledgment. We gratefully acknowledge support of the RCGI Research Centre for Gas Innovation, hosted by the University of São Paulo (USP) and sponsored by FAPESP São Paulo Research Foundation (2014/50279-4) and Shell Brasil. GRSA acknowledges the support of FAPESP (2011/00205-6) and CNPq (306917/2015-7).

References

1. Tognarelli, M., Taggart, S., Campbell, M.: Actual VIV fatigue response of full-scale drilling risers: with and without suppression devices. In: Proceedings of the 27th International Conference on Ocean, Offshore and Arctic Engineering (ASME-OMAE 2008), p. 57046 (2008)
2. Taggart, S., Tognarelli, M.: Offshore drilling riser VIV suppression devices - what's available to operators?. In: Proceedings of the 27th International Conference on Ocean, Offshore and Arctic Engineering (ASME- OMAE 2008), p. 57047 (2008)
3. Assi, G.R., Bearman, P.W., Tognarelli, M.A.: On the stability of a free-to-rotate short-tail fairing and a splitter plate as suppressors of vortex-induced vibration. Ocean Eng. **92**, 234–244 (2014)
4. Silva-Ortega, M., Assi, G.: Flow-induced vibration of a circular cylinder surrounded by two, four and eight wake-control cylinders. Exp. Therm. Fluid Sci. **85**, 354–362 (2017)
5. Korkischko, I., Meneghini, J.R.: Experimental investigation of flow-induced vibration on isolated and tandem circular cylinders fitted with strakes. J. Fluids Struct. **26**(4), 611–625 (2010)
6. Freire, C.M., Meneghini, J.R., Gioria, R.S., Assi, G.R.S.: Comparison between the Koopman modes for the flow around circular cylinder and circular cylinder fitted with helical strakes. In: Theofilis, V., Soria, J. (eds.) Instability and Control of Massively Separated Flows. Fluid Mechanics and its Applications, vol. 107. Springer (2015)
7. Assi, G.R.S., Bearman, P.W., Carmo, B.S., Meneghini, J.R., Sherwin, S.J., Willden, R.H.J.: The role of wake stiffness on the wake-induced vibration of the downstream cylinder of a tandem pair. J. Fluid Mech. **718**(3), 210–245 (2013)
8. Assi, G., Bearman, P., Kitney, N., Tognarelli, M.: Suppression of wake-induced vibration of tandem cylinders with free-to-rotate control plates. J. Fluids Struct. **26**, 1045–1057 (2010)
9. Assi, G.R.S., Bearman, P.W., Meneghini, J.R.: On the wake-induced vibration of tandem circular cylinders: the vortex interaction excitation mechanism. J. Fluid Mech. **661**(10), 365–401 (2010)

10. Assi, G.R.S.: Mechanisms for flow-induced vibration of interfering bluff bodies. Ph.D. thesis, Imperial College London (2009)
11. Chaplin, J., Teigen, P.: Steady flow past a vertical surface-piercing circular cylinder. J. Fluids Struct. **18**(3), 271–285 (2003)
12. Williamson, C.H.K., Govardhan, R.: Vortex-induced vibrations. Ann. Rev. Fluid Mech. **36**, 413–455 (2004)
13. Blevins, R.D.: Flow-Induced Vibration, 2nd edn. Van Nostrand Reinhold, New York (1990)

A Practical Analysis of the Influence of Tool Positioning During High-Frequency Mechanical Impact Treatment on a Welded Joint

Hector O. Ruiz[1(✉)], Naoki Osawa[1], Adan Vega[2], and Luis C. De Gracia[1]

[1] Department of Naval Architecture and Ocean Engineering, Osaka University, Osaka 565-0871, Japan
hector_ruiz@naoe.eng.osaka-u.ac.jp
[2] Zoartech Consulting, Panama City 0819-03568, Panama

Abstract. The fatigue strength improvement of materials and welded structures calls for calls for preciseness. The advantages of removing or mitigate tensile residual stresses and using the beneficial compressive residual stresses created by thermal and mechanical post-weld treatments are already known in welding communities. One solution is to improve the fatigue strength of welded structures by mechanical impact post-weld treatments. By an efficient High-Frequency Mechanical Impact (HFMI) treatment a significant improvement on local work hardening, reduction of notches effect, and fatigue strength could by obtained.

It is well known that HFMI is widely adopted for ships and offshore structures to improve fatigue life of welded joints. However, numerous aspects of this method need to be clarified or further investigated. The present study focuses on the influence of the position of the tool during HFMI treatment by numerically investigating the peening response on a welded joint under different positions, using a symmetric T-Welded Joint Model, with different peening tool angles regarding to the web and different impact positions around the weld toe.

Keywords: High-Frequency Mechanical Impact (HFMI) · Fatigue strength improvement · Compressive residual stress

1 Introduction

Most of the structures that we use in our daily life are assembled using welding; this has been used for many years due to its efficiency, versatility and reliability. However, it is well known that welding decreases the fatigue life of materials. The benefit of removing potential threats such as tensile residual stresses and exploiting the beneficial compressive residual stresses created by mechanical treatments are widely known.

In the last two decades, International Institute of Welding (IIW) Commission XIII on Fatigue of Welded Components and Structures has been investigating on High-Frequency Mechanical Impact (HFMI) technology development and experimental studies involving HFMI. Commission XIII has coined the term HFMI as a generic term

V. A. J. E. Carreño Moreno et al. (Eds.): CIDIN 2019/COPINAVAL 2019, pp. 298–307, 2020.
https://doi.org/10.1007/978-3-030-35963-8_25

to describe several related technologies for improving the fatigue strength of welded structures.

In 2007, Hagensen through IIW Commission XIII, proposed a best practice guideline of post-weld treatment methods for steel and aluminum structures. In this guideline the authors classified hammer peening as residual stress modification techniques which induce compressive residual stresses at the weld toe or in the contact region, increasing the fatigue life of welded joints.

The benefits of this treatment comes by impacting cylindrical indenters with a component or structure with high-frequency (> about 90 Hz). Marquis (2013) concluded that "HFMI introduces compressive residual stresses, work hardening, and reduction of the notches effect at the weld toe are the three main contributions of this treatment".

HFMI-process has been investigated for several researchers such as Foehrenbach (2016) and Ernould (2018). In order to take into account dynamic effect, an explicit elastic-plastic finite element (EPFE) commercial code (ABAQUS) was utilized.

In 2018, Lennart and Ruiz examined the effect of Finite Element (FE) mesh and various parameters such as boundary conditions, friction coefficient, tool indentations, strain rate, yield stress, etc., through the peening response of a flat stress-free-plate. However, there are large variations in the actual improvements achieved, and the results obtained by various methods are not always ranked in a consistent manner. One explanation for the observed variations is the lack of standardization of the optimum method of application, but variations in the material, type of loading and type of test specimens may also have influenced the results. The effectiveness of the treatment also depends heavily on the skill of the operator. Therefore, there is a need to investigate the peening tool angle and the impact positions at the weld toe to understand their effects on peening response near the top surface and the thickness direction.

In previous study one of the author Ruiz (2018) used the explicit commercial code MSC. Dytran to validate the proposed numerical method with experimental measurements carried out by Foehrenbach (2017) on flat stress-free plates. In this study, same methodology, software, material modelling and HFMI-conditions are used to investigate the effect of the peening tool on residual stresses, using a symmetric T-Welded Joint Model, with different peening tool angles and impact positions around the weld toe.

2 Material

As Ruiz (2018), in this study is thought that a yield model that correctly represents the material behavior, including material hardening is indispensable.

The Chaboche's kinematic hardening law is adopted (Eq. 1).

$$J_2(\sigma - \alpha) = \sigma_o \tag{1}$$

Equation 2, shows the hardening law, back stress tensor:

$$d\boldsymbol{\alpha} = \sum\nolimits_{i=1}^{M} d\boldsymbol{\alpha}_i;\, d\boldsymbol{\alpha}_i = C_i \frac{d\varepsilon^p}{\sigma_o}(\boldsymbol{\sigma} - \boldsymbol{\alpha}) - \gamma_i \boldsymbol{\alpha}_i d\varepsilon^p \tag{2}$$

where, $\boldsymbol{\alpha}$ is the back stress tensor, C_i, γ_i material parameters, $d\varepsilon^p$ is equivalent plastic strain increment, M number of kinematic hardening components, and i component number. M = 2 is used in this study. σ_o is the yield stress, defined to account for combined isotropic-kinematic hardening and strain rate dependency as (Eq. 3):

$$\sigma_o = \sigma_{o,0} F(\dot{\varepsilon}^p) G(\varepsilon^p); F(\dot{\varepsilon}^p) = 1 + (\dot{\varepsilon}^p/H)^{1/\rho}, G(\varepsilon^p) = 1 + a(\varepsilon^p)^b \tag{3}$$

where, $\sigma_{o,0}$ is the initial yield stress, H and ρ are Cowper-Symonds strain hardening parameters, and a and b are isotropic strain hardening parameters. Material parameters are the same as those adopted in Ernould (2018), and are listed in (Table 1). In this table "H" and "a" are set to disable the isotropic part of hardening and strain rate effect.

Young's modulus E is 210GPa, and Poison's ratio ν is 0.3. In (Eq. 2), γ_1 = 2.19E2 and γ_2 = 1.07E2. For (Eq. 3), b = 0.6. ρ is 5.80 (Yasuda 2016), and negligibly small a = 1.0E-6 is adopted because the pure kinematic hardening is assumed.

Table 1. Material parameters (Ernould 2018)

E [GPa]	ν [-]	σ_o [MPa]	C1 [MPa]	γ1 [-]	C2 [MPa]	γ2 [-]
210	0.3	435	8971.8	218.65	12654.88	106.98

3 Numerical Analysis

Mechanical behavior during HFMI treatment is analyzed using the commercial code MSC.Dytran (2017). The material model is implemented in a user's subroutine. First, the influence of tool angle is investigated. Then, the effect of impact position at the weld toe is analyzed.

HFMI simulations can be described either displacement controlled simulations (DCS) or force controlled simulations (FCS). By FCS impact velocity or impact force has to be prescribed, but for manually treated cases it is difficult (Ruiz 2018). In this study indentation and impact period were assumed constant, therefore, DCS is chosen.

4 Influence of Tool Angle and Impact Position

4.1 Workpiece and FE Model Boundary Conditions

The FE model is built of hexahedron brick elements. All nodes of the workpiece lower surface are prevented from motion, and nodes of the symmetric surface are prevented from transverse motion (Y-direction), see (Fig. 1). Peening tool with a pin radius of 2 mm to carry out all simulations is used.

Figure 1 shows the FE-model and tool which is composed of a conical pin and a guide. Rigid shell elements are used to model this tool. Pin is modeled as a rigid body. Element size in the peening region is 0.2 mm as was found necessary by (Lennart 2018) and (Ruiz 2018).

4.2 Material and Peening Parameters

The non-linear kinematic hardening material stated before is adopted. Material parameters are shown in (Table 1).

HFMI analysis is carried out for a length of 10 mm in X-direction, with an indentation of 0.2 mm in Z-direction, and peening impact pitch of 0.4 mm. A peening frequency of 100 Hz is used.

Figure 2 shows the position of the elements located at the weld toe, in the thickness direction where the results are obtained taking a stress average of each line of element at the same depth, in order to minimize numerical noise and plotted in the following sections.

4.3 Peening Response Due to Tool Angle

The peening tool has an angle with the web of 0°, 15° and 30° as shown in Fig. 1. Peening parameters are described in the Sect. 4.2. The comparison of the weld toe profile (transversal RS) due to different peening angles at the middle of the peening path is shown in (Fig. 3). The following can be seemed:

- With an angle of 0° tensile stress is induced near the top surface of the weld toe and the weld toe shape is very irregular.
- An angle of 15° induces compressive stress near the top surface and through the thickness direction. Weld toe shape is smoother than the case with an angle of 0°.
- Using an angle of 30° smaller compressive stresses than the case with an angle of 15° are obtained near the top surface, however, higher compressive stresses in the thickness direction. Finally, weld toe shape is very smooth.

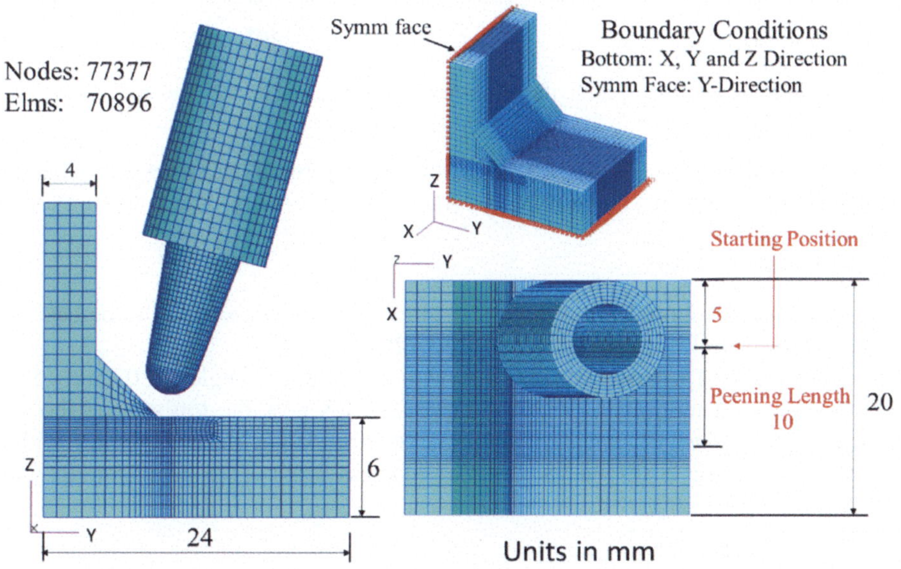

Fig. 1. FE-Model, boundary conditions and peening length.

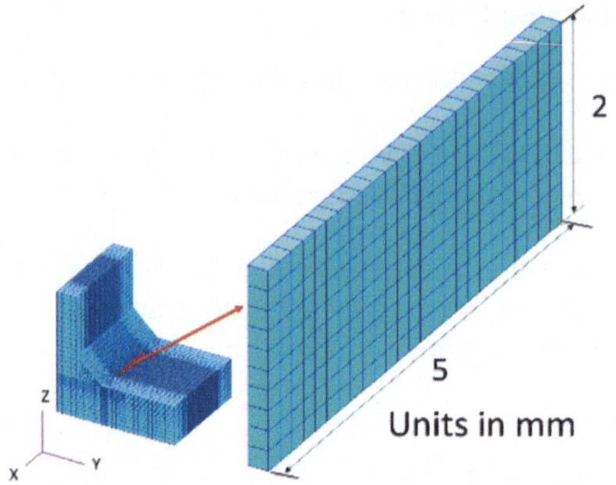

Fig. 2. Elements analyzed in the thickness direction.

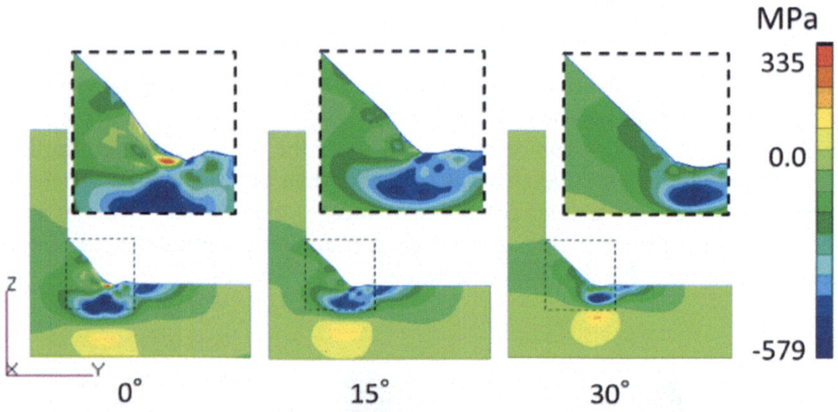

Fig. 3. Weld toe profile due to different tool angles (0°, 15° and 30°).

Longitudinal and transversal RS (S_{XX} and S_{YY}) through the thickness direction from the peening path are shown in, (Figs. 4 and 5) respectively. The following can be observed:

On longitudinal RS:

- Longitudinal RS show high compressive stress near the top surface, same behavior was obtained by Ernould (2018) and Ruiz (2018).
- Similar trend and values were obtained with peening tool angles of 0° and 15°.
- With a peening tool angle of 30° compressive RS near the top surface increase about 12%. However, smaller compressive stress than those obtained with a peening tool angle of 0° and 15°, from a thickness depth of about 0.3 mm to about 1 mm.

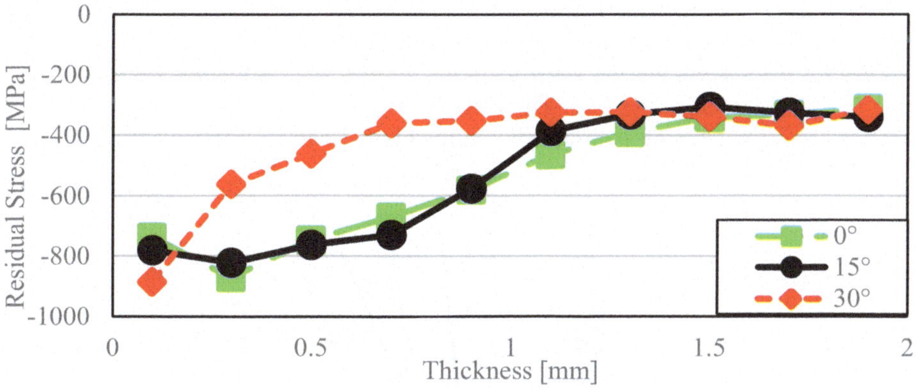

Fig. 4. Longitudinal RS profile in the thickness direction.

On transversall RS:

- Compressive RS near the top surface is comparable to the yield stress level of the material. Similar results were obtained experimentally by Leitner (2017).
- Similar trend and values were obtained with peening tool angles of 0° and 15°.
- Fluctuations of stresses near the top surface may be observed with a peening tool angle of 30°.

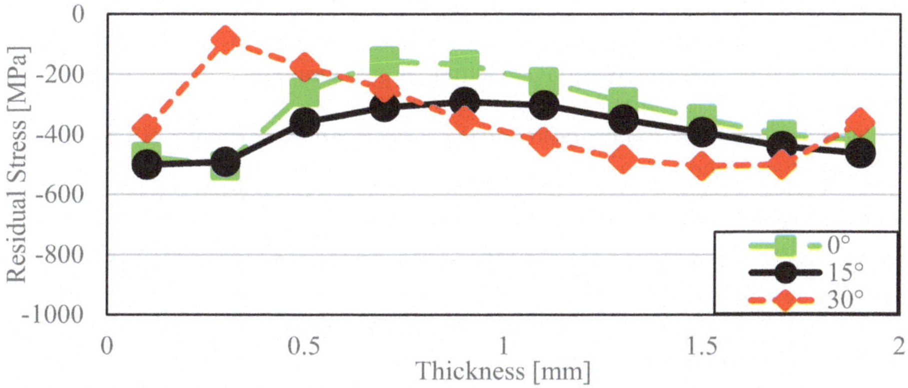

Fig. 5. Transversal RS profile in the thickness direction.

4.4 Peening Response Due to Impact Position

The influence of the impact position is analyzed by using the symmetric stress-free T-Joint model shown in (Fig. 1). In total three cases are carried out as shown in

(Fig. 6). Where a is located 1 mm from the weld toe over the filler metal and b is located 1 mm from the weld toe over the base metal.

Peening parameters are the same as those in Sect. 4.2. All cases are using a tool angle of 15° regarded to the web. The comparison of S_{XX} and S_{YY} through-thickness direction from the peening path are shown in (Figs. 7 and 8) respectively.

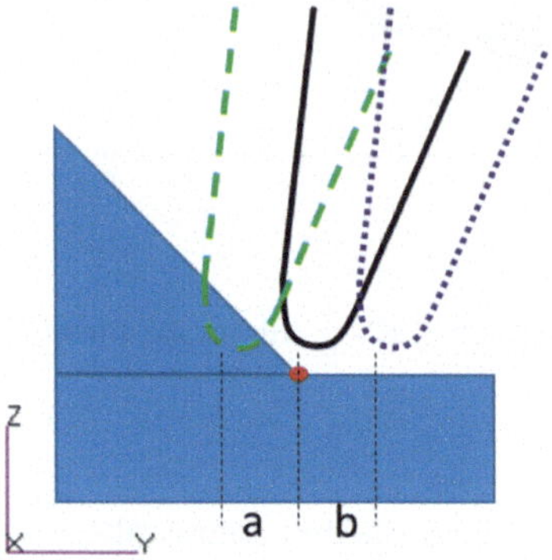

Fig. 6. Transversal RS profile in the thickness direction.

Comparison of longitudinal RS (S_{XX}) shows that:

- When the pin impacts inside the filler metal fluctuations of stresses near the top surface is observed.
- When the pin impacts at the weld toe smooth change of stresses through the thickness direction are observed. Getting high compressive stress near the top surface.
- When the pin impacts outside or in the base metal compressive RS are smaller through the thickness direction.

Comparison of transversal RS (S_{YY}) shows that:

- When the pin impacts inside the filler metal fluctuations of stresses near the top surface is observed.
- When the pin impacts at the weld toe smooth change of stresses through the thickness direction are observed.
- When the pin impacts outside or in the base metal sudden change of compressive RS near the top surface is observed.

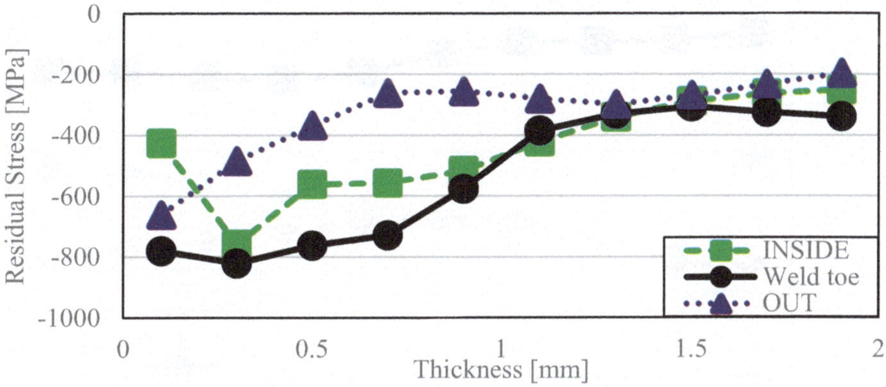

Fig. 7. Longitudinal RS profile in the thickness direction due to impact position.

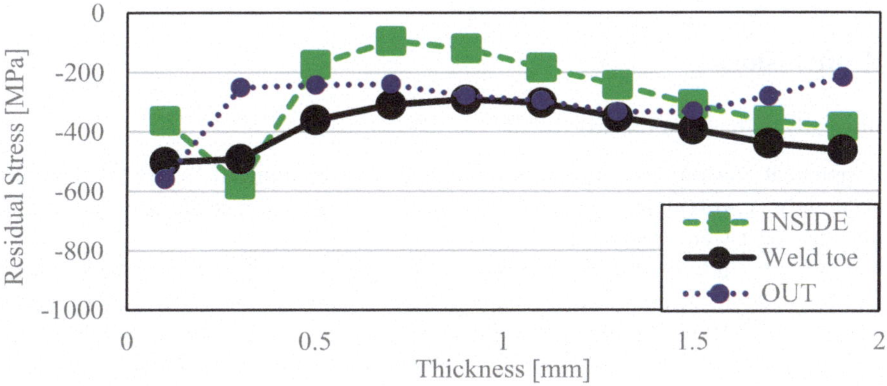

Fig. 8. Transversal RS profile in the thickness direction due to impact position.

Figure 9 shows the three normal stress components and Von Mises after HFMI-treatment. It shows high compressive RS in the longitudinal direction (S_{XX}), same behavior is shown by (Foehrenbach 2016) and (Ernould 2018). It should be noted that each normal component is compressive which is the main objective of this post-weld treatment technique. Furthermore, Von Mises stress increases to the yield stress level with consideration of hardening, similar behavior was found by Leitner (2017).

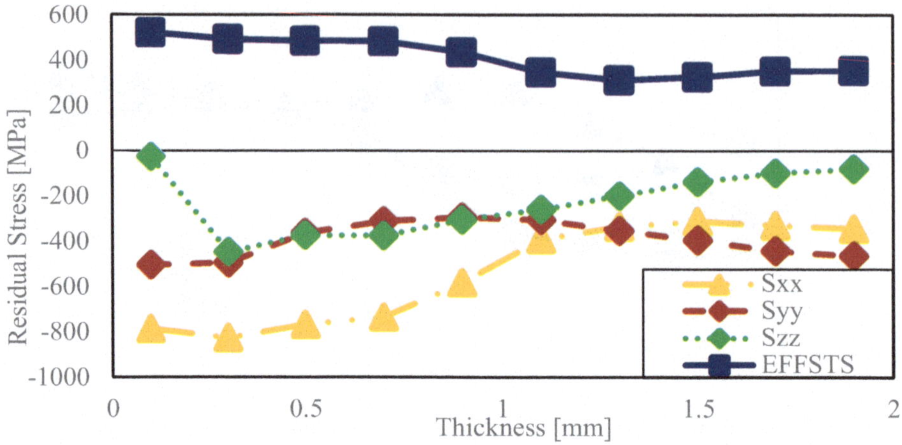

Fig. 9. Normal stress comparison and Von Mises after HFMI.

5 Conclusions

In this study, it has demonstrated that;

- A practical analysis of compressive residual stresses induced by HFMI-treatment, accurately represent the expected behavior, getting compressive residual stresses after the post-weld treatment.
- The present study of different peening tool angles showed that same behavior and similar residual stresses close the top surface are obtained. However, smooth changes of RS in the thickness direction are observed with an angle of 15° regarding to the web. Furthermore, a peening tool angle of 0° induces tensile RS near the top surface and irregular weld toe shape on a T-joint. Finally, a peening tool angle of 30° induces smaller compressive RS near the top surface.
- The study of peening response due to impact position showed significant sensitivity. It was found that when the pin impacts the filler metal (inside case) shows sudden changes of RS close to the top surface. When the pin impacts the base metal (outside case) shows smaller compressive RS in the thickness direction. Getting the smoothly behavior when the impact position is at the weld toe.

6 Future Works

This study provides the basis to carry out a practical numerical simulation of HFMI. However, it is recommended to investigate the stability of residual stress under common variable loading conditions on marine structures.

References

Hagensen, P.J., Maddox, S.J.: IIW Recommendations on Post Weld Fatigue Life Improvement of Steel and Aluminum Structures. Woodhead Publishing Ltd. (2013)

Marquis, G.B., Barsoum, Z.: A guideline for fatigue strength improvement of high strength steel welded structures using high-frequency mechanical impact. Proc. Eng. **66**, 98–107 (2013)

Marquis, G.B., et al.: Fatigue strength improvement of steel structures by high-frequency mechanical impact: proposed fatigue assessment guidelines. Weld. World **57**, 803–8022 (2013)

Foehrenbach, J., Hardenacke, V., Farajian, M.: High-Frequency Mechanical Impact Treatment (HFMI) for The Fatigue Improvement: numerical and Experimental investigations to describe the condition in the surface layers. Weld. World **60**(4), 749–755 (2016)

Ernould, C., et al.: Application of different simulation approaches to numerically optimize high-frequency mechanical impact (HFMI) post-treatment process. In: IIW Annual Assembly 2017, Shanghai (2018)

Lennart, J., et al.: Committee V.3 materials and fabrication technology. In: Proceedings of the 20th International Ship and Offshore Structures Congress (ISSC 2018), Progress in Marine Science and Technology, vol. 2, pp. 143–191 (2018)

Ruiz, H., et al.: Stability of compressive residual stress introduced by HFMI technique. In: International Conference on Offshore Mechanics and Arctic Engineering, ASME: Materials Technology: V004T03A019, vol. 4 (2018). https://doi.org/10.1115/omae2018-77887

MSC Software: MSC DYTRAN2017 Reference Manual (2017)

Yasuda, A., Rashed, S.: Efficient collapse analysis of ductile components subjected to impact loads. In: Proceedings of 35th International Conference on Ocean, Offshore and Arctic Engineering (OMAE2016), paper OMAE2016-54480 (2016)

Leitner, M., et al.: Stability of high-frequency mechanical impact (HFMI) post-treatment induced residual stress states under cyclic loading of welded steel joints. Eng. Struct. **143**, 589–602 (2017)

Fiberglass Stiffened Panels: Estimation of Natural Vibration Frequency

Franklin J. Domínguez Ruiz(✉) and Luis Manuel Carral Couce

Universidade da Coruña - UDC, A Coruña, Spain
jdominguez@tecnavin.com,
{f.j.dominguez.ruiz,lcarral}@udc.es

Abstract. In the shipbuilding industry, the application of fiberglass composite laminate panels has increased in the decades. In the preliminary design stage, it is convenient to use simplified formulas to estimate the first natural frequency of a stiffened panel that can be compared with the recommended ones by the Germanischer Lloyd's [3]. In this work, it is proposed a formula to performed this estimation considering panel features, aspect ratio and the number of stiffeners.

The behavior of fiberglass stiffened panels was studied through Finite Element Analysis method considering 684 models where principal dimensions, aspect ratios, thicknesses, and stiffeners were combined. Based on these numerical results, a posynomial formula has been derived that allows estimating the natural frequencies of fiberglass stiffened panels. As a result of this investigation, a set of coefficients are proposed, which are only applicable within a range of aspect ratios. Some examples with different aspect ratios and dimensional variations of the panels and stiffeners are included to validate the proposed formula. The estimation error is in average 5% when compared to LR recommendation, that is considered acceptable in the preliminary design stage.

Keywords: Natural frequency · Composite lamination · Fiberglass · Finite Element

1 Introduction

Most of the Classification Societies have set technical rules and formulas to estimate the natural frequency of panels, beams, and stiffener [1, 5, 8]. However, they do not provide a general formula to estimate the frequency of the composite panels with stiffeners. In the preliminary design stage, it is required to estimate natural frequencies of stiffened panels, that can be used to make a vibration analysis using the finite element method.

In the literature, there are several formulas available for the estimation of the natural frequencies of panels and stiffeners by different methods. For example, Constantinescu and Rigo [3] did a theoretical analysis on vibrations in rigid panels and beams, Liu and Zhao [7] considered the vibration of sandwich panels with the plate theory. Seung Cho, Vladimir, and MukChoic [11] predicted the natural frequency of panels with openings using rigid structures and different boundary conditions. Macchiavello and Tonelli [9]

V. A. J. E. Carreño Moreno et al. (Eds.): CIDIN 2019/COPINAVAL 2019, pp. 308–320, 2020.
https://doi.org/10.1007/978-3-030-35963-8_26

performed a detailed vibration analysis with Finite Elements Method to identify the possible resonances conditions in the structure of a pleasure craft. Although, the mentioned calculations required a lot of computational time; it is possible to analyze the structures as individual components only.

In the preliminary structural design stage, it would be desirable to have a simple formula to estimate the natural frequency of the stiffened panels to avoid possible resonances conditions with the propulsion system [4]. For this reason, the goal of this investigation is to develop a formula to estimate the first natural frequency of fiberglass composite panels using a posynomial regression due to this formula offered the best correlation.

The following procedure has been considered for the deduction of this formula applicable in fiberglass composite panels:

- To develop a database considering different dimensionless parameters of stiffened panels members.
- To calculate the first natural frequency of each panel considered using Finite Element.
- To calculate the frequency of the simple composite panels (without reinforcement), using the Leissa formula [6] modified by the frequency coefficient N,
- To calculate the frequency of stiffener with an associated plate.
- To calculate the frequency of a composite panel with stiffener, taking the regressions based on the aspect ratio. The terms of the final formula include 8 terms, for which we have fitted the posynomial coefficients for each aspect ratio.

2 Variables and Formulas Considered for Posynomial Regression

For the development of the posynomial regression, the panel is considered as with stiffener clamped at their ends. The following main variables were including:

- Number of stiffeners, n
- Base width of the stiffener section, b_w
- Aspect ratio, r_a
- Simple panel width, M_i
- Longitudinal radius of gyration, R_M
- Transversal radius of gyration, R_L
- Equivalent modulus of the sandwich panel laminate, E_{PL}
- Equivalent modulus of the stiffener with the associated plate, E_R
- Equivalent Inertia of the panel, I_{PL}
- Equivalent Inertia of the stiffener with associated plate, I_R
- Laminate panel Thickness, t_{PL}
- Stiffener Thickness, t_R
- Cross section area of the stiffener, A_R
- Cross section area of the panel, A_{PL}

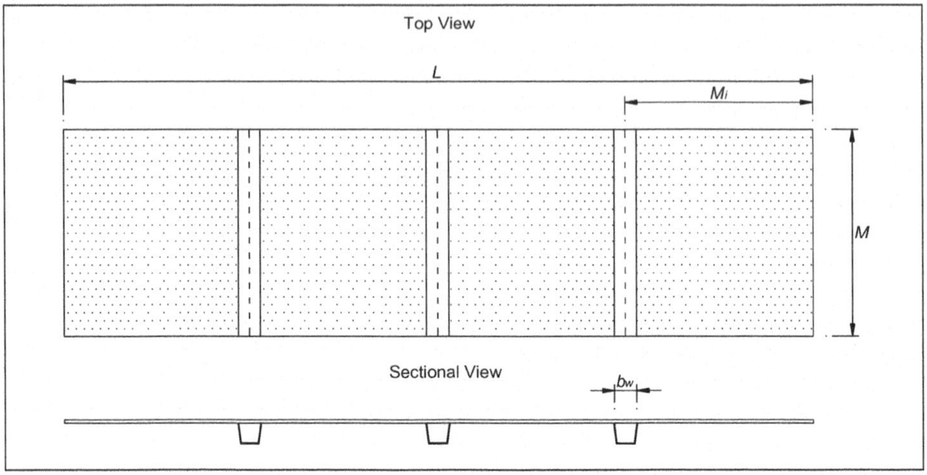

Fig. 1. Panel graph with 3 stiffeners

The *dimensions of the stiffened panel* are defined according to simple panel dimensions, being: L, the panel width and M, the length, see Fig. 1:

$$L = (n+1) \cdot M_i \tag{1}$$

$$M = r_a \cdot M_i \tag{2}$$

The *radius of gyration*, R, is defined according to Eq. (3), where I is equivalent inertia of the panel and A is its cross-section area. For the longitudinal radius of gyration, it only was considered: the thickness of the panel t_{PL} and length M.

$$R = \frac{I}{A} \tag{3}$$

$$I = \frac{EI_{eq}}{E_{eq}} \tag{4}$$

Because the panels are made of composite material, the inertias have been calculated using the equivalent stiffness modulus EI_{eq} and equivalent Young's modulus E_{eq}, as indicated in Eq. (4).

The *relation m",* is the relationship between the product of the modulus of elasticity, the cross-sectional area of the panel and the one of the panel with stiffener, defined in Eq. (5).

$$m'' = \frac{m_{PL}}{m_{PL} + m_R} \tag{5}$$

where

$$m_{PL} = E_{PL} \cdot A_{PL} \qquad m_R = E_R \cdot A_R \tag{6}$$

The resulting posynomial formula consists of 8 terms that have allowed relating all the geometric variables of the stiffened panel. The variables have been grouped in such a way that the regression obtained has 5% error. The terms that intervene in the posynomial formula are:

- $\ln(L/R_M)$
- $\ln(E_{PL} \cdot L \cdot t_{PL})$
- $\ln(n+1)$
- $I_R \cdot E_R \cdot t_R / I_{PL} \cdot E_{PL} \cdot t_{PL}$
- $M/(M_i - b_w)$
- m''
- R_L/M
- $\ln(E_{PL} \cdot M \cdot t_{PL})$

In the proposed posynomial formula, *f*" is the relation between the simple composite panel natural frequency (without stiffener) and the natural frequency of panel with stiffener; see Eq. (4). The panel frequency has been estimated using the modified Leissa formula, see Sect. 3.1. The frequency of the panel with stiffener used in the database has been obtained by means of Finite Elements.

$$f'' = \frac{f_{PL}}{f_{PLR}} \tag{7}$$

3 The Natural Frequency of a Simple Composite Panel (Without Stiffener) Using the Modified Leissa Formula

The simple panel natural frequency is defined by the empirical formula proposed by Leissa [6]:

$$f_{PL}^2 = \frac{\pi^4 \cdot D \cdot K}{a^4 \cdot \rho \cdot N} \tag{8}$$

$$D = \frac{E_{PL} \cdot t_{PL}^3}{12 \cdot (1 - \nu^2)} \qquad K = 12 + 8 \cdot \left(\frac{a}{b}\right)^2 + 12 \cdot \left(\frac{a}{b}\right)^4$$

Where f_{PL} is the panel frequency in Hz, D is the panel flexural rigidity, ρ is the mass density of the panel in Kg/m^3, K and N are the regresional natural frequency coefficients, a width of the panel in meters; b is length of the panel in meters, E_{PL} is Young's modulus in N/m^2, t_{PL} is the thickness in meters and ν is Poisson's ratio. Originally, the value of N has been proposed for a Poisson's ratio of 0.25, with the condition of edges clamped.

3.1 Modification of the Regression Coefficient N for Several Poisson's Ratio Values

Due to the Poisson's ratio in Eq. (8) was defined as ν = 0.25, a posynomial equation has been proposed to modify the frequency coefficient N. For this reason, it has been generated a database for different panels, the composite laminate was varied, see Table 1, in such a way, as to obtain variations in the Poisson's ratio. The frequency f_{PL} has been obtained by means of a Finite Elements Analysis.

Table 1. Data for estimation of the regression frequency coefficient, N

t_{PL} (m)	a (m)	b (m)	E_{PL} (N/m^2)	ρ_{PL} (kg/m^3)	D	K	f_{PL} (Hz)	ν	N
0.03945	3	0.75	1.44E+09	369.34	7473.1	3212	143.80	0.115	3.78
0.0331	3	0.75	1.70E+09	412.56	5259.8	3212	125.00	0.143	3.15
0.0257	3	0.75	2.18E+09	489.88	3179.5	3212	101.85	0.178	2.42
0.04003	3	0.75	1.62E+09	383.04	9017.3	3212	153.28	0.192	3.84
0.03368	3	0.75	1.92E+09	429.30	6421.0	3212	134.36	0.221	3.23
0.02628	3	0.75	2.44E+09	509.65	3950.6	3212	109.87	0.257	2.47
0.0406	3	0.75	1.80E+09	398.33	10782.4	3212	163.27	0.260	3.92
0.03425	3	0.75	2.12E+09	445.48	7760.3	3212	143.25	0.289	3.28
0.02685	3	0.75	2.69E+09	528.57	4852.9	3212	118.14	0.325	2.54

For this regression, the following variables are considered:

- Rigidity module ratio, E_{PL}/E_N
- Density ratio, ρ_{PL}/ρ_N
- Poisson's ratio, v

Where E_N y ρ_N is Young's modulus and density of the core (balsa) respectively.

In Eq. (9), the posynomial formula for the natural frequency coefficient N is proposed according to the aforementioned variables, and its coefficients are presented in Table 2.

$$N = a_0 \cdot \left(\frac{E_{PL}}{E_N}\right)^{R_1} \cdot \left(\frac{\rho_{PL}}{\rho_N}\right)^{R_2} \cdot v^{R_3} \tag{9}$$

Table 2. Posinomiales coefficient of the frequency coefficient N

a_0	R_1	R_2	R_3
0.202	2.270	−4.623	−0.156

4 Posynomial Formula for the First Natural Frequency of Panel with Stiffener

4.1 Database

For the development of the natural frequency formula of composite panels was used stiffener type Omega; its dimensions are shown in Table 3. The database is made with 684 variations of the stiffened panel. For each sample, it has been varied the following parameters: panel thickness, number of stiffeners, stiffener size, aspect ratio, and width of the panel. The combinations used in the present work are:

- Thickness of the panel: 26, 33 and 40 mm
- Number of stiffener: from 1 to 3
- Aspect ratio: 0.5, 1.0, 1.5, 2.0 and 3.0
- Thicknesses of stiffener: 5.2 mm and 8.1 mm,

Figure 2 show the combinations used to obtain all the 684 variations. These range of variables were selected because they are the most common in superstructures of yachts with the length between [30, 50] meters.

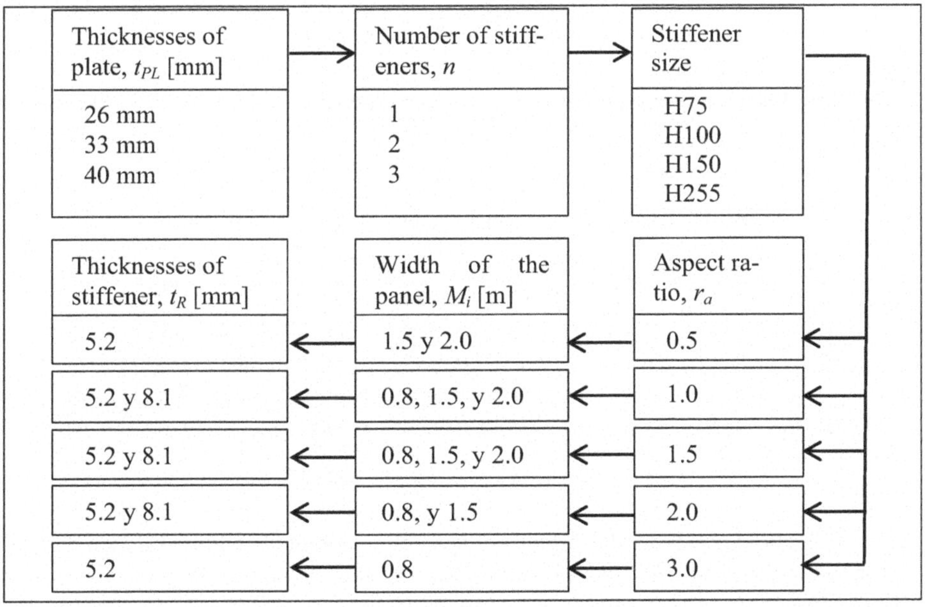

Fig. 2. Scheme for database development.

Table 3. Omega type stiffener used in the database, measured in mm.

Stiffener	High	Base width	Crown width	Thickness	Laminate
H75	75	86	67	5.22	3Mat450 + 2WR600
H100	100	92	76	5.22	3Mat450 + 2WR600
H150	150	130	100	5.22	3Mat450 + 2WR600
H255	255	220	170	8.11	4Mat450 + 3WR800

4.2 Proposed Formula for First Natural Frequency Estimation of a Stiffened Panel

From the proposed Eq. (7), we obtain the natural frequency of a composite panel with stiffener (10), with the dimensionless relation *f''* estimated by the posynomial Eq. (11) and f_{PL} through Eq. (8).

$$f_{PLR} = \frac{f_{PL}}{f''} \tag{10}$$

$$f'' = \frac{a_0 \ln\left(\frac{L}{R_M}\right)^{R_1} \ln(E_{PL} \cdot L \cdot t_{PL})^{R_2} \ln(n+1)^{R_3} \left(\frac{I_R E_R t_R}{I_{PL} E_{PL} t_{PL}}\right)^{R_4} \left(\frac{M}{M_i - b_w}\right)^{R_5}}{m''^{R_6} \left(\frac{R_L}{M}\right)^{R_7} \ln(E_{PL} \cdot M \cdot t_{PL})^{R8}} \tag{11}$$

The resulting posynomial coefficients are presented in Table 4 for different aspect ratios.

Table 4. Results of posinomial coefficients for stiffened panels

a	a0	R1	R2	R3	R4	R8	R5	R6	R7
0.5	3.38E+00	0.1534	0.279	−0.0678	−0.0121	−0.2619	0.2975	−0.0167	0.8773
1	8.56E+01	0.8897	−3.438	−0.1360	−0.0879	−0.4413	0.6148	−0.1149	−1.4514
1.5	5.25E+03	1.8488	−6.447	−0.3907	−0.2143	−0.8327	1.1601	−0.2302	−2.6866
2	4.40E+04	0.9168	−3.477	−0.5481	−0.2431	0.1677	1.2911	0.0323	1.0962
3	4.46E+15	1.8628	−13.689	−0.6240	−0.3964	0.0000	2.8021	0.0079	0.0000

Figure 3 corresponds to the graph between the estimated natural frequency relationship using the modified Leissa formula, Eq. (8) and the natural frequency obtained by Finite Elements. These results are compared with the frequency obtained by the posinomial formula *f''*, Eq. (11). Also, an error of about 5% was obtained between the estimated results of *f''* using the posynomial formula.

For cases with an aspect ratio value not found in Table 4, it is recommended to use linear interpolation to estimate f''.

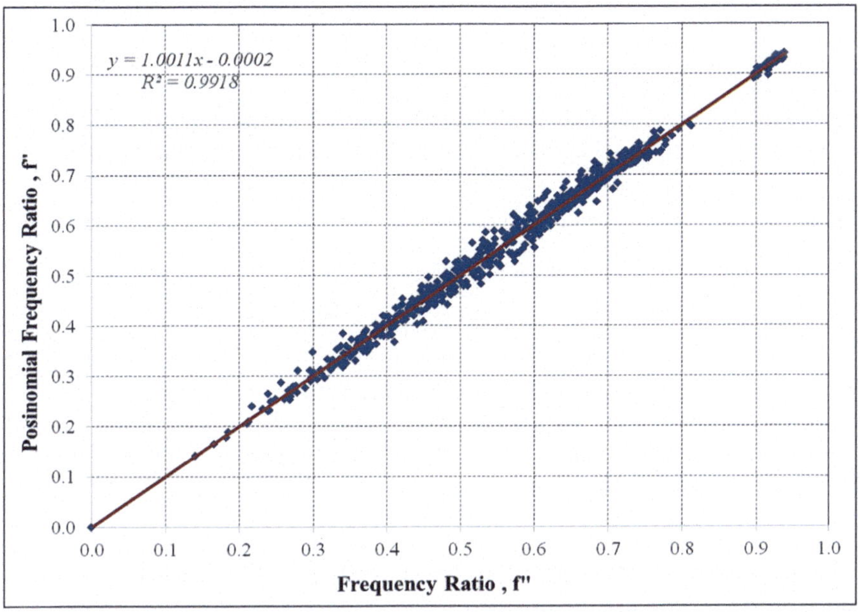

Fig. 3. Graph of Posynomial Frequency Ratio using Eq. (8) vs. natural Frequency ratio using FEM.

5 The Formula for Natural Frequency Estimation of a Stiffener with Its Associated Panel

The formula to estimate the natural frequency of a stiffener with its associated panel with clamped ends is presented in Eq. (15). This formula has been deduced from the Eqs. (12a) where *m* is mass, *c* is damping and *k* is rigidity.

The equation of the frequency without damping, *c* is [2]:

$$m \cdot x''(t) + c \cdot x'(t) + k \cdot x(t) = 0 \tag{12a}$$

$$x''(t) + \omega^2 \cdot x(t) = 0, \text{being c} = 0 \tag{12b}$$

where

$$\omega^2 = \frac{k}{m} \tag{12c}$$

Where *k* is deduced from the formula of bending of beams, obtaining:

$$k = \frac{384 \cdot E \cdot I}{l^3}$$

The mass [10] is represented by:

$$m = \frac{A \cdot \gamma \cdot l}{g} \tag{13}$$

Replacing the previous equations in (12) and equalizing Eq. (12) with (14), we obtain the natural frequency equation of a stiffener with its associated panel.

$$\omega^2 = (2\pi)^2 f_R^2 \tag{14}$$

$$f_R = \frac{1}{2\pi}\sqrt{\frac{384 \cdot E_R \cdot I_R \cdot g}{A_{RPA} \cdot \gamma \cdot l^4}} \tag{15}$$

Where E_R is Young's modulus of the stiffener with associated plate in N/m^2, I_R is inertia of the stiffener with associated panel in m^4, A_{RPA} is transversal area of the stiffener with associated panel in m^2, l is length not supported in m, γ is the specific weight of the stiffener in N/m^2, and g is gravity in m/s^2.

6 Comparison of Results Obtained with the Proposed Natural Frequency Formula

A side panel of a yacht with fiberglass superstructure [12, 13] has been taken as a reference to validate the proposed formula for the natural frequency of a stiffened panel, Eq. (10), and for the natural frequency of a stiffener with its associated panel, Eq. (15). The panel selected as an example has stiffeners omega type of 125 mm high, which is not found in the database of the deduction of the formula.

The main characteristics of the example panel are:

Table 5. Example: Omega of H125, side panel 01

Calculation of the first natural frequency of a panel with stiffener			
Element	Variable	Unit	Magnitude
Omega type			H125
Number of stiffeners	n		2
The base width of the stiffener section	b_w	mm	114
Thickness of core	t_N	mm	25.4
The thickness of the panel	t_{PL}	mm	34.25
Thicknesses of stiffener	t_R	mm	5.22
Simple panel width	M_i	m	1.71
Aspect ratio	r_a		1.49
Width of the panel with stiffeners	L	m	5.13
Length of the panel	M	m	2.55
Cross section area of the panel	A_{PL}	mm^2	1.76E+05
Cross section area of the stiffener	A_R	mm^2	1.70E+03

(*continued*)

Table 5. (*continued*)

Calculation of the first natural frequency of a panel with stiffener			
Element	Variable	Unit	Magnitude
Mass density of the panel	ρ_{PL}	kg/m^3	445
Mass density of the core	ρ_N	kg/m^3	144
Poisson's ratio	ν		0.289
Equivalent modulus of the sandwich panel laminate	E_{PL}	N/mm^2	2125
Equivalent modulus of the stiffener with the associated plate	E_R	N/mm^2	3147
Inertia equivalent of the panel	I_{PL}	mm^4	3.75E+07
Inertia equivalent of the stiffener with associated plate	I_R	mm^4	7.17E+07
Transversal radius of gyration	R_L	mm	31.81
Longitudinal radius of gyration	R_M	mm	14.60
Variable 1	$\ln(L/R_M)$		5.862
Variable 2	$\ln(E_{PL}Lt_{PL})$	N	19.738
Variable 3	$\ln(n+1)$		1.099
Variable 4	$E_RI_Rt_R/E_{PL}I_{PL}t_{PL}$		0.431
Variable 5	$M/(M_i-b_w)$		1.596
Variable 6	m''		0.986
Variable 7	R_L/M		0.012
Variable 8	$\ln(E_{PL}Mt_{PL})$	N	19.038
Natural frequency of a simple composite panel	f_{PL}	Hz	12.38
Relation of posinomial frequency	f'		0.4897
Natural Frequency of panel with stiffener, estimated with the posynomial formula	f_{PLR}	Hz	**25.71**

The side panel analyzed is reinforced with an omega of H125 and an aspect ratio of 1.49. The value obtained from the frequency proposed with the posinomial formula is 25.71 Hz, while the frequency of similar panel obtained by the Finite Element analysis is 23.74 Hz (see Table 5), shown an error of 6%.

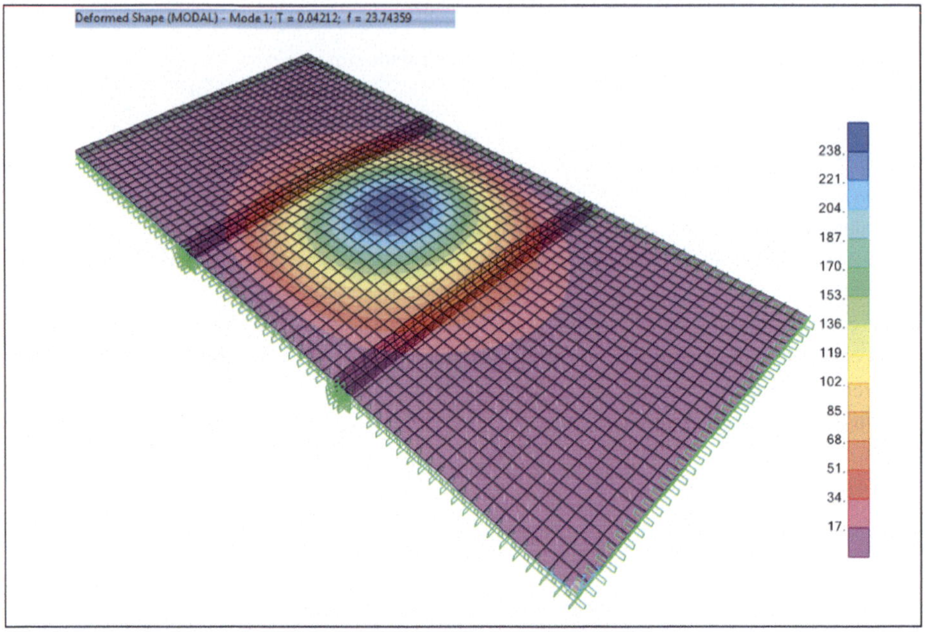

Fig. 4. Finite Element Analysis: Side panel - Omega H125, 1st natural frequency. For panels features, see Table 5

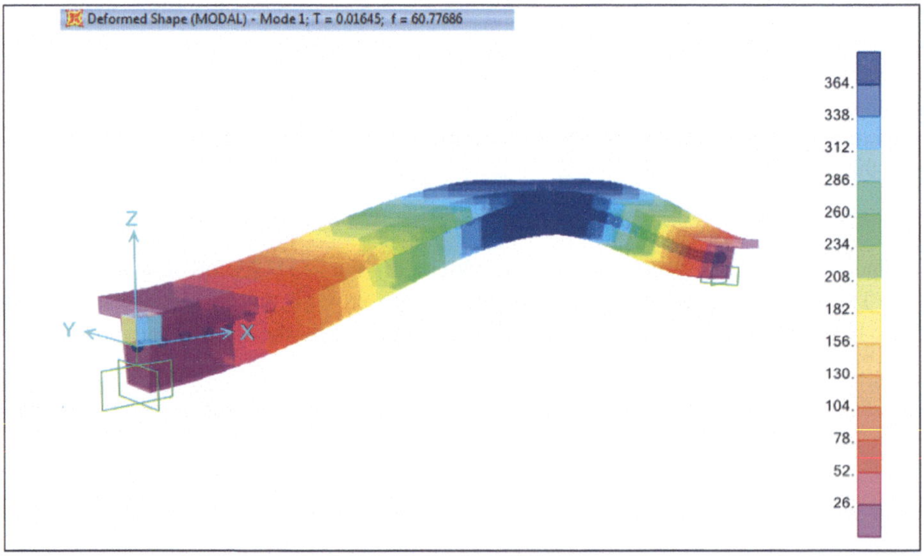

Fig. 5. Finite Element Analysis: Stiffener with associated plate, 1st natural frequency. For stiffened panels features, see Table 5

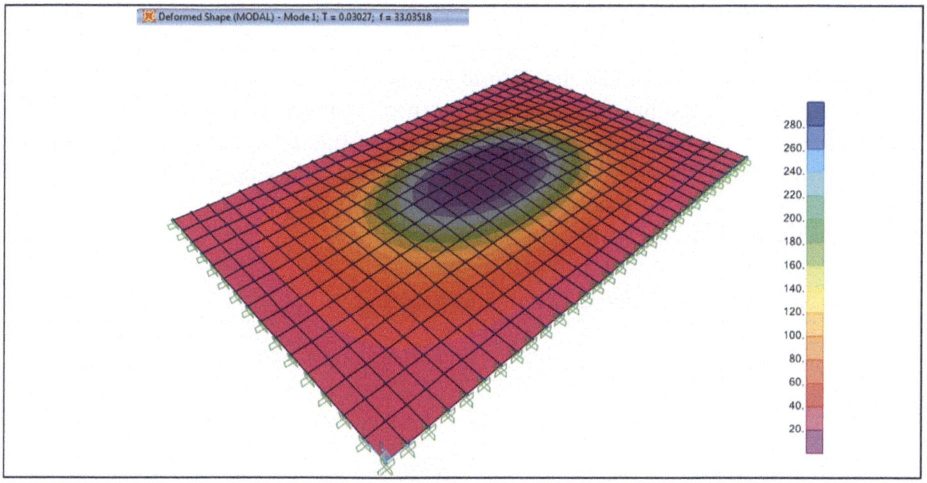

Fig. 6. Finite Element Analysis: Simple side panel, 1st natural frequency. For panels features, see Table 5

For the example shown in Figs. 4, 5 and 6, the Table 6 shows the estimated frequencies by formula, the frequency obtained with Finite Element analysis for the stiffened panel; the frequency of stiffener with its associated panel, and the frequency of simple panel (without stiffener). On this table is possible to observe that the estimation error is acceptable and that the frequencies obtained are out resonance range concerning to Propeller blade frequency.

Table 6. Results of frequency analysis of panel

Element	Estimated frequency (Hz)	FEM frequency (Hz)	Error (%)
Stiffened panel, f_{PLR} Ec.(11)	25.25	23.74	5.96
Stiffener with its associated panel, f_R Ec. (15)	58.38	60.78	4.10
Simple panel, f_{PL} Ec. (8)	32.88	33.03	0.47
Propeller blade frequency, f_{Blade}	50.90		

7 Conclusions

- The formula (8) to estimate the natural frequency of a simple composite panel (without stiffeners), with clamped edges, proposed in Leissa's formula for a Poisson's ratio of $\upsilon = 0.25$, has been modified to extend its range of application from υ [0.1 a 0.4] through the modification of frequency coefficient N [6].
- The formula (15) to estimate the frequency of a composite stiffener with an associated panel, with clamped edges, has been deduced based on the equation of

motion, it is expected that the formula has restriction on application interval or in the use of other materials.

- The formula (11) to estimate the natural frequency of a composite stiffened panel, with clamped edges has been proposed from a regression analysis. From the examples developed to validate the application of this formula, it is recommended to use it in the preliminary design stage within the following ranges:
  - The thickness of the panel: 26, 33 and 40 mm
  - Number of stiffener: from 1 to 3
  - Aspect ratio: 0.5, 1.0, 1.5, 2.0 and 3.0
  - Thicknesses of stiffener: 5.2 mm and 8.1 mm,

References

1. ABS: Guidance notes on Ship Vibration, April 2016
2. Avallone, E., Baumeister, T., Sadegh, A.: Standard Handbook for Mechanical Engineers, 11th edn. McGraw Hill, New York City (2006)
3. Constantinescu, A., Rigo, P.: Vibration assessment of beam structures and stiffened panels. Naval Architecture Research Unit (ANAST), University of Liège (2008)
4. Domínguez, F., Cali, M., Garcia, L.: Analysis of Ship Hull Girder Hydro-Vibration: excitation force induced for propulsion system and rudder interaction. In: 6th International Conference on Marine Structures (Marstruct 2017), Lisbon, Portugal, pp. 75–86, May 2017
5. Germanischer Lloyd's: GL Technology Ship Vibration, Hamburg (2001)
6. Leissa, A.: Vibration of plates, Ohi State University, National Aeronautics and Space Administration, SP-160, Washington, February 1970
7. Liu, Q., Zhao, Y.: Prediction of natural frequencies of a sandwich panel using thick plate theory, Department of Mechanical Engineering-Engineering Mechanics, Michigan Technological University, Houghton, MI 49931, USA (2001)
8. Lloyds Register: Hull Construction in Composite. Rules and Regulations for the Classification of Special Service Craft, July 2016
9. Macchiavello, S., Tonelli, A.: Pleasure vessel vibration and noise finite element analysis. In: 6th BETA CAE International Conference, Italy
10. Parson, W., Vorus, W., Richard, E.: Added Mas and Damping of vibrating propeller, Department of Naval Architecture and Marine Enginnering, The University of Michigan, October 1980
11. SeungCho, D., Vladimir, N., Tae, M.: Numerical procedure for the vibration analysis of arbitrarily constrained stiffened panels with openings. Int. J. Nav. Architect. Ocean Eng. **6** (4), 763–774 (2014)
12. Dominguez, J., Carral, L.: Hybrid joint between steel deck and fiberglass superstructure. In: Proceedings of the Multidisciplinary International Conference of Research Applied to Defense and Security MICRADS, April, vol. 94, pp. 284–295. Springer (2018)
13. Dominguez, J., Carral, L.: Superstructure desing: combination of fibreglass panel and tubular structure with naval steel hull. In: Proceedings of the 25th Pan-American Coference of Naval Enginnering - COPINAVAL 2017, pp. 81–91. Springer, July 2018

Synthesis of a Modified Polyurethane Paint with Better Properties of Corrosion Resistance Applicable to Vessels and Naval Artifacts

Gabriel Peluffo Martínez[1(✉)],
Karen Patricia Domínguez Martínez[2],
and Eduardo Espinosa-Fuentes[3]

[1] Escuela Naval de Suboficiales ARC "Barranquilla", Barranquilla, Colombia
gbrl.peluffo@gmail.com
[2] Cotecmar, Cartagena, Colombia
[3] Universidad Libre, Barranquilla, Colombia

Abstract. The present manuscript deals with the formulation of a polyurethane paint with better properties of resistance to corrosion, by using a linear polymer (expanded polystyrene) as a potential inhibitor used in addition to the components traditionally used in these formulations. The manufacturing methodology deals with a sequential mixing process, finalized with an experimental part of chemical saturation between normal polyurethane paint (PUN) and expanded polystyrene (PER). The Electrochemical Impedance Spectroscopy Analysis (EIS) indicated the resistance of the polyurethane paint film to the phenomenon of corrosion, applied on carbon steel substrate, showing different characteristics, compared to commercial paints. Finally, the benefits and applications of painting on boats and naval artifacts are exposed.

Keywords: Polyurethane paint · Modified · Polystyrene · Substrate carbon steel electrochemical impedance · Ships

1 Introduction

Since the days of cave paintings have been used as decorative and protective films. The Greeks and Romans first used lead carbonate "white lead" white pigment (toxic). The Egyptians and Greeks manufactured rubber based paints and pigmentations natural wax such as iron oxide and carbon black. In the Middle Age and the Renaissance it had strong growth artistic paints, industrial by this time had not expanded, due to lacking applicability of the time. The first synthetic resins as components of paints (nitrate and cellulose acetate) appeared during the First World War, currently there are many components of a paint which confer viscosity and plasticity and other properties [1].

The paintings are most suitable for the protection of the different materials used in construction and industry method. A painting may be defined as a dispersed system of finely dispersed solids in a fluid medium that meets as vehicle functions. This material is based on a binder substance, also called film-forming binder or immersed in a solvent to which is incorporated additives and plasticizers. In the case of organic solvent based

V. A. J. E. Carreño Moreno et al. (Eds.): CIDIN 2019/COPINAVAL 2019, pp. 321–333, 2020.
https://doi.org/10.1007/978-3-030-35963-8_27

paints, the carrier is a liquid solution (molecular dispersion of the polymeric material) suspending the pigment particles; during drying system a significant volumetric shrinkage of the film [2–4] is observed.

In the area of the paintings they stand out hundreds of researchers working on the development and improvement thereof, including highlights:

In 1956, Reilly and Orchin prepared and reported properties of films based on polyurethane polymers and studied protection aluminum surfaces [5].

In 1989, Chen et al studied the chemical properties and polyurethane applications on metal and plastic surfaces [6].

In 2004, Chattopadhyay and colleagues studied the thermal and flame retardancy stability to polyurethane films based on HMDI and MDI [7].

In 2009, Giudice and colleagues made several technologies for paint coating based on expanded polystyrene [8].

In 2013, Benitez and Velez obtained from an anticorrosive coating from recycled expanded polystyrene using limonene main [9] solvent.

In 2011, Shikata et al studied Polystyrene solutions cyclic monoterpenes found in essential oils of trees, in order to develop a reducing organic agent expanded polystyrene [10].

In 2014, Jing Yang and colleagues studied polyurethanes to combat the attack of fuel and marine biocorrosion [11].

The present research deals with the preparation of organic solvent based paints, using polyurethane polymers, expanded polystyrene commercially known as ICOPOR and light protective additives used in the polystyrene industry. It is expected that the paint to be synthesized will have physical and chemical properties comparable to commercial paints, and it is estimated that the film has favorable anticorrosive properties due to the hydrophobic nature of the ICOPOR material.

2 Methodology

2.1 Materials

For the base preparation of the polyurethane paint, the following materials were used:

- Thinner polyurethane (solvent): A solvent is a pure or mixed liquid, which is used to give fluidity to the paint before its application. In practice, the term solvent, solvent, and diluent are used interchangeably. Solvents are chemicals that can dissolve, suspend or extract other materials, usually without chemically changing any of the solvents or other materials. Solvents make it possible to process, apply, clean or separate materials. These work according to the principle of "like dissolves like", therefore, for a solvent to work, it must have chemical characteristics similar to the substance that is trying to dissolve. Water is also a solvent, which is described as "inorganic" [12].
- Cobalt: it is primarily an oxidation catalyst and therefore acts as a surface drying agent. Cobalt siccatives, due to their great activity, should preferably be added at the end of the manufacturing process.

- Calcium Octate: Calcium blotters have a weak drying action by themselves, but they are very used in combination with active drying agents for this reason it is known as secant auxilia. They are also used as a dispersant and wetting agent, due to this property, it is convenient to be added to the system during grinding.
- Zirconium Octate: It is a secant used generally in combination with cobalt, manganese and calcium siccatives. The zirconium secant makes it possible to reduce the amount of cobalt normally used without affecting the drying properties.
- Dispersant: it is an additive that is used to make a solute have distribution and dispersion in a solvent.
- Titanium dioxide: It is widely used as a white pigment in exterior paints because it is chemically inert, due to its great coating power, its opacity to UV light damage and its self-cleaning capacity. The dioxide has also been used as a bleaching and opaque agent in porcelain enamels, giving a final finish of great brightness, hardness and acid resistance.
- Expanded polystyrene: Cellular and rigid plastic material manufactured from the molding of pre-expanded polystyrene pearls or one of its copolymers, which has a closed cell structure and filled with air.

3 Preparation of Modified Polyurethane Paint Film

3.1 Preparation of PER Solution in Limonene

The solution of PER in the limonene solvent will be made in a temperature range of 30 to 60 °C, at a step of 10 °C. These mixtures will be stirred at 300 to 500 rpm. Solvent: PER binary mixing ratios will be made from 4:1 to 2:2. Taking into account the three variations, the combination number following a factorial design are listed in the following table (see Table 1). In the table the dilution ratio (a:b) is shown, for this relationship the corresponding dilution factor will be taken into account assuming that the additive volumes a/(a + b). Thus a 1:1 ratio corresponds to a factor 1/(1 + 1) = 1/2 or 0.5. A 1:4 ratio corresponds to a factor 1/(1 + 4) = 1/5 or 0.2 (López et al. 2014).

Table 1. Factorial design solution preparation (PER-Solvent)

Test No.	Temperature	Mixing revolutions	Solvent ratio:PER
1	30	300	4:1
2	30	300	3:2
3	30	300	2:2
4	30	400	4:1
5	30	400	3:2
6	30	400	2:2
7	30	500	4:1
8	30	500	3:2
9	30	500	2:2
10	40	300	4:1
11	40	300	3:2

3.2 Preparation of Modified Polyurethane Paint (PUM)

The PUM will be made by mixing the normal polyurethane paint (PUN). For the preparation of the polyurethane suspension, the following process was carried out: in a resistant container, 2.5 kg of resin was added, with the aid of a mixing machine, it was stirred for 10 min. 0.5 L of polyurethane solvent was then added. Maintaining agitation. Subsequently, 1.5 kg of titanium dioxide and 44 g of Dispersant are added under constant stirring. After homogenizing the mixture, 20 g of 12% cobalt octoate, 20 g of 24% zirconium octonate, 20 g of calcium octoate and 44 g of dispersant (second addition) are added. Finally, maintaining the agitation for one hour, then a saturated solution of (PER) was added, which was prepared with expanded polystyrene and lemon according to the factorial design (Table 1); this mixture was made by adding portions of 5 mL of solution (PER) until saturation of the normal PUN polyurethane paint was obtained, the percentage levels were calculated with the empirical formula of percentage.

$$\% = \frac{Solute\ mass}{Mass\ of\ dissolutionn} \times 100 \qquad (1)$$

3.3 Electrochemical Impedance Spectroscopy (EIS) Analysis

As the phenomenon of corrosion process electrochemical nature, and shape it is possible to predict their behavior study by electrochemical techniques. Electrochemical impedance spectroscopy is a non-destructive technique that allows, through small electrical disturbances applied to the sample, to evaluate its performance against corrosion. This technique is particularly useful for evaluating anticorrosive organic coatings, since it produces some characteristic parameters protective coating ability in short time intervals. The most important parameters for the protective capacity of an antifouling coating are organic capacitance and electrical resistance. The first is related to the ability of the coating to absorb water.

The second relates to the barrier property of the coating to prevent the diffusion of aggressive species. In principle, an electrochemical process may be modeled by elements of electric circuits such as resistors, capacitors and inductors. The facility to model a process using electrical circuits corrosion is one of the main advantages of using the impedance technique. The values and shapes of the curves allow us to differentiate between different paintings behaviors studied, or between the same paint applied to different substrates. The electrochemical cell used in this studied was three-electrode cell, which has the working electrode plate painted with Sample White Paint developed by ACR Naval Academy whose effective area is 38.5 cm^2, as auxiliary electrode a platinum wire is used as reference electrode an electrode of silver/chloride Silver (Ag/AgCl). The test solution is a solution of NaCl at 4% and the equipment used is a Gamry potentiostat 600. Each paint system was initially assessed after one hour stabilization potential and then the same system is evaluated at 7 days and 15 days to observe what happens to the painting in time.

At the end of the specimen images after the electrochemical assay it is presented as auxiliary electrode a platinum wire is used as reference electrode and a silver electrode/silver chloride (Ag/AgCl). The test solution is a solution of NaCl at 4% and the equipment used is a Gamry potentiostat 600. Each paint system was initially assessed after one hour stabilization potential and then the same system is evaluated at 7 days and 15 days to observe what happens to the painting in time. At the end of the specimen images after the electrochemical assay it is presented as auxiliary electrode a platinum wire is used as reference electrode and a silver electrode/silver chloride (Ag/AgCl). The test solution is a solution of NaCl at 4% and the equipment used is a Gamry potentiostat 600. Each paint system was initially assessed after one hour stabilization potential and then the same system is evaluated at 7 days and 15 days to observe what happens to the painting in time. At the end of the specimen images after the electrochemical assay it is presented.

Each paint system was initially assessed after one hour of stabilization of the potential and the same system after 7 days and 15 days to observe what happens with paint over time is evaluated. At the end of the specimen images after the electrochemical assay it is presented. Each paint system was initially assessed after one hour of stabilization of the potential and the same system after 7 days and 15 days to observe what happens with paint over time is evaluated. At the end of the specimen images after the electrochemical assay it is presented.

4 Results and Analysis

4.1 Mounting of the Specimens

For the present assay performed assemblies presented in Figs. 1 and 2. Two samples White paint on metal substrate naked were analyzed and encoded as: M13 and M15. Another two samples on galvanized steel substrate were analyzed and coded as: M16 and M24.

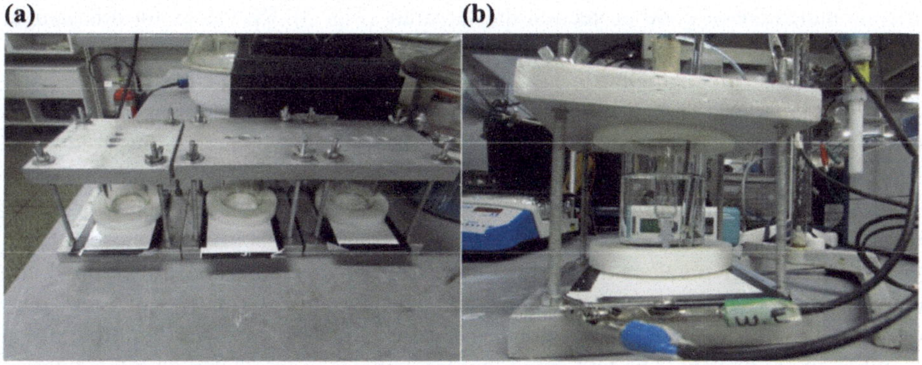

Fig. 1. Assemblies for testing EIS. Source: Laboratorio CIDEMAT. a. Detail of the cells used. b. Cells with the solution and electrodes

4.2 Analysis of Paintings by Electrochemical Impedance Spectroscopy (EIS)

4.2.1 Paints Applied on Carbon Steel Substrate

The results of the impedance are analyzed from graphs or diagrams of Nyquist linking imaginary impedance (-Z Imaginaria) with real impedance (Z actual).

In Fig. 2 the impedances of the M13 samples in time are compared, with the reference sample that obtained at the start of the test after one hour in NaCl solution.

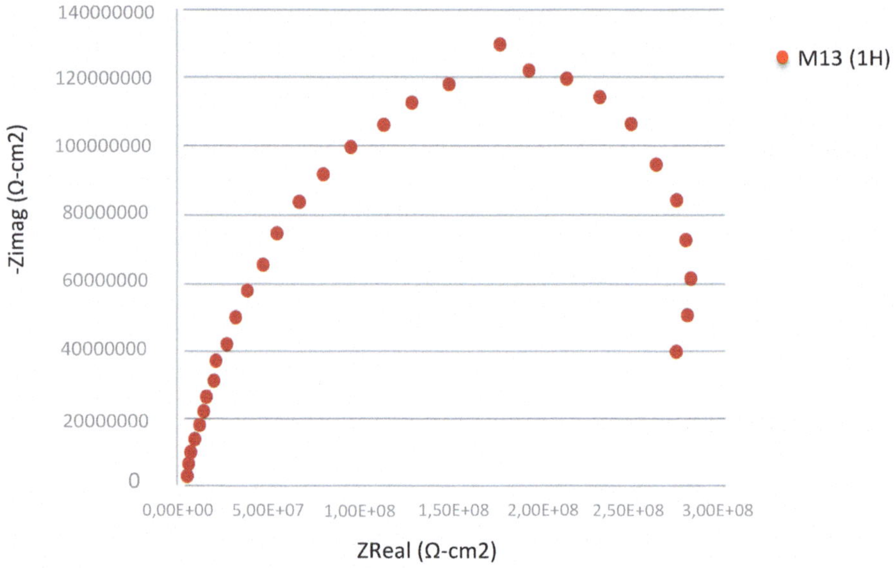

Fig. 2. Sample 13 (1 h) Carbon steel substrate Source: Laboratorio CIDEMAT

For the first paint applied on bare steel carbon sample it corresponds to M13 (1 h) an open arc high resistance film (Rp) which is close to high frequency values occurs which is indicative of good protection the substrate (Fig. 3). Rp values for this sample are near 0.3 GOhm.cm^2, although in practice a painting unaffected can reach values up to 10 GOhm cm^2, (1 Giga Ohm = 1 $\times$ 109) with an arc that is not closed. As the paint system will deteriorate much less semicircle is formed and other forms associated with diffusion processes, oxidation and formation of species which are observed even slightly in the diagram of Fig. 3 for curves M13 (7 days) and M13 (15 days).

Compared with the previous sample, it can be said that the initial behavior of the sample M15 - (1 h) Fig. (4) is very similar, with a large capacitive arc closes at low frequencies taking Rp values near 2.5 GOhm.cm^2, (Fig. 5) Contrary to the above, the behavior of the sample curves to 7 days and 15 days this completely different phenomenon. In curve M15 - (7 days) a capacitive arc much lower than the initial and a second open arc present. The first arc at low frequencies is associated with the external paint system and the second high associated intermediate reaction products on the electrode surface frequencies. This behavior of the impedance indicates the initial

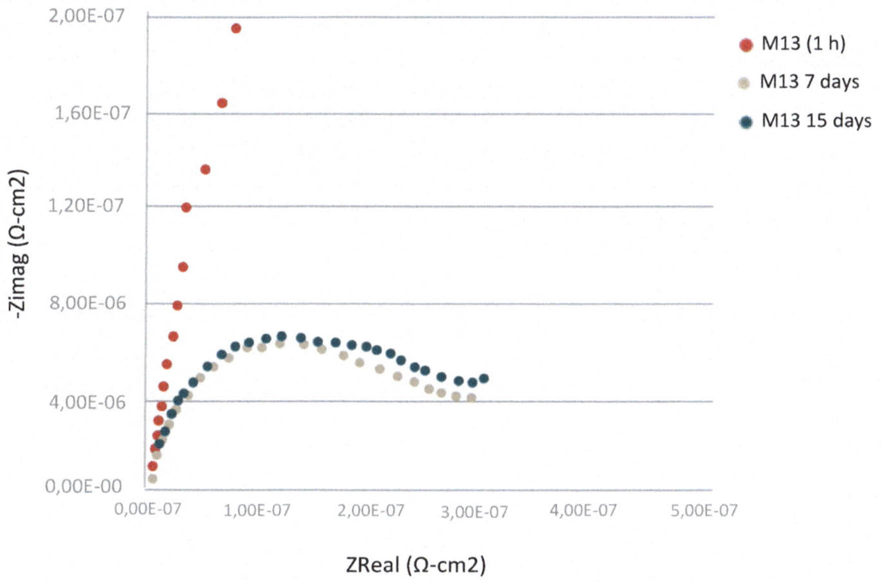

Fig. 3. EIS analysis using samples for M13 on carbon steel. Initial M13-1 h (Red), M13-7 days (Grey), M13-15 days (Blue)

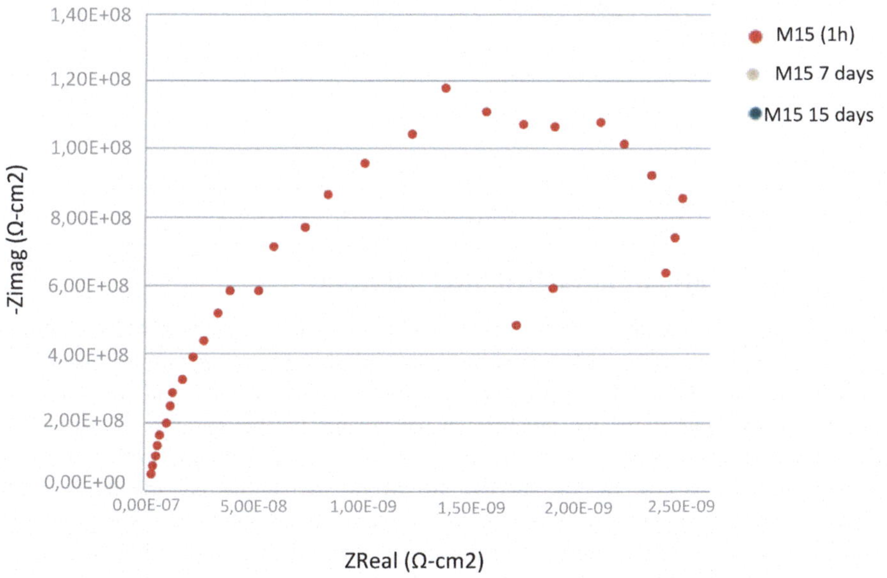

Fig. 4. Diagram sample 15 (1 h) in carbon steel substrate. Source: Laboratorio CIDEMAT

existence of defects in the coating which allow contact between the electrolyte and metal substrate. The resistance associated with the coating of this curve is well below the initial sample and is close to 1.17 × Rp = 105 O.cm^2, (0.17 MOhm.cm^2). Meanwhile curve M15 - (15days) also has a capacitive arc at high frequencies and tries to form a second arc at low frequencies thus presenting a very similar behavior to the previous curve. The value Rp for the last curve is 0.70 × 105 O.cm^2, (0.07 MOhm. cm^2). This result corroborates what has been observed in the sample, which shows as many defects at the end of blistering evaluations.

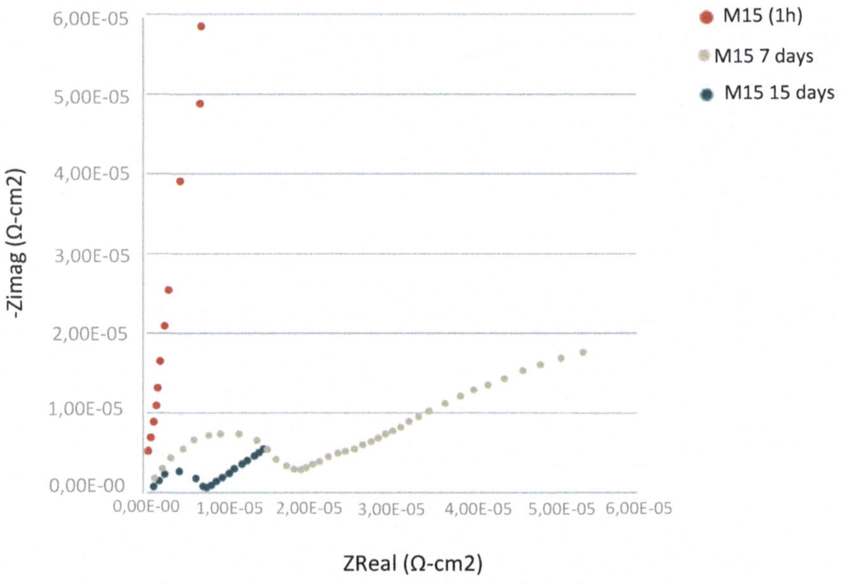

Fig. 5. EIS for analysis by M15 shows on carbon steel. Initial M15-1 h (Red), M15-7 days (Grey), M15-15 days (Blue).

4.3 Applied Paints on Galvanized Steel Substrate

Paints applied on galvanized substrate correspond to the M16 and M24 shown. In Fig. 6 the impedances of the M16 sample (1 h) with those obtained after 7 and 15 days are compared. In M16 (1 h) shows two capacitive arcs, one of small diameter, high frequency associated with the phenomena occurring in the paint is observed; and another large at low frequencies characteristic of processes occurring in the galvanized substrate. The large size of the second arc M16-1 h sample is indicative that the galvanized surface phenomena even no active corrosion of galvanized substrate. With the passing of time this second arc tends to become smaller, indicating the onset of corrosion processes and/ or processes of reaction intermediates on the zinc surface (Fig. 7).

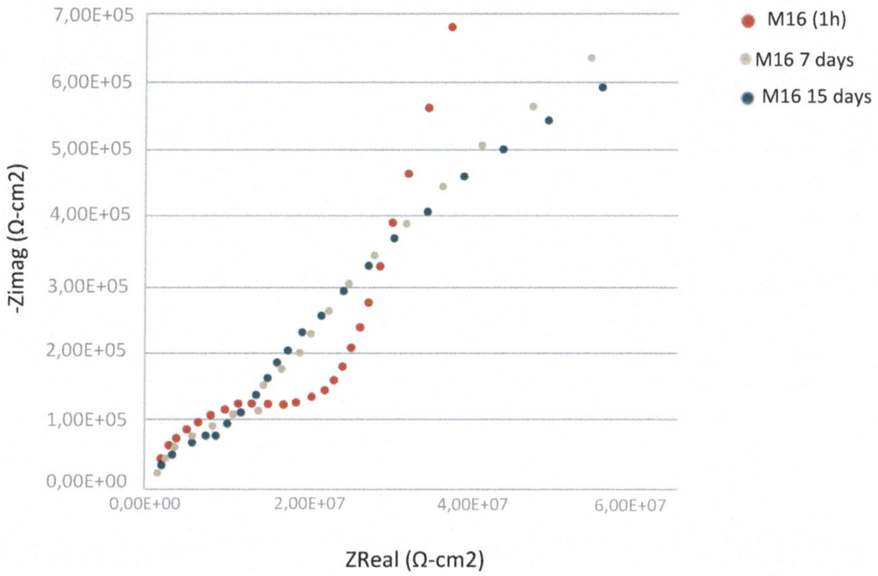

Fig. 6. Initial M16-1 h (Red), M15-7day (Gray), M15-15 (Blue) all at high frequency in galvanized steel substrate. Source: Laboratorio CIDEMAT

The arcs at high frequency (Fig. 6) that represent the processes in the painting have similar values and tend to be kept in the same order of magnitude with time, which implies that from the first hours of evaluation the painting shows certain defects in the coating that allow contact between the electrolyte and the metal substrate. The resistance associated to the initial coating M16 - (1 h) is of the order of 0.3 MOhm. cm^2 being comparatively with the carbon steel samples three orders of magnitude smaller. After several days of exposure, the Rp values are of the order of 0.13–0.15 $MOhm.cm^2$ for the curves at 7 days and 15 days respectively. On the other hand, the resistances associated with the phenomena that occur in the galvanized surface fall from 1.2×107 $O.cm^2$ (large red arc) to 3.07×106 $O.cm^2$ (gray and blue arcs) as shown in (Fig. 7), in which the large red arc closes to reach smaller values as a consequence of the corrosion processes in the galvanized substrate.

For the M24 sample representing the second galvanized substrate, the behavior of the impedance of the sample at one hour (M24-1 h) is very different, since the paint system applied to this plate shows a large capacitive arc indicating a high rp typical of a good coating. At low frequencies the M24-1 h sample tends to form a second typical of what occurs in a galvanized not yet experience shows that corrosion phenomena arc. Contrary to the previous sample for this sample the two arcs are quite large and tend to merge into one (Fig. 8). The resistance values of the film are in the order of 0.13 $GOhm.cm^2$.

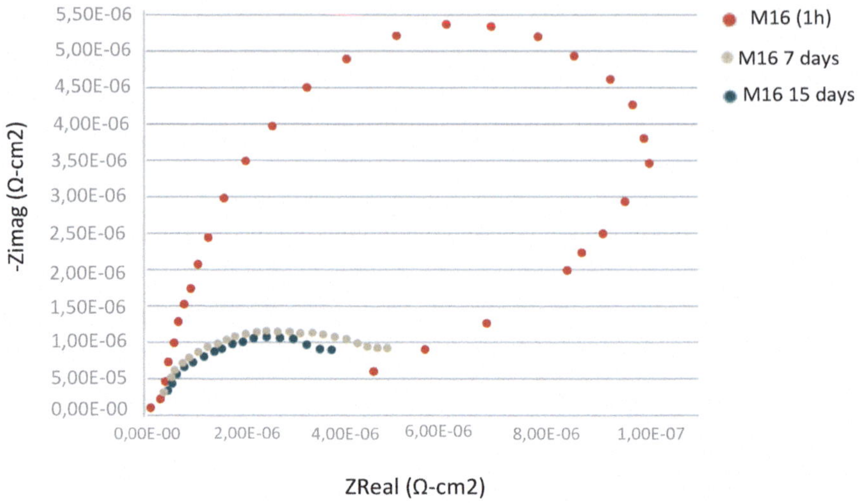

Fig. 7. Initial M16-1 h (red), M15-7 days (Gray), M15-15 days (Blue) all low frequency substrate galvanized steel

With the passage of time are formed two arcs is smaller: The first high frequencies associated with paint and the second at low frequencies associated with the phenomena of corrosion of galvanized substrate. As in the previous graphs showing M24-7days and M24-15days have similar values and tend to be conserved in the same order of magnitude over time, the shape of the curves very similar to each other (Fig. 9). The resistance associated with the coating of this sample after several days of exposure is in the range of 1.4 to 1.6 MOhm.cm^2, decreases dramatically relative to sample value one hour, which implies a strong involvement of the paint and corrosion processes galvanized substrate.

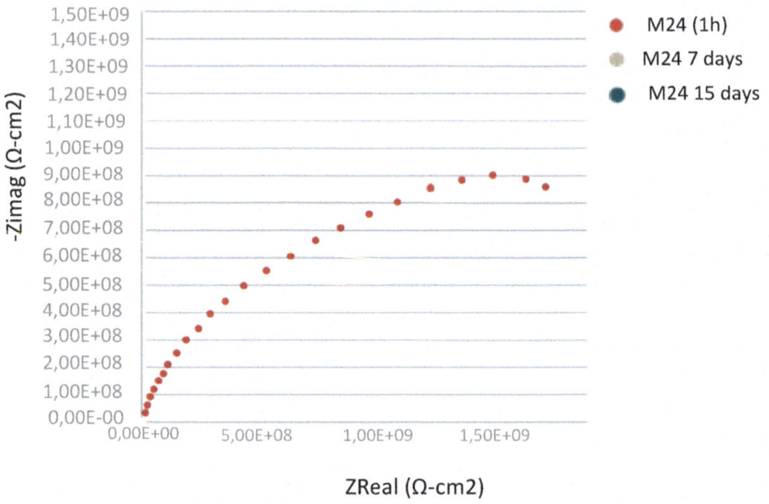

Fig. 8. Initial sample 24 (1 h) in galvanized steel subs. Source: Laboratorio CIDEMATtrate.

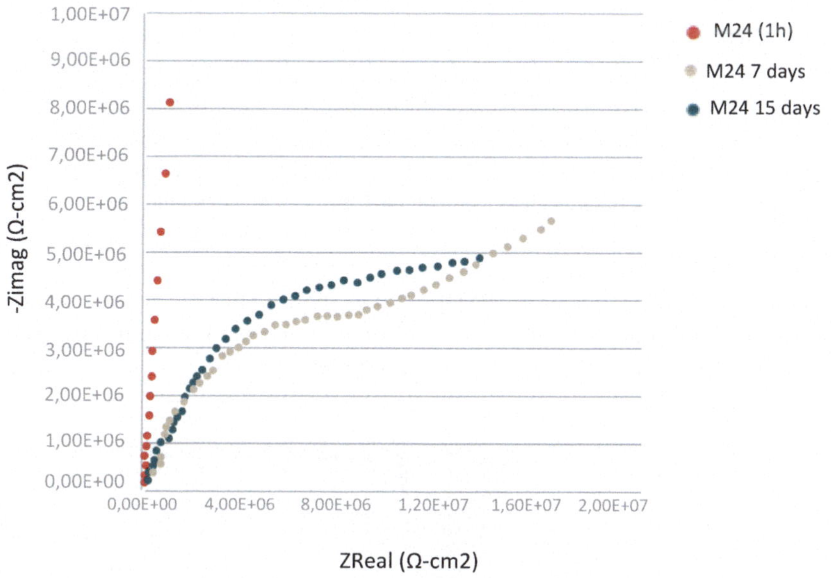

Fig. 9. Inicial M24 - 1 h (Red), M24-7day (Gray), M24-15days (Blue)

The analysis of the curves and the determination of their values are made using equivalent electrical circuits similar to those observed in Fig. 10 with the purpose of obtaining quantitative parameters of the properties of the coating applied to the samples. These circuits are modeled according to the characteristics of the system. For example, a simple circuit with two resistors and a capacitor (Randles) allows simulating a semicircle, while a more complex one of three circuits in cascade simulates a painting with galvanized substrate. Table 2 summarizes the calculated values for Rp of the previous samples.

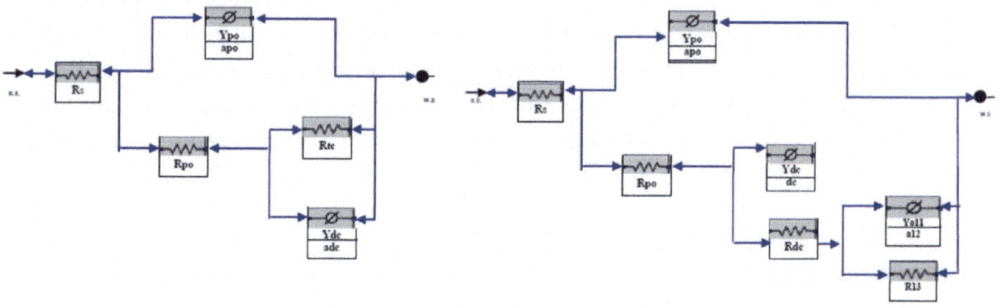

Fig. 10. Equivalent circuits that allow to adjust the impedance curves

Table 2. Summary of Rp values for the four samples. Source Laboratorio CIDEMAT

Sample	Period		
	1 h	7 days	15 days
M13	0.3 (GOhm.cm2)	25 (MOhm.cm2)	25 (MOhm.cm2)
M16	2.5 (GOhm.cm2)	0.17 (MOhm. cmtwo)	0.07 (MOhm.cm2)
M16	0.3 (MOhm.cm2)	0.13 (MOhm. cm2)	0.15 (MOhm.cm2)
M24	0.13 (GOhm.cm2)	1.6 (MOhm. cm2)	1.4 (MOhm.cm2)

In conclusion, the results of impedance curves indicate that all samples are affected in varying degrees, and the sample is less affected is the M13, although it is possible that a future said sample present corrosion problems.

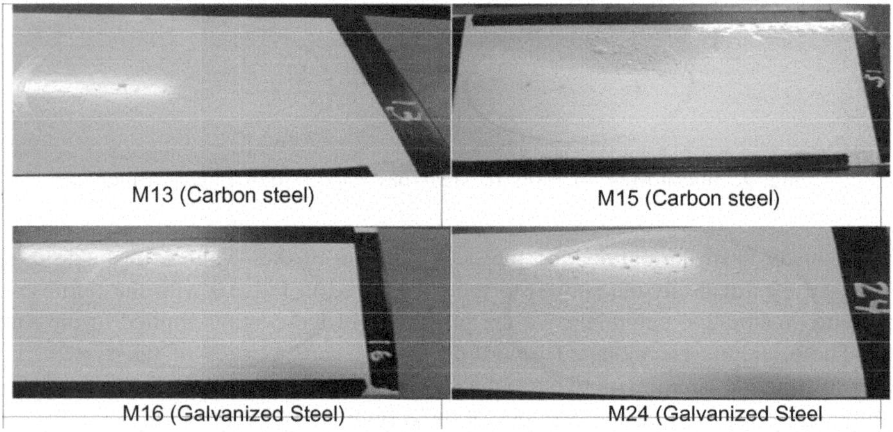

Fig. 11. Samples of carbon steel and galvanized steel, Source: Laboratorio CIDEMAT

Although the curves obtained by EIS have different behaviors depending on whether the substrate is carbon steel or galvanized steel, if you can find some common ground on all systems from their relationship against blistering. The images in Fig. 11 show the results of picture after disassembling the cell greater density of blisters of the substrates of the M16, M24 and M15 samples, the substrate in carbon steel M13 is evident.

5 Conclusions

The Electrochemical Impedance Spectroscopy analyzes allow us to conclude that the sample with the best performance is M13, applied on a carbon steel substrate. The open capacitive arc of the impedance of the sample M13 exposed to a few hours immersion in the electrolyte is characteristic of a good barrier system, which says well of the correct application of the coating and good initial performance of it; however, the study carried out does not allow to determine for how long said coating may grant protection to the substrate. The rest of the samples present parameters that indicate a poor performance of the protection system or the initiation of corrosion processes in the area of the interface between metal and paint, phenomena that are physically evident from the strong blistering that the samples present afterwards of 15 days of exposure in NaCl solution.

One of the most difficult plastic waste to eliminate is the PER because of its characteristics, its disposition is very complex and entails environmental hazards. Expanded polystyrene, being one of the most used materials in everyday life, has become a high pollutant to the environment, is a material that is not very biodegradable, so it is essential to make better use of it. In this research the expanded polystyrene was used as a component of a polyurethane paint improving its properties, det of protection against corrosion, and within the technological prospective of innovation and technological development can be used in the nautical finishes of the dead work, and super structure of the afloat units of the National Navy.

References

1. Clemitson, I.: Castable polyurethane elastomers: Taylor & Francis Group, LLC Boca Raton, FL (2008)
2. Baker, J.W., Holdsworth, J.B.: The mechanism of aromatic side-chain reactions with special reference to the polar effects of substituents. Part XIII. Kinetic examination of the reaction of aryl isocyanates with methyl alcohol. J. Chem., 713–726 (1947). https://doi.org/10.1039/JR9470000713
3. Baker, J.W., Gaunt, J.: Ibid, 9 (1949)
4. Edwards, P.A., Striemer, G., Webster, D.C.: Synthesis, characterization and self-crosslinking of glycidyl carbamate functional resins. Prog. Org. Coat. **57**(2), 128–139 (2006). https://doi.org/10.1016/j.porgcoat.2006.08.002
5. Reilly, C.B., Orchin, M.: Product and process development: preparation and properties of polyurethane coatings. Editorial Ind. Eng. Chem. **48**(1), 59–63 (1956)
6. Chen, H., Fan, Q., Chen, D., Yu, X.: J. Appl. Polym. Sci. **79**, 295 (2001)
7. Chattopadhyay, D.K., Webster, D.C.: Thermal stability and flame retardancy of polyurethanes. Prog. Polym. Sci. **34**(10), 1068133 (2009)
8. Giudice, C.A., Pereyra, A.M.: Paints and coatings technology components, formulation, manufacture and quality control. Edutecne, Buenos Aires (2009)
9. Benitez, I.J., Vélez, J.A.: Preparation of an anticorrosive coating from expanded polystyrene recycling. Undergraduate thesis, University of Cartagena (2013)
10. Lopez, D. et al.: Treatmant of expander polystyrene using solvent Green. Appl. Res. J. 8(1), 1–9 (2014). https://revistas.upb.edu.co/index.php/investigaxionesaplicadas. Accesed 22 June 2017
11. Shikata, S.: Dissolution of polystyrene into cyclic monoterpenes present in tree essential oils. J. Mater. Waste Manag. Cycles **13**, 127–130 (2011)
12. Yang, W.: Prog. Polymer Sci. **39**, 1017–1042 (2014)
13. Talbert, R.: Paint Technology Handbook. CRC Press/Taylor & Francis Group, New York (2008)

Methodology for the Early-Stage Design of Electrical Power Systems in a Future Generation of Colombian Warships

Gabriela Salas-Berrocal[1(✉)] and Richard Bucknall[2]

[1] The Science and Technology Corporation for the Development of the Naval, Maritime and Riverine Industries (COTECMAR), Cartagena, Colombia
gsalas@cotecmar.com
[2] Faculty of Engineering, University College London, London WC1E 7JE, UK

Abstract. This paper proposes a methodology for the development of concept designs for naval electrical power systems, based on the latest standards for marine and naval electrical installations, as the IEEE Std. 45.3-2015. To achieve this methodology, the authors review the relevant aspects involved in the design of the Shipboard Power Systems (SPS), including the components and main features of the SPS, as well as the most common network topologies. Finally, potential areas of improvement for the implementation of the proposed methodology in the design of a future generation of Colombian warships are presented.

Keywords: Electrical power systems · Naval ship design · Shipboard Power Systems

1 Introduction

The design of a naval ship is a complex and iterative process, taking place in several phases. It starts with what is usually referred in the literature as concept exploration and concept design [1–6]. During these initial phases, important decisions regarding key specifications of the vessel such as speeds, ranges, and configuration of the mission systems are adopted [7]. Consequently, these decisions have a deep impact on the selection of the electrical power and propulsion system architectures.

Regarding electrical energy requirements, a modern warship is comparable to a small town constrained to a space smaller than a typical sports stadium [8]. Hence, the Electrical Power System (EPS) is a very complex but essential part of a naval vessel, considering its influence in the feasibility and affordability of given ship design. Moreover, the way the components of the electrical power system are interconnected, which is the network topology, can make a difference between two ship designs in terms of number and size of equipment to be used, survivability, reliability, and efficiency [9]. For this reason, in recent years there has been a renewed interest in the study of EPS and network topologies for warships.

The knowledge of the main elements and characteristics of the SPS in warships is the baseline for its design. However, it is also required a clear methodology to

V. A. J. E. Carreño Moreno et al. (Eds.): CIDIN 2019/COPINAVAL 2019, pp. 334–346, 2020.
https://doi.org/10.1007/978-3-030-35963-8_28

undertake the design of the electrical power systems during the early stages of a naval project, which has traditionally relied on initial electrical power estimates made by Naval Architects.

This paper presents an approach to the early-stage design of EPS, through a simplified methodology to integrate elements of the electrical engineering into the traditional workflow of the naval ship concept design, aiming to the development of a future generation of Colombian warships.

2 Elements Involved in the Design of the Electrical Power Systems for Warships

Nowadays, the trend in the design of warships is going towards an increasing degree of electrification [10, 11]. Figure 1 illustrates the relationship between the full load displacement and the installed electrical power for a set of surface combatants with mechanical propulsion, identified in a parametric survey conducted by Salas and Vásquez [12]. It shows that the rate of increase of electrical power is higher in ships less than 25 years of commissioning, which suggests that modern naval vessels are more "energy dense" compared to older ships. This trend is the result of the rising number of consumers, driven by the use of advanced communication systems, sensors, and weapons [13–15]. Moreover, propulsion systems in surface combatants are migrating from fully-mechanical arrangements to hybrid and all-electric systems.

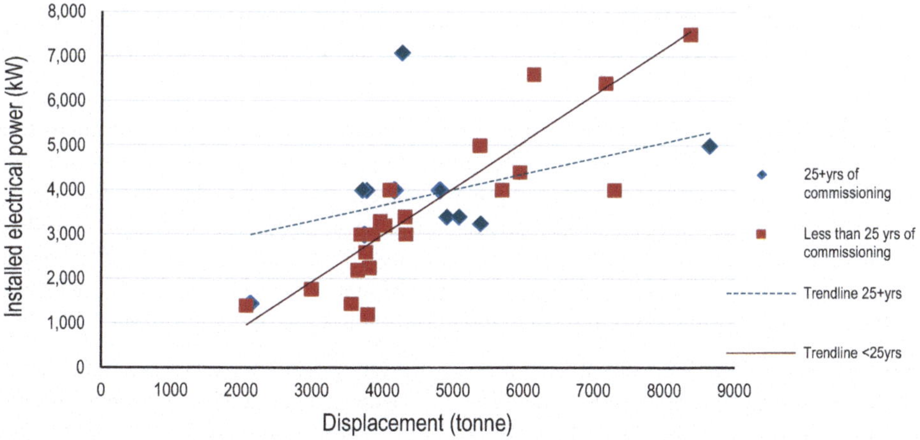

Fig. 1. Survey of installed electrical power vs. displacement in surface combatants with mechanical propulsion [12]

To satisfy the power/energy demand of all the above-mentioned consumers it is necessary to have power generating sources, a distribution system, power conversion stages to provide different voltages and frequencies as required by the final loads, all of these linked through automated monitoring, control, and protection systems. This group of components constitute the EPS of the vessel, known in the literature as Shipboard Power System (SPS) [16].

The SPS is a backbone of a naval vessel due to its strong interaction with most of the systems and sub-systems of the ship and particularly, with sensors, weapons and other mission-related equipment. As stressed by Amy [13], "transport of a military mission, payload or cargo is the fundamental function of a (naval) ship. The ship's power system is what makes this happen. Without a power system, a ship is just a barge". Moreover, the EPS is one of the key drivers of cost and performance in a naval vessel. As a result, the ship designer must be familiarised with the main concepts of the SPS, which are summarised in the next sections.

2.1 Components of an SPS

Figure 2 depicts the main types of elements in an SPS. These components can be differentiated by its functionality in the vessel: Power generation, power distribution, power conversion, energy storage, electrical power control, and loads [17, p. 14]. Each element is briefly described, as follows.

Power Generation. The power generation component converts an energy source, such as fuel, into electrical power. Traditionally, this conversion process has been carried out on board through a mechanical prime mover coupled to a rotating AC alternator. Typical prime movers for rotating power generation are shown in Fig. 3; these include technologies powered by fuels as the diesel engines and gas turbines, as well as the shaft generators, driven by the propulsion train and wind-driven turbines. On the other hand, electrical power can also be generated by static machines as the Fuel Cells (FC) and Photovoltaic Cells (PV).

Power Distribution. The power distribution system is formed by the equipment that transfers the electrical power from the generating sources to the final consumers: Switchboards, load centres, motor control centres, power panels, and the cables interconnecting these elements. Furthermore, it also includes the switchgear used to protect and isolate electrical equipment. The components of the power distribution system are supervised and controlled via Power Management System (PMS), which is in charge of monitoring, reconfiguring, and coordinating the operation of the SPS [18].

Power Conversion. The power conversion equipment allows adapting the characteristics of a power supply (voltage, AC or DC, frequency, waveform, power quality) per the requirements of an end consumer or a distribution sub-system. The most common power conversion equipment are transformers, but this group also contains Variable Speed Drives (VSD), soft-starters, Active Harmonic Filters (AHF), inverters, rectifiers, power supplies, and so on.

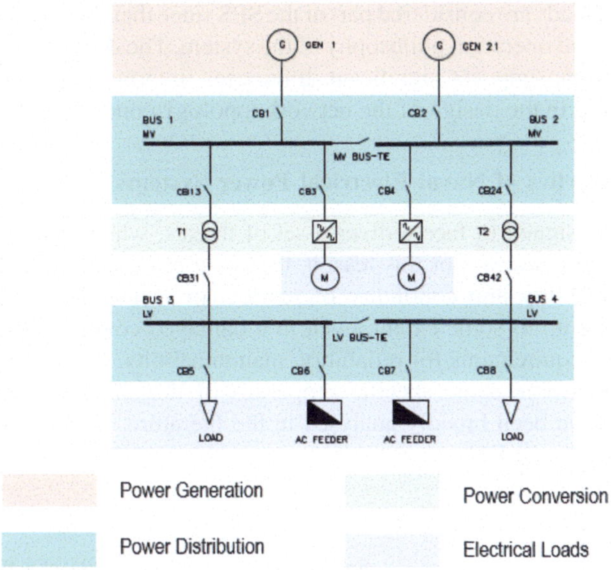

Fig. 2. Simplified single line diagram

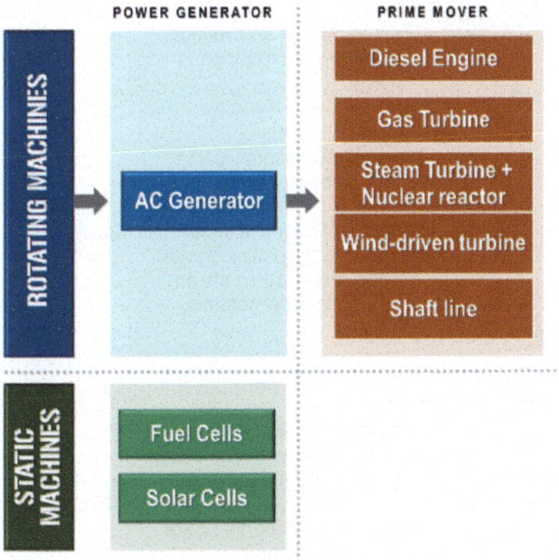

Fig. 3. Types of electrical power generation in ships

Energy Storage. Energy storage systems are needed to provide uninterrupted power to critical consumers, as well as to that equipment susceptible to power failures and shutdowns. Moreover, nowadays the energy storage systems are being used to supply peak power in support to the power generation elements of the vessel, in order to maximise their fuel efficiency. The most common energy storage technology in naval vessels is the battery, but new designs are also implementing other technologies as the ultracapacitors and flywheels.

Loads. The final loads are considered part of the SPS since their nature play an important role in the design and operation philosophy of the system. The distinction between 'vital' and 'non-vital' loads supposes significant differences in terms of redundancies, which must be considered in the design of the network topology concept [13].

2.2 Main Properties of Naval Electrical Power Systems

Naval ships are designed to face a diverse set of threats, which can potentially cause them some level of damage. For this reason, the naval electrical power systems aim to ensure the survivability and continuity of service, in such a way that the ship can continue operating up to some extent, even in a damaged condition [9].

The stringent requirements for reliability, maintainability, and efficiency that naval ships possess determine some specific properties of the shipboard electrical power systems, which have been broadly analysed in the literature. Figure 4 summarises the core properties of Naval EPS.

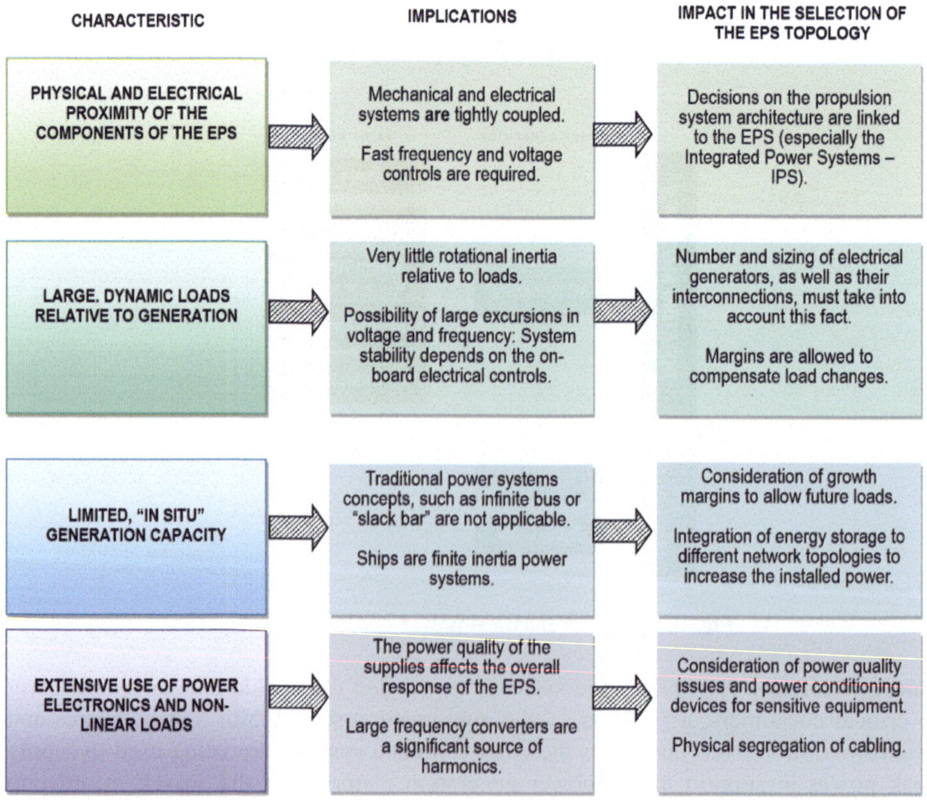

Fig. 4. Summary of the main characteristics of naval power systems and their implications

2.3 Electrical Network Topologies Used in Ship Design

The components of an SPS can be interconnected following different network topologies. This architecture influences the size and number of elements, as well as other characteristics such as efficiency, emissions, survivability, reliability, and fuel consumption, to mention some of them. As argued by Amy [13], the power system architecture generally remains unaltered during the lifetime of the ship; hence, its selection has impacts in both short-term (design and construction) and long-term (lifecycle costs). This section presents the key features of the most commonly used electrical network topologies in surface combatants, according to the literature and the standard IEEE 45-2015.

Radial Architecture. In a radial distribution system, the electrical loads, grouped in load centres and switchboards, are fed directly from a single connection point. Indeed, the term "radial" indicates that loads radiate outward from the switchboards, arranged in several distribution levels from the main power generating sources up to the end consumers. In this architecture, the power flows from the main distribution centres through feeder cables that interconnect the various distribution levels [18, 19]. The radial architecture is the most common philosophy used in the design of SPS due to its simplicity. However, a fully radial distribution system is unable to provide the level of reliability and survivability demanded by a warship. For this reason, the implementation of redundant circuits for vital and mission-critical loads is compulsory according to the regulations of all the Classification Societies.

Zonal Architecture. The concept of the zonal electrical distribution system (ZEDS) was introduced to overcome the limitations of the radial architecture due to the increasing number of vital and mission-critical loads in modern warships. In a typical zonal architecture, the ship is divided into sections, called zones, which are interconnected by two main buses that longitudinally run throughout the ship at port and starboard sides. To increase the survivability of the vessel, these longitudinal buses must be installed as far as possible from each other and in the less vulnerable places of the ship (for example, below the structural box girders) to prevent that both buses become unavailable at the same time during a combat situation [9, 13, 17, 20–22].

Hybrid Architectures. According to the standard IEEE 45-2015 [17], some architectures are combinations of radial and zonal elements. These arrangements are considered hybrid architectures and the most common topologies found in the literature are the ring bus, breaker-and-a-half, and meshed grid [23].

3 A Methodology for the Concept Design of a SPS in Warships

One of the primary motivations for the development of a methodology for the concept design of naval electrical power systems is depicted in Fig. 5: Design decisions taken in initial stages of design have a smaller impact on cost. For this reason, there is a recent trend to incorporate the design of the SPS earlier in the design process.

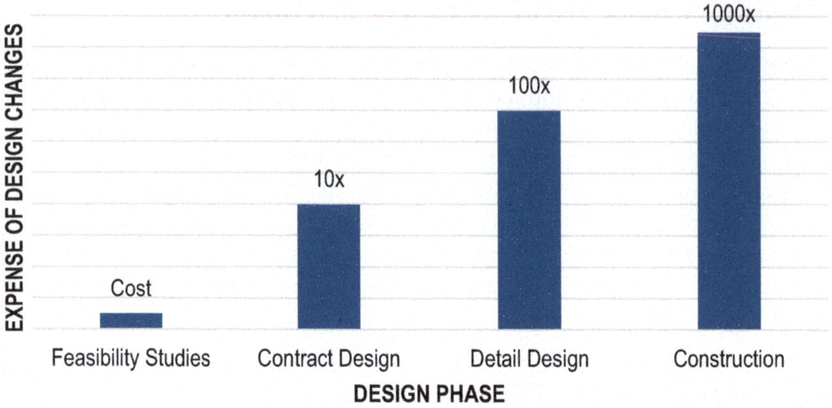

Fig. 5. Cost of design changes during different naval ship design phases [24]

The following methodology attempts to enhance the integration of the electrical power systems into the trade-off studies undertaken in the early-stages of a naval project, through the characterisation of the required inputs and expected outputs for the processes involved in the workflow.

Figure 6 shows an adaptation of the traditional design spiral, in which ship design is represented as an iterative process [5]. This approach has been applied by McCoy [11] to the SPS design, explaining a sequence of processes that are undertaken during concept design. Furthermore, [13, 25], and [17] discuss the foremost considerations in the design of naval EPS, providing an overview of the design process from the systems engineering perspective.

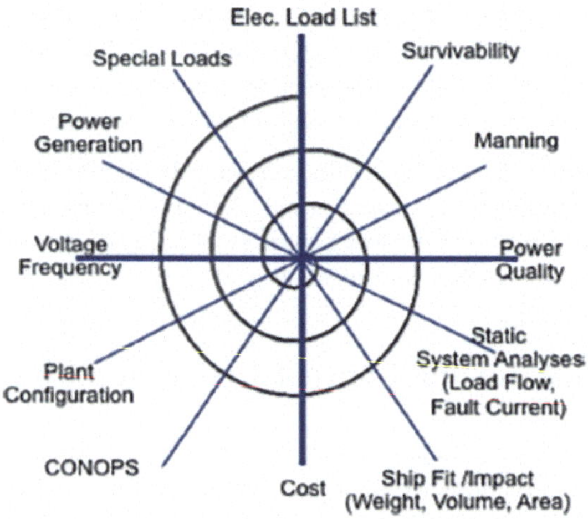

Fig. 6. Traditional design spiral process applied to shipboard electric plant design [11].

Taking the literature findings as a starting point, a methodology to integrate the concept design of the SPS into the initial stages of a naval project is proposed. Figure 7 illustrates this methodology as a block diagram and the main aspects of each step are explained below.

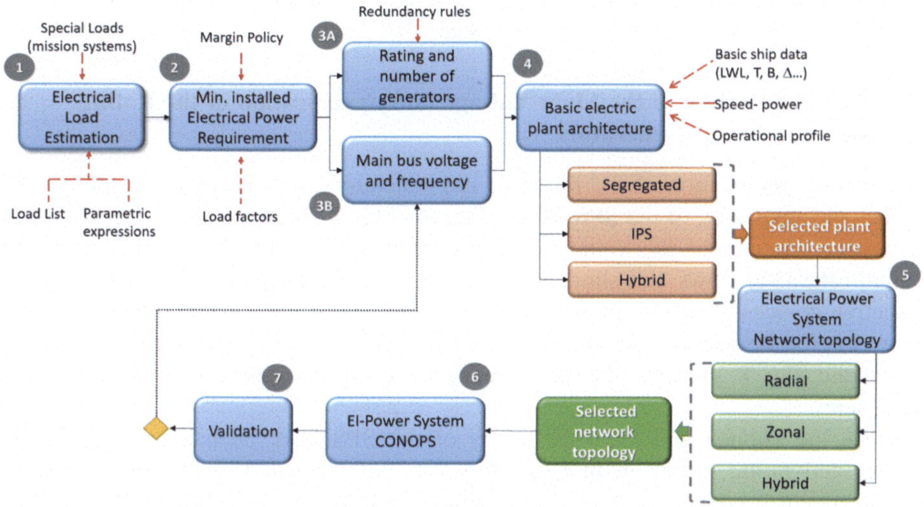

Fig. 7. Block diagram of the proposed methodology for the concept design of electric power systems in warships

3.1 Step 1: Electrical Load Estimation

The first step is to undertake the electrical load estimation. Typically, the load estimation starts with an electric load list of the vessel [26]. However, sometimes during the early stages of design there are not enough data to elaborate a complete list for the electrical load analysis. Moreover, in the case of a new class of warship, it is very likely to face some difficulties in the estimation of the electrical power requirement due to the lack of reference information of a comparable ship. Hence, a first approximation to the Maximum Functional Load (MFL) of the vessel, can be achieved through parametric expressions. Morales and Salas [27] applied the MIT XIII-A mathematical model to obtain an estimation of the installed electrical power for the case study of the Strategic Surface Platform (PES, for the acronym in Spanish). It is worth mentioning that for naval combatants the MFL must be evaluated for the combat condition.

The results of the estimations performed by using parametric models can be validated through a benchmarking (parametric survey) of similar vessels. In addition to the MFL estimated for the platform, another input required for this step is total power consumption of the special loads, including all mission systems.

3.2 Step 2: Minimum Installed Electrical Power Requirement

Once an initial estimate of the connected load is achieved (step 1), then the minimum installed electrical power can be determined, allowing margins for design and growth in accordance with the Navy and Shipyard margin policies, as well as allowances for variations in load through load factors [17, p. 36]. Moreover, the installed electrical power requirement drives the selection of the rating and number of generators (step 3A) and the main bus voltage and frequency (step 3B).

3.3 Step 3: Selection of the Main Voltage and Specification of the Rating and Number of Generators

Prior to the specification of the rating of the generators, it is necessary to define the voltage and frequency of the SPS. According to [11], the total installed electric power drives the selection of the voltage and frequency of the power system in order to maintain the load flows and short circuit current within the permissible limits of the protective devices. Table 1 summarises the guidelines of the standard IEEE 45-2002 [28, p. 22] for the selection of voltage and system type in accordance with the installed power and Fig. 8 presents the possible options for the selection of voltage and frequencies.

Once the voltage and main frequency of the SPS have been decided, then the number and rating of the main generators can be selected. For this purpose, it is recommended to apply the "N + 1" redundancy rule for the selection of the required number the generator sets. This rule establishes that the electrical power generation plant must be able to supply the MFL with one generator down, which is an allowance for maintenance [11, 17]. Consequently, the sizing of the generators can be performed through (1).

$$P_{GEN} = \frac{kW_{MFLM}}{(n-1) * 0{,}9} \tag{1}$$

Where: kW_{MFLM} is the MFL with margins; n is the number of generators; 0,9 is the load factor, indicating that each generator can only be loaded up to 90% of its rated capacity during normal operating conditions.

Table 1. Recommended voltage levels for AC power distribution in ships [28]

Electrical installed power	Generated voltage (V)	Utilisation voltage (V)
<100 kW	120	115
100 kW to 1 MW	230, 240	220, 230
1 MW to 10 MW	450, 480, 600, 690	440, 460, 575, 660
>10 MW	2.400; 4.160; 6.600	2.300; 4.000; 6.000

3.4 Step 4: Basic Electric Plant Architecture

The fourth step is to select the basic electric plant architecture, which is related to the integration of the propulsion system with the electrical power system. Three possibilities for this integration are recalled by literature: segregated, integrated (IPS), and hybrid systems (diesel-electrical configuration).

The decision of which basic plant architecture is suitable for each warship depends on different factors, including the operational profile, project constraints (cost, weight, volume), and technology readiness and availability for a particular Navy. For instance, [11, 14, 21, 29] explain the advantages of the IPS over the traditionally segregated architectures. Nevertheless, a new trend in the design of small surface combatants are the hybrid architectures, as outlined by [30–32], and [33], due to its benefits in fuel efficiency at low speeds.

When it comes to the application of the methodology, it is important to mention that if IPS or hybrid architectures are selected, then it is required to have additional information for the specification of the electrical components of the propulsion system; this is mainly the speed-power curve, the basic ship data, and the operational profile.

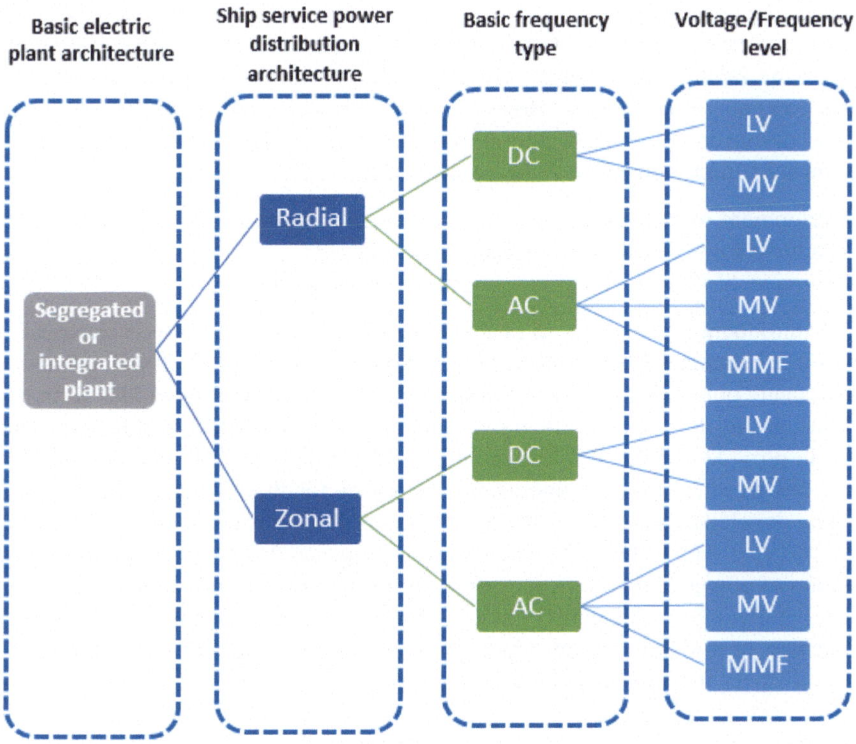

Fig. 8. Electric plant architecture taxonomy [11]

3.5 Steps 5 and 6: Definition of the Electrical Network Topology and EPS-CONOPS

After the basic plant architecture has been chosen, the next step is the selection of the electrical network topology, considering the options of radial, zonal, and hybrid architectures (covered in Sect. 2.3). Finally, the preparation of the Concept of Operations for the Electrical Power System (EPS-CONOPS) and the validation of the results are undertaken.

The EPS-CONOPS is introduced in the IEEE 45.3-2015 (IEEE, 2015) as a statement of the required performance of the electrical power system based on the expected use of the vessel.

4 Conclusions and Further Work

The traditional approach to concept design, where the selection of the propulsion and electrical power system architecture was carried out after initial estimations, conducted from the Naval Architecture perspective, is not suitable for the design of modern warships. Instead of this, the literature proposes the integration of some elements of the EPS design into the early-stages of the naval projects, to achieve a balanced solution in compliance with the functional requirements of the vessel. The proposed methodology identifies the main steps to undertake this task, as well as the required inputs and expected outputs of the process. However, since this an initial attempt to introduce the electrical design in the concept design of a future generation of Colombian warships, some areas of improvement have been identified:

- Development of metrics to assess each topology in terms of physical characteristics (weight, volume), cost, survivability, and reliability.
- Integration of these metrics into an effectiveness model.

The introduction of metrics to evaluate the effectiveness of a given electrical configuration is beneficial in the context of a "set-based design" [4, 7, 34]. In a "set-based design", many potential solutions are generated, but feasible configurations form only a small part of this design space; therefore, metrics for the electrical architectures will be helpful to provide additional information to perform the verification and validation of feasible design alternatives. Some useful references for the further work are found in [35], who developed a model to support the trade-off studies during concept design for the US Navy, as well as [36, 37], and [38] who proposed diverse methodologies for the evaluation of power generation concepts using metrics such as reliability and quality of service. Finally, although the use of metrics during concept design provides guidance in the selection of a given network topology, it is important to bear in mind that as the project advances the EPS must be tested to ensure its robustness.

Acknowledgements. Mrs. Salas-Berrocal would like to thank CeiBA Foundation and the Government of Bolivar for the support provided for the development of this project and her studies at UCL. Finally, the views expressed in this paper are that of the authors and do not necessarily represent the views and opinions of COTECMAR and the Colombian Navy.

References

1. Andrews, D.J.: A comprehensive methodology for the design of ships (and other complex systems). Proc. Math. Phys. Eng. Sci. **454**(1968), 187–211 (1998)
2. Brown, A., Mierzwicki, T.: Risk metric for multi-objective design of naval ships. Naval Eng. J. **116**(2), 55–72 (2004)
3. Brown, A., Salcedo, J.: Multiple-objective optimization in naval ship design. Naval Eng. J. **115**(4), 49–62 (2003)
4. Doerry, N., et al.: Using set-based design in concept exploration. Presentado en SNAME Chesapeake Section Technical Meeting, Arlington, VA, USA (2014)
5. Lamb, T.: The ship design process. In: Ship Design and Construction, vol. 1–2. Society of Naval Architects and Marine Engineers (SNAME) (2003)
6. Mandel, P., Chryssostomidis, C.: A design methodology for ships and other complex systems. Philos. Trans. Roy. Soc. Lond. A Math. Phys. Eng. Sci. **273**(1231), 85–98 (1972)
7. Chalfant, J.: Early-stage design for electric ship. Proc. IEEE **103**(12), 2252–2266 (2015)
8. Chryssostomidis, C., Cooke, C.M.: Space reservation for shipboard electric power distribution systems. In: 2015 IEEE Electric Ship Technologies Symposium (ESTS), pp. 187–192 (2015)
9. Skjong, E., Volden, R., Rodskar, E., Molinas, M., Johansen, T., Cunningham, J.: Past, present and future challenges of the marine vessel's electrical power system. IEEE Trans. Transp. Electrification **PP**(99), 1 (2016)
10. Lai, K., Illindala, M.S.: Graph theory based shipboard power system expansion strategy for enhanced resilience. IEEE Trans. Ind. Appl. **54**(6), 5691–5699 (2018)
11. McCoy, T.J.: Integrated power systems - an outline of requirements and functionalities for ships. Proc. IEEE **103**(12), 2276–2284 (2015)
12. Salas, G., Morales, C.: Estimación de la Potencia Eléctrica demandada por la Plataforma Estratégica de Superficie. Gerencia de Diseño e Ingeniería (COTECMAR), Cartagena, Concepto Técnico CT-ARC-3138-310-01 (2014)
13. Amy, J.V.: Considerations in the design of naval electric power systems. In: 2002 IEEE Power Engineering Society Summer Meeting, vol. 1, pp. 331–335 (2002)
14. Thongam, J.S., Tarbouchi, M., Okou, A.F., Bouchard, D., Beguenane, R.: All-electric ships - a review of the present state of the art. In: 2013 8th International Conference and Exhibition on Ecological Vehicles and Renewable Energies (EVER), pp. 1–8 (2013)
15. Wang, J., Sumner, M., Thomas, D.W.P., Geertsma, R.D.: Active fault protection for an AC zonal marine power system. IET Electr. Syst. Transp. **1**(4), 156–166 (2011)
16. Sabatucci, L., Simone, G.D., Cossentino, M.: Shipboard power system reconfiguration: a self-adaptation exemplar. In: 2018 IEEE/ACM 4th International Workshop on Software Engineering for Smart Cyber-Physical Systems (SEsCPS), pp. 49–52 (2018)
17. IEEE SA-45.3-2015: IEEE recommended practice for shipboard electrical installations – systems engineering. IEEE, July 2015
18. IEEE Std 45.1-2017: IEEE recommended practice for electrical installations on shipboard–design, pp. 1–198. IEEE (2017)
19. Tidd, C.N.: Hardware model of a shipboard zonal electrical distribution system (ZEDS): alternating current/direct current (AC/DC). Massachusetts Institute of Technology, Cambridge (2010)
20. Ciezki, J.G., Ashton, R.W.: Selection and stability issues associated with a navy shipboard DC zonal electric distribution system. IEEE Trans. Power Deliv. **15**(2), 665–669 (2000)
21. Doerry, N.: Naval power systems: integrated power systems for the continuity of the electrical power supply. IEEE Electrification Mag. **3**(2), 12–21 (2015)
22. Doerry, N.: Zonal ship design. Naval Eng. J. **118**(1), 39–53 (2006)

23. Agnello, L., Cossentino, M., Simone, G.D., Sabatucci, L.: Shipboard power systems reconfiguration: a compared analysis of state-of-the art approaches. In: Proceedings of the Smart Ship Technology, London, p. 9 (2017)
24. Rigterink, D.T.: Methods for analyzing early stage naval distributed systems designs, employing simplex, multislice, and multiplex networks. Ph.D. thesis, University of Michigan, Ann Arbor, United States (2014)
25. Jurkiewicz, D.J., Chalfant, J., Chryssostomidis, C.: Modular IPS machinery arrangement in early-stage naval ship design. In: 2013 IEEE Electric Ship Technologies Symposium (ESTS), pp. 121–127 (2013)
26. USN Naval Sea Systems Command: Design data sheet (DDS 310): Electric power load analysis (EPLA) for surface ships (2012)
27. Morales, C., Salas, G.: Installed electric power estimation and analysis of diesel generator set alternatives during the concept exploration phase for next generation of colombian navy combat ships. Presentado en XXIV Congreso Panamericano de Ingeniería Naval, Transporte Marítimo e Ingeniería Portuaria, Montevideo, Uruguay (2015)
28. IEEE 45-2002: IEEE recommended practice for electrical installations on shipboard. IEEE (2002)
29. Zivi, E.: Design of robust shipboard power automation systems. Ann. Rev. Control **29**(2), 261–272 (2005)
30. Buckingham, J.: Hybrid drives for naval auxiliary vessels. Presentado en Pacific 2013 International Maritime Conference: The Commercial Maritime and Naval Defence Showcase for the Asia Pacific, pp. 437–446. ACT: Engineers Australia, Barton (2013)
31. Kwasieckyj, B.: Hybrid propulsion systems: efficiency analysis and design methodology of hybrid propulsion systems. Delft University of Technology, Delft (2013)
32. McCoy, T., Zgliczynski, J., Johanson, N.W., Puhn, F.A., Martin, T.W.: Hybrid electric drive for DDG-51 class destroyers. Naval Eng. J. **119**(2), 83–91 (2007)
33. Simmonds, O.J.: Advanced hybrid systems and new integration challenges. Presentado en 2016 International Naval Engineering Conference (INEC) (2016)
34. Singer, D.J., Doerry, N., Buckley, M.E.: What is set-based design? Naval Eng. J. **121**(4), 31–43 (2009)
35. Chalfant, J.S., Chryssostomidis, C.: Analysis of various all-electric-ship electrical distribution system topologies. In: 2011 IEEE Electric Ship Technologies Symposium, pp. 72–77 (2011)
36. Zhang, T.: A method for evaluating power generation concepts considering the optimal concept of operations. ProQuest Dissertations Publishing (2013)
37. Stevens, B., Dubey, A., Santoso, S.: On improving reliability of shipboard power system. IEEE Trans. Power Syst. **30**(4), 1905–1912 (2015)
38. Ali, H., et al.: Cross-platform validation of notional baseline architecture models of naval electric ship power systems. In: 2011 IEEE Electric Ship Technologies Symposium, pp. 78–83 (2011)

Evaluation of the Damage in Composite Materials Modified with Viscoelastic Layers for the Hull of Boats Subjected to Slamming Impacts

Patrick Roger Townsend Valencia[1(✉)],
Juan Carlos Suárez Bermejo[2(✉)], Paz Pinilla Cea[2(✉)],
and Estela Sanz Horcajo[2(✉)]

[1] Escuela Superior Politécnica del Litoral ESPOL, Guayaquil, Ecuador
ptownsen@espol.edu.ec
[2] Universidad Politécnica de Madrid UPM, Madrid, Spain
{juancarlos.suarez,paz.pinilla,
estela.sanz.horcajo}@upm.es

Abstract. A laboratory methodology is proposed, which allows to evaluate the evolution of intralaminar and interlaminar damage, for the hull of vessels built with reinforced polymer and fiberglass - (GFRP) modified viscoelastically, subjected to low energy impacts reproducing the blow of slamming during your navigation at half speed. The method proposes the reproduction of the phenomenon with the equipment designed for this purpose, analyzing the energy absorbed and the energy returned based on the results obtained by vertical impacts of the same energy level. Interlaminar and intralaminar delaminations are visually appreciated and correlated with the results obtained through the use of image analysis software. The deformations have been evaluated with strain gauges supported by the respective mathematical formulation. The advances in the use of new generation viscoelastic layers and their future perspectives are shown.

Keywords: Slamming · GFRP · Impact · Boats · Delamination

1 Introduction

Slamming is an important event during the navigation of the ship. It is presented as the impact of the ship in navigation with bow waves. This generates potential energy due to the impact between the bottom of a vessel's hull and the surface of the water. This deformation occurs in pulses of very short duration of the order of a few milliseconds, which act on a very small surface and which causes high peaks of pressure.

The severity of this impact and its damage to the spacecraft is so unpredictable that the Captains must take precautions and reduce speed so as not to cause additional damage during the voyages. This is aggravated in the case of high-speed planning hull vessels. The importance of slamming on the safety and integrity of marine vehicles has

V. A. J. E. Carreño Moreno et al. (Eds.): CIDIN 2019/COPINAVAL 2019, pp. 347–356, 2020.
https://doi.org/10.1007/978-3-030-35963-8_29

motivated a continuous research process. It is known that this phenomenon was initially studied by Von Karmman in 1929 [1].

The complexity of the phenomenon occurs when the bottom of the ship, the fluid impacts the surface of the hull and the surface of the sea which have an angle difference. This generates that in the contact region, the water expands at high speed, independent of the speed of the ship. This phenomenon was analyzed by Wagner in 1932 [2]. Kabsengerg [3] compiles the research carried out in recent years, and confirms that the study of this problem is pending to be resolved, therefore more experimentation should be done, improve mathematical models and try models or reproductions at scale.

The application of composite materials: Glass Fiber Reinforced Plastic - GFRP in boats was initially developed as a need to obtain lighter vessels, resistant to corrosion, resistance to the environment, ability to mold structures with complex geometries, good mechanical strength with relation to its low weight, ease of repair and excellent durability, Selvaraju [4] and Recuero [5].

The use of new "out of autoclave" pre-impregnated materials - OoA for oven curing in GFRP structures - requires special consideration in the study of the dissipation of energy from cyclic impacts such as slamming. The impacts, during the milliseconds that the deformation lasts, leave residual damage in the laminate that accumulates in the material at the micro structural level.

In the case of GFRP vessels, the phenomenon of slamming has the particularity that the impact of the sea is converted into energy, which is damped in the composite material, and can produce different levels of damage. This results in a reduction of the strength and rigidity of the composite material, creating a problem of a complex nature to study [6, 7].

Numerous types of tests are carried out based on the principle of energy conservation, in which an impactor is launched from a height, on the test piece to be studied, falling freely with an acceleration equal to gravity (g).

According to the research of Lopes and Seresta [8] this technique provides very complete results on the behavior of kinetic energy, when it becomes potential energy and when it is transferred to the sample, it produces a damage that is transmitted on the surface of the sample. GFRP materials, when subjected to this type of impact test, dissipate the energy received through the laminate, which, depending on its intensity, results in different types of damage.

At the moment of impact at low energy, the composite panel begins to flex elastically, generating internal stresses and deformations, product of the response to the different energies that are transferred from the impactor. This is a fatigue effect, a cyclic event in which the composite material reaches the elastic limit or damage threshold.

At that moment of the material, due to the energy of the impact, the structure presents intralaminar cracks and, depending on the orientation of the fibers due to the deflection of the panel, tension and compression tensions are generated causing the panel to break. The fatigue damage in the GFRP laminates always adopts a ladder shape in which the delaminations are linked to the interlaminar cracks.

The reproduction considered at real scales, allows through interlaminar separation planes, to know the amount of energy returned that does not become damage and to have an idea of what happens with the GFRP as time progresses versus the number of impacts applied. And if the amount of kinetic energy dissipated during the deformation should be mitigated to reduce the damage to the laminate, it is essential to modify this laminate to decrease resin cracking, fiber separation and delamination which dramatically affect the overall strength and the integrity of the structure.

When the panel is modified by inserting viscoelastic layers, this energy dissipates protecting the laminate from damage. This occurs because the design of the viscoelastic sheets is related to the three-dimensional Hooke Law, restricting the viscoelastic in two of its directions within a capsule, this is compressed in only one direction. This causes the impact energy to be returned in a large percentage of its totality.

The present work shows an experimental methodology used to reproduce the phenomenon of slamming in the laboratory and are compared with the results of similar panels of GFRP impacted at low energy.

2 Reproduction of the Slamming

The proposed method reproduces the slamming phenomenon by applying a cyclic pressure load on test pieces, similar to those designed for the bottom of the vessels. For this purpose, pre-impregnated GFRP laminated samples of 270 × 270 mm have been used, with laminated layers typical of planning hull boat with 13-meter long, "V" bottom and 25 knots of speed. It is a laminate of 9 layers, to which a viscoelastic sheet has been inserted (45°/−45°/90°/viscoelastic/−45°/45°/0°/90°/−45°/+45°). The panel was cured out of autoclave in the oven. (OoA materials, Out-of-Autoclave) as shown in Fig. 1.

Fig. 1. Cured panels OoA type

A detail of the viscoelastic sheet is shown in Fig. 2. The design of the impact face, the face opposite to the impact and the detail of how the viscoelastic is inserted into the sheet is observed. Elastomers with their property of Poisson constant close to 0.5 allow them large elongations and absorption of energy when they are restricted, and they are the key to the design of the viscoelastic sheet. In such a way that encapsulating the elastomer inside a rigid polymer would allow playing with the weight of the sheet, its adhesion capacity and its thickness. Considering that it will be placed in the hull of the vessels and that its benefits are between reducing the weight of the ship by placing fewer layers after the viscoelastic and being able to adapt to the curvature of the ship. A hexagonal design is proposed, which allows to create a set of cells that are grouped with each other when they are manufactured and can form a set.

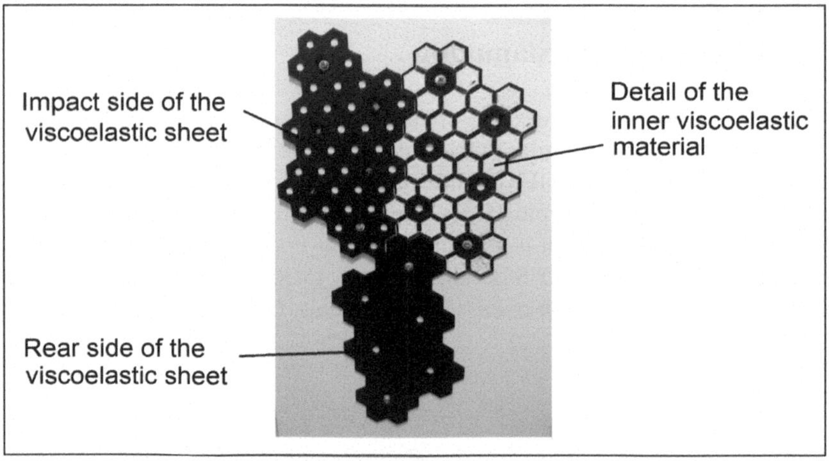

Fig. 2. Details of the viscoelastic sheet

For the manufacture of the outer capsule, acrylonitrile butadiene styrene or ABS of 3 mm was used, which is a very impact resistant plastic composed of three blocks, acrylonitrile, butadiene and styrene, which is why it is terpolymer. Acrolonitrile gives rigidity, resistance to chemical attacks, hardness and stability at high temperatures. Butadiene provides toughness for low temperatures and impact resistance, and styrene gives it mechanical strength, rigidity, brightness and hardness.

For the internal elastomer, a linear thermoplastic polyurethane or TPU was used, which among its properties has high resistance to wear, abrasion, traction and tear. It has very good cushioning capacity and is tenacious, excellent elastic recovery.

To reproduce the slamming, the following specific characteristics were considered: the non-linear character in which the domain of time predominates and not the frequency, the impacts are random, and the hydro-elastic effect is to be disregarded considering water as an elastic solid in which the impact is returned to the equipment. The slamming loads are considered as an increase in the value of the design load estimated by empirical data based on the classification societies.

The equipment manufactured for the impact test by slamming in the laboratory of the Polytechnic University of Madrid ETSIN Navales [9], consists of an electric motor, an eccentric impact cam, a rev counter, and a rigid frame for adjusting the test tube. Figure 3 show the slamming equipment.

Fig. 3. Impact with an eccentric cam

The slamming reproduction equipment has the electric motor-variator unit, connected by an axis supported on two bearings. The steel cam that is fitted with key to the shaft, has lightening in its mass so that it is balanced with the center of the shaft and avoid inertial loads due to its eccentric shape.

The installed drive works at a frequency of 200 to 320 RPM. At this speed, the motor-cam assembly is more efficient and the vibrations produced by the equipment when turning do not affect the experiment. Rubber gaskets were placed on the bases of the bearings that supported the shaft. These joints dissipate the reaction energy of the cam against the panel, with the purpose of not affecting the lateral load on the variator motor assembly. The design of the cam causes the deformation and relaxation of the panel to press gradually on the contact face, avoiding point loads. A micro switch was adapted to one side of the axis to count the cycles.

The equipment was embedded in the middle of two steel supports, to hold the frame where the panel to be tested is located. The panels were mounted on screwed presses to prevent translation and rotation at their edges. This allows that during the test, its boundary conditions are embed, with all its degrees of freedom restricted.

One of the characteristics of the equipment designed for the test is the adjustment distance of the cam. This is regulated by moving the frame on guide's perpendicular to the axis of the system. The regulation of this distance is directly related to the unitary deformation on one of its faces.

In the equipment it is possible to deform between 0 to 2050 μm/m corresponding to the minimum and maximum distance of the specimen in contact with the cam. To be able to control the distance, it is convenient to place the panel and strain gauge near the impact area, and using a finite element model, to correlate the pressure in the area of the impact.

The cam is firmly supported by two bearings, which allows no displacement and all the energy of the cam remains in the panel to hit. To estimate the contact surface of the cam with the panel corresponding to the slamming pressure, it was calculated using a finite element model. The slamming is an event of pressures, by which the area to analyze, corresponds only to the contact with the cam.

3 Evaluation of the Damage

Slamming reproductions were performed at pressures of 263 kN/m^2, over 2.1×10^5 cycles, it was clearly observed a delamination on the face contrary to the impact in the Fig. 4.

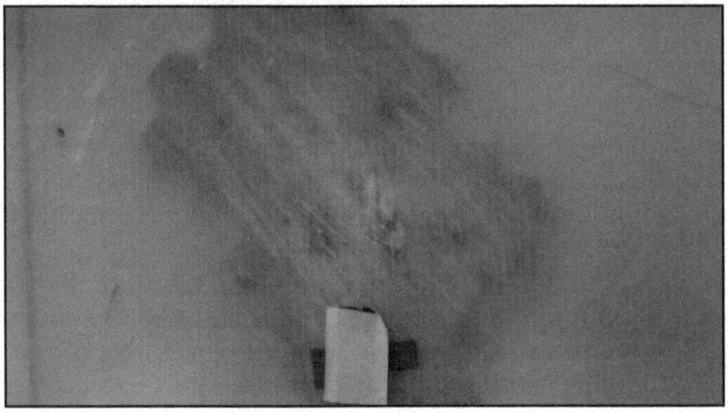

Fig. 4. Evolution of slamming damage to 2.1×10^5 impact cycles of the cam at 263 kN/m^2

To evaluate the damage produced by the slamming, the evolution of the damage has been compared as vertical impacts to low energy, which are governed by the formulation presented in [9].

From the results presented by the author in reference [10], this behavior of the damage is equivalent to that of a vertical impact due to weight loss made with an energy of 10 J. The formulation for evaluating the energy absorbed is presented in Eq. (1).

$$E_a = E_o - (|E| - |K|) \tag{1}$$

It is expressed that the energy absorbed Ea during the impact, is in relation to the initial energy Eo which is the impact of slamming, less the energy that returns the material product of its restitution. This energy that returns the material is the difference between the energy of impact E, and the kinetic energy K due to the displacement of the panel during the deformation.

The vertical impact equipment has an accelerometer that takes data from the deceleration of the impactor, while breaking the layers of the panel.

Figure 5 shows the results of accelerations obtained after impacting 10 J of energy, on an unmodified and a modified panel, to establish a comparison. These results correspond to the data obtained with an accelerometer installed in the impactor, to test the deformation and the restitution of the impacted panel.

The accelerations due to the effect of the viscoelastic layer are more uniform, which indicates that the subsequent layers absorb minimal damage. In the unmodified panel, sudden acceleration changes clearly indicate material breakage losing stiffness and may further deformation.

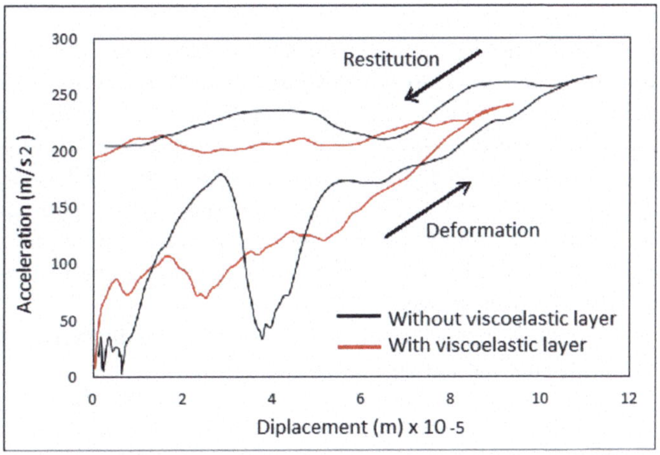

Fig. 5. Acceleration versus displacement of an unmodified and modified panel

From equation [1] converting the accelerations into energy, in Fig. 6 the variation of energy absorbed during the impact is compared with the kinetic energy, on an unmodified and a modified panel, to establish a comparison. It is observed that while receiving more kinetic energy to the panel during impact, it is absorbed by it.

On the 3 J of kinetic energy developed by the impactor while deforming the panel, there is a variation in the energy absorbed. The modified panel maintains its ability to absorb energy while the unmodified panel decreases in its ability to absorb energy.

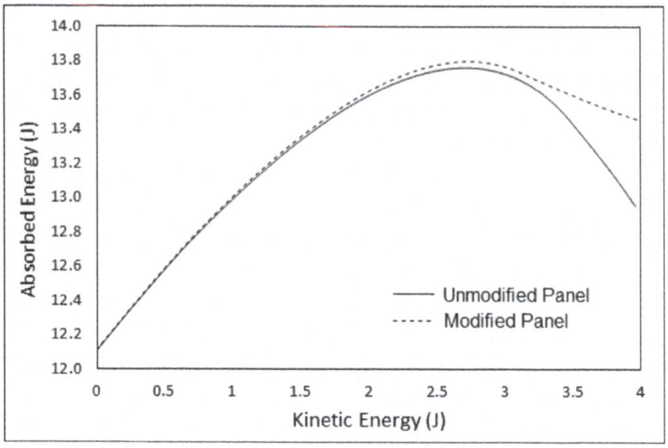

Fig. 6. Energy absorbed by the impact at 10 J versus the kinetic energy contributed during the impact

Fluorescent penetrant liquid was used for the evaluation process of the impacted panels. A 0.5 mm drill was drilled so that the liquid enters through the interlaminar and intralaminar delaminations, and can be observed under fluorescent light. With solvent, the ink is removed from the outside, and it does not have the ability to penetrate.

When exposed to fluorescent light, it is observed how the ink has penetrated and has left marked the cracks and faults of the material inside. With diamond disc equipment, the transversal sections of the specimens were cut in the impact area, lubricating oil was used in such a way that burns did not occur in the cut and the composite material was not destroyed.

Based on this comparison, the panel tested at 2.1 × 105 impact cycles with viscoelastic layer, has the internal damage as shown in Fig. 7.

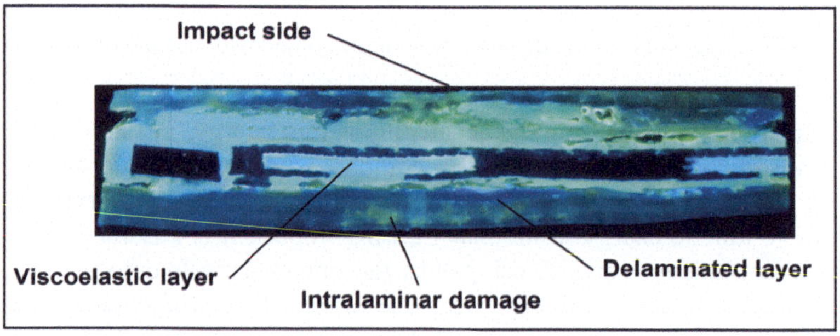

Fig. 7. Intralaminar damage observed under fluorescent light

It is observed that the damage has caused the matrix to break, while in the area of the panel subjected to greater internal tension, the layers have delaminated.

4 Conclusions

4.1. The methodology allows reproducing the slamming fatigue event. This occurs on the contact area of the GFRP panels with the cam. The laminations of the layers could be observed under fluorescent light and compared with the equivalent damage produced in the panels, when subjected to cyclical impacts.

4.2. The tests carried out and the experimental results obtained have allowed observing several internal modes of failure within the laminate of the composite material. This has allowed to predict the appearance of meso-meshes and micro-cracking in the matrix.

4.3. The prediction of damage using micromechanical models in composite materials can be locally accurate in the description of failure initiation at critical points. Microscopy has shown that the equipment allows to differentiate the damage that goes from interlaminar to intralaminar, to measure the degradation of the rigidity throughout the application of the cycles of the load and the rupture of the laminate at pressures lower than its static limit, which they are always related to plastic deformation according to the results. In the microstructure observed for low cycle tests, the increase in temperature affects the dissipation of the stresses in the material. It cannot be estimated because the friction of the cam during the tests raises the temperature more than this level because the reproduction equipment of slamming is not adiabatic.

4.4. At this level of pressures with the slamming reproductive equipment, it is physically observed how in a short interval of time the dynamic of redistribution of effort occurs. This explains the presence of discontinuities by the percentage of damage produced by cycle. The tests went from low energy slamming impacts in which the microcracks are aligned interlaminarly, to medium energy level impacts, in which the interlaminar damage also produces intralaminar damage. This can be confirmed by microscopic observation SEM. Fragmentation of the fiber, produced at the apex of a macro-crack, leads to an increase in the fracture toughness of the compound and an additional dissipation of energy.

4.5. The method of reproduction proposed to study the phenomenon of slamming uses a dry pressure to study the appearance of damage within the material and its evolution. This can be improved by developing research that considers including the viscous effect of water, because this induces structural damping. It will also act as a lubricant to decrease the heating of the plate and the cam.

4.6. Designing a reading system using an accelerometer installed on the face opposite to the impact, would allow to evaluate the behavior of the laminate during each cycle of slamming. The acceleration in time shown in Fig. 5 is the laboratory demonstration that the viscoelastic material absorbs impact energy, because the panel is restored in a better way showing fewer variations in the curve. That is, the smaller variation of peaks indicates that fewer layers broke under the viscoelastic sheet.

4.7. The ability of the viscoelastic to absorb energy is demonstrated by the results of Fig. 6, the unmodified panel loses its ability to absorb energy before the modified panel. Both panels initially behave in the graph in a similar way when the impactor breaks the upper layers. Then they begin to behave differently when the impactor tries to deform the viscoelastic sheet. As the impactor immediately breaks the lower layers of the unmodified panel, it quickly decreases its ability to absorb energy. This means, that more energy is converted to damage to the unmodified panel.

4.8. The future of the shipbuilding of boats made from GFRP is in the correct insertion of the viscoelastic layers to protect the ships from the destructive load of slamming. Effectively with the presented methodology, it is observed that the useful life of the glider vessels is prolonged, since the laminate of the hull of the ship is protected by the absorption of energy. This reduces the appearance of micro cracks that increase the flexibility of the panel, rendering it unusable.

4.9. In the experimentation to verify the effectiveness of the insertion of the viscoelastic sheet, it was possible to appreciate that both panels tested under the same conditions, had a significant variation in the microdeformations at the end of the tests. The modified panel was less flexible in a large proportion than the unmodified panel. Esto se verifica con la observación bajo la luz fluorescente. La Figura 7, muestra claramente que las capas protegidas por la lámina viscoelástica, reciben poco daño. El viscoelástico absorbió la energía y la devolvió.

4.10. This will be directly related to the useful life of the vessels, since the pressures tested correspond to very severe slamming events, tested with the purpose of observing the damage produced.

References

1. Von Karman, T.: El impacto de los flotadores de hidroaviones durante el aterrizaje. Nota técnica 321, Washington D.C. (1929)
2. Wagner, H.: Uber stoss-und gleitvergange an der oberflache von flussigkeiten. Zeitschrift für angewandte mathematik und mechanik **12**(4), 193–235 (1932)
3. Kapsenberg, B.: Slamming or ships: where are we now? Philosofical Trans. Roy. Soc. **369**, 2892–2897 (2011)
4. Selvaraju, S., Ilaiyavel, S.: Applications of composites in marine industry. J. Eng. Res. Stud. (JERS) **2**, 89–91 (2011). E-ISSN0976-7916
5. Recuero, A.: Refuerzo de estructuras con composites avanzados. Informes de la Construcción **49**(452) (1997)
6. Hayman, B.: Response of sandwich structures to slamming and impact loads. In: Composite Materials in Marine Structures, vol. 2. Cambridge University Press, Cambridge (1993)
7. Qin, Z., Batra, R.C.: Local slamming impact of sandwich composite hulls. Int. J. Solids Struct. **46**, 2011–2035 (2009)
8. Lopes, C., Seresta, O.: Low-velocity impact damage on disperse stacking secuence laminates. Part I: experiments. Compos. Sci. Tenchol. **69**(2009), 926–936 (2009)
9. Townsend, P., Suárez-Bermejo, J.C., Pinilla-Cea, P., Sanz-Horcajo, E.: Reduction of slamming damage in the hull of high-speed crafts manufactured from composite materials using viscoelastic layers. Ocean Eng. **159**, 253–267 (2018)
10. Townsend, P., Suárez Bermejo, J.C.: Paneles GFRP modificados con capas viscoelásticas para mejorar la respuesta frente a impactos de baja velocidad. In: XII Congreso Nacional de Materiales Compuestos - MATCPMP 2017, Donostia-San Sebastián (2017)
11. Kim, Y.J., Yoshitake, I.: Composite hull structures subjected to wave-induced slamming impact. J. Reinf. Plast. Compos. **33**(1), 3–13 (2014)

Numerical Model to Analyze an Artificial Reef

Maria Isabel Lamas Galdo[1(✉)], Patrick Roger Townsend Valencia[2], Maria Jesus Rodriguez Guerreiro[1], Laura Castro Santos[1], and Luis Manuel Carral Couce[1]

[1] Universidade da Coruña, A Coruña, Spain
isabellamas@udc.es
[2] Escuela Superior Politécnica del Litoral (ESPOL), Guayaquil, Ecuador

Abstract. Many seas in the world are productive but limited to a few pelagic species, highly influenced by environmental conditions. This dependency makes necessary to develop solutions in order to promote wealth and biodiversity in productive regions. One of the recommended solutions to enrich the seas is the implantation of artificial reefs. These reefs mimic the environment and do not alter the preexistent equilibrium. Current studies about artificial reefs have focused on experimental observations. An alternative to these experimental analyses is CFD (Computational Fluid Dynamics). The main advantage of numerical simulations is that these provide much more information at a lower cost and less time. In this regard, this paper presents a numerical simulation of an artificial reef in order to analyze the most relevant hydrodynamic characteristics.

Keywords: Artificial reef · CFD · Computational fluid dynamics · Plankton

1 Introduction

Fishing and shellfishing in Galicia (Spain) constitute an important socioeconomic sector. Fishing techniques have evolved and the fishing sector has been adapting to new conditions. The Galician fleet operates all kinds of resources: long (waters of third countries and international), community and nearby waters (Spanish waters, inland waters). The coastal areas of Galicia, especially estuaries, have been recognized at regional, national, and international level due to their uniqueness and importance as population support, productive activities, biodiversity and resource. Its estuaries are ecosystems of great biological production. Due to the human activity (fishing, shellfishing, urbanization of their surroundings, recreational use, etc), estuaries need protection mainly because they develop some marine species that occupy first stages in the food chain [1]. It is possible to improve the systems in order to allow future sustainable development of resources through instruments as licensing, control gear, quotas of species, effort quotas, etc. [2–4]. Particularly, an important measure recommended to deal with the loss of structural complexity of communities and alteration of food chains is the implantation of artificial reefs [5]. Artificial reefs can provide shelter species associated with rocky bottoms, sandy or next, and where the presence of algae and

V. A. J. E. Carreño Moreno et al. (Eds.): CIDIN 2019/COPINAVAL 2019, pp. 357–364, 2020.
https://doi.org/10.1007/978-3-030-35963-8_30

other organisms allows greater availability of food [6–9], thus recreating habitats that improve the biodiversity of marine ecosystems.

In order to design an artificial reef, it is very important to know its performance. Although experimental studies have revealed many key points of artificial reefs, much is still unknown regarding the behavior of the fluid flow. Recently, CFD has become a powerful tool to study this phenomenon. By solving the Navier-Stokes equations, CFD provides information about the pressure and velocity fields. Numerical simulations are very useful to complement traditional testing and experimentation, providing faster time and lower costs. Unfortunately, only a few numerical simulations about artificial reefs are available in the literature. Su et al. [10, 11] developed several works to characterize the fluid flow around and inside an artificial reef. Düzbastılara and Şentürkb [4] calculated the mobilizing forces and drag coefficients. Jiao et al. [12] analyzed reefs composed by tubes and compared their experimental results with PIV measurements. Liu et al. [13, 14] and Jiang et al. [15, 16] also developed PIV measurements to compare CFD simulations.

In this paper, an artificial reef was designed. In order to analyze the flow around and inside the reef, a tri-dimensional CFD model was developed using the open software OpenFOAM. This paper begins with an introduction of the proposed design. The numerical method and results are then briefly described and finally the conclusions of this work and areas for future research are presented.

2 Design of the Artificial Reef

The experimental reef is shown in Fig. 1. It consists of 4 concrete blocks. Each block, $2 \times 2 \times 2$ m, has 6 holes in order to provide nests for crustaceans. This modular design was proposed in order to facilitate production, transportation and installation. These reefs will be installed in Galicia (Spain), at 10–20 m depth in a sea 7.5–8.4 pH.

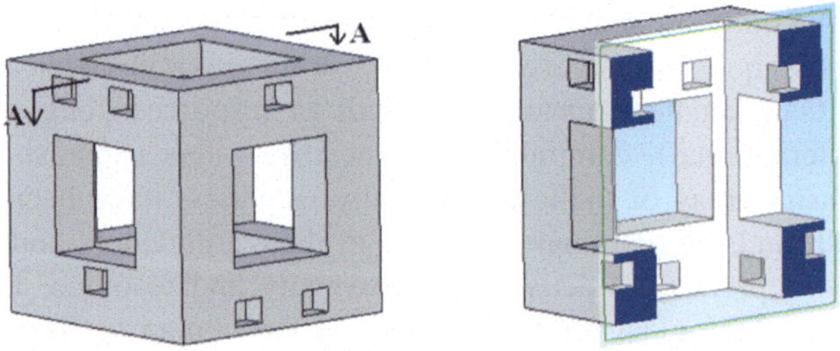

(a) (b)

Fig. 1. Design of the artificial reef. (a) 3D (b) A-A section.

3 Numerical Implementation

The fluid flow around and inside this reef was studied numerically. The velocity and pressure fields were characterized.

3.1 Computational Mesh

The computational domain is shown in Fig. 2. It consists of a 8 × 16 × 32 m block with the artificial reef inside, as can be seen in the middle plane indicated in Fig. 2(b). Several simulations were carried out in order to simulate an infinite medium, verifying that these dimensions are too long to neglect any boundary effect on the flow. Besides, several mesh sizes were simulated in order to check mesh quality. The size mesh was refined near the reef in order to resolve the fields as they travel downstream.

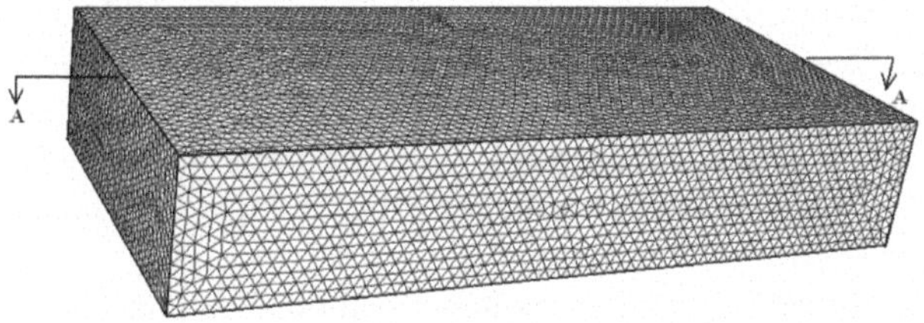

(a)

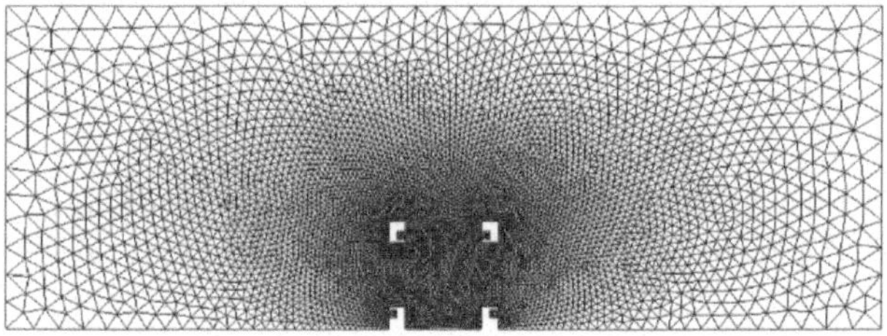

(b)

Fig. 2. (a) Computational mesh; (b) A-A section.

3.2 Boundary Conditions

The boundary conditions are indicated in Fig. 3. The left side was imposed as inlet. The upstream influx velocity was set as 1 m/s. The downstream right side was set as outlet. The reef and seafloor sides were set no slip and the remain sides were set zero gradient in order to simulate an infinite liquid medium.

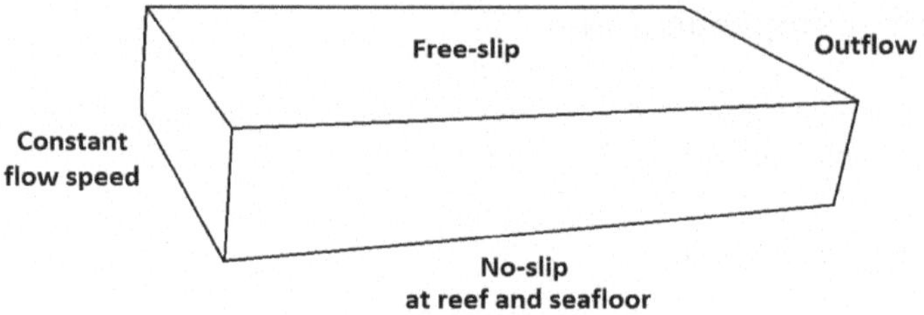

Fig. 3. Boundary conditions.

The initial conditions were free-stream velocity and atmospheric pressure.

3.3 Numerical Methodology

Continuity and momentum equations were solved using the CFD software employed was OpenFOAM. It was chosen because it is an open code and thus allows manipulating the governing equations in order to adapt them to each particular problem. A new OpenFOAM solver was developed using C++ programming language. Both OpenFOAM and CFD experience was accredited by the authors of the present work through previous papers [17–26].

OpenFOAM is based on the finite volume method. The pressure-velocity coupling was treated using the PISO (Pressure Implicit Splitting of Operators) procedure. The equations were discretized by the QUICK interpolation and the temporal treatment was solved by an implicit method.

3.4 Validation

This CFD model was validated with experimental results obtained by Jiao et al. [12]. These authors used PIV to analyze the flow around tube artificial reefs, as shown in Fig. 4.

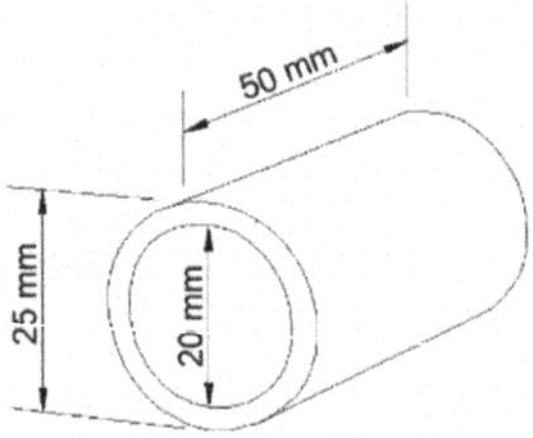

Fig. 4. Geometry of the tube employed to constitute artificial reefs in [12].

Figure 5 illustrates the length of the affected back region against the free-stream velocity. In this case, the affected region was determined as the region in which the vertical component of the velocity is equal to or greater than 10% inlet velocity. As can be seen, coherent results are provided by the CFD model developed in the present work, which indicates its adequacy.

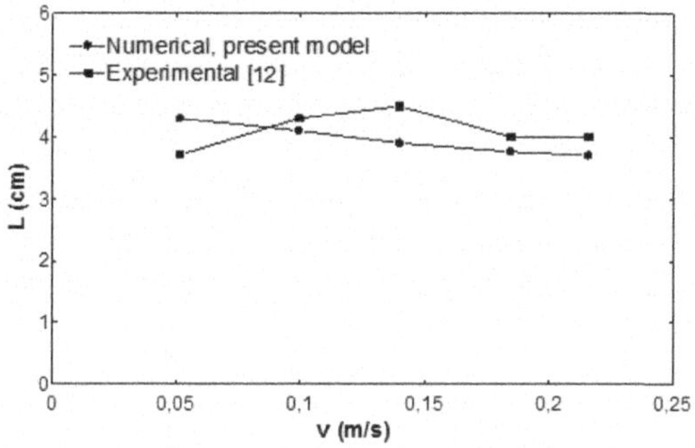

Fig. 5. Length of affected back region obtained numerically and experimentally.

Other comparisons between CFD models and experimental measurements can be consulted in the literature [13–16]. These studies and the present one validate CFD simulations to analyze artificial reefs.

4 Results and Discussion

The velocity field at the middle plane is indicated in Fig. 6. As can be seen, the free stream velocity is perturbed due to the presence of the reef. The reef increases the flow velocity inside it but decreases it behind. The surface flow velocity at the reef and seafloor is zero. The velocity at the downstream side is almost zero. Thus, the reef has an important influence on the free-stream fluid flow.

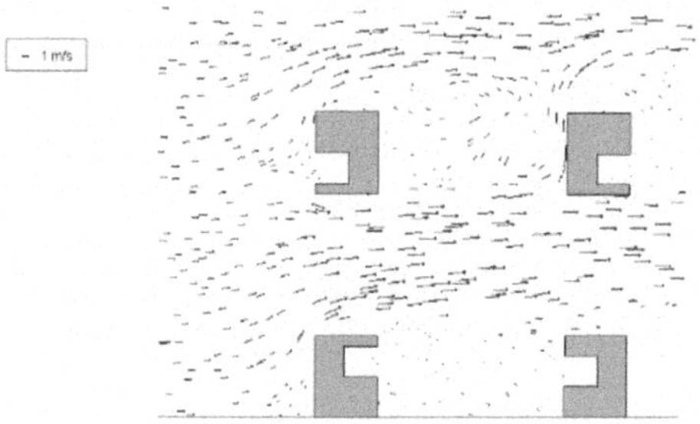

Fig. 6. Velocity field.

In order to have an adequate design, it is primordial to preserve circulation along the whole reef. A proper circulation is necessary to guarantee supply of nutrients. Hydrodynamics provides information about the water circulation and velocity in the interior and exterior of the reef. As can be seen in Fig. 6, the velocity field provided by the CFD analysis shows a high velocity in the interior. This interior velocity is guarantees the renewal of favorable nutrients. In this particular design of artificial reef, the function of the central hole is to make circulation possible inside the reef, while the smaller holes provide nests. Regarding the internal secondary holes, these experiment a reduction of velocity. This reduction of velocity is extremely negative in the transportation of nutrients. Nevertheless, it was verified that this velocity is enough to guarantee nutrients.

Another aspect that must be analyzed through hydrodynamics is sedimentation. According to Fig. 6, there are no risk of sedimentation in the proposed design. For this reason, the design proposed can be considered acceptable, as well as the size of the central and secondary holes.

5 Conclusions

This paper proposes a CFD model to simulate the fluid flow around and inside an artificial reef. These numerical results are useful for marine biologists and ocean engineers since the flow pattern can be characterized.

In order to be adequate, the design of an artificial reef must be adapted to the hydrodynamic characteristics. The hydrodynamic model predicts the velocity field. Since the food supply is directly related to circulation, the velocity field also predicts the nutrient supply. The model was used to study the hydrodynamic behavior. It was found that the velocity field appropriate inside the reef, guarantying nutrient supply. The zones with lower velocities are the secondary holes, but the values provided by the CFD model are considered adequate.

This initial paper can be used in the future to study aspects such as the design of a reef. In this sense, CFD is a useful tool to adapt the reef model geometry for the adequate circulation of nutrients along the cavities.

Another application of the present model is to analyze the interaction between several reefs instead a single one. The size of the circulation opening links the interior of the reef with the exterior. This size is related with the separation distance between several reefs and the separation between consecutive reefs must be such that the velocity field is regulated between them.

Another further work is to analyze the orientation of the reef according to the predominant currents and the effect of a change in the direction of the current.

Acknowledgements. The authors gratefully acknowledge the financial support from the regional government of Galicia, Xunta de Galicia, through the project CN-10MMA003CT.

References

1. González-Garcés, A., Vila Martín, F., Antón Álvarez Salgado, X.: La ría de Vigo. Una aproximación integral al ecosistema marino de la Ría de Vigo. Instituto de Estudios Vigueses, Vigo, Spain (2008)
2. Bellan-Santini, D., Lacaze, J.C., Poizat, C.: Les biocénoses marines et littorales de Méditerranée. Synthèse, menaces et perspectives. Collection Patrimoines Naturels, vol. 19. Múseum National d'Histoire Naturelle, Paris (1994)
3. Corral, J.: Las rías Gallegas como ecosistema. Consellería de Pesca Marisqueo e Acuicultura. Xunta de Galicia. Santiago de Compostela, Spain (1991)
4. Düzbastılara, F.O., Şentürkb, U.: Determining the weights of two types of artificial reefs required to resist wave action in different water depths and bottom slopes. Ocean Eng. **36** (12–13), 900–913 (2009)
5. Seaman Jr., W., Buckley, R.M., Polovina, J.J.: Advances in knowledge and priorities for research, technology and management related to artificial aquatic habitats. Bull. Mar. Sci. **44** (2), 527–532 (1989)
6. Grove, R.S., Wilson, C.A.: Introduction to aquatic habitat enhancement. Bull. Mar. Sci. **55** (2–3), 265–267 (1994)
7. Pickering, H., Withmarsh, D., Jensen, A.: Artificial reefs as a tool to aid rehabilitation of coastal ecosystems: investigating the potential. In: Reef Design and Materials Workshop, Southampton (1998)
8. Pratt, J.R.: Artificial habitats and ecosystem restoration: managing for the future. Bull. Mar. Sci. **55**(2–3), 268–275 (1994)
9. Ramos-Esplá, A.A., Guillén, J.E., Bayle, J.T., Sánchez-Jérez, P.: Artificial anti-trawling reefs off Alicante, South-Eastern Iberian Peninsula: evolution of reef block and set designs. In: Jensen, A., Collins, K.J., Lockwood, A.P.M. (eds.) Artificial Reefs in European Seas, pp. 195–218. Kluber Academic Publishers, London (2000)
10. Su, D.T., Liu, T.L., Ou, C.H.: Numerical investigation into effects of seabed topography on flows in and around artificial reefs. Fish. Sci. **74**, 236–254 (2008)
11. Liu, T.L., Su, D.T.: Numerical analysis of the influence of reef arrangements on artificial reef flow fields. Ocean Eng. **74**, 81–89 (2013)
12. Jiao, L., Yan-xuan, Z., Pi-hai, G., Chang-tao, G.: Numerical simulation and PIV experimental study of the effect of flow fields around tube artificial reefs. Ocean Eng. **134**, 96–104 (2017)
13. Liu, Y., Guan, C.T., Zhao, Y.P., Cui, Y., Dong, G.H.: Numerical simulation and PIV study of unsteady flow around hollow cube artificial reef with free water surface. Eng. Appl. Comput. Fluid Mech. **6**, 527–540 (2012)
14. Liu, Y., Zhao, Y.P., Dong, G.H., Guan, C.T., Cui, Y., Xu, T.J.: A study of the flow field characteristics around star-shaped artificial reefs. J. Fluids Struct. **39**, 27–40 (2013)
15. Jiang, Z., Liang, Z., Liu, Y., Tang, Y., Huang, L.: Particle image velocimetry and numerical simulations of the hydrodynamic characteristics of an artificial reef. Chin. J. Ocean Limnol. **31**, 949–956 (2013)
16. Jiang, Z., Liang, Z., Huang, L., Liu, Y., Tang, Y.: Characteristics from a hydrodynamic model of a trapezoidal artificial reef. Chin. J. Ocean Limnol. **32**, 1329–1338 (2014)
17. Lamas, M.I., Rodríguez, J.D., Rodríguez, C.G., González, P.B.: Three-dimensional CFD analysis to study the thrust and efficiency of a biologically-inspired marine propulsor. Pol. Marit. Res. **18**(1), 10–16 (2011)

18. Lamas Galdo, M.I., Rodríguez Vidal, C.G., Rodríguez García, J.D., Fernández Quintás, M. V.: Modelo de mecánica de fluidos computacional para el proceso de barrido en un motor Otto de dos tiempos. DYNA **86**, 165–172 (2011)
19. Lamas, M.I., Rodríguez, C.G.: CFD analysis of the scavenging process in the MAN B&W 7S50MC two-stroke diesel marine engine. J. Ship Res. **56**(3), 154–161 (2012)
20. Lamas, M.I., Rodríguez, C.G.: Numerical model to study the combustion process and emissions in the Wärtsilä 6L 46 four-stroke marine engine. Pol. Marit. Res. **20**(2), 61–66 (2013)
21. Lamas, M.I., Rodríguez, C.G., Aas, H.P.: Computational fluid dynamics analysis of NOx and other pollutants in the MAN B&W 7S50MC marine engine and effect of EGR and water addition. Int. J. Marit. Eng. **155**(Part A2), A81–A88 (2013)
22. Lamas, M.I., Rodríguez, C.G., Rodríguez, J.D., Telmo, J.: Internal modifications to reduce pollutant emissions from marine engines. A numerical approach. J. Nav. Arch. Mar. Eng. **5** (4), 493–501 (2013)
23. Lamas, M.I., Rodríguez, C.G., Telmo, J., Rodríguez, J.D.: Numerical analysis of emissions from marine engines using alternative fuels. Pol. Marit. Res. **22**(4), 48–52 (2015)
24. Lamas, M.I., Rodríguez, C.G., Redo, J.M.: Numerical model to study the valve overlap period in the Wärtsilä 6L 46 four-stroke marine engine. Pol. Marit. Res. **19**(1), 31–37 (2012)
25. Lamas, M.I., Rodríguez, C.G.: Numerical model to analyze NOx reduction by ammonia injection in diesel-hydrogen engines. Int. J. Hydrog. Energy **42**, 26132–26141 (2017)
26. Lamas, M.I., de Dios Rodriguez, J., Castro Santos, L., Manuel Carral, L.: Effect of multiple injection strategies on emissions and performance in the Wärtsilä 6L 46 marine engine. A numerical approach. J. Clean. Prod. **216**, 1–10 (2019)

Application of a General Pattern of Colonization of a "Verde-PROARR"-Artificial Reef in the Ría de Ares-Betanzos (Spain)

Maria Jesus Rodriguez Guerreiro[1], Jose Carlos Alvarez-Feal[2], José Ángel Fraguela-Formoso[2], Laura Castro Santos[2], and Luis Manuel Carral Couce[2](✉)

[1] Departamento de Química, Escuela Politécnica Superior de Ferrol, Universidad de A Coruña, 15403 Ferrol, Spain

[2] Departamento de Ingeniería Naval e Industrial, Escuela Politécnica Superior de Ferrol, Universidad de A Coruña, 15403, Ferrol, Spain

L.carral@udc.es

Abstract. The Galician estuaries (rías) do not manage to increase the fishing yield, in spite of the fact that they are ecosystems of great biological production. Therefore, one of the measures to be taken is the implantation of artificial reefs (AR). They attract fish from neighboring areas and/or increase the area colonized by epifaunal species. An important task is to plan the layout of the different modules that will make up a reef to obtain the greatest biological and economic profitability. Therefore, it is necessary to predict equilibrium states between benthic and nektonic communities in different geographical areas.

In the present work, a possible general pattern of colonization of an artificial reef is established, during the first five years of its development in the Ría de Ares-Betanzos. It is based on the official statistics of artisanal fishing, of commercial species of the benthic, demersal and pelagic communities unloaded in the fish markets of the area under study.

Keywords: Artificial reef · Colonization · Benthic, demersal and pelagic community

1 Introduction

The implantation of artificial reefs involves the development of new biological communities that provide the biomass necessary to feed commercially important species, in order to achieve greater fishing performance [1, 2].

In our study [3], we conducted an analysis of the evolution of artisanal fisheries in Galicia based on the discharges of the most important commercial species groups. In this study, we conclude the need to increase the performance of commercial fishing, in general, throughout the Galician coast.

Therefore, the present work analyses the data on the evolution of artisanal fisheries in the Ría de Ares-Betanzos, during the last 5 years, based on the official statistics

V. A. J. E. Carreño Moreno et al. (Eds.): CIDIN 2019/COPINAVAL 2019, pp. 365–372, 2020.
https://doi.org/10.1007/978-3-030-35963-8_31

available from PescadeGalicia, in order to demonstrate the need to implement our first reef in the mentioned estuary object of study.

The PROARR research project has allowed us to have modules of AR of production and protection of the green type. It has been sought to have an effective repopulating agent of marine ecosistems and the recovery of artisanal fishing with its adaptation to the geomorphology of the Galician estuaries (rías) [3]. The results of the project have been the subject of a patent [4].

Thus, a hypothetical study on the possible colonization in the supposed reef located in the Ría de Ares-Betanzos has been developed based on the studies of colonization predictions in AR of several authors [5–7]. This study takes into account on the one hand, the fact that all the species named in this work can be developed in depths up to 50 m and, on the other hand, the feeding habitats of the main species caught.

2 Study Area

The Ría de Ares-Betanzos, consists of an area of 72 km^2, and it is located between the estuaries (rías) of A Coruña and Ferrol. It is a double estuarine system [8], with a volume of 0.75 km^3 and a length of 19 km: the estuary of the Eume river and the estuary of the Mandeo river (Fig. 1).

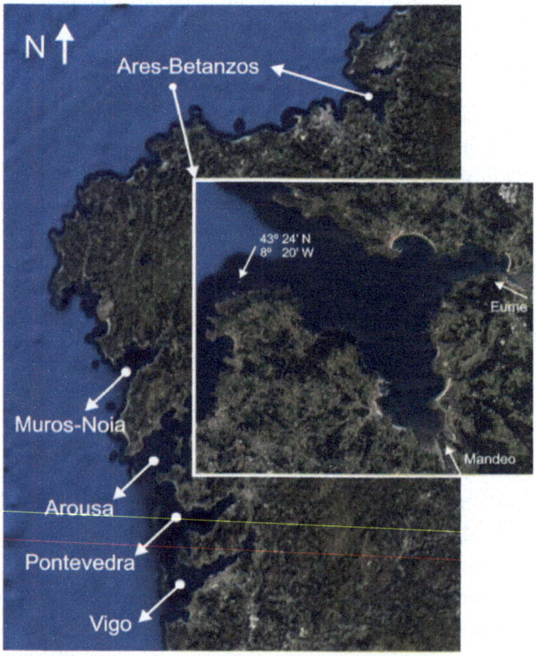

Fig. 1. Situation of the Ares Betanzos estuaries (rías). Own elaboration.

These rivers flows have seasonal variations with an average value of 16.5 and 14.1 m^3/s, respectively [9]. The two sectors are joined in a confluence zone that is connected to the Atlantic Ocean through a mouth of 40 m of depth and 4 km wide.

The current speed is driven by the tide and is usually <10 cm s^{-1} due to the low contributions of the rivers [10].

3 Evolution of Artisanal Fishing in the Ría de Ares-Betanzos

The data offered according to the official statistics available from Pescadegalicia of the Consellería do Mar, based on the unloading of some of the most important commercial species in the area under study (2012–2017), has been analysed.

For many groups of species, in the last five years the volume landings have followed a decreasing trend (Figs. 2 and 3).

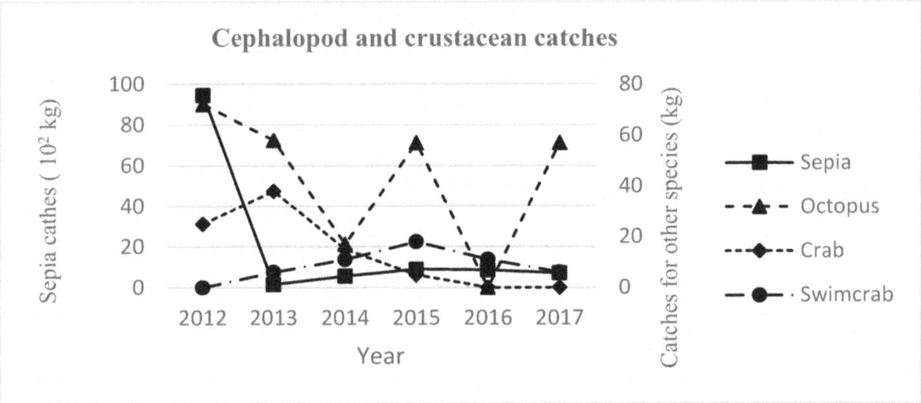

Fig. 2. Landings in kg of cephalopods and crustaceans of the fish markets corresponding to the Ría de Ares-Betanzos in the period 2012–2017. The figure shows a continuous decline in fishing in kg over time for the four species studied. Cephalopods: octopus (Octopus vulgaris) their catches have decreased by 21% and sepia (Sepia officinalis) their catches have decreased by 92.5%. Crustaceans: Crab (Maja squinado) its catches have decreased in a 100% and velvet swimcrab (Liocarcinus puber), its catches have been reduced in a 67%. Source: Own preparation considering [11].

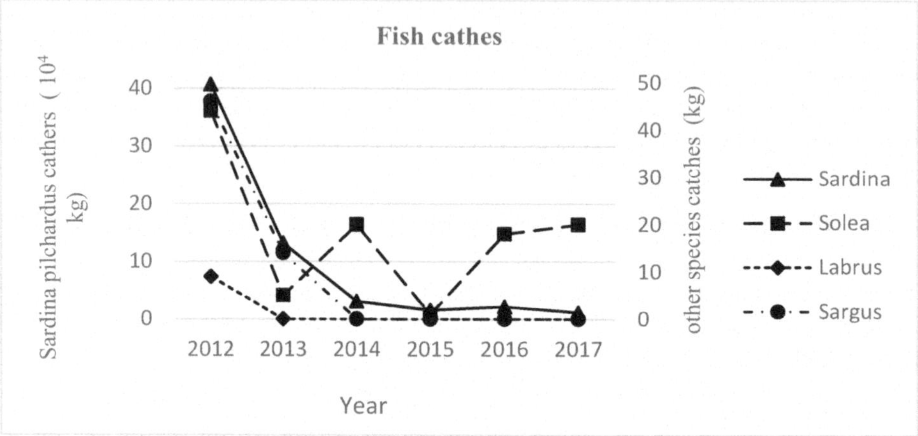

Fig. 3. Landings in kg of fish from the fish markets corresponding to the Ría de Ares-Betanzos in the period 2012–2017. The figure shows a continuous decline in fishing in kg over time for the four species studied. Sardine (Sardina pilchardus) its catches have decreased by 97%; flounder (Solea vulgaris) their catches have decreased by 54.5%; ballan wrasse (Labrus barylta) its catches have decreased by 100% and white seabream (Diplodus sargus) its catches have decreased by 100%. Source: Own preparation. Considering [11].

4 First Year of Colonization in the AR

The trophic chains begin in the nutrients that are dissolved in the water of the sea in small amounts and distributed in a non-uniform way, giving rise to cyclical variations in their abundance. Of these nutrients, it is mainly nitrates and phosphates that condition the production of phytoplankton, so these algae take nutrients from the sea and take advantage of solar energy to synthesize sugars (photosynthesis). In the Galician stuaries, phytoplankton abundance is variable, with two maximums in spring, autumn and minimum in winter. This minimum is due, on the one hand to the cold waters and on the other hand to the great agitation of the waters that does not allow the phytoplankton to remain suspended in the photic zone. In this way, the main group of microscopic algae that constitute phytoplankton in the Rías de Galicia are diatoms [12].

In the food chain phytoplankton serves as food for zooplankton, wich is made up of a great variety of fish larvae, cephalopods, mollusks and fundamentally crustaceans, being the copepods, amphipods and euphausians the most representative.

The zooplankton will be the food base of the plant-eating fish in the food chain. Both the plant-eating fish and the zooplankton will serve as food for the predatory fish and fish and these for the seabed fish.

The depositions of all these organisms, as well as the organisms themselves at their death, go to the seabed, where they will be decomposed into the basic elements (nutrients) by the bacteria. Therefore, there is a concentration of nutrients in the seabed, far from the photic zone, which by means of marine currents rise to the surface of the water, closing the cycle continuously (Fig. 4).

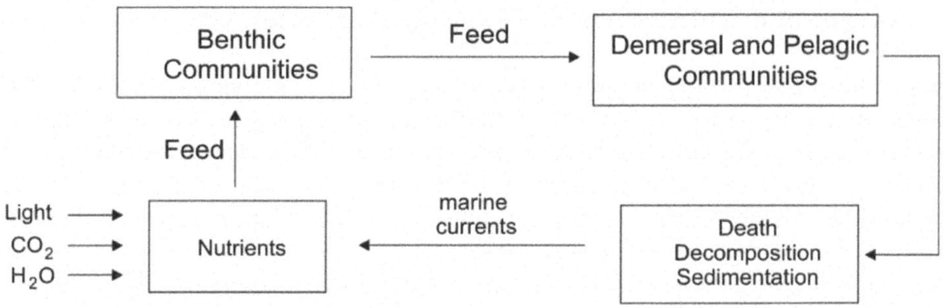

Fig. 4. Development cycle of benthic, demersal and pelagic communities. Own source.

It can be foresee that the first colonization of the benthic community, based on studies on the colonization of an AR [7], will be carried out by hydroids, polychaetes, mollusks (Mytilus galloprovincialis) and crustaceans (Balanus perforatus, Pollicipes pollicipes). The number of plantonic larvae will have its maximum in spring and autumn. The first colonizations are expected around the holes, cavities and corners of the AR more exposed to the movement of water, later the larger surfaces of the module will be colonized [13].

Juveniles of the demersal and pelagic community, for example of the type of Chelon labrosus, Coris julis, Sardina pilchardus, Solea vulgaris, Symphodus melops and Mullus surmuletus), are expected around the reef in this period. Their feeding is developed in Table 1.

Table 1. Main feeding of the possible fish species that will be able to colonize the reef during the first 5 years of its installation. Own source.

Fish species	Feeding					
	Mollusks	Crustaceans	Polychaetes	Plankton	Algae	Fishes
Chelon labrosus	x	x			x	
Coris julis	x		x	x		
Sardina pilchardus				x		
Solea vulgaris	x		x			
Symphodus melops	x	x	x			
Mullus surmuletus		x				
Dipodus sargus	x	x			x	
Solea lascaris	x		x		x	
Labrus bergylta	x	x				
Sparus aurata	x	x			x	
Trachurus trachurus		x				x
Scomber scombrus						x
Dicentrarchus labrax		x				x
Belone belone						x
Micromesistius poutassou		x				x
Phycis physis	x	x				x

5 Second and Third Year of Colonization in the AR

During this period, the mussels will dominate the benthic community, reaching 13,000 individuals/m^2 [7]. This colonization will influence the settlement of other epibenthic species, since the sediment and mud particles produced by the mussels produce a soft seafloor that gives rise to the development of bottom species that serve as food for the cephalopods (Octopus vulgares, Sepia officinalis, Loligo vulgaris), crustaceans (Pollicipes pollicipes, Maja esquinado) and fishes (Dipodus sargus, Solea lascaris, Labrus bergylta, Solea vulgaris and Sparus aurata). Colonization in this period will be based on species that feed especially on mussels (Table 1).

6 Fourth and Fifth Year of Colonization in the AR

On the one hand, sedimentation of mussel mud will further reduce hard substrates and promote new species of soft substrates. On the other hand, the development of the pelagic community that feeds on mussels, will probable cause these mollusks to disappear gradually. In certain studies [7], a decrease in values of 3,000 individuals/m^2 has been observed in this period.

Thus, in general, the following colonization is expected at this stage: cephalopods (Octopus vulgaris and Loligo vulgaris), crustaceans (Maja squinado and Liocarcinus puber) and fish, greater than those that exist in the reef due to previous colonizations (Trachurus trachurus, Scomber scombrus, Dicentrarchus labrax, Belone belone, Micromesistius poutassou and Phycis physis). The diet in this period is mainly based on crustaceans and fish (Table 1).

7 Conclusions

The According to our study on the evolution of artisanal fishing in Galicia in the markets of the entire Galician coast [3], it has been observed that for many groups of species, in the last ten years the landings in volume have followed a trend. This has resulted in a reduction of 51.5% in cephalopods, 53% in crustaceans, 29% in bivalves and 3% in fish.

The groups of species of major importance for our study that have decreased their catches in the last six years are cephalopods and crustaceans. Therefore, the valuation of the date on the catches of a representative example of cephalopods, crustaceans and fish in the Ría de Ares-Betanzos has made during the last six years. We observed that we found the justification for the implantation of our reef.

Most of our species under analysis have decreased their catches in percentages higher than 50%, reaching half of those studied, the disappearance of 100%.

According to our general pattern of colonization studied, the temporal evolution of a benthic community of reefs follows a slow and gradual process towards a state of equilibrium. The slow process from the initial colonization, especially by the species Mytilus galloprovincialis to the stability, is due to the speed of the changes in the atypical hard substrate of an artificial reef located in a soft seabed area. Under these conditions, it is subject to high sedimentation rates [13, 14] that will finally define the benthic, demersal and pelagic community that can survive.

Taking into account the results obtained on the improvement of biomass production with the installation of artificial reefs [15, 16], we expect similar results for our estuary under study.

Acknowledgements. The work carried out is part of the research project "Design and production of an artificial reef module for the protection and rehabilitation of coastal ecosystems in the autonomous community of Galicia". CN-10MMA003CT; Program INCITE Programs Sectoriais PEME I + D and R & D SUMA Xunta de Galicia.

References

1. Bombace, G.: Introduction générale sur le théme récifs artificiels. FAO Fish. Rep. **357**, 29–41 (1986)
2. Relini, G., Orsi Relini, L.: Artificial reefs in the Ligurian Sea (Northwestern Mediterranean): aims and results. Bull. Mar. Sci. **44**(2), 743–751 (1989)
3. Carral Couce, L., Alvarez-Feal, J.C., Tarrio-Saavedra, J., Rodriguez Guerreiro, M.J., Fraguela, J.Á.: Social interest in developing a green modular artificial reef structure in concrete for the ecosystems of the Galician rías. J. Clean. Prod. **172**, 1881–1898 (2018)
4. Carral Couce, L., Rodríguez Guerreiro, M.J., Fraguela Formoso, J.A., Álvarez-Feal, J.C., Filgueira Vizoso, A.: Green artificial reef PROARR: repopulation of coastal ecosystems and waste recycler of the maritime industries. In: Proceedings of the 25th Pan-American Conference of Naval Engineering – COPINAVAL. Springer, Heidelberg (2019c)
5. Ardizzone, G.D., Gravina, M.F., Belluscio, A., Somaschini, A.: Colonization and disappearance of Mytilus galloprovincialis Lam. on an artificial habitat in the Mediterranean sea. Estuar. Coast. Shelf Sci. **43**(6), 665–676 (1996)
6. Ardizzone, G.D., Belluscio, A., Somaschini, A.: Fish colonisation and feeding habits on a Mediterranean artificial habitat. In: The Responses of Marine Organisms to they Environments, Proceedings of the 30 th European Marine Biological Symposium, pp. 265–275 (1997)
7. Ardizzone, G., Somaschini, A., Belluscio, A.: Prediction of benthic and fish colonization on the Fregene and other Mediterranean artificial reefs. In: Artificial Reefs in European Seas, pp. 113–128. Springer, Dordrecht (2000)
8. Asensio, I., Grajal, M.: Rasgos morfológicos y sedimentológicos de la ría de Betanzos (A Coruña, Galicia, España). Cuad. Lab. Xeol. Laxe **2**, 197–208 (1981)
9. Sánchez-Mata, A., Glémarec, M., Mora, J.: Physico-chemical structure of the benthic environment of a Galician ria (Ria de Ares-Betanzos, north-west Spain). J. Mar. Biol. Assoc. United Kingdom **79**(1), 1–21 (1999)
10. Piedracoba, S., Álvarez-Salgado, X.A., Labarta, U., Fernández-Reiriz, M.J., Gómez, B., Balseiro, C.: Water flows through mussel rafts and their relationship with wind speed in a coastal embayment (Ría de Ares-Betanzos, NW Spain). Cont. Shelf Res. **75**, 1–14 (2014)
11. Xunta de Galicia: Consellería de Pesca (2019). http://www.pescadegalicia.gal/. Accessed 25 Dec 2018
12. Bao, R., Rey, J., Varela, M.: Cuad. Lab. Xeol. Laxe **14**, 133–158 (1989)
13. Ardizzone, G.D., Pelusi, P.: Modificazioni di una prateria di Posidonia oceanica (L.) Delile nel Medio Tirreno sottoposta ad attivitá di pesca a strascico. Nat. Sicil. **3**, 509–515 (1982)

14. Bombace, G., Fabi, G., Fiorentini, L.: Artificial reefs and mariculture: the Italian experiences. In: Proceeding ECOSET, Japan International Marine Science and Technology Federation, vol. 2, pp. 830–835 (1995)
15. Antsulevich, A., Laihonen, P., Vuorinen, I.: Employment of artificial reefs for environmental maintenance in the Gulf of Finland. In: Artificial Reefs in European Seas, pp. 319–329 (2000)
16. Ramos-Esplá, A.A., Guillén, J.E., Bayle, J.T., Sánchez-Jérez, P.: Artificial anti-trawling reefs off Alicante, South- Eastern Iberian Peninsula: evolution of reef block and set designs. In: Artificial Reefs in European Seas, pp. 195–218. Springer, Dordrecht (2000)

Virtual Laboratory with Algodoo for the Determination of the Metacentric Height of a Rectangular Vessel

Roberto A. Del Carpio Minaya[1(✉)] and Yalmar Ponce Atencio[2(✉)]

[1] Programa de Doctorado en Computación, Universidad Nacional del Altiplano, Puno, Peru
rdelcarpio1969@hotmail.com

[2] Universidad Nacional José María Arguedas, Andahuaylas, Apurímac, Peru
yalmar@gmail.com

Abstract. In the present work, a virtual laboratory using the Algodoo App was implemented in order to determine the metacentric height of a rectangular Ship in 2D. Experiments were carried out with different rectangular Ships, of different densities and geometries, which floated on a density fluid of fresh or salt water. Vertically and horizontally sliding pieces of different weights were placed on each Ship, causingtilt on the Ship, which could be measured for determining the stability gradient. Following, dispersion graphs of the vertical gravity center position versus the stability gradient were made, and with a linear data adjustment, the Ships transverse stability metacenter was determined. Finally, with a statistical treatment, the Ships metacentric height and its respective interpretation of the stability were determined. It was concluded that the method could serve as an option for instruction in understanding the stability of Ships and this method could serve as an option for instruction in understanding the stability of ships and in suggesting the future work with more complicated ship geometries.

Keywords: Algodoo · Metacentric height

1 Introduction

Fluid Mechanics (FM) has become a fundamental tool in various areas of technology study and modern science. In many areas of engineering, FM plays a very important role, for example, in the aerodynamic design of structures shape (such as hydraulic dams), propeller blades, turbine blades, Ship hulls, bodywork of aircraft and cars. Though, the analytical study of fluid problems is very elegant and precise, it should be noted that, in most fluid studies, it is practically impossible to obtain analytical solutions due to the immense complexity of the involved systems, due to this it's necessary to use numerical methods in order to get solutions which must be computed in very short times, achieving greater effectiveness for the application in current technology. Thus arises a branch of the FM called Computational Fluid Dynamics (CFD), which studies the appropriate numerical approximations that should be used in the physical equations of FM. These CFD methods allow to properly calculate the velocities, pressures, densities and internal forces of the particles that make up the mass of the fluid. In some studies of

V. A. J. E. Carreño Moreno et al. (Eds.): CIDIN 2019/COPINAVAL 2019, pp. 373–390, 2020.
https://doi.org/10.1007/978-3-030-35963-8_32

CFD, it's adequate to view on the computer screen the results of calculations in real time, and for that there are powerful computational tools such as graphics cards and the possibility to compute general purpose calculations in Graphical Processing Units (GPU), achieving an important link between the CFD and the Computer Graphics (CG), which results in a very interesting study alternative when we want to simulate fluids, since you can get conclusions that would not be easily discerned by other methods. Now, for the study of the buoyancy and stability of Ships, it is necessary to do an experimental study based on calculations of gravity centers, buoyant centers, measurement of heights, displacement of weights, etc. All this procedure involves working sometimes in unfavorable situations, such as unstable waters that make the results of the experiment unpredictable and because the unreliability of the measuring instruments used. This is why the connection between the CFD and CG provides a good alternative to study the buoyancy and stability of Ships, but in controllable virtual environments. For this purpose it is necessary to use Ships which have not too complicated shapes, in sucha way to do a computational implementation that allows to get adequate and viable results which could be used in the design and construction of Ships. The simplest option is to study the flotation of a rectangular box, this due to its geometry and symmetric characteristics allow an easy computational implementation. That is why this work aims to provide a simple computational technique for the analysis of the buoyancy and stability of rectangular Ships and thus set a precedent to improve the technique and determine the stability and buoyancy of geometrically more complicated Ships.

2 Preliminaries

2.1 Fluid Simulation

The equations that govern the movement of an incompressible fluid are given by Navier-Stokes [1] equations. The conservation of mass and momentum are given by

$$\frac{D\rho}{Dt} = 0 \tag{1}$$

$$\frac{D\vec{v}}{Dt} = -\frac{1}{\rho}\nabla P + \mu\nabla^2\vec{v} + \vec{g} \tag{2}$$

where ρ, $\vec{v}$, P, μ, $\vec{g}$ are respectively the density, speed, pressure, coefficient of dynamic viscosity and acceleration of gravity.

To solve numerically the Navier-Stokes equations, the Smoothed Particle Hydrodynamics (SPH) method can be used. This is a technique for the simulation of Lagrangian fluids, which tracks the material particles that make up the fluid and calculates its physical properties at each moment of time.

In this approach, fluids are represented by a set of particles $i \in [1..N]$, that is, the tracking of each particle i that makes up part of the fluid that contains N particles. Each particle i has position $\vec{r}_i$ and mass m_i. The SPH method defines how to calculate the value of any physical property in an arbitrary position $\vec{r}$ in space, performing a soft interpolation, on the set of all nearby or neighboring particles j, it means that the properties of a particle can be obtained based on the known properties of neighboring particles that surround it.

In this way the density $\rho_{(\vec{r}_i)}$ in a fluid element [2] is computed with the following equation.

$$\rho_{(\vec{r}_i)} = \sum_j m_j W_{\left(\vec{r}_i - \vec{r}_j, h\right)} \tag{3}$$

where m_j is the mass of the neighboring particles and $W_{\left(\vec{r}_i - \vec{r}_j, h\right)}$ is a smoothing function called *kernel* that has a radius of action or length of smoothing h, the mathematical form of the *kernel* depends on the type of fluid to be studied.

For the calculation of the fluid pressure and viscosity forces, the Eqs. (4) and (5) are used respectively [3]

$$\vec{f}_i^{\text{pressure}} = -\sum_j m_j \frac{P_i + P_j}{2\rho_j} \nabla W_{\left(\vec{r}_i - \vec{r}_j, h\right)} \tag{4}$$

$$\vec{f}_i^{\text{viscosity}} = \mu \sum_j m_j \frac{\vec{v}_j - \vec{v}_i}{\rho_j} \nabla^2 W_{\left(\vec{r}_i - \vec{r}_j, h\right)} \tag{5}$$

where, P_i is the pressure in the particle i, P_j is the pressure in the particle j, $\vec{v}_i$ is the velocity of the particle i, $\vec{v}_j$ is the velocity of the particle j, and μ is the fluid viscosity.

Notice that for the calculation of the pressure force we use the gradient of a kernel and for the calculation of the viscosity force we use the Laplacian of a kernel.

2.2 Boat Simulation

To simulate the movement of the boat on the fluid we can use the equations of the rigid bodies mechanics, which basically computes the rigid body displacement and rotation.

If we consider M the total mass of the rigid body, the acceleration $\vec{a}_{cm}$ of the rigid body, at its center of mass, will be given by

$$\vec{a}_{cm} = \frac{\vec{F}_{rigid}}{M} \tag{6}$$

where the force $\vec{F}_{rigid}$ is the total sum of all the forces applied on the rectangular Ship considered as a rigid body. This linear acceleration is enough to calculate the displacement of each of the particles that make up the rigid body.

To calculate the rotation of the rigid body we need to calculate the torque of the system [4] first $\vec{\tau}_{rigid}$ by the equation

$$\vec{\tau}_{rigid} = \sum_k (\vec{r}_k - \vec{r}_{cm}) \times \vec{f}_k^{total} \tag{7}$$

where $\vec{r}_{cm}$ is the position of the mass center of the rigid body and $\vec{r}_k$ is the position of each particle of the rigid body and $\vec{f}_k^{total}$ is the sum of the external forces applied to each particle k that makes up the boat.

Then we must make a calculation of angular momentum by

$$\vec{L}_{rigid} = \sum_k m_k(\vec{r}_k - \vec{r}_{cm}) \times (\vec{v}_k - \vec{v}_{cm}) \tag{8}$$

where m_k is the mass of each particle of the rigid body, $\vec{v}_k$ is the linear velocity of each particle and $\vec{v}_{cm}$ is the linear velocity of the center of mass of the rigid body. There is a relationship between the inertial tensor I with angular velocity $\vec{\omega}$ of the rigid body as follows

$$\vec{L}_{rigid} = I\vec{\omega} \tag{9}$$

The inertia tensor is determined according to the geometry of the rigid body. With the calculated angular velocity, the way in which the rigid body rotates on the fluid is finally determined.

2.3 Stability Criteria for the Boat Simulation

For the calculation of the stability of a boat, the keel (K), the gravity (G), the buoyant center (B) and the metacenter (M) shown in Fig. 1 are taken as main points.

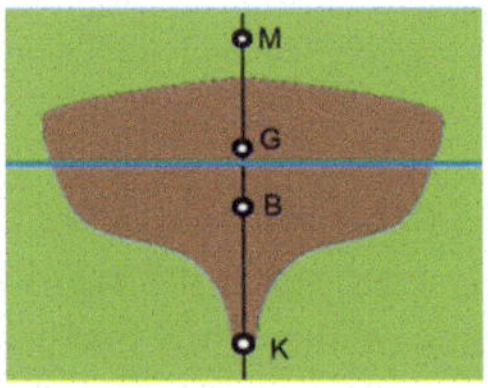

Fig. 1. Main reference points.

With these points the following distances are calculated:

Height of the center of gravity (KG): is measured from the plane of the keel (K) or baseline to the center of gravity (G) of the boat.

Metacentric height (GM): is measured from the center of gravity (G) to the metacenter (M).

Metacentric Radio (BM): is measured from the buoyant center (B) to the metacenter (M).

With these distances, the following equality can be calculated:

$$KM = GM + KG \tag{10}$$

The different equilibria that the boat can adopt are:

- *Stable equilibrium or positive stability:* It occurs when KM > KG.
- *Unstable equilibrium or negative stability:* It occurs when KM < KG.
- *Indifferent equilibrium or null stability:* It occurs when KM = KG.

3 Implementation

3.1 Transverse Stability of a Rectangular Ship

The cross-sectional area is rectangular and to calculate its value we consider the measurements of the beam and the prop of the boat (Fig. 2).

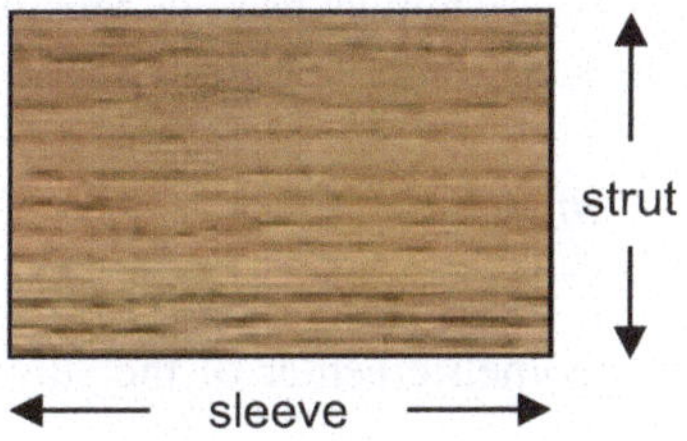

Fig. 2. Sleeve and strut in a boat.

$$transversal\,area = (sleeve) \cdot (strut) \tag{11}$$

To calculate the surface density, we must consider the value of the mass to be used in the following relationship

$$superficial\,density = \frac{m_{vessel}}{transversal\,area} \tag{12}$$

To determine the center of gravity G we must consider a reference system XZ where the origin of the system is in the lower part of the strut and in the middle part of the sleeve (Fig. 3).

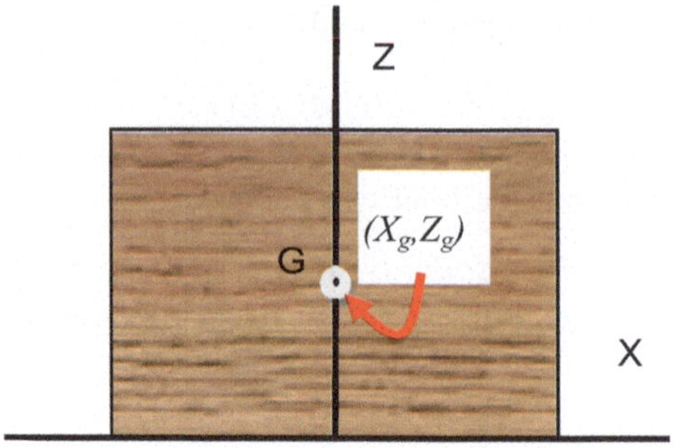

Fig. 3. Cross section gravity center.

The center of gravity is in the geometric center of the rectangular cross-sectional area and will have the coordinates (X_g, Z_g) given by

$$(X_g, Z_g) = \left(0, \frac{strut}{2}\right) \tag{13}$$

To calculate the draft of the boat we must use the density of the water where it will float, this can be fresh water or salt water. The draft will be given by

$$draft = \frac{m_{vessel}}{\rho_{water}(sleeve)} \tag{14}$$

Buoyant center B is in the geometric center of the submerged area (Fig. 4).

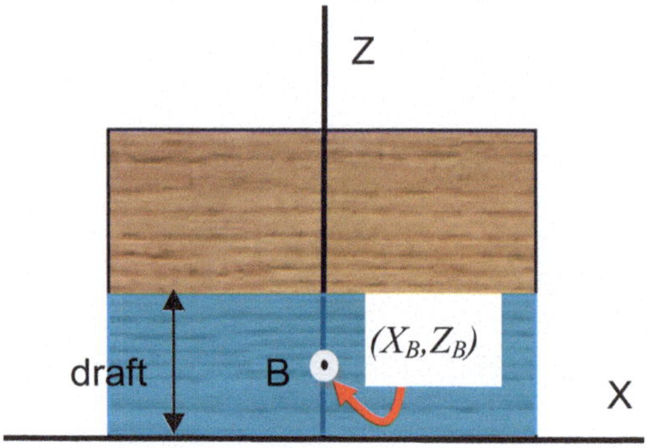

Fig. 4. Buoyant or floating center.

The buoyant center [5] will have the coordinates given by

$$(X_B, Z_B) = \left(0, \frac{draft}{2}\right) \tag{15}$$

By laterally placing a certain mass on the boat, the center of gravity will move towards the mass (Fig. 5).

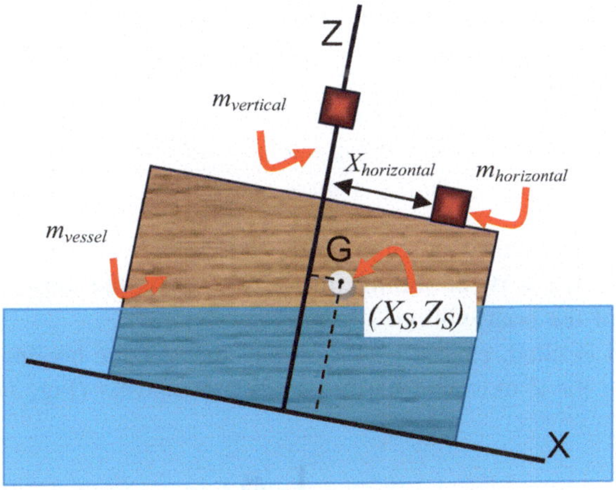

Fig. 5. Horizontal displacement of G.

The location of the center of gravity as well as the buoyant center will be modified due to the heel formed [6]. To calculate the new horizontal position of the center of gravity we use

$$X_s = \frac{m_{horizontal} X_{horizontal}}{m_{vessel} + m_{horizontal} + m_{vertical}} \tag{16}$$

Where $m_{horizontal}$ is the horizontal sliding mass, $m_{vertical}$ is the vertical sliding mass, $X_{horizontal}$ is the horizontal position of the horizontal sliding mass.

By vertically placing a certain mass on the boat, the center of gravity will move towards the mass (Fig. 6).

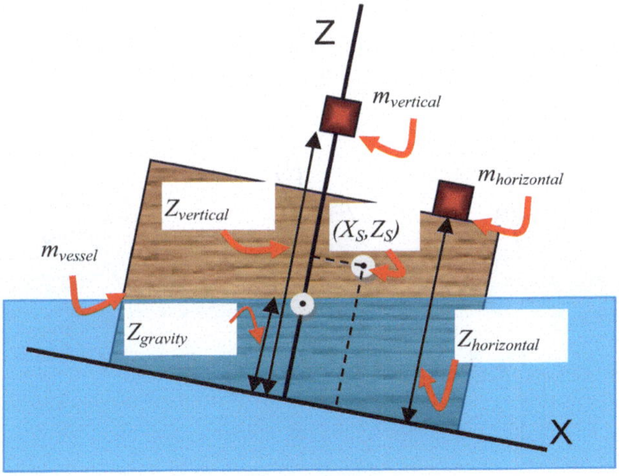

Fig. 6. Vertical displacement of G.

As in the previous case, the location of the center of gravity and the buoyant center will be modified, due to the list formed [6]. To calculate the new vertical position of the center of gravity we use

$$Z_s = \frac{m_{vessel} Z_{gravity} + m_{horizontal} Z_{horizontal} + m_{vertical} Z_{vertical}}{m_{vessel} + m_{horizontal} + m_{vertical}} \tag{17}$$

where $Z_{horizontal}$ is the vertical position of the horizontal sliding mass, $Z_{vertical}$ is the vertical position of the vertical sliding mass, $Z_{gravedad}$ is the vertical position of the center of gravity of the boat without slip masses.

When the boat is tilted, caused by the sliding masses, the heeling angle θ, formed by the vertical and the Z axis, can be measured successively (Fig. 7).

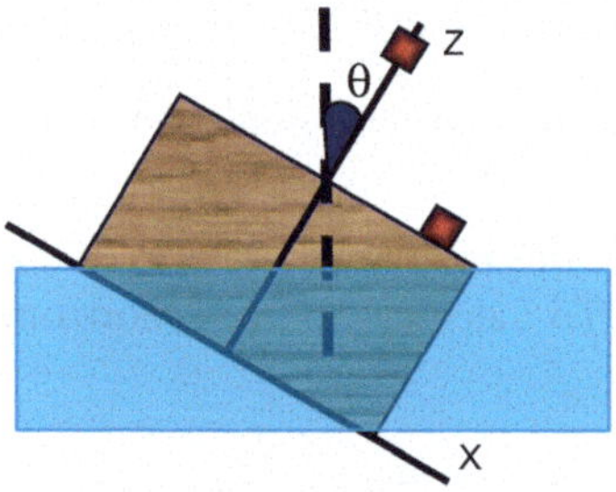

Fig. 7. Heeling angle

To measure the stability of a Ship, the stability gradient is defined as the variation of X_S related to heeling angle θ.

$$grad = \frac{dX_S}{d\theta} \equiv \frac{X_S}{\theta} \tag{18}$$

In an experiment we can place the horizontal and vertical sliding masses in different positions, in order to obtain different heeling angles (Fig. 8).

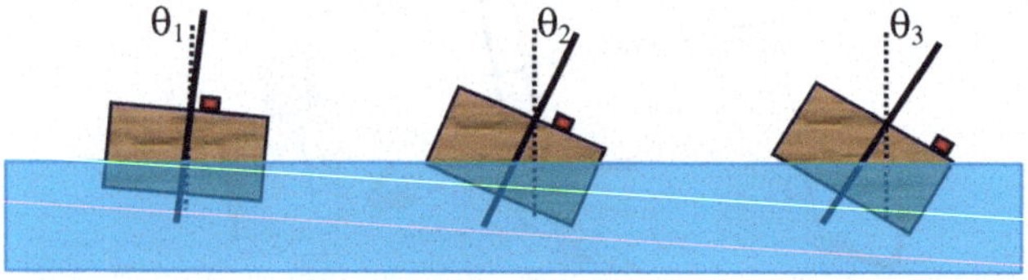

Fig. 8. Variation of heel angles by weight

Then we can record the measurements made in a table as shown in Fig. 9.

$X_h = X_{h1}$			$X_s = X_{s1}$		
Z_v	Z_{v1}	Z_{v2}	...	...	Z_{v5}
θ	θ_1	θ_2	...	...	θ_5
Z_s	Z_{s1}	Z_{s2}	...	...	Z_{s15}
grad	$grad_1$	$grad_2$	...	...	$grad_5$

Fig. 9. Variables to be used in the experiments.

We make a scatter chart Z_s versus *grad* and we perform a linear adjustment as shown in Fig. 10.

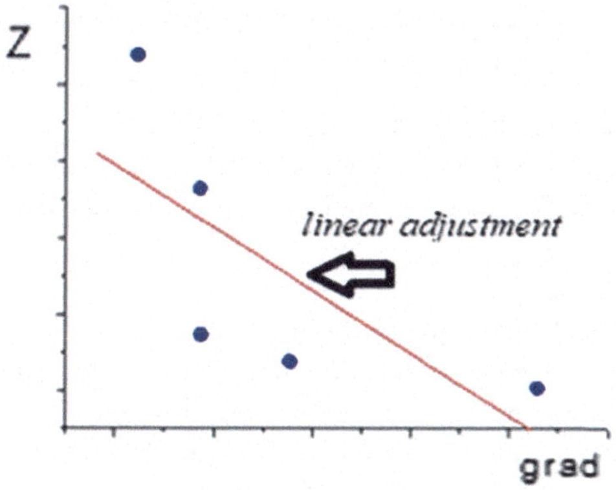

Fig. 10. Dispersion and linear data adjustment

As a result of the adjustment we will obtain a linear equation of the form

$$Z = A - B \cdot grad \tag{19}$$

where A and B are coefficients obtained from the adjustment.

The zero stability gradient provides the location of the metacenter of transverse stability, this is fulfilled at the intersection of the line of linear adjustment with the vertical axis Z, then compute the following equations

$$Z = A - B \cdot (0) \tag{20}$$

$$KM = A \tag{21}$$

For the calculation of metacentric height we use [7]

$$GM = KM - KG \tag{22}$$

where KG is the height of the center of gravity of the boat. We use the stability criterion:

GM > 0 (boat with stable equilibrium)
GM = 0 (boat with neutral balance)
GM < 0 (boat with unstable equilibrium).

3.2 Virtual Laboratory with Algodoo

Algodoo is a 2D physics simulator software for educational purposes. This software includes program tools which can create polygons, cutter, brush, gear, fixer, rotator, laser pointer, etc. It also has characteristics that allow control and modify the speed, forces, density, collision, pressure, buoyancy, refraction and others in order to simulate complex situations. In the case that we are interested in, Algodoo allows users to perform very real fluid simulations using the SPH method in any implementation. The steps that were done for the implementation of a virtual laboratory using Algodoo in the simulation of a fluid and a rectangular object floating in it were:

- **Implementation of a vessel, fluid and a ship (boat)**
  With polygon tools and rectangle of Algodoo we have proceeded to create a Vessel, fluid and a Ship. With the material tool, the mass and density were assigned to the Ship.
- **Float tests of the boat**
  We proceeded to perform some flotation tests of the Ship. First the boat was released from a certain height on the fluid and it was expected that it would come to rest displaying an adequate behavior of stable buoyancy (Fig. 11). Next, a movement of the fluid was caused with the moving tool, and the boat again displayed acceptable flotation with the wave oscillation (Fig. 12).

Fig. 11. Simulation of a boat falling on a fluid.

Fig. 12. Boat simulation moving in the fluid.

- **Tilt tests of the boat**
  We proceeded to perform some tests with weights on the boat to determine the degree of inclination it experiences, achieving an adequate behavior for the measurement of heel angles in the experiments (Fig. 13).

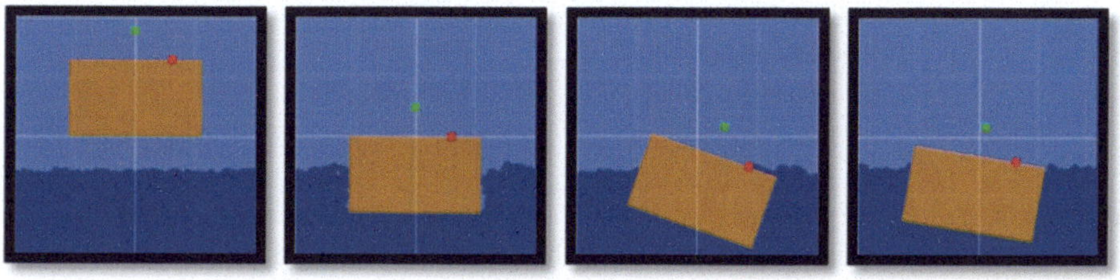

Fig. 13. Simulation of tilt of the boat with weights.

- **Measurement of heel angle and position of the weights**
  For these experiments it is necessary an adequate measurement of the angle of heeling of the boat, as well as the position of the sliding weights, for this the Script tool was used to modify the *angle* and *position* entries in the dialogue window.

4 Results

A rectangular Ship model with the data of Table 1 was implemented and the stability calculations were carried out.

Table 1. Data for the model

Description	Symbol	Data
Sleeve	D	2.22 m
Strut	P	1.30 m
Heel	L	13.40 m
Transversal section area	A	2.886 m^2
Transversal area mass	m_a	1.00 kg
Water density	ρ_w	1.0000 kg/m^2
Ship density	ρ_a	0.3465 kg/m^2
Vertical gravity center	Zg	0.65 m
Horizontal gravity center	Xg	0.00 m
Horizontal sliding mass	m_h	0.05 kg
Vertical sliding mass	m_v	0.10 kg
Horizontal position of m_h	X_h	0.20 m; 0.30 m; …
Vertical position of m_h	Z_h	1.30 m
Horizontal position of m_v	X_v	0.00 m
Vertical position of m_v	Z_v	1.50 m; 1.70 m; …

Calculations for X_h = 0.20 m
We use the displacements of the weights shown in Table 2 and perform the calculations of the stability gradient.

Table 2. Calculations with X_h= 0.20 m

$X_h = 0.20\ m$			$X_s = 0.008695652\ m$		
$Z_v\ (m)$	1.50	1.80	2.10	2.50	3.50
α	2.74°	3.92°	4.81°	6.37°	8.35°
$Z_s\ (m)$	0.752173913	0.778260870	0.804347826	0.839130435	0.926086957
grad	0.003173596	0.002218279	0.001807828	0.001365095	0.001041395

Next, we make a scatter plot with the results of the center of gravity versus the stability gradient and perform a linear adjustment as shown in Fig. 14.

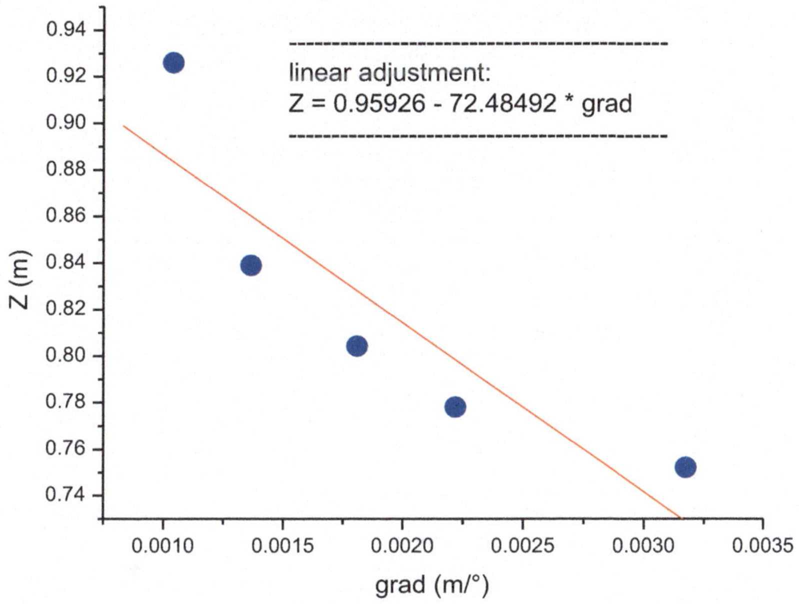

Fig. 14. Dispersion and linear adjustment of data for $X_h = 0.20$ m.

The equation obtained for the linear adjustment is:

$$Z = 0.95926 - 72.48492\,grad$$

Where the zero stability gradient provides the metacenter location of transverse stability, then

$$KM = 0.95926\,\mathrm{m}$$

Calculations for X_h = 0.30 m

We use the displacements of the weights shown in Table 3 and perform the calculations of the stability gradient.

Table 3. Calculations with X_h = 0.30 m

$X_h = 0.30\,m$			$X_s = 0.013043478\,m$		
$Z_v\,(m)$	1.50	1.70	1.90	3.00	4.00
α	2.47°	4.71°	6.98°	7.04°	10.71°
$Z_s\,(m)$	0.752173913	0.769565217	0.786956522	0.882608696	0.969565217
$grad$	0.005280760	0.002769316	0.001868693	0.001852767	0.001217878

Then we made a scatter plot with the results of the center of gravity versus the stability gradient and we made a linear adjustment as shown in Fig. 15.

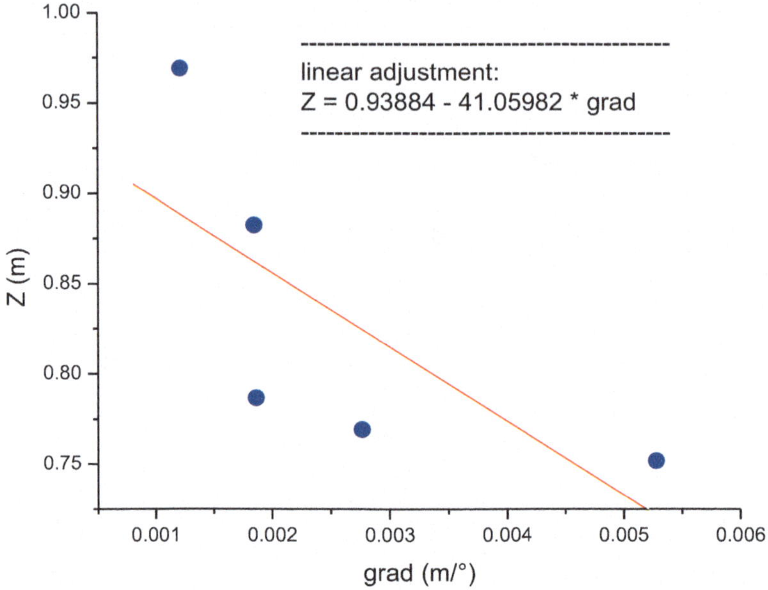

Fig. 15. Dispersion and linear adjustment of data for $X_h = 0.30$ m.

The equation obtained for the linear adjustment is:

$$Z = 0.93884 - 41.05982\, grad$$

Where the zero stability gradient provides the metacenter location of transverse stability, then

$$KM = 0.93884\,\mathrm{m}$$

Calculations for $\mathbf{X_h = 0.40}$ m

We use the displacements of the weights shown in Table 4 and perform the calculations of the stability gradient.

Table 4. Calculations with $X_h = 0.40$ m

$X_h = 0.40\ m$			$X_s = 0.017391304\ m$		
$Z_v\ (m)$	1.50	1.70	2.00	3.50	4.00
α	5.40°	6.38°	7.06°	12.60°	15.44°
$Z_s\ (m)$	0.752173913	0.769565217	0.795652174	0.926086957	0.969565217
grad	0.003220612	0.002725910	0.002463358	0.001380262	0.001126380

Next, we make a scatter plot with the results of the center of gravity versus the stability gradient and perform a linear adjustment as shown in Fig. 16.

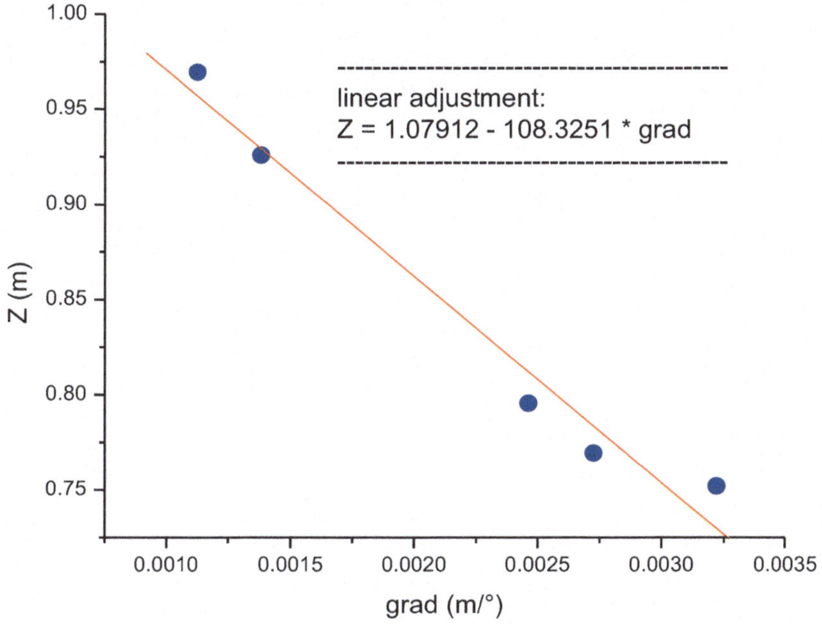

Fig. 16. Dispersion and linear adjustment of data for $X_h = 0.40$ m.

The equation obtained for the linear adjustment is:

$$Z = 1.07912 - 108.3251\,grad$$

Where the zero stability gradient provides the metacenter location of transverse stability, then

$$KM = 1.07912\,\mathrm{m}$$

Calculations for $X_h = 0.50$ m

We use the displacements of the weights shown in Table 5 and perform the calculations of the stability gradient.

Table 5. Calculations with $X_h = 0.50$ m

$X_h = 0.50\,m$			$X_s = 0.02173913\,m$		
$Z_v\,(m)$	1.50	1.70	1.90	2.50	3.00
α	5.35°	6.91°	7.00°	8.02°	8.95°
$Z_s\,(m)$	0.752173913	0.769565217	0.786956522	0.839130435	0.882608696
$grad$	0.004063389	0.003146039	0.003105590	0.002710615	0.002428953

Next, we make a scatter plot with the results of the center of gravity versus the stability gradient and perform a linear adjustment as shown in Fig. 17.

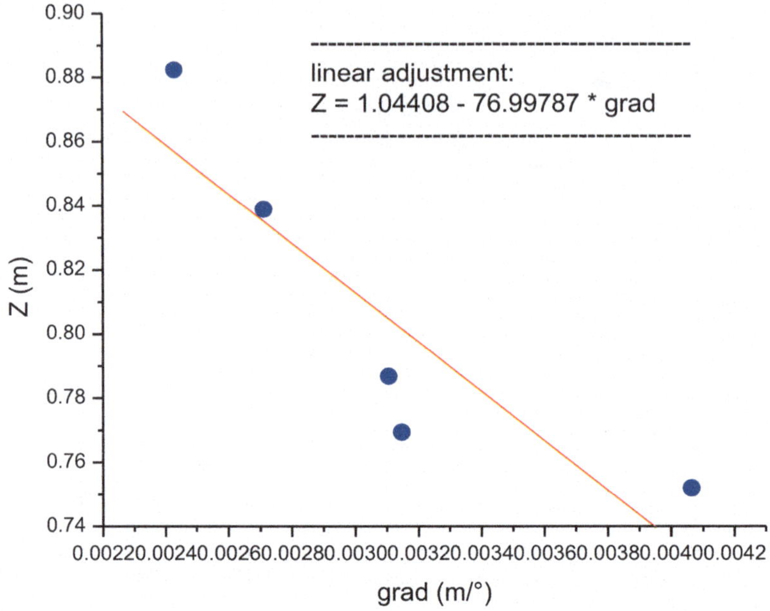

Fig. 17. Dispersion and linear data adjustment for X_h = 0.50 m.

The equation obtained for the linear adjustment is:

$$Z = 1.04408 - 76.99787\,grad$$

Where the zero stability gradient provides the metacenter location of transverse stability, then

$$KM = 1.04408\,\mathrm{m}$$

Calculations for $\mathbf{X_h}$ = 0.60 m

We use the displacements of the weights shown in Table 6 and perform the calculations of the stability gradient.

Table 6. Calculations with X_h = 0.60 m

$X_h = 0.60\,m$			$X_s = 0.026086957\,m$		
$Z_v\,(m)$	1.50	1.70	2.00	3.00	4.00
α	5.85°	9.11°	9.40°	11.19°	17.41°
$Z_s\,(m)$	0.752173913	0.769565217	0.795652174	0.882608696	0.969565217
$grad$	0.004459309	0.002863552	0.002775208	0.002331274	0.001498389

Next, we make a scatter plot with the results of the center of gravity versus the stability gradient and perform a linear adjustment as shown in Fig. 18.

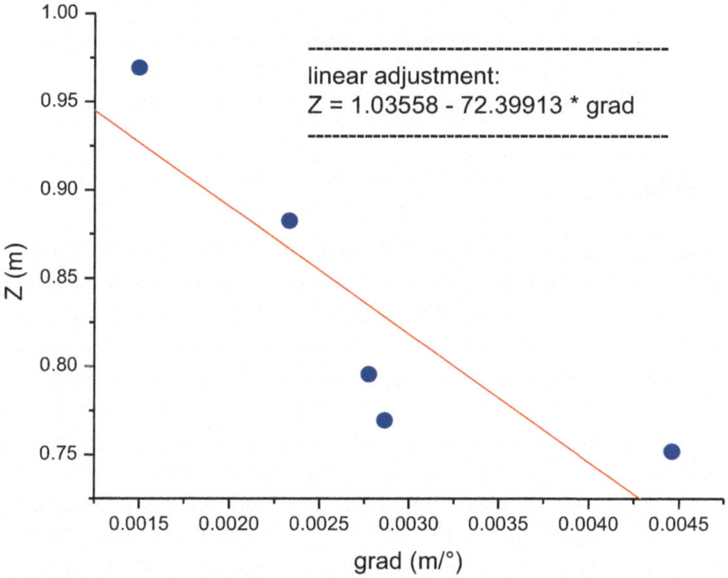

Fig. 18. Scattering and linear adjustment of data for $X_h = 0.60$ m.

The equation obtained for the linear adjustment is:

$$Z = 1.03558 - 72.39913\, grad$$

Where the zero stability gradient provides the metacenter location of transverse stability, then

$$KM = 1.03558\,\mathrm{m}$$

Calculation of Average Metacentres

With the different results found previously we calculate the average cross-section stability metacenter for the boat

$$\langle KM \rangle = 1.011376\,\mathrm{m}$$

Finally, we round the average and we will have the experimental metacenter of our boat

$$\langle KM \rangle = 1.011\,\mathrm{m}$$

According to Table 1 the vertical center of gravity is 0.65 m, then

$$GM = KM - KG = 1.011\,\text{m} - 0.650\,\text{m}$$

$$GM = +0.361\,\text{m}$$

Therefore, as GM > 0 it could be said that it is a stable Ship.

5 Conclusions

It was possible to implement a virtual laboratory for the study of fluid and object floating on it, with the help of Algodoo software, which allows the use of the Smoothed Particle Hydrodynamic method for fluid simulation. Likewise, a rectangular Ship was floated on fluid, manipulating its physical properties as density, and for the ship case its geometrical measurements. With this virtual laboratory, One can perform experimental tests to determine the stability of a rectangular Ship. When performing the calculations for the experimental metacentric height (GM), a height of +0.361 m was obtained for the model, and since the stable Ships must have a GM > 0, we conclude that the model studied is stable. Therefore, it is possible to perform experiments in a virtual controlled environment for the determination of buoyancy and stability of rectangular Ships. In addition, this research shows that it is possible to continue with the experiments and carry out an analysis of the stability and buoyancy of Ships with more complex shapes.

References

1. Takahiro, H., Seiichi, K., Yoichiro, K.: Smoothed particle hydrodynamics in complex shapes. In: Proceedings of the 23rd Spring Conference on Computer Graphics, Budmerice, Slovakia (2007)
2. Solenthaler, B., Schläfli, J., Pajarola, R.: A unified particle model for fluid–solid interactions: research article. In: Computer Animation and Virtual Worlds (2007)
3. Müller, M., Charypar, D., Gross, M.: Particle based fluid simulation for interactive applications. In: ACM SIGGRAPH/Eurographics Symposium on Computer Animation (2003)
4. Baraff, D.: An introduction to physically based modeling: rigid body simulation/unconstrained rigid body dynamics. In: SIGGRAPH 1997 (1997)
5. Boris, L., Guerrero, B.: Equilibrio de los Cuerpos Flotantes. Humboldt Marine Training (2011)
6. Forcano, C.P.: Análisis de la Estabilidad de un Buque durante la Operación de Descarga de una Carga Pesada. Facultad de Náutica de Barcelona/Universitat Politécnica de Catalanya (UPC), Barcelona – España (2014)
7. Van Dokkum, K., Koomen, K., Ten Katen, H., Pinkster, J.: Ship Stability, 3rd edn. Enkhuizen, DokMar (2008)

Optimization of the Life Cycle in the Warships: Maintenance Plan and Monitoring for Costs Reduction

Antonio Fernández Jove[1], Antonio Mackinlay[1], and José María Riola[2(✉)]

[1] Grupo Fernández Jove, Torrelavega, Spain
tony@fjove.com, amac-cas@fernandezjove.com
[2] Universidad Politécnica de Madrid, Madrid, Spain
josemaria.riola@upm.es

"My logisticians are a humorless lot ... they know if my campaign fails, they are the first ones I will slay"

(Alexander the Great)

Abstract. By its own meaning, a warship is a complex and expensive system, with a long acquisition cycle and a very long operational life. Therefore, the optimization of the lifecycle is increasingly important. One should not think only about the building or acquisition cost, but also about the cost through the lifecycle. The Spanish Navy (Armada) defines, on one hand, the lifecycle engineering as the set of the required activities in order to adapt the hull and the armament systems to their operational requirements, the configuration management and the definition, assessment and upgrade of the equipment support through their entire operational life. And on the other hand, the logistic support as the set of actions in order to provide the means required by their units/assets and systems to accomplish their missions. The marine industry around the warship world is immersed in the digitalization of all the logistics processes, which mainly refers to the maintenance process. In this paper we intend to reflect how the maintenance methods of the units/assets are to be based on the monitoring of equipment and how their centralized control is to be based on a detailed Maintenance Plan that would help to their preservation during their lifecycle. And also, reflect how the auxiliary marine industry, as part of the industrial technological base, may help to ease such maintenance. The methodology of Reliability Centered Maintenance (RCM) aims to determine the operational maintenance based upon the operability of a system or equipment. This methodology, which has been adopted by the Armada, is of special importance considering the incorporation of the Center of Supervision and Data Analysis (CESADAR) to obtain the required 4.0 level.

Keywords: Lifecycle · Warships · Maintenance · RCM · CESADAR

"Amateurs discuss tactics, professional soldiers study logistics"

(Tom Clancy)

V. A. J. E. Carreño Moreno et al. (Eds.): CIDIN 2019/COPINAVAL 2019, pp. 391–401, 2020.
https://doi.org/10.1007/978-3-030-35963-8_33

1 Introduction

The word "*logistics*" comes from the Greek "*logistikos*", which stands for to know how to calculate. This word was already used in Greece, in 489 BC, and it was defined as the "*doing something logical*". But normally, the first conception of modern logistics is attributed to Baron Jomini, who in his text "*Précis del art de la guerre*", refers to a theory of war supply and distribution of troops. Since then, there have been multiple examples of how logistics may define course of the war.

It is easy to be in General Fuller's shoes when the III Reich was involved in the Eastern Mediterranean scenario. Thus, in the campaign waged in Libya and Egypt, where from the local soil only camel dung and water could be obtained, each kilogram of food, ammunition or liter of fuel consumed by the troops had to be transported by sea from Italy. The distance to travel between Tripoli and Alexandria was 3,000 km after the loss of Benghazi due to the O'Connors offensive. Without a railway line, except for a few narrow lines, they had to be crossed by a single road, via Balbia, which ran parallel to the coast, and was always an easy target for the enemy aviation. The Tripoli port facilities could handle a maximum of five merchant ships in a month. So, how many troops and how far can they operate if your supply is through a port with an offloading capacity of 45,000 tons per month and if you have to transport the supplies across a road without a railroad network of about 3,000 km?. Logistics showed us the course of history in advanced (Fig. 1).

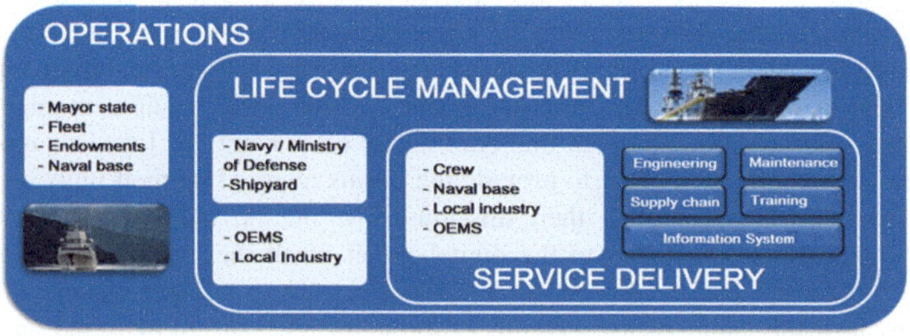

Fig. 1. Collaboration with the industry (Source: Navantia)

If we understand the maintenance of a warship as a combination of technical, administrative and management actions [2], in Spain, the General Directorate of Armament and Material (DGAM) is the governmental body responsible for the planning and development of the Ministry of Defense (MoD) policy with respect to armament systems and material. The *Armada*, which is responsible of the ship's maintenance, is based in the Logistic Support Headquarters (JAL), where the administration and control of the material resources is performed. Therefore, in the new build of naval ships, the acquisition of systems which will be integrated into the ship is tailored to the ship design and operational capabilities based upon the *Armada*'s high level requirements (REM). Most of the Integrated Logistics Support (ILS) policy is written for the acquisition of new warships or systems in order to meet a specific mission solution. The new systems will not be integrated unless the required logistics support is in place. ILS for naval programs [5] is defined as a disciplined approach that pushes the product design and the support solution in order to optimize the supportability with minimum cost. The fundamental objective of ILS is to provide the life cycle support, therefore, a process where design logistics, planning and budgeting activities are well integrated, is to be followed. According to ILS documents, we use elements such as maintenance planning, manpower and personnel, tests, training devices, documentation, packaging, storage and transportability, facilities, reliability, maintainability and computer resources support.

Our initial required maintenance cycle for the warship is a programmed maintenance service of six weeks every two and a half years, a docking period of two months every five years and a modernization period of forty weeks in the 15th year. The suppliers need to provide information with respect to any additional periods required for their equipment, which will be shorter than the ones indicated before, and will be assessed in order to determine if the requested immobilization periods are compatible with planned maintenance tasks which require immobilization of the ship. The *Armada* breaks down the levels of maintenance intervention into four steps: the first one is to be undertaken by the own ship's crew onboard; a second and a third ones are to be undertaken at the designated facilities (i.e. not onboard the ship) based upon the Admiral of the Fleet and the Admiral of Logistic Support's decision; and a fourth one for which the tasks will be executed by the supplier or authorized companies to overhaul and perform renewal tasks beyond the capacity of the lower maintenance levels.

The JAL presented the publication [1] with the fundamentals of the Reliability Centered Maintenance (RCM) following the guidelines from the U.S. Navy [6] which methodology is followed to its full implementation. This publication shows three phases: a first phase, the "*methodology and data preparation*" identifies the necessary means for the generation of predictive tasks that will be integrated into management systems such as the GALIA logistic application; a second phase, "*integration and adaptation to the current maintenance program*" in which the analyses of the equipment identified in the first phase are undertaken; a third phase, "*migration to the new RCM program and implementation*" which defines the maintenance tasks based on the RCM and GALIA adaptation. It seems unanimous to conclude that the benefits of the RCM makes it a valid tool in organizations such as the Navies, which have highly demanding needs, in terms of quality and quantity, and in which the economic aspects prevail. However, a correct application requires a great amount of time dedicated to management. In the figure below, the complete support cycle is shown.

For the RCM implementation the suppliers need to provide a list including the reliability and maintainability data for the whole equipment and main components subject to repair tasks as well as their expected service life and the following three important mean times: between failures (MTBF), to repair (MTTR) and between overhauls (MTBO) (Fig. 2).

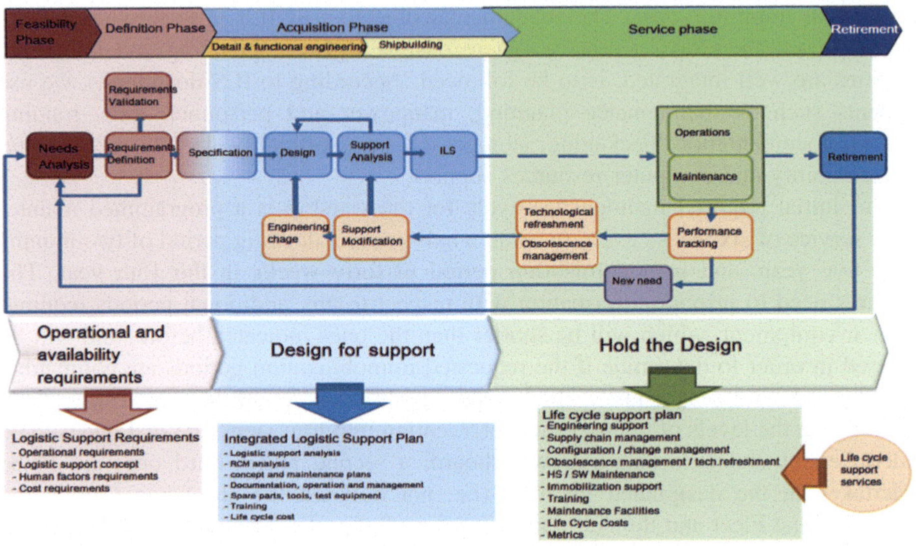

Fig. 2. Logistic support cycle. (Source: Armada)

2 Auxiliary Industry

With reference to the maintenance of ships, we usually consider four types: corrective, preventive, predictive and RCM, which are differentiated by the objectives to be achieved, the available resources and/or the type of need to attend. The appropriate combination of these methodologies will allow the optimization of resources [4]. However, RCM may be considered as the application of all of them, since it assigns to each system whatever is considered most convenient (from the on board first step of maintenance to the last step).

The auxiliary industry is the center around which all the naval/marine support and its associated maintenance gravitates. And the two pillars that we would highlight are, the required NATO standardization of products, such as the use of standardized DIN valves, which allows to ease the support to the life cycle of the ship in terms of maintenance tasks and spare parts (a National Stock Number - NSN - is allocated to these spares by means of the NATO codification system). And the automation and monitoring of the equipment and systems' operation, since having information about the condition of the installed equipment helps to achieve maintenance based on the condition as well as predictive and preventive maintenance. In order to detect and prevent failures, it is really necessary to have a good monitoring tool. Monitoring systems are responsible for controlling the technology used by an organization such as hardware, networks, operating systems or applications, in order to analyze their operation and performance, and to alert or detect about possible problems. A good monitoring system allows to monitor devices, infrastructures and applications. It prevents incidents and if these incidents occur, they are detected quicker, which saves money and time.

Monitoring is a huge and complex activity, which is not a fixed concept, but it depends on, as we advised, the needs of each organization. But monitoring systems have normally a number of common features such as system alerts, notifications by different means (so the message is received by the right people), graphic visualization, analysis in real time, production of reports, record available, possibility of installing plug-ins or distinction by user type (Fig. 3).

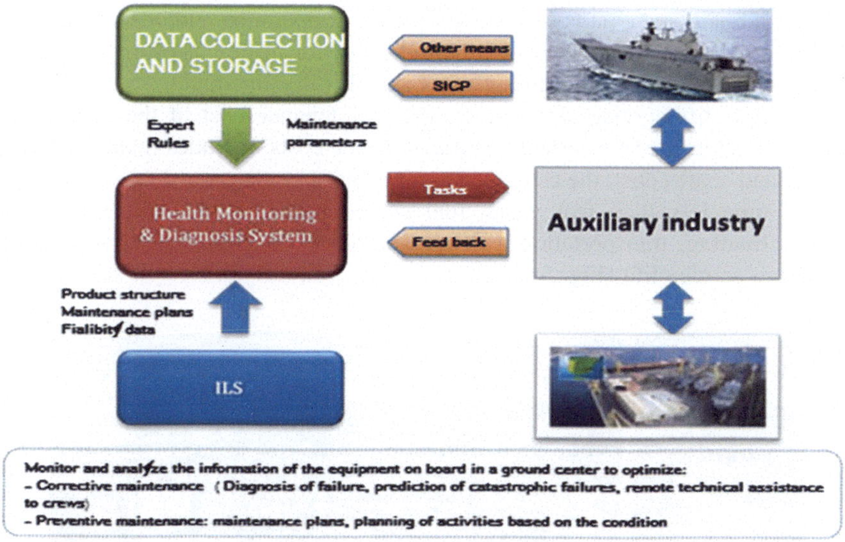

Fig. 3. Maintenance plans. (Source: author)

Monitoring is the first brick in the 4.0 digital processes and R & D activities are the tool for the naval auxiliary industry to achieve the desired digitization objectives. The required data provided by the suppliers are the basis of the optimization of the maintenance plans in the GALIA application. Whole Life Cost (WLC) includes the identification of costs but not limited to the initial acquisition and development costs but also the costs related to operation, support and disposal through the entire life of the system or equipment. Therefore, the supplier will understand that WLC is one of the major parameters in the design and selection of any equipment or systems under consideration (Fig. 4).

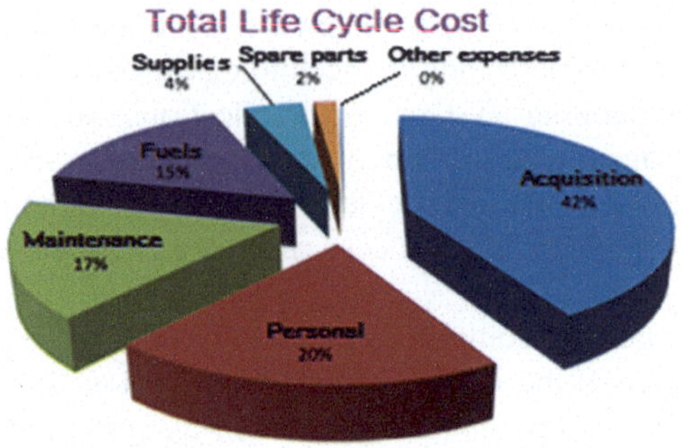

Fig. 4. Estimated warship lifecycle costs. (Source: author)

The supplier must supply the WLC data for the system and equipment as procurement cost of equipment (which includes: installation, set up, commissioning, trials, etc.), procurement cost to cover the maintenance requirements for the life of equipment, the cost of on board spare parts to allow sixty days of operation and onshore spare parts for the first year of maintenance as required, the cost for insurance of major spare parts and rotary pool spares, initial training costs and ILS documentation.

For a ship's life service of 35 years, the maintenance cost of the equipment, as preventive, corrective and programmed maintenance costs, includes maintenance labor man hours and spares cost, the estimated equipment upgrade costs due obsolescence in periods of time less than five years, fuel consumption and lube oil consumption, continuous training, transportation and handling costs, electrical power consumption and optimum service life (Fig. 5).

Fig. 5. Auxiliary naval industry. (Source: Fernández Jove)

Suppliers provide a complete set of data, equipment breakdown and spare parts with the necessary information as element description, units, technical and physical characteristics, part number, supplier reference number, drawings, recommended on board spares quantity, NATO stock number, and different codes (delivery delay, expire date, order…), dimensions, weight package, unit price value, remarks, etc.

3 SICP (IPMS)

The Integrated Platform Management System (SICP/IPMS) is responsible for monitoring sensors and actuators of the systems onboard the platform. It consists of a centralized control that allows to monitor the operation of equipment. Since the former implementations, the SICP has been in continuous evolution in several directions; it has incorporated the most up to date technology available in the market at every moment, it has increased and upgraded the capabilities in order to cope with the increasingly number of signals provided by the equipment and systems on board, it has improved its internal performance in terms of speed of response, quality of presentation of information, ability to process signals, etc., it has integrated new applications that ease the operation of the ship and also it has provided tools that support new concepts of sustainability.

As an example, the SICP of the multi-purpose amphibious assault ship-aircraft carrier *Juan Carlos I* broke records since it implements the management of more than 45,000 signals in real time. For the acquisition, processing, distribution and presentation of all this information, the SICP includes data acquisition units, local substations based on programmable automata and operator positions. All the information is distributed through a redundant fiber optic Giga Ethernet network and finally the data is transferred by satellite to CESADAR (Fig. 6).

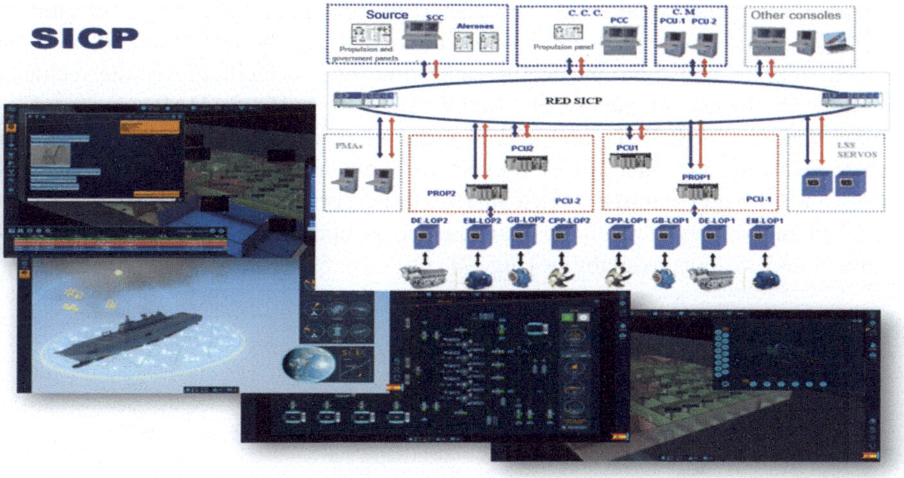

Fig. 6. SICP properties. (Source: Navantia)

The management of all this information is transformed into "knowledge" of the real condition of the ship, which allows the operators to execute the necessary control actions to operate of the ship as appropriate. The SICP concept is developed around the paradigm of knowledge management and information systems. The SICP supervises and controls the propulsion, the power generation and electrical distribution and the operation of all the auxiliaries and breakdowns of systems on board. In addition to the standard functions of fire detection, flooding detection, NBC pollution, firefighting and ballast control, HVAC control, monitoring and control of access and sealing and incident tracking, the SICP damage management module incorporates a series of tools and functions which help to improve the accomplishment of the ship's mission, such as a vessel stability calculator both under normal and damage conditions, a 3D map which shows the status of the alarms, dynamic cards of immediate action that allows, with a single action of an operator from a console, to execute all the recommended actions to fight against incidents such as fire, flood, NBC pollution, etc.

4 CESADAR

Considering the *Armada*'s size, it must remain at the spearhead of the available technologies to provide the necessary superiority in combat, and this includes the digital technologies as the 4.0 heart. We can summarize the interests of the *Armada* with the following ideas: integration of all ships in a network "*Fleet 4.0*", monitoring of the onshore systems in order to optimize their predictive maintenance, remote control and maintenance, cyberphysical systems, analysis of the data generated on board, automated vessel that allows a crew reduction, works on board with augmented reality and production of spare parts with 3D printing technology.

The platform of the naval ships has been strengthened in terms of operations and the required maintenance activities, the diagnostic centers and the use of information technologies. This has been made possible due to having the operational data of the deployed ships in real time and at the same time due to facilitate remote technical assistance. The Center of Supervision and Data Analysis (CESADAR), is where the recorded data at the SICP is analyzed in order to reduce maintenance costs and to optimize the tasks and intervals of preventive maintenance. From the data analysis, ships are accordingly advised at sea in real time, corrective maintenance tasks are undertaken in port and preventive maintenance is updated in order to upgrade the equipment and systems availability (Fig. 7).

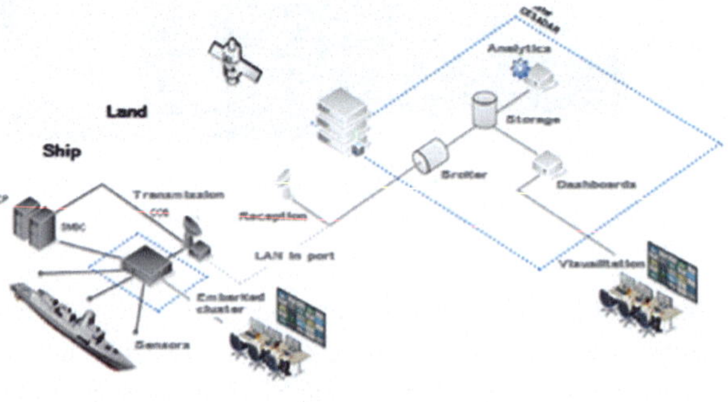

Fig. 7. CESADAR information data transfer. (Source: Armada)

The operational data of the critical systems, such as the main propulsion engines/motors and/or the gas turbines, is saved in the SICP and transmitted via satellite when the ships are at sea. In addition to the dynamic data, the CESADAR receives data related to the sea water, hydraulic fluids etc. This information is integrated into the maintenance database and the analyses are the guide for their predictive maintenance. This organization (CESADAR) comprises six centers, a central one and five local branches. More than thirty ships are covered, comprising Frigates, Mine-hunters, Hydrographics, Amphibians, Patrol vessels, supply ships or LHD. Due to their own idiosyncrasy as warships, the operation, security of information and control systems are critical factors to support the decision. A cybersecurity effort is also considered, which focuses not only on classified systems and information, but also has a strong dependence in all areas: operations, logistics, platform control, combat systems, navigation, etc.

The technologies which are included in the 4.0 industrial revolution have been described in an article published in the Harvard Business Review by Porter et al. [4]: "Smart, connected products require companies to build and support an entirely new technology infrastructure. This technology stack is made up of multiple layers, including new product hardware, embedded software, connectivity, a product cloud consisting of software running on remote servers, a suite of security tools, a gateway for external information sources, and integration with enterprise business systems". From the authors [3], data that has been appearing since the first phases of design. Around this database will appear the applications that control the components life cycle. The key is that all devices, sensors and components are directly connected to the IoT network.

The core of a 4.0 warship is the digital twin, a digital representation of a ship with associated processes and systems, based on continuous data collection. So, these processes are presented digitally. Instead of the arrangement of a physical test cycle, the processes can be tested and validated easily and quickly. It seems that some areas for potential improvements that can be calculated by a digital twin are tasks as efficient thrusters or bulb, hull performance, propeller cleaning, the optimal operation of machinery systems, verification of ship performance, visualization of the changes effect, etc. We highlight the importance of augmented reality within the first maintenance step. The virtual and augmented reality programs in the shipbuilding environment currently offer an intuitive user interface that incorporates navigation tools and capabilities to facilitate navigating the naval/marine designer (Fig. 8).

"The line between disorder and order lies in logistics..." (Sun Tzu)

Fig. 8. Shipping 4.0. (Source: Haugesundkonferansen)

5 Conclusions

- Ship lifecycle efficiency optimization is a holistic process of identifying and planning value-generating activities to optimize ship performance and lower costs over the warship lifecycle. It starts in the construction design or analyzing the true vessel condition. Based in these analysis, operation and maintenance can be optimized.
- The auxiliary naval industries are the necessary tool for the Navies in order to accomplish the required level of maintenance. They must provide a level of digitization, enough for the Navies to meet their 4.0 expectations.
- R & D activities are the tool for the auxiliary naval/marine industry in order to achieve the desired digitization objectives.
- The actual capacities of the on board SICPs and the analysis capability of the CESADAR are the other two pillars to be integrated with the auxiliary naval/marine industries in order to achieve the required optimization of costs reduction in the processes of the construction and maintenance plans.

References

1. Española, A.: Mantenimiento basado en fiabilidad en la Armada. Volumen I y II. Servicio de publicaciones de la Armada (2006)
2. DoD Directive 5000.01, The Defense Acquisition System
3. Muñoz, J.A., Pérez, R.: La integración de la inteligencia artificial en los procesos y metodología de diseño naval. 57 Congreso de Ingeniería Naval e Industria Marítima, Valencia (2018)

4. Porter, M.E., et al.: How smart, connected products are transforming competition. Harvard Bus. Rev. **92**, 64–88 (2014)
5. US Navy. SECNAVINST 4105.1C. Independent logistics assessment and certification requirements (2012)
6. US Navy. Reliability Centered Maintenance (RCM) Handbook (2007)

Environmental Impact of Ship Emissions Based on AIS Big Data for the Port of Rio de Janeiro

Maricruz A. Fun Sang Cepeda(✉), Gabriel P. Monteiro, João V. M. de Oliveira Moita, Luiz Felipe Assis, and Jean-David Caprace

COPPE, Federal University of Rio de Janeiro, UFRJ, Rio de Janeiro, Brazil
{maricruzcepeda, joaov, felipe, jdcaprace}@oceanica.ufrj.br, gabriel_premolimonteiro@poli.ufrj.br

Abstract. Automatic Identification System (AIS) data records a huge quantity of information regarding the safety and security of ships and port facilities in the international maritime transport sector. However, this big database is not only useful for the security of ships operations and port facilities. It can also be helpful for other important functions in maritime traffic such as reducing environmental impacts, improve the logistics and analyses compliance with current International Maritime Organization (IMO) regulations. This study develops an analytical approach to quantify the impacts of ship emissions in the Guanabara Bay of Rio de Janeiro (Brazil) using AIS database as well as life cycle assessment (LCA) tool. The paper describes a method in two steps. First, the inventory of ship emissions is evaluated and geolocated with AIS data through the assessment of fuel consumption calculated for each individual vessel. Then, the impact of the emissions is assessed with the ReCiPe LCA method that translates emissions into a limited number of environmental impact scores by means of so-called characterization factors. The results show that the proposed methodology is efficient to estimate the environmental impact of ship emissions over the Rio de Janeiro Port area. We suggest that quantifying the number of emissions from ships in order to fulfil IMO regulations and reduce the health impacts of people who are living in surrounding areas of high maritime traffic is important for decision makers and for the maritime authorities to improve their strategies.

Keywords: Ship emissions · Environmental impact · Automatic Identification System · Bigdata · Marine traffic · ReCiPe method · LCIA · LCA

1 Introduction

The emergence of Big Data Analytics due to the data availability, storage capacity and increasing capability have become tools for to stakeholders driver in mitigating future uncertainty in all type industries [1]. Due to that, Big Data is used in countless applications.

V. A. J. E. Carreño Moreno et al. (Eds.): CIDIN 2019/COPINAVAL 2019, pp. 402–412, 2020.
https://doi.org/10.1007/978-3-030-35963-8_34

In the marine industry, Big Data is used in some studies related to smart ports, shipping, traffic, operations, hydrodynamic, emissions, etc. [2], and [3].

Moreover, Automatic Identification System (AIS) is a mandatory automatic tracking system used on ships and by vessel traffic services (VTS) for identifying and locating vessels by electronically exchanging data with other nearby vessels throughout the world. AIS system required to be installed on ships by the International Maritime Organization (IMO) and the Maritime Safety Administration in several countries [4]. With the popularization of AIS technology and massive navigation, logs recording ship activities can be collected [5].

The data collection allowed to develop numerous studies focusing on the analysis of ship traffic based on AIS. Although, only a smaller portion of monitorable operational data available is used for different purposes than maritime traffic [2].

In a previous study, Cepeda et al. reviewed 41 papers about emissions calculations since 2009. It is observed that in most of the cases the calculation of emissions is focusing ports in Europe and Asia [6]. Figure 1 presents a stacked bar chart of the number of papers encountered in the literature where the AIS data are used to perform predictions of ship marine traffic emissions. The quantity of papers is plotted per years and the color of the stacked bars refers to if whether the authors of the papers carried out social cost impacts. It appears that the authors of the papers do not assessed the environmental impact and the life cycle assessment in the reviewed studies. In addition, the maximum occurrence of the papers corresponds to the last 5 years.

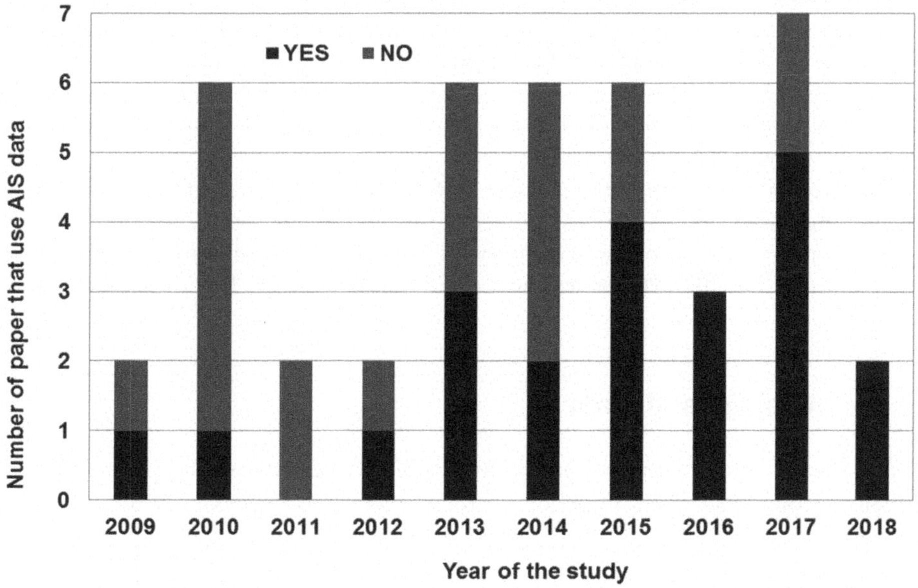

Fig. 1. The number of publications using AIS data for air emissions calculation.

LCA is an environmental management tool that helps to evaluate the effects of a product on the environment over the entire period of its life [7]. Emissions and consumption of resources can be evaluated at every stage of the life cycle. The emergence of Big Data Analytics is a tool that helps us to better understanding the LCA.

According to the literature review, the evaluation of the environmental impact in commercial vessels has been analyzed in some case studies.

Strazza et al. evaluates the use of methanol in Solid Oxide Fuel Cells (SOFC), as auxiliary power systems for commercial vessels, through LCA and the comparison with the conventional auxiliary power system [7].

Chatzinikolaou and Ventikos present a case study on a Panamax oil tanker, life cycle emission inventory has been developed using the LCA framework and results are presented and discussed accordingly [8].

Another study that uses LCA in the analysis of ship hull maintenance strategies is the paper by Want et al. which can provide information to shipyards and owners to facilitate reliable long-term maintenance decisions and demonstrate LCA as an appropriate tool for that purpose [9].

In shipping, the LCA for the quantification of the environmental impacts is applicable in all the life phases of the ship, including in the final phase called scrapping. Ko and Gantner present distribution of added value per impact on the environment over the life cycle of a ship, dividing this into three stages: production, operation and end of life. An important conclusion of their study is that most of the economic benefit is gained during the use phase and remains in Europe, which reflects the purpose of the existence of the ship. Due to the long use of the ship, most of the environmental impact is also generated in the operational phase but is emitted worldwide, reflecting again the importance of studying emissions from ships [10].

The use of LCA is not common in emission studies and that it has been studied in the maritime area in isolation. On the other hand, the use of bigdata in conjunction with LCA has not been done.

Therefore, it is appropriate to add one-step to the study of the emissions inventory, which is the evaluation of the environmental impact through the LCA.

The purpose of the paper is to quantify the impacts of ship emissions using the AIS database as well as life cycle assessment (LCA) tool inside the Bay of Rio de Janeiro (Brazil).

2 Methodology and Data

The degree of detail of the bottom-up methodology used allows the presentation of emissions disaggregated by ships, by operation modes, as explained in the following section.

2.1 Case Study

The area of the study is an oceanic bay called Guanabara Bay (GB), located on Southeast Brazil in the state of Rio de Janeiro, between 22°40′S and 23°00′S latitude and between 043°00′W and 043°18′W longitude. The port of Rio de Janeiro is the second largest in the country after the port of Santos. The bay has an area of approximately 384 km^2, including islands. On its western shore lies the city of Rio de Janeiro and fifteen other municipalities, with a population of 12 million people in 2017 based on the Brazilian Institute of Geography and Statistics data (IBGE).

The data source used comes from Automatic Identification System (AIS) data and from the other databases as Marine traffic or IHS Maritime.

AIS is a mandatory collision avoidance system required to be installed on ships by the International Maritime Organization (IMO) and the Maritime Safety Administration in several countries to locate most of the vessels throughout the world.

International voyaging ships with a Gross Tonnage (GT) of 300 or more, passenger ships of all sizes, domestic vessels with a GT of 200 or more travelling in coastal waters, and inland ships with a GT of 100 or more, are all required to be equipped with AIS. This type of vessels is classified in class A. Class B vessels are the small ships that are not required to comply with SOLAS conventions (recreational vessels, fishing vessels of less than 15 m, etc.), but these voluntarily adapt to AIS system [4].

The base station of our research group is named UFRJ-COPPE located at Technology Centre of Federal University of Rio de Janeiro [6]. The main table in the database contains 196 different fields extracted from the messages. The area covered is about 2091 km^2, the average range of the signal reception is 7.21 NM with a maximum of 56.23 NM, the average of records received in one hour are 32 distinct vessels, and 15958 positions, see Fig. 2.

The AIS data provides efficiency to assist ships in collisions avoidance as well as ports and maritime authorities in traffic monitoring ensuring better surveillance of the sea. However, it also has other potential utilization, such as the estimation of emissions with real data.

In this study, four months of AIS data, from January until April 2018, were utilized to estimate the emission inventory over the area of the study. During the time period of the research 2681 vessels were registered, of which 1432 are those that are studied. The vessels that presented less than 500 AIS position reports in the DB has been disregarded in this study, which corresponds to 47% of vessels discarded after data treatment.

The distribution of the type of ships of the 1432 vessels studied is 48% cargo vessels, 23% other vessels, 12% tankers, 6% cargo vessels hazardous category A, 5% tugs, 2% passenger and military vessels, and 1% fishing and sailing vessels, see Fig. 3.

In a previous study the database was using the first two months of 2018 and the environmental impact study was omitted [6]. The current study doubles the number of records analyzed in months and adds the quantification of environmental impact.

The environmental impacts and related aspects were evaluated through life cycle assessment (LCA): a quantitative method to assess the environmental performance throughout the entire life cycle of a product. In practical terms, the life cycle of a ship is composed of ship production, ship operation, and ship end of life [10]. In this study, the evaluation of environmental impact is in the operational phase of the cycle.

The ReCiPe method has been used including hierarchic, egalitarian and individualist perspectives, with the aim of identifying potential distortions in the LCA results that may ultimately affect the evaluation of the environmental impact.

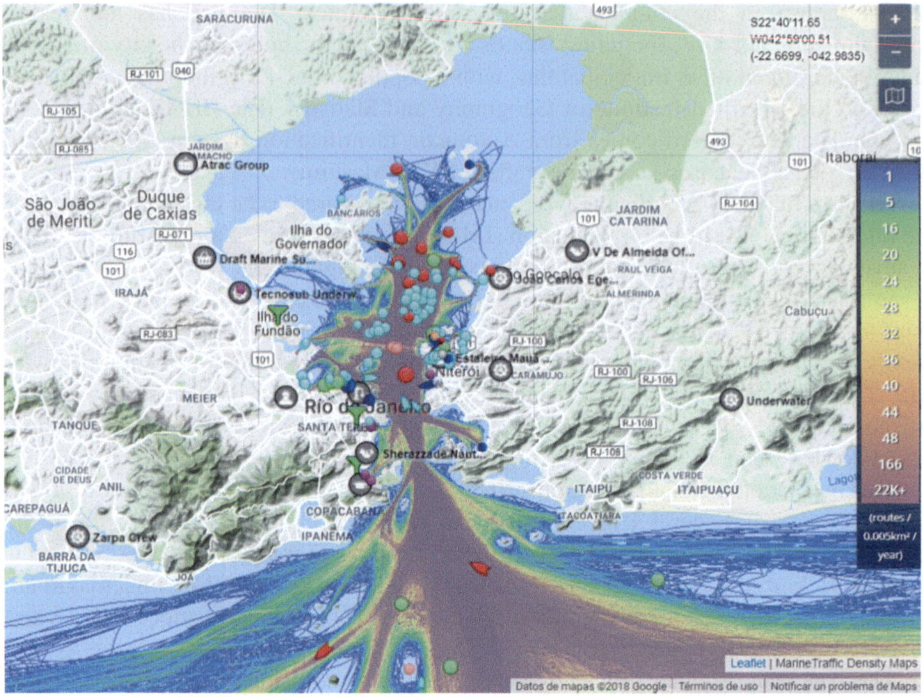

Fig. 2. View of Guanabara Bay included the Rio de Janeiro Port with AIS data and the base stations, source: Marine Traffic December-2018

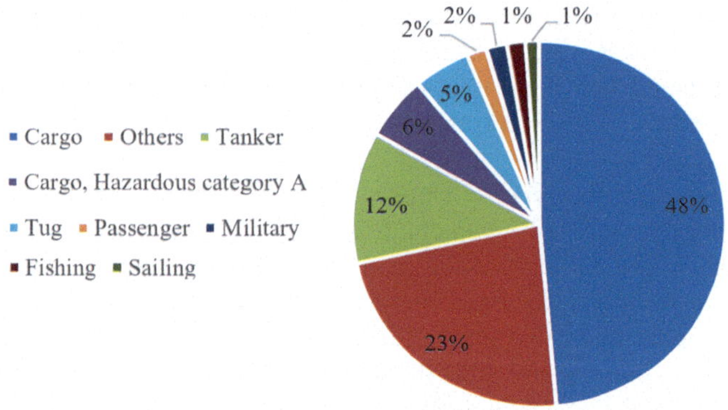

Fig. 3. Distribution of the type of ships studied

2.2 Model

The workflow of the method developed in this paper consists of the following steps: data collection, data store, data treatment, estimation of ship emissions, application of ReCiPe 2008 method, analysis of results.

The quantification of ship CO_2, SO_2, NO_X, PM_{10}, and $PM_{2.5}$ emissions have been calculated between two report positions of a vessel based on Eq. 1 developed by Goldsworthy and Goldsworthy [11] and [12]. This formulation depends mainly on the installed power of the ship engines, type of the fuel used as well as of the load factor of the engine. However, these data are not provided by the AIS data. It is acquired from other databases or regressions calculated for each type of ship [6].

$$E_{i,j,k,l} = P_j \times LF_{j,l} \times T_{j,k.l} \times \frac{EF_{i,j,k}}{10^6} \tag{1}$$

$$\boldsymbol{LF_{j,l} = (AS/MS)^3} \tag{2}$$

The Total emission of pollute i from engine j using fuel type k during operation mode l (tons) is $E_{i,j,k,l}$. P_j is the installed power for engines j (kW); $LF_{j,l}$ is the Load factor for engine j during operation mode l (%); $T_{j,k.l}$ is the operating time for engine type j, using fuel type k during operation mode l (h); AS is the Actual Speed (knots); and MS is the Maximum Speed (knots).

The emissions factors used in this study are taken from Fan et al. [14] considering the machine type as ME and the oil type as RO for all the ships. For each engine, the corresponding emission factors were applied, as described in Table 1. The auxiliary engines used for generating energy onboard has been disregarded in the present study due to the difficulty to obtain the correct installed power of this type of equipment.

Table 1. Emission factors (CO_2, SO_2 and NO_X, PM_{10}, and $PM_{2.5}$) for pollute and fuel type for each engine type (g/kWh) [13] and [14].

Machine type	Engine type	Oil type	CO_2	SO_2	NO_X	PM_{10}	$PM_{2.5}$
ME	Slow Speed Diesel (SSD)	Residual Oil (RO)	622	10.30	18.10	1.378	1.22
ME	Medium Speed Diesel (MSD)	Residual Oil (RO)	686	11.31	14.00	1.193	1.22
ME	High Speed Diesel (HSD)	Residual Oil (RO)	686	11.31	12.7	0.65	0.50

LCA is classified in different levels, one such example is the development of midpoint-oriented and endpoint-oriented methods for life cycle impact assessment (LCIA). The method ReCiPe 2008 for LCIA has been used. ReCiPe provides a recipe to calculate life cycle impact category indicators. The acronym also represents the initials of the institutes that were the main contributors to this project. ReCiPe 2008 comprises two sets of impact categories with associated sets of characterization factors. Eighteen impact categories are addressed at the midpoint level, and at the endpoint level, most of these midpoint impact categories are further converted and aggregated into the following three endpoint categories. Figure 4 sketches the relations between the life cycle impact (LCI) parameter (left), midpoint indicator (middle) and endpoint indicator (right) [15].

CO_2 is in the climate change (CC) midpoint level, SO_2 and NO_X are in the terrestrial acidification (TA) midpoint level, PM_{10}, and $PM_{2.5}$ are in particulate matter formation (PMF) midpoint level. At the endpoint level, most of these midpoint impact categories are further converted and aggregated into the following two endpoint categories: damage to human health (HH) and damage to ecosystem diversity (ED).

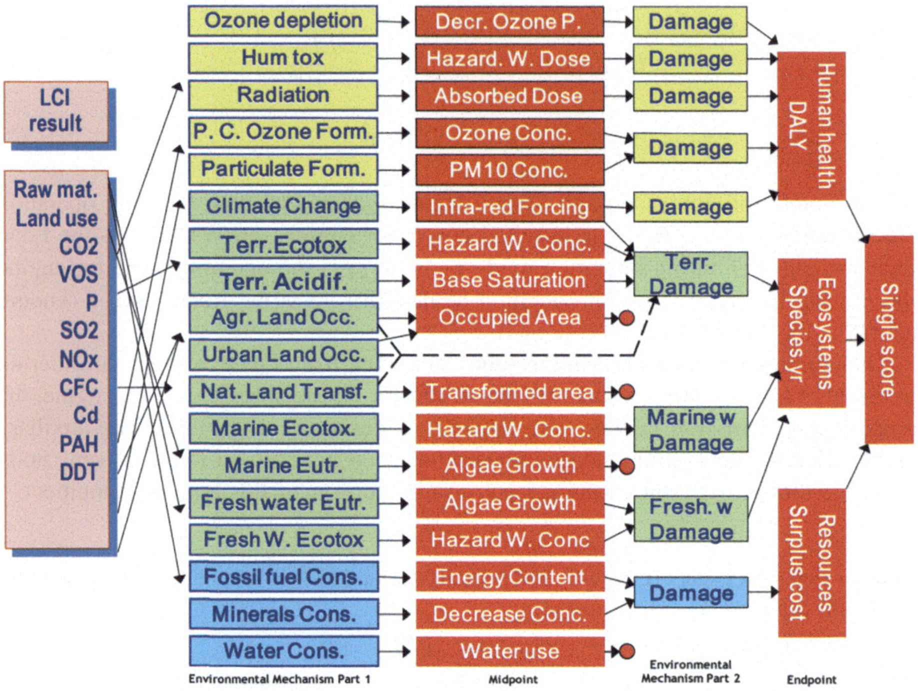

Fig. 4. Relationship between LCI in the different level of impact (environmental impact, midpoint and endpoint) [15].

ReCiPe presents three value perspectives: individualist (I), hierarchist (H), and egalitarian (E), which differs from each other in terms of magnitude and time scales of environmental issues. Individualist (I) perspective assumes a short-term interest (e.g. for climate change the time horizon is 20 years) and optimistic technological breakthroughs in the future. Hierarchist (H) perspective assumes an intermediate time frame (e.g. 100 years' time horizon for climate change) and common policy principles. Egalitarian (E) perspective is more precautionary as it assumes the longest period (e.g. 500 years' time horizon for Climate change) and pessimistic development scenarios [15].

3 Results

Total estimated emissions from ships for January until April 2018 as well as an estimation of the annual average are presented in Table 2. The CO_2 emissions are the most important with over than 43457 tons per year followed by NOx and SO_2 emissions. The AIS data allowed plotting a high-resolution geographical characterization of emissions.

Table 2. Total of emission due to marine traffic in Rio de Janeiro in tons, comparing the estimations by the previous study [6] with the actual results

Period	CO_2	SO_2	NO_X	PM_{10}	$PM_{2.5}$
January until April 2018	14485.7	238.8	291.4	23.1	23.0
Annual average previously study	40208.4	666.0	887.4	77.4	76.8
Actual annual average	43457.1	716.5	874.2	69.4	68.9

Table 2 shows the differences between the previous study about emissions inventory with the actual study. The different is about +8% in CO_2 and SO_2 emissions, about −1.5% in NO_X emissions and about −10.3% in PM_{10} and $PM_{2.5}$ emissions.

The heat maps of the quantitative assessment of the emissions are illustrated in Fig. 5. These maps have been constructed using Google maps API through a solution provided by Raffael Vogler in (www.joyofdata.de). The API calculates the heat map based on the contribution of each point in 50 pixels of distance. The map's maximum intensity is fixed at 2.5 t and is represented by the red color. Color gradient follows the default order: light green, yellow, orange and red, representing roughly 25%, 50%, 75% and 100% or more of the maximum intensity. The peak of the emissions is observed in the south part of the bridge between Rio de Janeiro downtown center and Niteroi municipalities.

The assessment of the emission impacts on the population of the surrounding municipalities is in of the scope of this study. The LCIA is present in Table 3.

The LCIA was calculated using the factors of characterization and normalization available in the LCA-ReCiPe website (rivm.nl/en/life-cycle-assessment-lca/recipe), based on the total emissions estimated.

Of all the 18 mid-points, only 6 were directly affect by the considered pollutants, and only 2 of the 3 end-points. Each pollute contribute in at least one of these mid-points and were converted to the equivalent unit when necessary, following the ReCiPe's factors.

The resultant impact in Human health is measured in DALY (Disability-Adjusted Life Year), which is a measure of overall disease burden, expressed as the cumulative number of years lost due to ill-health, disability or early death. In short, the DALY sum up the years lived with disability and the years of life lost. Moreover, the DALY is an established term in medical world. The resultant impact in ecosystem diversity, it's on the other hand measured in species loss per year.

Table 3. LCIA of actual annual average emission due to marine traffic in Rio de Janeiro

Endpoint	Midpoint	Individualist	Hierarchist	Egalitarian	Unit
Human health	Global Warming	1,2	13,4	181,1	DALY
	Fine particulate matter formation	14,4	78,2	78,2	
	Photochemical ozone formation	0,3	0,3	0,3	
Ecosystem diversity	Global Warming	7,7E−03	4,1E−02	3,6E−01	Species lost per year
	Photochemical ozone formation	3,77E–02	3,77E–02	3,77E–02	
	Acidification	7,29E−02	7,29E−02	7,29E−02	

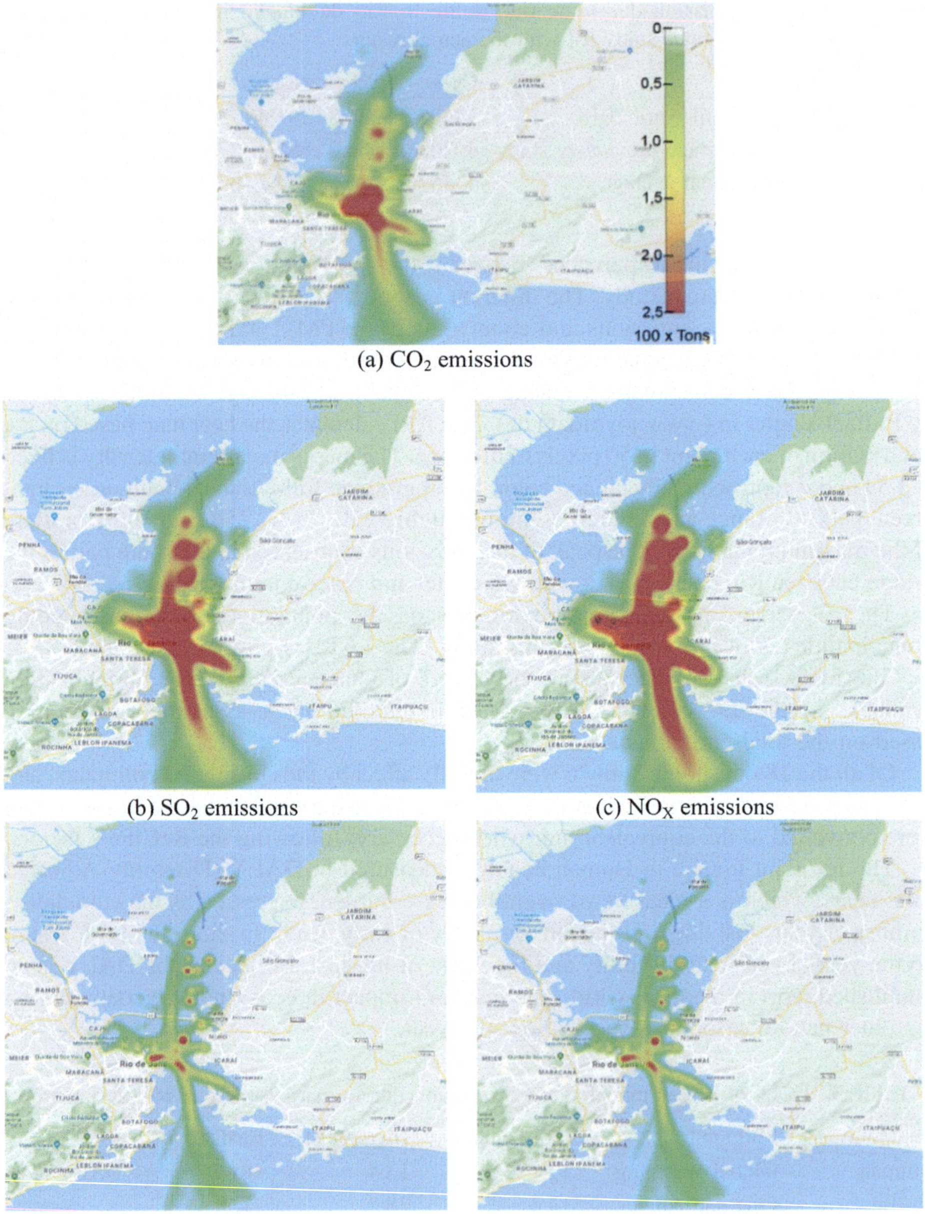

(a) CO_2 emissions

(b) SO_2 emissions

(c) NO_X emissions

(d) PM_{10} emissions

(e) $PM_{2.5}$ emissions

Fig. 5. Distribution of emissions in tons per year around Rio de Janeiro (Guanabara Bay)

4 Conclusions and Recommendations

The Rio de Janeiro Guanabara Bay, one of the busiest ports of Brazil, has a great environmental and socio-economic importance for the region of the study. Its current state of environmental degradation including by GHG emissions pose risks to the human populations of its surroundings, who use its waters for pleasure, transportation, or for their livelihood.

This study focusses on the assessment of the emissions due to marine traffic base on 4 months AIS data (January until April 2018). The major findings of this study, which is the first ship emission inventory for this zone, may be summarized as follows: Total estimated emissions from ships for January to April are 14485.7 tons of CO_2, 238.8 tons of SO_2, 291.4 tons of NO_X, 23.1 tons of PM_{10} and 23.0 tons $PM_{2.5}$.

Continuously storing AIS data will allows us soon to better understand the distribution of ship emissions in the Rio the Janeiro Bay. The main differences between the previous study about emissions inventory with the actual study is in CO_2 emissions, and PM_{10} emissions; therefore, it is necessary to take more attention, when an inventory of this type is carried out, in the CO_2 emissions that are the ones of greater relevance.

The implementation of estimation of life cycle impact assessment is an accomplished achievement quantifying the environmental impact and damages to human health (HH) as well as damages to ecosystem diversity (ED).

The biggest impact is in HH endpoint in Global Warming and Fine particulate matter formation, the first one is due to the CO_2 emissions, and the last one is due to the $PM_{2.5}$ emissions. Therefore, these two emissions should be prioritized in the future.

The present study shows the impacts of Marine emissions in the port of Rio de Janeiro through the Disability-Adjusted Life Years (DALY) reaching respectively 1.2 (I), 13.4 (H) and 181.1 (E) for Global Warming and 14.4 (I), 78.2 (H) and 78.2 (E) for fine particulate matter formation.

However, special attention should be paid to the construction of consistent databases about the ship engines installed power for both main propulsion and auxiliary power units.

Acknowledgements. This study was financed in part by the Coordenação de Aperfeiçoamento de Pessoal de Nível Superior - Brasil (CAPES) - Finance Code 001.

References

1. Keane, A., Brett, P.O., Ebrahimi, A., Gaspar, H.M., Garcia Agis, J.J.: Preparing for a digital future - experiences and implications from a maritime domain perspective. In: Proceedings of 16th Conference on Computer and IT Applications in the Maritime Industries, COMPIT 2017, Cardiff (2017)
2. Abbasian, N.S., Salajegheh, A., Gaspar, H., Brett, P.O.: Improving early OSV design robustness by applying 'Multivariate Big Data Analytics' on a ship's life cycle. J. Ind. Inf. Integr. **10**, 29–38 (2018)

3. Bons, A., van Wirdum, M.: Big data and (Inland) shipping: a sensible contribution to a strong future. In: Proceedings of 15th International Conference on Computer and IT Applications in the Maritime Industries, COMPIT 2016, Lecce (2016)
4. IMO: Guidelines for the Installation of a Shipborne Automatic Identification System (AIS). IMO, London (2003)
5. Huang, L., Weng, Y., Geng, X., Zhou, C., Xiao, C., Zhang, F.: Estimation and spatio-temporal analysis of ship exhaust emission in a port area. Ocean Eng. **140**, 401–411 (2017)
6. Cepeda, M.F.S., Monteiro, G.P., de Oliveira Moita, J.V., Caprace, J.D.: Estimating ship emissions based on AIS big data for the port of Rio de Janeiro. In: Proceedings of 17th Conference on Computer Applications and Information Technology in the Maritime Industries (COMPIT), Pavone (2018)
7. Strazza, C., Borghi, A.D., Costamagna, P., Traverso, A., Santin, M.: Comparative LCA of methanol-fuelled SOFCs as auxiliary power systems on-board ships. Appl. Energy **87**, 1670–1678 (2010)
8. Chatzinikolaou, S.D., Ventikos, N.P.: Holistic framework for studying ship air emissions in a life cycle perspective. Ocean Eng. **110**(Part B), 113–122 (2015)
9. Wang, H., Oguz, E., Jeong, B., Zhou, P.: Life cycle cost and environmental impact analysis of ship hull maintenance strategies for a short route hybrid ferry. Ocean Eng. **161**, 20–28 (2018)
10. Ko, N., Gantner, J.: Local added value and environmental impacts of ship scrapping in the context of a ship's life cycle. Ocean Eng. **121**, 317–321 (2016)
11. Goldsworthy, L., Goldsworthy, B.: Modelling of ship engine exhaust emissions in ports and extensive coastal waters based on terrestrial AIS data–an Australian case study. Environ. Model Softw. **63**, 45–60 (2015)
12. Entec UK Limited: Quantification of emissions from ships associated with ship movements between ports in the European Community (2002)
13. Chen, D., Zhao, Y., Nelson, P., Li, Y., Wang, X., Zhou, Y., Lang, J., Guo, X.: Estimating ship emissions based on AIS data for port of Tianjin, China. Atmos. Environ. **145**, 10–18 (2016)
14. Fan, Q., Zhang, Y., Ma, W., Ma, H., Feng, J., Yu, Q., Yang, X., Ng, S.K.W., Fu, Q., Chen, L.: Spatial and seasonal dynamics of ship emissions over the Yangtze River Delta and East China Sea and their potential environmental influence. Environ. Sci. Technol **50**, 1322–1329 (2016)
15. Goedkoop, M., Heijungs, R., Huijbregts, M., De Schryver, A., Struijs, J., van Zelm, R.: ReCiPe 2008 a life cycle impact assessment method which comprises harmonised category indicators at the midpoint and the endpoint level. Ruimte en Mileieu, Netherlands (2013)

Proactive Maintenance in the Azimuth Marine Propulsion: SCHOTTEL Condition Monitoring Solutions

Edgard Mulford(✉), Julio Carrasquilla, and Gonzalo Moreno

SCHOTTEL de Colombia SAS, Cartagena de Indias, Colombia
{edgard.mulford, julio.carrasquilla, gonzalo.moreno}@schottel.co

Abstract. This article exposes the importance of implementing strategies to improve the reliability of naval propulsion systems, minimizing the possibility of failures through programmed actions that seek to increase the lifetime of the systems, improve their efficiency and operational productivity, reduce time costs of inactivity and prevent costly interventions for corrective actions. The document present from the SCHOTTEL experience, the development of field activities such as the vibrations & oils analysis; electromechanical inspections; computer based tools (SCHOTTEL CONDITION MONITORING, S-COM) for machine condition monitoring for the early detection of faults, an effective diagnosis of the system condition and smart decision making; up to case of study that contemplate the operation of vessels based on reactive maintenance, vessels with preventive strategies and the success obtained by vessels with proactive maintenance programs in Latin America.

Keywords: Diagnosis · Strategy · Maintenance based on machine condition · Proactive maintenance · Propulsion

1 Introduction

Maintenance strategies have a direct impact on the safety, capacity and availability of every vessel. It is a subject of great importance for any ship owner, from commercial organizations to governments. Determining the ideal maintenance strategy is not a trivial exercise. An ideal maintenance strategy is intended to maintain the design purpose, achieve availability objectives, and be affordable and flexible enough to deal with uncertainty. That it be sustainable and backed by good learning from experience and operational assumptions.

In general terms, for some ship owners and/or ship operators, maintenance has historically been considered a financial burden, rather than as a mechanism to deliver a safe, reliable and high quality service. However, the benefits of developing and implementing a robust and systematic maintenance strategy have gained greater acceptance over the years, resulting in reductions in unnecessary downtime, as well as increases in the operational capacity of vessels [1].

V. A. J. E. Carreño Moreno et al. (Eds.): CIDIN 2019/COPINAVAL 2019, pp. 413–423, 2020.
https://doi.org/10.1007/978-3-030-35963-8_35

Although the naval industry is complex, conservative and restricted in relation to the merchant, maritime and commercial industry; the constant search for best practices based on the experience of manufacturers, ship owners and operators; as well as the current technological advances, are leading to the development of increasingly complex diagnostic methods capable of implementing successful maintenance and support strategies, thus allowing optimizing the management of the fleet in terms of costs and reliability of the equipment.

This document will give you an overview of proactive maintenance strategies based on propulsion systems and their benefits.

2 Wear and Fail

Service vessels such as tugboats, pushers, tankers, ferries, OPV (offshore patrol vessel) among others, are designed for long duration; however, many of the operational loads that interfere in the structural and operational components of the system are unique and directly affect the reliability of them. Unfortunately, these loads (see Fig. 1) are not equivalent for each type of propeller since almost all vessels operate differently and even each propeller in the same boat has specific demands.

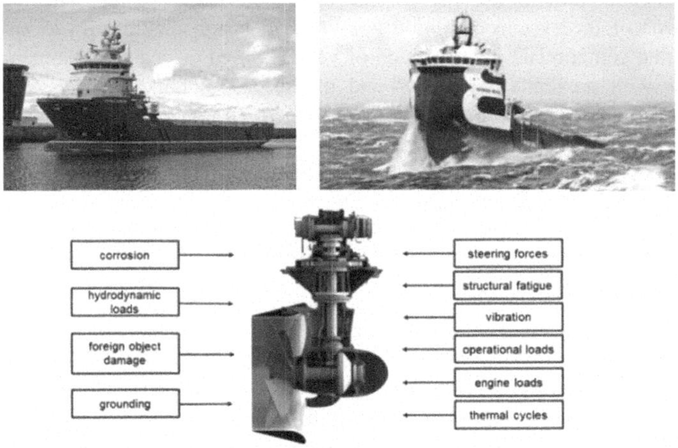

Fig. 1. Different modes of operation (left, PSV Havila Aurora in calm conditions; right, PSV Bourbon Mistral in rough conditions; center, loads in main propulsion) [2].

During its useful life, the thrusters are exposed to a slight and continuous wear from the first start-up after manufacturing and ending in the worst case with total deterioration. Wear is a form of depreciation that occurs even when equipment is used competently, with care and with proper maintenance.

In general, there are different degradation steps that affect the propeller's reliability (see Fig. 2):

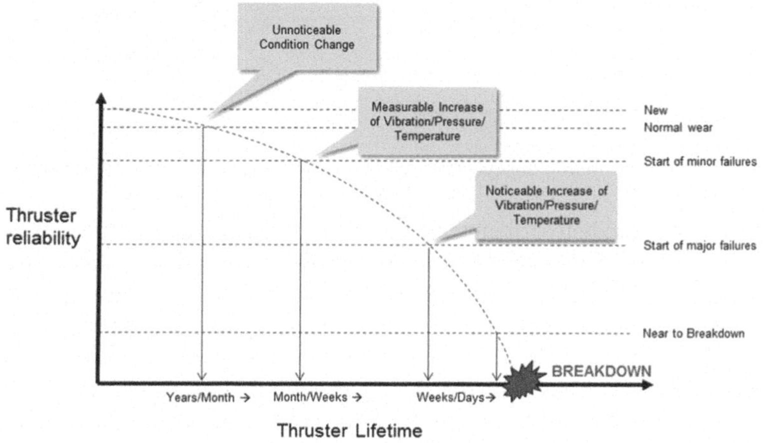

Fig. 2. Relationship of the different degrees of degradation suffered by an azimuth thruster in its useful life in relation to reliability [2].

- When condition changes are imperceptible.
- When there is a measurable increase in conditions (vibrations, temperatures, pressures).
- When there is a noticeable increase in these conditions.

Depending on the intensity of the operation and the maintenance measures taken, the advance of the propeller wear is defined. The slower it degrades, the longer the service life. However, you never know exactly when a mechanical failure will actually occur.

3 Proactive Maintenance

From a general point of view, there are two maintenance approaches. One with reactive approach and the other proactive. The choice of method depends on the particularity of the systems and especially on operating factors directly related to the impact of the mechanical failure. The cost of mitigating the failure must be balanced effectively with the financial, operational and safety costs generated by the failure [1].

Although in practice, in many cases due to culture, budget or ignorance, reactive practices are used, it is experimentally proven that a proactive maintenance plan generates the capacity to prolong the useful life of the units and prevent a collapse complete and unexpected of the system (see Fig. 3).

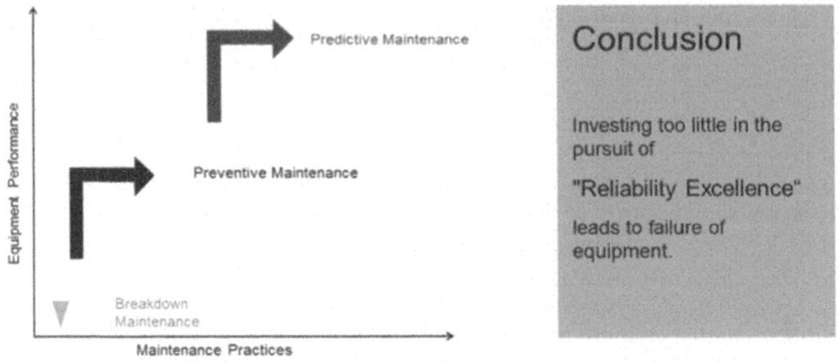

Fig. 3. Proactive maintenance practices based on the performance of the device.

Preventive and predictive maintenance are the basis of current proactive maintenance practices. A proactive approach is characterized according to this methodology:

Detection: the failure is identified when it occurs (or is about to occur). The objective is to determinate quickly the actions to safeguard the operation and avoid further damage. *Forecast:* the fault is predicted within a given period of time. The probability of failure is usually included with this prediction. *Diagnosis:* the defective component and/or root cause of the fault is identified. Then, through empirical, experimental models or computational tools (artificial intelligence and expert systems among others) efficient solutions are generated [3].

A combination of methodologies can allow the optimization of maintenance programs and improve availability by reducing downtime; however, the methodologies used should be subject to continuous or periodic monitoring to ensure that they remain optimized [1].

3.1 Preventive Actions

In naval propulsion systems, normally scheduled maintenance does not really consider the individual operation of the propeller. These are planned based on experiences with existing units or based on similar applications. Depending on the application, oil changes, seals or bearings, are scheduled according to maintenance programs and vary between 5 and 15 years. The objective is to keep the vessel working and usually performing minor actions on a regular basis or when necessary. More complex activities such as the change of a bearing have an even more serious effect from the interruption of work or even the use of dry dock.

3.2 Predictive Actions

During the execution of predictive activities, it is sought to gather as much data as possible of important parameters during similar conditions of operation of the propeller and make a trend. Mainly parameters such as *vibration, temperature, saturation of water in oil, levels and particles in oil, speed and steering angles of the propeller* are recorded.

This data should be analyzed and compared with existing service data. Based on experience and predefined technical limitations. The resulting trends can be used to prevent future failures in the propeller. The trend will also be used to make a forecast based on the actual conditions of the components.

Therefore, surprises can be avoided and if a replacement is needed, a necessary service action can be scheduled at the most convenient time for the operator. Visual and electronic inspection, check of alarms and operating parameters, non-invasive measurements, scheduled tests, performance data are analyzed and reported.

Vibration Analysis. Frequency of individual faults can be defined by measuring the vibration and the analysis of the signal (see Fig. 4).

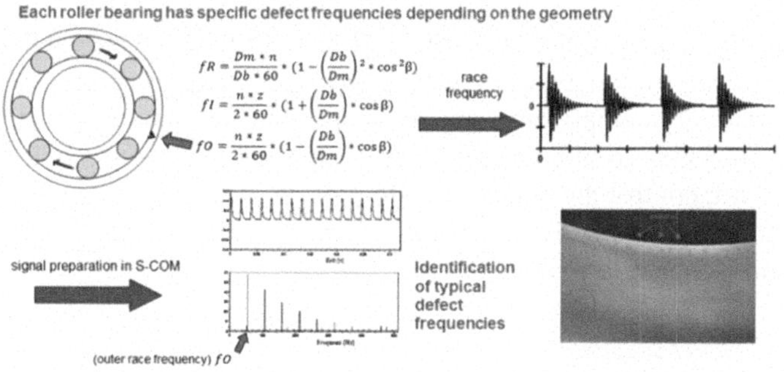

Fig. 4. Detection of bearings defects [3].

Each rotating part, for example the bearings in the unit, creates specific frequencies. Any change in these frequencies can show wear or possible damage, which can be detected at an early stage. By analyzing the frequency, for example, it is possible to identify if a defect in the bearing is in the outer or inner race. These defect frequencies are mainly based on the known geometry of the bearings that are used (ISO 18436-2).

Oil Analysis. The particles in the oil (hydraulic/lubricant), contamination, wear of metals and additives, are always an early indication of failure; or the water content in it, which helps detect possible damage to the seals (ISO 4406-1999).

Treatment and Data Analysis. All measured values are stored in a data logger.

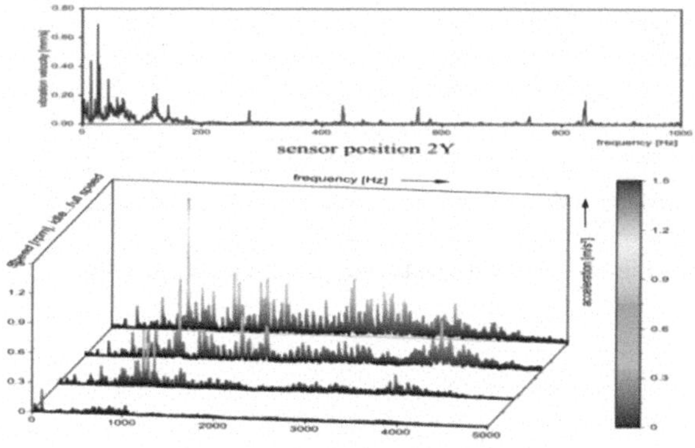

Fig. 5. Spectra of vibration analysis.

The received data is analyzed and compared with a professional spectrum analysis and analysis software as shown in Fig. 5, to identify trends over time.

Report. On a regular basis, the analyst generates a report and highlights unexpected deviations, generates conclusions, status reports, recommendations and maintenance proposals, spare parts and parts.

4 S-COM

Diagnostic systems such as S-COM (SCHOTTEL CONDITION MONITORING)® operate independently of the alarm system of any ship. A series of sensors are installed on the propeller or its components and these measure relevant parameters. They are located in the machine room and are easily accessible to service personnel. The monitoring system, control the actual condition of the propeller, including not only the condition of the oil but also the condition of the bearings in particular. In this way, predictive maintenance is possible (Fig. 6).

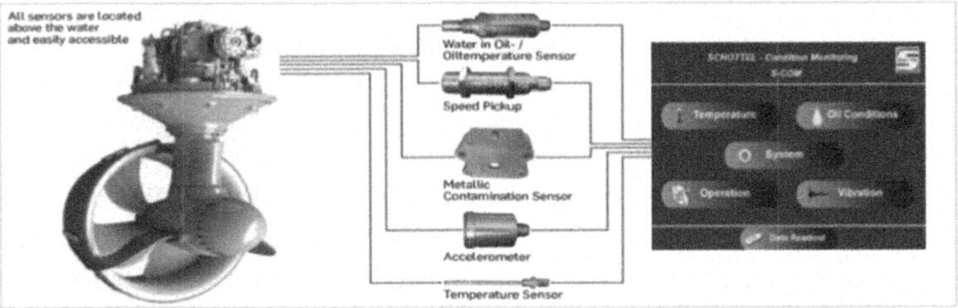

Fig. 6. S-COM (SCHOTTEL Condition Monitoring System)®.

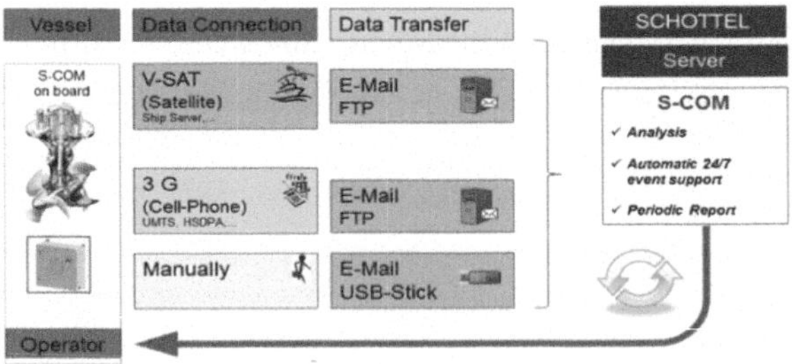

Fig. 7. S-COM (Data analysis and transmission).

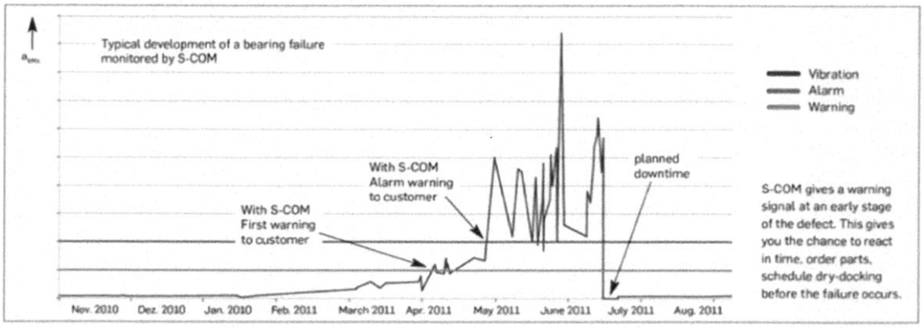

Fig. 8. S-COM (Example of bearing tendency).

Depending on the requirements of the operator, the stored data can be sent manually for validation or through a permanent connection through the Internet, which allows periodic and constant validation (see Fig. 7). For each parameter, different security levels are predefined. A level of "warning" will inform quickly before a defect becomes a problem. This ensures the attention of that peak or specific component and allows a careful follow-up, as can be represented in Fig. 8.

If a negative development of this trend is observed, the level of "alarm" is reached. Although a safe operation of the propeller can be guaranteed. Once the "alarm level" is achieved, the operator receives a notice about the findings and is informed if there are possible quick solutions by itself or not. Additionally, specific maintenance recommendations or appropriate repair tips are defined.

Systems like the S-COM monitor the propellers 24/7 and after the analysis described by the remote diagnosis system, the operator receives: Periodic reports and the status of each propeller (for example, every 3, 6, 12 months); On-order on-demand or automatic reports of events when the level of "alarm" was reached.

5 Trend

5.1 Fleet Management Optimization

Digitization and automation have a massive effect on the maritime industry. The new technologies are committed to significantly improve the operation of the boat in many different ways. The combination of sensor/machine data and the accessibility of this information through the internet is the central technological core in the digital age (Fig. 9).

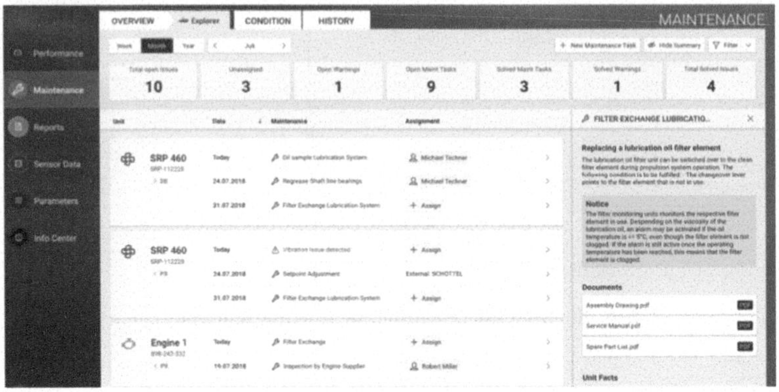

Fig. 9. SCHOTTEL fleet management.

Technological solutions of fleet management allow a better understanding of the operation of the boat, the state of their equipment and the maintenance profile. In relation to propulsion systems, these developments allow:

- Better understanding of the condition of the equipment.
- Faster reaction time in case of problems.
- Increase in equipment and vessel activity time.
- Programming maintenance tasks.
- Summary of all maintenance tasks for the entire fleet.
- Maintenance history
- The combination of tasks and manuals to ensure proper maintenance.
- Maintenance based on conditioning.

6 Case Studies

In many cases the operational efficiency, maintenance costs and reliability of the propulsion systems of the boats respond clearly to the maintenance routines to which the operator is programmed to execute.

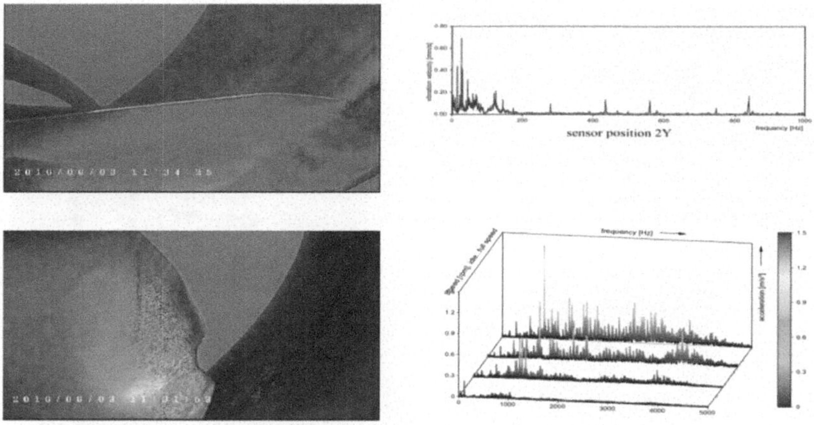

Fig. 10. Case 1, Detection of imbalance of the propeller.

Figure 10 illustrates the operation of the vessel based on reactive maintenance routines. In this case after the failure occurred, a diagnosis of machine condition confirmed unbalance in the propeller. An early detection would have reduced the reach of the dam by at least 2 days for the balancing of the propellers and consequently the costs due to unforeseen events, as well as possible propagation of bearing failures.

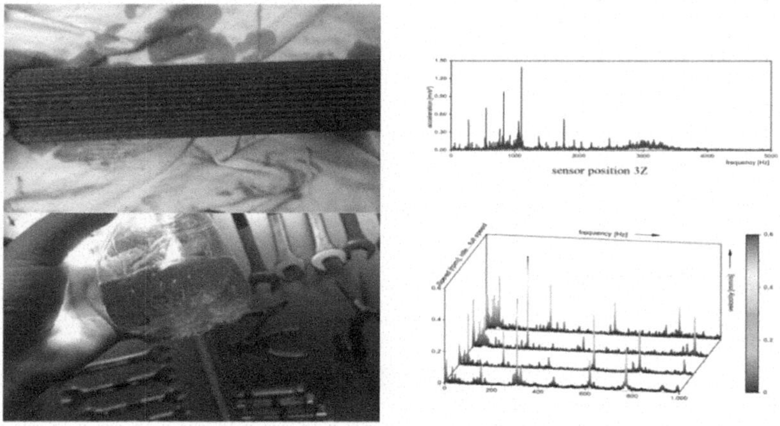

Fig. 11. Case 2, Detection of bearing wear.

Figure 11 represents the case of execution of preventive maintenance with early detection of bearing wear. With which it was possible to comply with the monitoring of trends, effectively plan the dike work and limit the operation effort of the propeller.

Fig. 12. Case 3, Detection of unbalance in propeller and bearing failures.

On the other hand, Fig. 12 represents the case in which predictive diagnoses were made, in which although possible modes of failure were detected in the system, the vessel continued to operate without taking into account recommendations according to novelties found in vibration analysis.

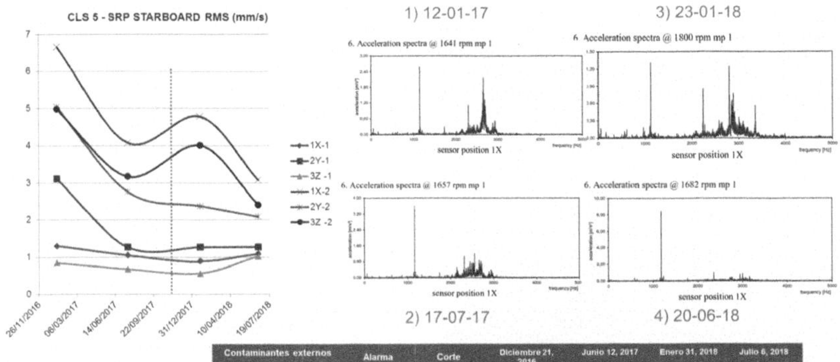

Contaminantes externos	Alarma	Corte	Diciembre 21, 2016	Junio 12, 2017	Enero 31, 2018	Julio 6, 2018
Silice (Si)	10.00	25.00	2	1	2	24

Fig. 13. Case 4, Analysis evolution of proactive fleet vibrations.

The case showed in Fig. 13 and 14 respectively, show the time evolution of the fleets that base their maintenance on proactive actions. In the case, Fig. 14 in the period shown, the inactivity (in days) of the boats due to dry docks and downtime was reduced to less than 50%. As well as the increase in the activity of the fleet in terms of availability.

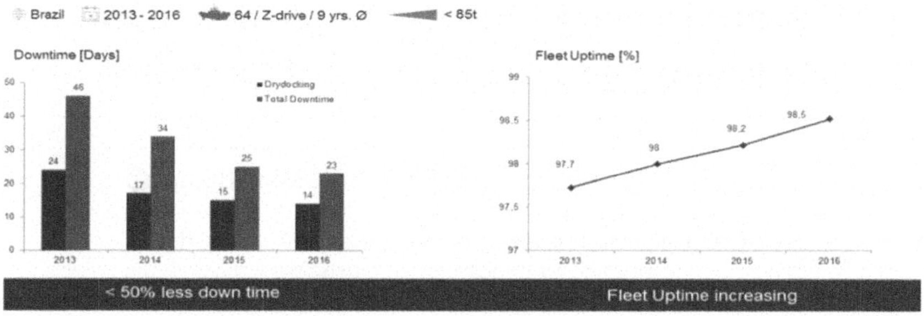

Fig. 14. Results of proactive fleet activity.

Approximately 27% reduction in maintenance costs after 24 months of performing proactive maintenance; more continuous and stable cash flow; prolongation in time between scheduled dry docks and reduction of more than 25% of the orders of non-scheduled corrective services; These are some of the indicators that support the importance of the implementation of these proactive maintenance strategies.

7 Conclusions

In maritime propulsion, the great challenge is to reduce maintenance costs and increase the reliability of the equipment. To do this, determine when there is a problem, obtain a diagnosis of condition (good or bad); if it is bad, know how bad it is to avoid catastrophic failures, know faults with specific problems, forecast the useful life and how much longer the equipment could work without Failure risk is the goal of selecting the best maintenance strategy.

That is why the development and implementation of maintenance routines allows studying the temporal evolution of the operating parameters of the propulsive systems and associate them with the evolution of the failures, to determine the period of time of the mechanical fault.

Failure is going to take an important relevance and thus to plan all the interventions and spare parts with enough time to act against the failure is the best way to maximize the reliability of the equipment.

References

1. Tomlinson, N.: What is the ideal maintenance strategy? A look at both MoD and commercial shipping best practice. IEng MIET MIMarEST. BMT Defence Services, UK (2013)
2. Zeller, F.: Preventive maintenance methods for heavy duty azimuthing thrusters. In: Dynamic Position Conference, 13–14 October 2015. Dynamic Position Committee, Marine Technology Society (2015)
3. Ferreira, L., et al.: A pilot for proactive maintenance in Industry 4.0. In: Conference Paper, CISTER Research Centre in Real Time and Embedded Computing Systems, CISTER-TR-170407 (2017)

Essential Factors for the Implementation of Sustainable Port Logistics Operations

Delmo Alves De Moura[1](✉) and Rui Carlos Botter[2]

[1] Federal University of ABC, Santo André, São Paulo, Brazil
delmo.moura@ufabc.edu.br
[2] University of São Paulo, São Paulo, Brazil
rcbotter@usp.br

Abstract. Promoting an environmental management system for each port is a necessary condition to seek the implementation of practical measures that contribute to so-called sustainable ports. Public policies focused on this area, involving companies that use private and public port terminals and state-owned companies that manage port operations, is a differential that can positively contribute to the search for sustainability and use of renewable and clean energy in port operations.

Having a port integrated with its surroundings, which knows how to live harmoniously with the city that surrounds it, generating wealth and not denigrating the environment, is a great challenge for humanity.

Some world ports located in Europe and the United States can serve as a reference model for other ports around the world. These ports use renewable energy in their transport, storage and handling operations. They use renewable energies to load and unload ships, minimize or eliminate the use of fossil fuels in their day-to-day operations, and manage the entire chain of entrances and ships at their terminals, as well as other modes integrated to ports.

This study will analyze some of the experiences of international ports in the use of the design of sustainable logistic operations and report on the role of each player in the quest to meet environmental aspects and promote a harmonious environment between ports and their surroundings, contributing to the low emission of gases contributing to the world more logistically sustainable.

Keywords: Green port · Renewable energy · Port operations · Environmental management system

1 Introduction

1.1 Sustainable Logistics Systems in Port Operations

Many ports around the world are deploying sustainable logistics systems in their port operations. For this, it is necessary to know the current state of a port and the role of the Port Authority in wanting to achieve logistically sustainable operations. Involving all actors who act directly and indirectly in port operations and defining the part of each in the new sustainable operations system is an essential condition for the success of the business.

V. A. J. E. Carreño Moreno et al. (Eds.): CIDIN 2019/COPINAVAL 2019, pp. 424–433, 2020.
https://doi.org/10.1007/978-3-030-35963-8_36

Several factors interfere in achieving an international sustainable port certification, such as the port's relationship with the local community, if it interferes negatively in the region, such as noise (noise pollution), air pollution, use of non-renewable energy (fossil fuel), if the port pollutes the soil, water etc. [1, 2].

The self-diagnosis method is a system to assist managers in periodically reviewing their environmental management programs and thereby assess how much a port is engaged in a sustainable operations system.

Each port should, therefore, determine its significant environmental aspects to focus its time and efforts on issues with the high potential for environmental impact, providing the greatest assurance that the environment will be protected and also encouraging efficient and resources. The Guidelines for the Self-Diagnosis Method have defined significant environmental aspects that were developed by the EcoPorts Foundation in 2004.

Noise: According to the Self-Diagnosis Method, noise is an unwanted sound. Sound pollution has increased dramatically in ports and is considered one of the priorities in environmental management. Ports with large movements usually work 24 h a day and 365 days a year.

The source of noise pollution in ports is the traffic of ships, trucks and equipment used to move goods/products. The vessels contribute to noise pollution from engines/propulsion engines, auxiliary engines, propellers, heating and air conditioning systems. Road traffic includes passenger cars, trucks and tractors used at the terminals. The loading operations refer to the noise of machines such as quay cranes, pumps, among others. Underwater noise can be generated by equipment during dredging operations. Construction activities can contribute to creating noises and vibrations, mainly generated by construction equipment, truck traffic, workboats and other similar sources [3].

It is noticeable that loud noise has serious medium and long-term consequences for humans and a port focusing on a sustainable system direct and indirect collaborators, as well as residents around the port, the local community, etc. If noise in port operations is a disturbing factor for humans, one of the first consequences that a port may suffer is a likely community complaint, triggering lawsuits and legal proceedings against the Port Authority. High noise is a major cause of emotional and psychological disorders, which cause loss of productivity, lack of concentration, poor performance in normal activities of a human being, and can cause serious health problems. Noise can also interfere with marine animal life and can aggravate a port ecosystem. Therefore, with the development of ships with greater transport capacity, as well as increased transport flow in ports, noise must be well managed, otherwise it will affect humans and marine life.

Ship noise can cause harm to human health and it was therefore necessary to establish a regulation requiring ships to be built with a focus on reducing noise on board and also when they are docked in ports for the protection of shipboard personnel; on the continent. Thus, the IMO (International Maritime Organization) defined a regulation of the International Convention for the Safety of Life at Sea (SOLAS), in 2012, requiring that in the construction of ships this noise factor should be reduced as much as possible, avoiding damage to human health. A ship's design must provide that several areas of a vessel do not cause hearing disturbance to its crew, or even to other people working in ports or local residents of a port.

The European Community has defined an Environmental Noise Assessment and Management Directive to prevent the problem-causing effects of ship-induced noise when docking, loading and unloading operations and disembarking in ports. The purpose of this legislation is focused on this item, noise, and bring greater comfort in port operations, avoiding discomfort in humans and marine animals, preserving a healthy and sustainable environment in port operations. This also means that these measures have been expanded to the other items of a transport logistics chain, the other modes and equipment, aiming at reducing noise levels.

Measures: Measures to combat noise pollution should be taken by port authorities. The adoption of low noise equipment, the installation of soundproofing seals, the switch to electric vehicles and port machines or the limitation of working hours can contribute to considerably reduce the noise produced. Reducing speed limits for vessels or allowing shore-side electricity to replace the power generated on board can also help to reduce noise levels in the port area. Also, workers should receive protective equipment, such as earplugs.

Indicators of environmental policies that ports must have to start the process for a green certification [4].

- Have an environmental policy;
- Environmental policy should be communicated to all relevant stakeholders/ stakeholders;
- Environmental policy should be communicated to all employees;
- Environmental policy should be publicly available on the port portal and
- The policy should include reference to the identification and control of port activities.

Having these indicators to assess the current status of a port is an essential condition for the sustainable port assessment system.

Environmental indicators guide the objectives to which a port wants to achieve. They must be measurable so that the port can evaluate its performance over a given period and with this, there may be parameters of comparison, to achieve goals. The objectives should be aligned with the port environmental policy [5–7].

2 Methodology

The methodology used in this work was the bibliographic review on the researched topic. The bibliographic review is a research stage that provides scientific support for the research, based on articles and scientific material on the Green Port theme, bringing knowledge subsidies. The theoretical reference was primarily based on international and documentary periodicals. Field research was also used in this work since data were collected from people who contributed to the understanding of the problem studied, the application of sustainable operations in European ports. Therefore, the exploratory research was used in general for this study, involving the bibliographical survey, interview with people with practical experiences with the problem researched and analysis of examples that contribute to the understanding of the problem.

3 Factors that Strengthen the Definition of Relevant Indicators

The management of a sustainable port should define qualitative and quantitative indicators to evaluate the environmental management in its operational processes and with this to evaluate which points need more attention for specific improvements and to reach measurable goals of sustainable logistic performance. Therefore, some important indicators are [8–11]:

- The port should set objectives for improving environmental management;
- The port should set goals for its purposes;
- The objectives and goals must be communicated to all actors in the process and
- The port should have quantitative objectives

Indicators of the Environmental Monitoring Program

- The port must have an environmental monitoring program;
- The port should identify ecological indicators to monitor environmental performance trends and
- The port must have an inventory of Significant Environmental Aspects.

Indicators of Environmental Training and Awareness

- The Port Authority must have an environmental training program for its employees;
- Environmental training should be appropriate to employee activities;
- All employees should be aware of the importance of compliance with environmental policy standards;
- All employees should be aware of the possible environmental impacts of their work activities and
- All employees should be aware of their responsibility to comply with the environmental policy and the management.

Indicators of environmental communication

- The port should publish an Environmental Report available to anyone;
- The port should release actual data by which the public can assess the trend of their environmental performance;
- Procedures should be in place to communicate environmental information internally among key environmental personnel and
- Procedures should be in place to exchange port environmental information with stakeholders/stakeholders, including external parties.

Indicators of environmental legislation

- The port should have an inventory of relevant environmental legislation and regulations related to its liabilities and responsibilities;
- There should be procedures to maintain and update the inventory and
- There must be methods to deal with non-compliance with internal and external standards.

Environmental budget indicators

- The port should have a budget specifically for environmental protection;
- Report the amount of funding for the environmental training of employees;
- Report the amount of funding allocated to control environmental impacts and
- Report the amount of funding allocated to emergency response and prevention.

Factors related to energy sources

On-site fossil fuel combustion includes stationary sources such as machinery, cranes, heating or cooling; and mobile sources essentially owned vehicles such as cars or boats. Electricity consumption comprises the electricity used for illumination of ports and heating and lighting of port buildings. Also includes the use of electricity from cranes, headlights or for other purposes.

To reduce CO_2 emissions, the demand for non-renewable energy needs to be lower. For this, efficient energy management is a crucial strategy to be achieved through redesigning processes, changing employee behavior and converting to greener technology.

Resource Consumption Indicators

- Define the percentage of energy sources of consumption, of total energy;
- Define the total annual energy consumption per energy source;
- The port should have a program to increase energy efficiency and
- The port should report how much energy is saved due to improvements in energy efficiency.

The Carbon Footprint is a measure of the total amount of greenhouse gas emissions (GHG) that is directly and indirectly caused by activity. The Carbon Footprint accounts for all six Kyoto GHG emissions: carbon dioxide (CO_2), methane (CH_4), nitrous oxide (N_2O), hydrofluorocarbons (HFCs), perfluorocarbons (PFCs) and sulfur hexafluoride (SF_6). Carbon dioxide emissions governed by the United Nations Framework Convention on Climate Change (UNFCC) and the subsequent Kyoto Protocol.

Carbon Footprint Indicators

- The port must measure or estimate its carbon footprint;
- The port should take steps to reduce its carbon footprint;
- The port should have the total annual emissions of greenhouse gas (GHG) emissions (Carbon Footprint) and
- The port should have the information on annual emissions of greenhouse gases (GHG) of direct emissions.

Based on the Self-Diagnosis System (SDM) noise is an unwanted sound in the process of operations. This noise comes from mechanical and industrial activities during daily operations at the ports [4].

Ship traffic is a noise/noise generator in the ports; there is also the road traffic that corroborates with this item in port operations. About ships, propulsion machinery and auxiliary engines are the major villains of noise in ports. The propeller, heating system, ventilation, and air conditioning contribute to boosting noise levels in ports. Trucks moving in ports also help increase port noise. Trucks moving in ports also help increase port noise: the operations of unloading of ships, loading, handling and transport in ports, where machinery such as dock cranes, pumps, terminal tractor, etc. are used elements that make a lot of noise in port operations [4, 12–14].

Sound pollution is an important and very significant factor when it comes to port environmental management, and port authorities should give full attention to this factor [4].

Noise indicators:

- Noise level at the terminals (day and night).
- Maximum noise level at the terminals.
- Frequency of noise measurements.
- Frequency of verification of the noise zoning map.

Waste Management

The self-diagnostic method system defines as waste any liquid or solid substance that is discarded during the routines of ports. The waste can come from ships that dock at terminals, from port equipment used in operations, from support equipment in port operations, from construction works and port reforms etc. [4].

Indicators of solid waste

- Total annual port waste collected by type.
- Annual amount of recycled solid waste.
- Annual amount of recycled liquid port waste.
- Annual amount of recycled non-hazardous industrial waste and
- Existence of separate containers for the collection of port waste.

4 Key Factors in the Green Port Establishment

In addition to the ten green port guidelines, based on different researches could develop a framework with key factors in the green port establishment. The establishment of a green port is first to split into three distinct areas. These are the Environmental Quality, the Environmental Construction and the Resource Management; from these three areas, it is possible to see the key factors that collaborate with the implementation of the green port concept [15–17].

- Environmental Quality: Carbon dioxide; Water quality; Land use and Environmental management.
- Environmental Construction: Expansion of planting; Green building; Comprehensive E-service and Wireless network in port.
- Resource Management: Material Selection; Management of waste; Water resource; Energy use and Transportation.

There are three environmental standards mainly used at ports; these are The International Organization for Standardization (ISO) 14001, the Eco-Management and Audit Scheme (EMAS) Regulation and the Port Environmental Review System (PERS). ISO 14001 do not provide any specific indicator to implement, but it gives examples that can be considered. However, both EMAS and PERS provide core EPIs to be used. PERS is the only protocol specific for ports.

The authors [6] noticed that it was essential to have a standard methodology that could provide specific results to any port, as long as each port is different and need EPIs to ensure environmental protection and sustainable development. Through the TEIP tool, it is possible to obtain the most appropriated indicators to use in a specific port. This can be obtained from a user-friendly online tool for any port, inland port or seaports. The advantages of the TEIP are not only the acquisition of the right indicators, but it can also provide information about the port to exchange knowledge between different ports.

The basis of the EMS implementation has its origin back in 1998 with the ECO-PORT project "Towards A Sustainable Transport Network". The establishment of EMS had as its main objective the creation of basic conditions to develop sustainable management in European ports. The ISO 14001 was the first standard to open doors for the EMS implementation, and nowadays in Europe, it is a prerequisite in the certification process to obtain an EMAS, which is one of the major standards [9].

The authors [9] still highlight the complexity of the variables at a port, and how difficult it can be to understand all of them in a complex ecosystem. As mentioned by [18]: "Seaports are complex, and dynamics entities", and [9] add that it is barely possible to an individual to try to completely understand it. For this reason, it was important to replace such variable by indicators in a limited number that could simplify the analysis.

Different countries in the World are deploying their green port concepts in different ways, and this is what [19] bring with an example through their study comparing four major ports in the world, these are: Port of Antwerp in Belgium, Port of Rotterdam in the Netherlands, Port of Shanghai in China and Port of Singapore in Singapore. Every port applies the green port concept through different ways, but these four ports apply practices that are equal or similar, and one example of practice adopted by these ports is the reward and punishment scheme, especially related to the need to get a better Low Sulfur Fuel Oil (LSFO).

Ships with emission levels above the allowed number may suffer the application of penalties, while ships with a good standard of emission may receive discounts at the port. One of the most known schemes of rewarding ships, that is being constantly implemented by several ports, is the Environmental Ship Index (ESI), which consists of a calculation method to measure ship's overall performance, and based on the results, discounts may be applied. For example, the Port of Antwerp offers a three-year discount on tonnages for companies that are investing in ESI. Also, when implementing the green port concept, the authors highlight that having a carbon footprint is a good start, and all the four ports bring their carbon footprint.

The Port of Rotterdam has an outstanding program that captures the carbon dioxide (CO_2) and stores it below the North Sea. This port is also known as the forefront port on the fight against the climate change. Together with the Port of Rotterdam, the Port of Antwerp also aims to reach an outstanding green status. Both ports prioritize the modal split of hinterland transport with barge transport and rail transport instead of road transport, in this term, they go straight to the Port of Shanghai and the Port of Singapore. Therefore those tendencies make it clear that Europe had started to implement the green port concept earlier than Asia. Nevertheless, the four ports focus on the same activities, prioritizing at first the ship traffic, followed by the cargo handling and storage, port expansion and port industrial activities.

The Port of Long Beach located in the United States is a model of green activities for several countries in the World. In 2005, this port formulated the first green policy, which consisted of initiatives that could positively contribute to the environment [20]. Different actions are applied by this port, and among these initiatives, it is possible to mention three of them, these are:

- The utilization of low-sulfur fuel in the main engines, as well as the incentive to the use of Liquefied Natural Gas (LNG).
- The Green Flag Program: The vessels arriving in Long Beach were encouraged to reduce their speed when coming closer to the port, to reduce air pollution. If the vessel keeps that practice for 12 months, it gets better dock charges and also the Long Beach Clean Port flag.
- The Clean Truck Program: This program consists of the prohibition of trucks manufactured before 1989. With the implementation of this program in 2008, the pollution levels were successfully reduced.

5 Conclusion

Port authorities, particularly in the relevant countries of the European Community and the United States and Canada, became familiar with the components of Environmental Management Systems (EMS) and the industry adopted well-established methodologies for port environmental management, such as the Self Diagnosis Methodology (SDM) or the Port Environmental Review System (PERS).

It is not simple to make the transition from a current port infrastructure, which uses, for example, fossil fuel in its day-to-day operations, to the deployment and development of clean and sustainable renewable energy use. Having public policies and involving all the actors that operate in the segment of port activities is an essential condition for achieving progress with this system. Understand the current reality and be willing to face the new technological challenges and a level that requires a lot of work, planning, and involvement in various areas of Port management. Evaluating possible market alternatives for renewable energy use and how these factors can impact port business is a considerable challenge.

Having all the information of the procedures of the current state of a port, as well as which practices related to the current environmental management, are essential conditions for the beginning of the work of future implantation of green port concepts. Knowing in detail all current port energy use, having key performance indicators that portray the current reality of the energy consumption of operations, having indicators of current environmental management is key. Having information about the consumption related to each operation and the current environmental management indicators will allow a port to have real conditions to analyze possible green logistics solutions and, thereby understand which points need improvements and innovations to achieve logistically sustainable operations.

Therefore, to implement a green port system, the high port administration must be committed to this system of sustainable logistics operations. Otherwise, the difficulty of implementing a model in this direction will become difficult and practically failure.

Acknowledgment. Project 2015/00277-8, sponsored by FAPESP, a Foundation for Research Support of the State of São Paulo.

References

1. Acciaro, M., Ghiara, H., Cusano, M.I.: Energy management in seaports: a new role for port authorities. Energy Policy **71**, 4–12 (2014)
2. Chiu, R.-H., Lin, L.-H., Ting, S.-C.: Evaluation of green port factors and performance: a fuzzy AHP analysis. Math. Probl. Eng. **2014**, 12 p. (2014). Article ID 802976. https://doi.org/10.1155/2014/802976
3. Davarzani, H., Fahimnia, B., Bell, M., Sarkis, J.: Greening ports and maritime logistics: a review. Transp. Res. Part D **48**, 473–487 (2016)
4. Duran, M.P.: Identification and selection of environmental performance indicators (EPIS) for use in the management of European seaports. A thesis submitted for the degree of Master of Philosophy (MPhil) in Cardiff University, School of Earth and Ocean Sciences, Cardiff University, February 2012
5. Lirn, T.-C., Wu, Y.-C., Chen, Y.J.: Green performance criteria for sustainable ports in Asia. Int. J. Phys. Distrib. Logist. Manag. **43**(5/6), 427–451 (2013)
6. Puig, M., Pla, A., Seguí, X., Darbra, R.: Tool for the identification and implementation of Environmental Indicators in Ports. Ocean Coast. Manag. **140**, 34–45 (2017)
7. Roh, S., Thai, V.V., Wong, Y.D.: Towards sustainable Asean port development: challenges and opportunities for Vietnamese ports. Asian J. Shipp. Logist. **32**(2), 107–118 (2016)
8. Pavlic, B., Cepak, F., Sucis, B., Peckaj, M., Kandus, B.: Sustainable port infrastructure, practical Implementation of the green port concept. Therm. Sci. **18**(3), 935–948 (2014)
9. Peris-Mora, E., Orejas, J.M.D., Subirats, A., Ibánez, S., Alvares, P.: Development of a system of indicators for sustainable port management. Mar. Pollut. Bull. **50**, 1649–1660 (2005)
10. Puig, M., Wooldridge, C., Casal, J., Darbra, R.M.: Tool for the identification and assessment of Environmental Aspects in Ports (TEAP). Ocean Coast. Manag. **113**, 8–17 (2015)
11. Puig, M., Wooldridge, C., Darbra, R.M.: Identification and selection of Environmental Performance Indicators for sustainable port development. Mar. Pollut. Bull. **81**, 124–130 (2014)
12. Sciberras, E.A., Zahawi, B., Atkinson, D.J.: Reducing shipboard emissions – assessment of the role of electrical technologies. Transp. Res. Part D **51**, 227–239 (2017)
13. Seguí, X., Puig, M., Quintieri, E., Wooldridge, C., Darbra, R.: New environmental performance baseline for inland ports: a benchmark for the European inland port sector. Environ. Sci. Policy **58**, 29–40 (2016)
14. Yu, H., Ge, Y.-E., Chen, J., Luo, L., Tan, C., Liu, D.: CO_2 emission evaluation of yard tractors during loading at container terminals. Transp. Res. Part D **53**, 17–36 (2017)
15. Tang, J., McNabola, A., Misstear, B., Caulfield, B.: An evaluation of the impact of the Dublin Port Tunnel and HGV management strategy on air pollution emissions. Transp. Res. Part D **52**, 1–14 (2017)
16. Walker, T.R.: Green Marine: an environmental program to establish sustainability in marine transportation. Mar. Pollut. Bull **105**, 199–207 (2016)

17. Yang, Y.-C.: Operating strategies of CO_2 reduction for a container terminal based on carbon footprint pespective. J. Clean. Prod. **141**, 472–480 (2017)
18. Bichou, K., Gray, R.: A critical review of conventional terminology for classifying seaports. Transp. Res. Part A **39**, 75–92 (2005)
19. Lam, J., Notteboom, T.: The greening of ports: a comparison of port management tools used by leading ports in Asia and Europe. Transp. Rev. **34**(2), 169–189 (2014)
20. Maritz, A., Yeh, S.-P.: Innovation and success factors in the construction of green ports. J. Environ. Prot. Ecol. **15**(3A), 1255–1263 (2014)

Implementation of Hydrodynamic Protuberances in the Design of Stabilizing Blades in a Human-Powered Submarine

Bryant E. Castañeda, Jorge Romero, Mariana Silva-Ortega(✉), Aldo Barradas, and Mariano A. Hernández

Department of Naval Architecture, Universidad Veracruzana, Boca del Río, Veracruz, Mexico
marsilva@uv.mx

Abstract. The humpback whale is one of the marine mammals with great capacity for maneuverability despite its large size; this is due to the peculiar shape of its fins and long in proportion to the length of its body and it shows undulations in the front. Searching for optimization of naval structures, engineers have tried to implement these protuberances to try to achieve the effects granted by the characteristics of the fins, it is estimated that the tubercles or undulations can provide less resistance to the advancement of the body. "Arcangello" is the submarine developed by the naval engineering education program of the Universidad Veracruzana. The project is designed and realized for participation in an international submarine race. In order to optimize the designs, undulations have been incorporated to the stabilizing blades. In this investigation, the submarine is analyzed between models with different designs of stabilizers. The corresponding analyses of the submarine's models was carried out using software CFD (computational fluid dynamics), will be compared different measures as coefficient of advance and torque in search of the most functional and optimal design.

Keywords: Humpback whale · Maneuverability · Tubercles/protuberances · Propeller · Stabilizing blades · Human-powered submarine · Amplitude · Wavelength · Aspect ratio · Computational fluid dynamics · Torque · Thrust · Efficiency

1 Introduction

Within marine mammal groups, humpback whales are studied due to physiological characteristics adaptable to naval artifacts.

Being a species with a robust and elongated body, weighing around 35 tons and a length of up to 17 m; this specimen plays a high performance at sea, outstanding in terms of maneuverability reflected in rotational movements. This capacity is due to its characteristic type of fins, being unique in their type, along the leading edge they have a series of rounded tubers that function as a means of flow control, which improves the behavior of the body through the water (Medrano and Urbán 2002).

V. A. J. E. Carreño Moreno et al. (Eds.): CIDIN 2019/COPINAVAL 2019, pp. 434–443, 2020.
https://doi.org/10.1007/978-3-030-35963-8_37

Through finite element analysis, several studies, it has been demonstrated that the presence of these bumps has the main effects of decreasing drag and increasing the elevation line caused by the angle of attack induced by these protuberances (Fish et al. 2011).

Although its body is plump and very large it has an outstanding capacity for maneuverability, achieved because its fins are considerably long in relation to the length of its body, about 1/3 of the total length and a series of protuberances, along its fin.

These tubercles have been implemented in the naval field on control surfaces such as ailerons, keels, blades and rudders; and many devices like turbines and propellers. (Hansen et al. 2009; Ibrahim and New 2015) (Fig. 1).

Fig. 1. Humpback whale acrobatic jump. Image from Pacific Whale Foundation.

1.1 Problem Description

ISR (International Submarine Races) is an international race of human-powered submarines every two years, held at David Taylor Model Basin, Carderock Division of the Naval Surface Warfare Center, in Bethesda Maryland, United States. Teams from different countries participate, one of them is the submarine "Arcangello" of the Universidad Veracruzana which seeks to improve the current model that participated in the last race held (2017).

The objective is to implement the protuberances of the humpback whale in the stabilizers and to demonstrate that they give a better efficiency to the submarine.

1.2 Submarine's Description

The submarine "Arcangello" is a design generated based on the shape of the tuna fish. The hull has a length of 3 m, a beam of 65 cm and a depth of 70 cm. Manufactured with fiberglass and aluminum reinforcements, it weighs approximately 60 kg.

Its propulsion system has a bicycle-type system composed of two axes connected by gears that configure the passage from 1 to 3 revolutions delivered by the pilot; the propeller is a model S type propeller version of 21 cm diameter, fixed pitch, made of aluminum T6061.

From the current model, optimization has been sought in the areas of design, speed and maneuverability. The stabilizing blades play a major role in the submarine's maneuverability and speed, for which an adaptation is needed to achieve a better performance; therefore, the fins of the humpback whale have been adapted to the new model of the submarine.

2 Methodology

2.1 Geometry of the Tubercles

In previous studies were determined and used measures of amplitude and wavelength:

- Bolzon et al. (2014) amplitude of 4.4% of the length of the stabilizer and a wavelength of 17.4% of the ratio of the profile.
- Shi et al. (2016) amplitude of 10% of the length of the stabilizer and a wavelength of 10% of the ratio of the profile.
- Ibrahim and New (2015) amplitude of 10% of the length of the stabilizer and a wavelength of 10% of the ratio of the profile.

With these values we determine the following profiles using the same length (.86 m) and profile ratio:

- A0L0: Amplitude 0% of the chord and wavelength of 0% of the ratio of the profile.
- A5L17: Amplitude 5% of the chord and wavelength of 17% of the ratio of the profile.
- A7L17: Amplitude 7% of the chord and wavelength of 17% of the ratio of the profile.

2.2 Stabilizing Blades

As previously mentioned, the approximate length of the pectoral fin of the humpback whale is one third of its total length, therefore, one third of the length of the submarine (2.6 m without the propeller) was used for the stabilizing blades, giving as a total a length of 0.86 m. The transverse section of the humpback whales fins is very irregular, to analyzes this section, it takes the transverse section with the maximum width as transverse section of the entire stabilizer, as a result the section has a width of 0.215 m; as a result we have an aspect ratio of 4.

For the main section, we chose the profile NACA 63 (4)-221 profiles with the pectoral fin of the humpback ball keeps a lot of similarity for the three different fins as illustrated in Fig. 2.

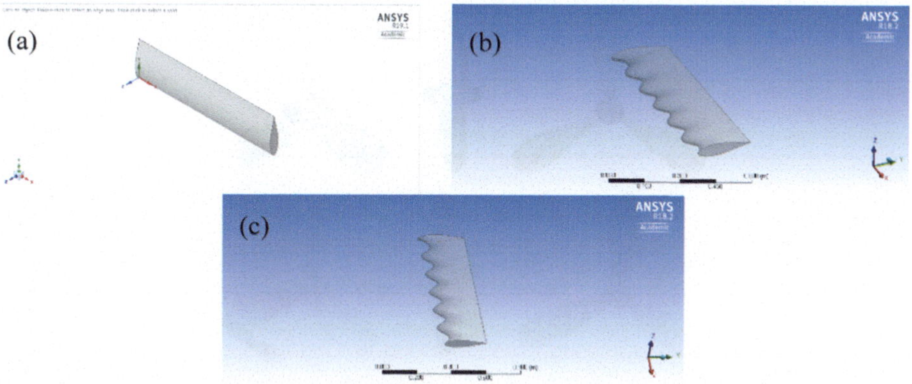

Fig. 2. (a) Model of the stabilizer fins: A0L0 (b) Model of the stabilizer fins: A5L17 and (c) Model of the stabilizer fins: A7L17.

2.3 Propeller

The propeller design was made using an open font code called Open Prop, and executable on MATLAB®. Open Prop can be used for design, analysis and optimization of propellers and turbines. The model is based in the lifting line theory, which is employed in parametric design codes for the U.S. Navy and commercial designers (Epps and Kimball 2003).

Open Prop can make geometry only using parameter in their data menu, those parameters are presented in Tables 1 and 2. The propeller is illustrated in Fig. 3.

Table 1. Open prop input data

Specifications	Value	Unit
Number of blades	3	
Angular velocity	195	RPM
Propeller diameter	0.65	m
Thrust required	736	N
Submarine velocity	3.5	m/s
Propeller hub	0.08	m
Fluid density	999.19	Kg/m^3

Table 2. Propeller performance

Specifications	Value	Unit
Advance ratio	1.65	
Torque	178.3	Nm
Power	3642	Watts

Fig. 3. A0L0 propeller views. (a) Output data in Open pro, (b) Models in Solid Works®.

In one study realized it was determined the requirements needed for the correct performance of the propeller:

- Thrust 736 N, being the force of resistance generated by the propeller, the thrust should withstand that force.
- Propeller hub of 8 cm to ensure propeller resistance.
- The density used is the density of the water at 15 °C, which is the temperature within David Taylor ModelBasin in summer.
- Angular velocity of 150 RPM.

The geometry files can be imported into the SOLID WORKS® Software to make a series of necessary modifications, obtaining the definitive model of the propeller.

Two more geometries were made, with the implementation of the protuberances in the submarine's propeller with the same amplitude and wavelength ratios, A5L17 and A7L17.

In the study the propeller was determined to generate less resistance force with the protuberances while in a static condition; when rotating the propellers with undulations they generate more drag than the smooth blades, and although they produce greater thrust is in very little quantity, a greater torque coefficient is obtained, which reduces the efficiency.

It is concluded that the most efficient helix and, therefore, the one that will be used in the following analyzes is with smooth blades or profile A0L0. Results coinciding with those of Ibrahim and New (2015) in its study of propellers with tubercles.

2.4 Numerical Simulation

Numerical simulations were analyzed as Fig. 4 is illustrated. Three different stabilizers fins with smooth propeller A0L0.

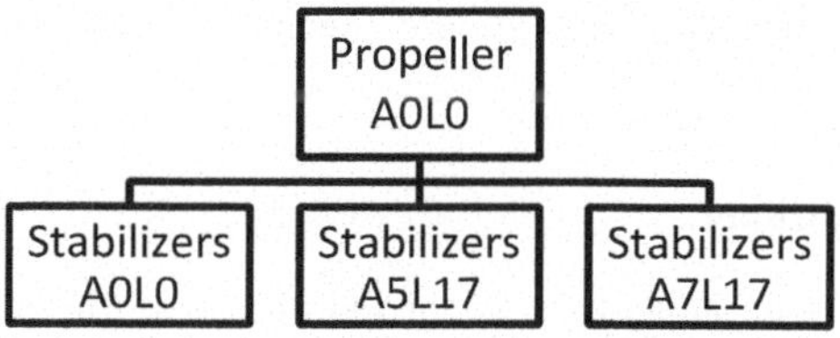

Fig. 4. Diagram of the numerical simulation to be analyzed

All models with smooth propeller (without protuberances) A0L0.

The comparison of the thrust coefficient, torque coefficient and efficiency will be made with the same advance ratio, using the following initial data and formulas:

(a) Advance ratio:

$$J = \frac{V_a}{N * D}$$

(b) Thrust coefficient:

$$K_t = \frac{T}{\rho * N^2 * D^4}$$

(c) Torque coefficient:

$$K_q = \frac{Q}{\rho * N^2 * D^4}$$

(d) Propeller efficiency:

$$\mu = \frac{K_t * J}{Kq * 2\pi}$$

- Va = Velocity inlet
- T = Thrust
- Q = Torque
- N = rps
- D = Propeller diameter
- P = Fluid density

3 Analysis

To determine the magnitudes of the drag force and torque, an analysis was made in CFD (Computational fluid dynamics), specifically the software ANSYS® 19.2 academic version, ANSYS® is a finite element solution software with which from a simulation numerical we can obtain the necessary unknowns to perform the comparison.

3.1 Boundary Conditions

For the different models an analysis was carried out using the same initial conditions; defining a cylindrical control region where the submarine is located. The fluid inlet at 1 m from the bow point with a radius of 4.5 m and ending 4.5 m behind the stern point called the fluid outlet. The region surrounding the submarine will be the flow of water to which the submarine is subjected.

The cylinder simulates the water flow of the test tank, since the forces produced on the surface of the hull are measured; the submarine is located in the center of the cylinder (Fig. 5).

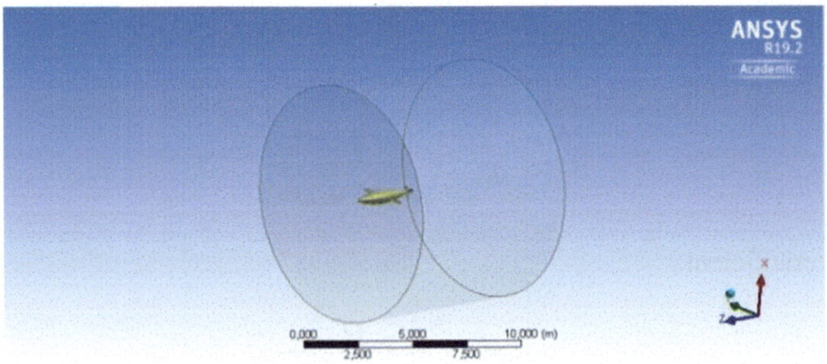

Fig. 5. Boundary conditions: cylinder non-uniform; 4.5 m of radius. Solid domain submarine body (yellow)

The governing equations for a fluid flow for a steady moving frame are (Ibrahim and New 2015)

Mass conservation:

$$\nabla \cdot \rho \vec{v}_r = 0$$

Momentum conservation:

$$\nabla \cdot (\rho \vec{v}_r \vec{v}) + \rho[\vec{\omega} \times \vec{v}] = -\nabla p + \nabla \cdot \bar{\bar{\tau}} + \vec{F}$$

Where:

ρ = Density (kg/m^3) = 998.2 (kg/m^3)
$\vec{v}$ = Absolute velocity (m/s)
$\vec{v}_r$ = Relative velocity (m/s)
$\vec{\omega}$ = Angular velocity (rpm)
$\bar{\bar{\tau}}$ = Viscous stress
$\vec{F}$ = Body Force.

3.2 Meshing

For meshing it was decided to use by default, the student version of ANSYS® having some operating limits, among them the number of nodes (520000), a fine or modified meshing exceeded the limit allowed by the license used.

Automatic mesh element of linear order, with element size of 1.1 m and medium smoothing. A fine mesh or manual with fixes requires a high number of elements, which student Ansys has as limit for analysis. The number of nodes and elements are shown in Table 3 (Fig. 6).

Table 3. Mesh data

Blade model with propeller	Number of nodes	Number of elements
A0L0	71735	389703
A5L17	86585	468245
A7L17	93090	502850

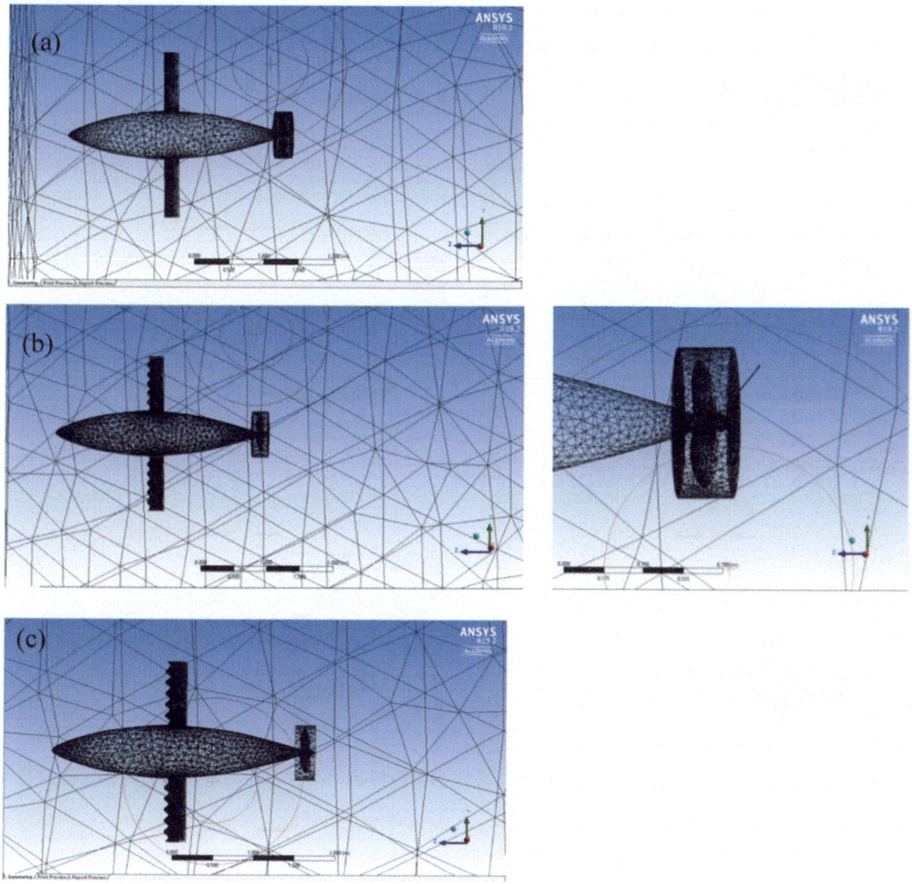

Fig. 6. Meshing: (a) Submarine A0L0 (b) Submarine A5L17 (c) Submarine A7L17

3.3 Setup

Once the meshing is done, the input data is configured, the flow of water that passes around the submarine enters at a velocity of 3.5 m/s, and as it is a dynamic analysis, the rotation velocity of the propeller is 195 RPM.

Table 4. Results

Model	Zone	Force	Torque	Kt	Kq	μ
A0L0	Propeller	1065.51	490.43	.565	.260	.57
A5L17	Propeller	1081.13	510.04	.58	.270	.572
A7L17	Propeller	1163.54	512.31	.617	.271	.59

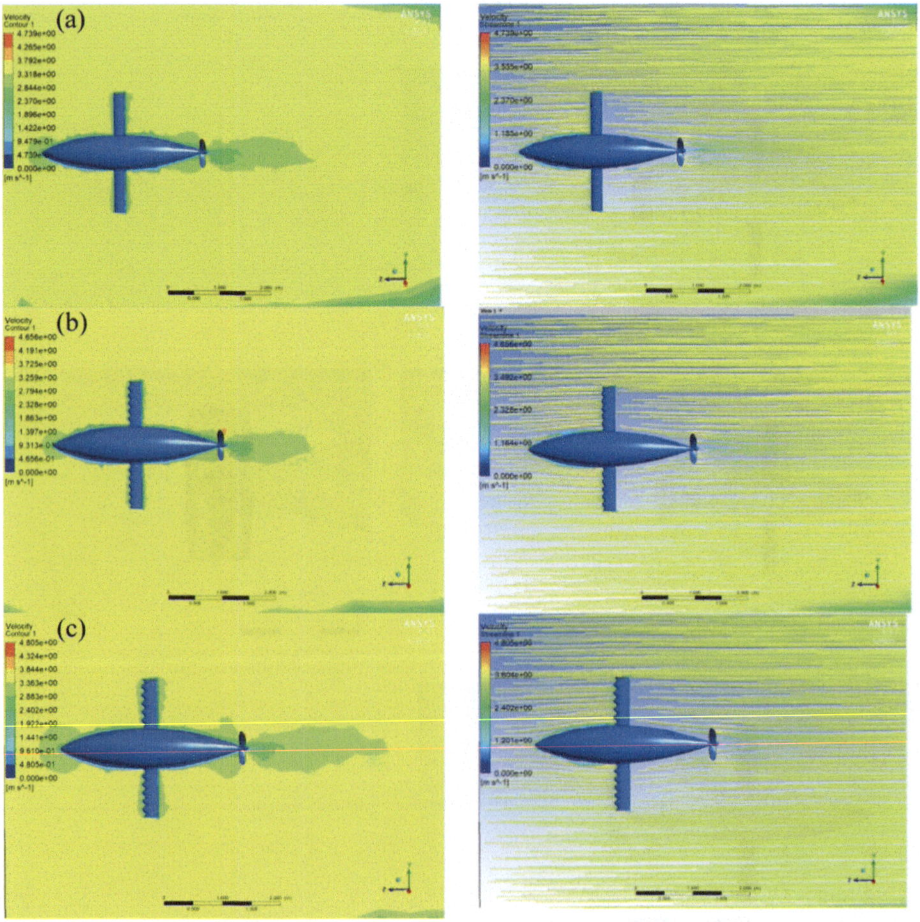

Fig. 7. Results with contour and Streamline (a) Submarine A0L0 (b) Submarine A5L17 (c) Submarine A7L17

3.4 Results

The results of the simulation are the drag forces and the torque and when included in the formulas outlined above are obtained: The results are shown in Table 4.

In the presence of tubercles in the leading edge allow the formation of vortices as same as the results of Hansen et al. (2009); Custodio (2007), although in this study it was not possible to determine how the leading edges alter the characteristics of the flux on the blade, if it is possible to appreciate the increase in velocity in the regions between tubercles, due to the low pressure regions as Fig. 7 is illustrated.

4 Conclusion

It can be concluded that the implementation of protuberances improves the behavior of the submarine, when used in the propellers they are not efficient but, in the stabilizers, they improve the behavior that they are smooth, this could be similar to conclusions of De Paula et al. (2016); Custodio (2007), especially about the increment of amplitude of the waves were enhance the effect of vortices and those decreases the drag forces in the surface.

References

Shi, W., Rosli, R., Atlar, M., Norman, R., Wang, D., Yang, W.: Hydrodynamic performance evaluation of a tidal turbine with leading-edge tubercles. Ocean Eng. **117**, 246–253 (2016)

Bolzon, M., Kelso, R.M., Arjomandi, M.: The effects of tubercles on swept wing performance (2014)

De Paula, A.A., Padilha, B.R.M., Meneghini, J.R., Paulo, S.: The airfoil thickness effect on wavy leading edge, pp. 1–43, January 2016

Custodio, D.: The Effect of Humpback Whale-Like Leading Edge Protuberances on Hydrofoil Performance. Worcester Polytechnic Institute, Worcester (2007)

Epps, B., Kimball, R.: OpenProp v3: open-source software for the design and analysis of marine propellers and horizontal-axis turbines (2013)

Ibrahim, H., New, T.H.: Tubercle modifications in marine propeller blades. In: 10th Pacific Symposium on Flow Visualization and Image Processing, pp. 1–11, June 2015

Hansen, K.L., Kelso, R.M., Dally, B.B.: The effect of leading edge tubercle geometry on the performance of different airfoils. In: ExHFT-7, pp. 1–9, June 2009

Fish, F.E., Weber, P.W., Murray, M.M., Howle, L.E.: The tubercles on humpback whales' flippers: application of bio-inspired technology. Integr. Comp. Biol. **51**(1), 203–213 (2011). Epub 15 May 2011

Medrano, L., Urbán, J.: La ballena jorobada en la Norma Oficial Mexicana 059-ECOL-1994. Comisión Nacional para el conocimiento y uso de la biodiversidad W024, pp. 69 (2002)

Author Index

A
Acosta Lopera, Oscar D., 134
Alvarez-Feal, Jose Carlos, 365
Alves De Moura, Delmo, 113, 424
Angulo, Marlis, 275
Assi, Gustavo R. S., 287
Assis, Luiz Felipe, 233, 402

B
Barradas, Aldo, 434
Beltrán Palomo, Publio, 43
Beltrán, Publio, 163
Botter, Rui Carlos, 113, 424
Bretas Rozo, Luiz Felipe, 233
Bucknall, Richard, 334

C
Cabrera Tovar, Jairo H., 134
Cali Y., E. Marcelo, 185
Caprace, Jean-David, 402
Carral Couce, Luis Manuel, 124, 308, 357, 365
Carralero Garcés, Romilio G., 153
Carrasquilla, Julio, 413
Castañeda, Bryant E., 434
Castro Santos, Laura, 357, 365
Crespi, Tommaso, 287

D
De Gracia, Luis C., 175, 298
de Oliveira Moita, João V. M., 402
de Troya-Calatayud, José, 104
Del Carpio Minaya, Roberto A., 373
Díaz-Cuadra, J. C., 163
Domínguez Martínez, Karen Patricia, 321
Domínguez Ruiz, Franklin J., 89, 308

E
Espinosa-Fuentes, Eduardo, 321

F
Fernández Jove, Antonio, 391
Filgueira-Vizoso, Almudena, 104
Fraguela-Formoso, José Ángel, 104, 365
Freiria, J., 79
Fuentes, David, 211
Fun Sang Cepeda, Maricruz A., 402

G
García Jaramillo, Lino Andrés, 89
Gómez, Pedro, 275

H
Hernández, Mariano A., 434

I
Insignares, Eduardo, 211

K
Kang, Kam Yu, 246

L
Lamas Galdo, Maria Isabel, 357
Leal, Luis D., 223
López-Arranz, Asunción, 104

M
Mackinlay, Antonio, 391
Martínez Villa, Pedro P., 153
Monteiro, Gabriel P., 402
Moreno, Gonzalo, 413
Mulford, Edgard, 413

V. A. J. E. Carreño Moreno et al. (Eds.): CIDIN 2019/COPINAVAL 2019, pp. 445–446, 2020.
https://doi.org/10.1007/978-3-030-35963-8

N
Naya, Salvador, 124
Nishimoto, Kazuo, 134
Núñez-Sánchez, M. J., 61

O
Osawa, Naoki, 175, 298

P
Patrón, Kevin E., 223
Peluffo Martínez, Gabriel, 321
Pérez Osses, Joel R., 258
Pérez-Rojas, Luis, 61
Philippi, E., 79
Piñeros Bello, Luis Andrés, 201
Pinilla Cea, Paz, 347
Podetti, Raúl Eugenio, 1, 11, 21
Podetti, Raúl Rafael, 1, 11, 21
Ponce Atencio, Yalmar, 373

R
Reusser Franck, Carlos A., 258
Riola, José María, 163, 391
Rodriguez Guerreiro, Maria Jesus, 357, 365
Rodríguez-Díaz, Aladar, 124
Romero, Jorge, 434
Ruiz, Hector O., 175, 298

S
Salas, Marcos A., 185
Salas-Berrocal, Gabriela, 334
Sampaio, Claudio M. P., 246
Sanz Horcajo, Estela, 347
Segovia Forero, Camilo Ernesto, 201
Silva-Ortega, Mariana, 434
Suárez Bermejo, Juan Carlos, 347

T
Tamaru, Hitoi, 175
Tarrío-Saavedra, Javier, 124
Till, Geoffrey, 30
Townsend Valencia, Patrick Roger, 89, 347, 357

V
Vasquez, Omar D., 223
Vega, Adan, 175, 298
Verma, Bharat, 211
Vieira, Daniel P., 246

The manufacturer's authorised representative in the EU is Springer Nature Customer Service Centre GmbH, Europaplatz 3, 69115 Heidelberg, Germany. If you have any concerns regarding our products, please contact ProductSafety@springernature.com

Printed and bound by CPI Group (UK) Ltd, Croydon, CR0 4YY
30/09/2026
02225458-0017